W0111920

Springer Monographs in Mathematics

Alberto A. Pinto • David A. Rand •
Flávio Ferreira

Fine Structures of Hyperbolic Diffeomorphisms

Springer

Alberto A. Pinto
University of Minho
Departamento de Matemática (DM)
Campus de Gualtar
4710 - 057 Braga
Portugal
aapinto@math.uminho.pt

David A. Rand
Mathematics Institute
University of Warwick
Coventry, CV4 7AL
UK
d.a.rand@warwick.ac.uk

Flávio Ferreira
Escola Superior de Estudos Industriais
e de Gestão
Instituto Politécnico do Porto
R. D. Sancho I, 981
4480-876 Vila do Conde
Portugal
flavioferreira@eseig.ipp.pt

ISBN 978-3-540-87524-6 e-ISBN 978-3-540-87525-3

DOI 10.1007/978-3-540-87525-3

Springer Monographs in Mathematics ISSN 1439-7382

Library of Congress Control Number: 2008935620

Mathematics Subject Classification (2000): 37A05, 37A20, 37A25, 37A35, 37C05, 37C15, 37C27, 37C40, 37C70, 37C75, 37C85, 37E05, 37E05, 37E10, 37E15, 37E20, 37E25, 37E30, 37E45

© 2009 Springer-Verlag Berlin Heidelberg

This work is subject to copyright. All rights are reserved, whether the whole or part of the material is concerned, specifically the rights of translation, reprinting, reuse of illustrations, recitation, broadcasting, reproduction on microfilm or in any other way, and storage in data banks. Duplication of this publication or parts thereof is permitted only under the provisions of the German Copyright Law of September 9, 1965, in its current version, and permission for use must always be obtained from Springer. Violations are liable to prosecution under the German Copyright Law.

The use of general descriptive names, registered names, trademarks, etc. in this publication does not imply, even in the absence of a specific statement, that such names are exempt from the relevant protective laws and regulations and therefore free for general use.

Cover design: WMXDesign GmbH, Heidelberg

Printed on acid-free paper

9 8 7 6 5 4 3 2 1

springer.com

In celebration of the 60th birthday of
David A. Rand

For

Maria Guiomar dos Santos Adrego Pinto

Bärbel Finkenstädt and the Rand kids: Ben, Tamsin, Rupert and Charlotte

Fernanda Amélia Ferreira and Flávio André Ferreira

Family and friends

Dedicated to Dennis Sullivan and Christopher Zeeman.

Acknowledgments

Dennis Sullivan had numerous insightful discussions with us on this work. In particular, we discussed the construction of solenoid functions, train-tracks, self-renormalizable structures and pseudo-smooth structures for pseudo-Anosov diffeomorphisms.

We would like to acknowledge the invaluable help and encouragement of family, friends and colleagues, especially Abdelrahim Mousa, Alby Fisher, Aldo Portela, Aloisio Araújo, Aragão de Carvalho, Athanasios Yannakopoulos, Baltazar de Castro, Bärbel Finkenstädt, Bruno Oliveira, Carlos Matheus, Carlos Rocha, Charles Pugh, Dennis Sullivan, Diogo Pinheiro, Edson de Faria, Enrique Pujals, Étienne Ghys, Fernanda Ferreira, Filomena Loureiro, Gabriela Goes, Helena Ferreira, Henrique Oliveira, Hugo Sequeira, Humberto Moreira, Isabel Labouriau, Jacob Palis, Joana Pinto, Joana Torres, João Almeida, Joaquim Baião, John Hubbard, Jorge Buescu, Jorge Costa, José Gonçalves, José Martins, Krerley Oliveira, Lambros Boukas, Leandro Almeida, Leonel Pias, Luciano Castro, Luis Magalhães, Luisa Magalhães, Marcelo Viana, Marco Martens, Maria Monteiro, Mark Pollicott, Marta Faias, Martin Peters, Mauricio Peixoto, Miguel Ferreira, Mikhail Lyubich, Nelson Amoedo, Nico Stollenwerk, Nigel Burroughs, Nils Tongring, Nuno Azevedo, Pedro Lago, Patricia Gonçalves, Robert MacKay, Rosa Esteves, Rui Gonçalves, Saber Elaydi, Sebastian van Strien, Sofia Barros, Sofia Cerqueira, Sousa Ramos, Stefano Luzzatto, Stelios Xanthopolous, Telmo Parreira, Vilton Pinheiro, Warwick Tucker, Welington de Melo, Yunping Jiang and Zaqueu Coelho.

We thank IHES, CUNY, SUNY, IMPA, the University of Warwick and the University of São Paulo for their hospitality. We also thank Calouste Gulbenkian Foundation, PRODYN-ESF, Programs POCTI and POCI by FCT and Ministério da Ciência e da Tecnologia, CIM, Escola de Ciências da Universidade do Minho, Escola Superior de Estudos Industriais e de Gestão do Instituto Politécnico do Porto, Faculdade de Ciências da Universidade do Porto, Centros de Matemática da Universidade do Minho e da Universidade do Porto, the Wolfson Foundation and the UK Engineering and Physical Sciences Research Council for their financial support. We thank the Golden Medal distinction of the Town Hall of Espinho in Portugal to Alberto A. Pinto.

Alberto Pinto
David Rand
Flávio Ferreira

Preface

The study of hyperbolic systems is a core theme of modern dynamics. On surfaces the theory of the fine scale structure of hyperbolic invariant sets and their measures can be described in a very complete and elegant way, and is the subject of this book, largely self-contained, rigorously and clearly written. It covers the most important aspects of the subject and is based on several scientific works of the leading research workers in this field.

This book fills a gap in the literature of dynamics. We highly recommend it for any Ph.D student interested in this area. The authors are well-known experts in smooth dynamical systems and ergodic theory.

Now we give a more detailed description of the contents:

Chapter 1. The Introduction is a description of the main concepts in hyperbolic dynamics that are used throughout the book. These are due to Bowen, Hirsch, Mañé, Palis, Pugh, Ruelle, Shub, Sinai, Smale and others. Stable and unstable manifolds are shown to be C^r foliated. This result is very useful in a number of contexts. The existence of smooth orthogonal charts is also proved. This chapter includes proofs of extensions to hyperbolic diffeomorphisms of some results of Mañé for Anosov maps.

Chapter 2. All the smooth conjugacy classes of a given topological model are classified using Pinto's and Rand's HR structures. The affine structures of Ghys and Sullivan on stable and unstable leaves of Anosov diffeomorphisms are generalized.

Chapter 3. A pair of stable and unstable solenoid functions is associated to each HR structure. These pairs form a moduli space with good topological properties which are easily described. The scaling and solenoid functions introduced by Cui, Feigenbaum, Fisher, Gardiner, Jiang, Pinto, Quas, Rand and Sullivan, give a deeper understanding of the smooth structures of one and two dimensional dynamical systems.

Chapter 4. The concept of self-renormalizable structures is introduced. With this concept one can prove an equivalence between two-dimensional hyperbolic sets and pairs of one-dimensional dynamical systems that are renormalizable (see also Chapter 12). Two C^{1+} hyperbolic diffeomorphisms that are smoothly conjugate at a point are shown to be smoothly conjugate. This extends some results of de Faria and Sullivan from one-dimensional dynamics to two-dimensional dynamics.

Chapter 5. A rigidity result is proved: if the holonomies are smooth enough, then the hyperbolic diffeomorphism is smoothly conjugate to an affine model. This chapter extends to hyperbolic diffeomorphisms some of the results of Avez, Flaminio, Ghys, Hurder and Katok for Anosov diffeomorphisms.

Chapter 6. An elementary proof is given for the existence and uniqueness of Gibbs states for Hölder weight systems following pioneering works of Bowen, Paterson, Ruelle, Sinai and Sullivan.

Chapter 7. The measure scaling functions that correspond to the Gibbs measure potentials are introduced.

Chapter refsmeasures. Measure solenoid and measure ratio functions are introduced. They determine which Gibbs measures are realizable by C^{1+} hyperbolic diffeomorphisms and by C^{1+} self-renormalizable structures.

Chapter 9. The cocycle-gap pairs that allow the construction of all C^{1+} hyperbolic diffeomorphisms realizing a Gibbs measure are introduced.

Chapter 10. A geometric measure for hyperbolic dynamical systems is defined. The explicit construction of all hyperbolic diffeomorphisms with such a geometric measure is described, using the cocycle-gap pairs. The results of this chapter are related to Cawley's cohomology classes on the torus.

Chapter 11. An eigenvalue formula for hyperbolic sets on surfaces with an invariant measure absolutely continuous with respect to the Hausdorff measure is proved. This extends to hyperbolic diffeomorphisms the Livšic-Sinai eigenvalue formula for Anosov diffeomorphisms preserving a measure absolutely continuous with respect to Lebesgue measure. Also given here is an extension to hyperbolic diffeomorphisms of the results of De la Llave, Marco and Moriyon on the eigenvalues for Anosov diffeomorphisms.

Chapter 12. A one-to-one correspondence is established between C^{1+} arc exchange systems that are C^{1+} fixed points of renormalization and C^{1+} hyperbolic diffeomorphisms that admit an invariant measure absolutely continuous with respect to the Hausdorff measure. This chapter is related to the work of Ghys, Penner, Rozzy, Sullivan and Thurston. Further, there are connections with the theorems of Arnold, Herman and Yoccoz on the rigidity of circle diffeomorphisms and Denjoy's Theorem. These connections are similar to the ones between Harrison's conjecture and the investigations of Kra, Norton and Schmeling.

Chapter 13. Pinto's golden tilings of the real line are constructed (see Pinto's and Sullivan's d-adic tilings of the real line in the Appendix C). These golden tilings are in one-to-one correspondence with smooth conjugacy classes of golden diffeomorphisms of the circle that are fixed points of renormalization, and also with smooth conjugacy classes of Anosov diffeomorphisms with an invariant measure absolutely continuous with respect to the Lebesgue measure. The observation of Ghys and Sullivan that Anosov diffeomorphisms on the torus determine circle diffeomorphisms having an associated renormalization operator is used.

Chapter 14. Thurston's pseudo-Anosov affine maps appear as periodic points of the geodesic Teichmüller flow. The works of Masur, Penner, Thurston and Veech show a strong link between affine interval exchange maps and pseudo-Anosov affine maps. Pinto's and Rand's pseudo-smooth structures near the singularities are constructed so that the pseudo-Anosov maps are smooth and have the property that the stable and unstable foliations are uniformly contracted and expanded by the pseudo-Anosov dynamics. Classical results for hyperbolic dynamics such as Bochi-Mañé and Viana's duality extend to these pseudo-smooth structures. Blow-ups of these pseudo-Anosov diffeomorphisms are related to Pujals' non-uniformly hyperbolic diffeomorphisms.

Appendices. Various concepts and results of Pinto, Rand and Sullivan for one-dimensional dynamics are extended to two-dimensions. Ratio and cross-ratio distortions for diffeomorphisms of the real line are discussed, in the spirit of de Melo and van Strien's book.

Rio de Janeiro, Brazil
July 2008

Jacob Palis
Enrique R. Pujals

Contents

1

Introduction

We study the laminations by stable and unstable manifolds associated with a C^{1+} hyperbolic diffeomorphism. We show that the holonomies between the 1-dimensional leaves are $C^{1+\alpha}$, for some $0 < \alpha \leq 1$, and that the holonomies vary Hölder continuously with respect to the domain and target leaves. Hence, the laminations by stable and unstable manifolds are C^{1+} foliated. This result is very useful in a number of contexts and it is used in all of the following chapters of the book. In general terms, it allows one to reduce many questions about 2-dimensional dynamics to questions about 1-dimensional dynamics.

We say that (f, Λ) is a C^{1+} *hyperbolic diffeomorphism* if it has the following properties:

(i) $f : M \to M$ is a $C^{1+\alpha}$ diffeomorphism of a compact surface M with respect to a $C^{1+\alpha}$ structure $\mathcal{C}_f$ on M, for some $\alpha > 0$.
(ii) Λ is a hyperbolic invariant subset of M such that $f|\Lambda$ is topologically transitive and Λ has a local product structure.

We allow both the case where $\Lambda = M$ and the case where Λ is a proper subset of M. If $\Lambda = M$, then f is Anosov and M is a torus (see Franks [41], Manning [74] and Newhouse [103]). Examples where Λ is a proper subset of M include the Smale horseshoes and the codimension one attractors such as the Plykin and derived-Anosov attractors.

1.1 Stable and unstable leaves

In this section and the rest of the introduction, we present some basic concepts and results in the research area of hyperbolic dynamics.

Definition 1 *An invariant set $\Lambda \subset M$ is* hyperbolic *for the map $f : M \to M$, if there is a continuous splitting decomposition $T_xM = E^s_x \oplus E^u_x$, with $x \in \Lambda$, and there exist constants $c > 0$ and $0 < \lambda < 1$ such that:*

(i) $Df(x)E^s_x = E^s_{f(x)}$ *and* $Df(x)E^u_x = E^u_{f(x)}$*; and*
(ii) $\|Df^n|E^s_x\| < c\lambda^n$ *and* $\|Df^{-n}|E^u_x\| < c\lambda^n$,

for all $x \in \Lambda$ *and* $n \in \mathbb{N}$.

Let d be a metric on M, and let $\Lambda \subset M$ be a hyperbolic set. For $x \in \Lambda$, we denote the *local stable* and *unstable manifolds* through x, respectively, by

$$W^s(x, \varepsilon) = \{y \in M : d(f^n(x), f^n(y)) \leq \varepsilon, \text{ for all } n \geq 0\}$$

and

$$W^u(x, \varepsilon) = \left\{y \in M : d(f^{-n}(x), f^{-n}(y)) \leq \varepsilon, \text{ for all } n \geq 0\right\}.$$

By Hirsch and Pugh [48], there exist constants $\varepsilon > 0$, $c > 0$ and $0 < \lambda < 1$ such that

(a) $d(f^n(y), f^n(x)) \leq c\lambda^n$, for all $y \in W^s(x, \varepsilon)$ and $n \geq 0$;
(b) $d(f^{-n}(y), f^{-n}(x)) \leq c\lambda^n$, for all $y \in W^u(x, \varepsilon)$ and $n \geq 0$;
(c) $T_x W^s(x, \varepsilon) = E^s_x$ and $T_x W^u(x, \varepsilon) = E^u_x$.

Furthermore, if f is C^r, then $W^s(x, \varepsilon)$ and $W^u(x, \varepsilon)$ are C^r embedded discs forming a C^0 lamination.

Let $f_\iota = f$ if $\iota = u$ or $f_\iota = f^{-1}$ if $\iota = s$. By the Stable Manifold Theorem (see Hirsch and Pugh [48]), the sets $W^\iota(x, \varepsilon)$ are respectively contained in the stable and unstable immersed manifolds

$$W^\iota(x) = \bigcup_{n \geq 0} f^n_\iota \left(W^\iota\left(f^{-n}_\iota(x), \varepsilon_0\right)\right)$$

which are the image of a $C^{1+\gamma}$ immersion $\kappa_{\iota,x} : \mathbb{R} \to M$, where we use ι to denote an element of the set $\{s, u\}$ of the stable and unstable superscripts and ι' to denote the element of $\{s, u\}$ that is not ι. An *open (resp. closed) full ι-leaf segment* I is defined as a subset of $W^\iota(x)$ of the form $\kappa_{\iota,x}(I_1)$ where I_1 is a non-empty open (resp. closed) subinterval in $\mathbb{R}$. An *open* (resp. *closed*) *ι-leaf segment* is the intersection with Λ of an open (resp. closed) full ι-leaf segment such that the intersection contains at least two distinct points. If the intersection is exactly two points we call this closed ι-leaf segment an *ι-leaf gap*. A *full ι-leaf segment* is either an open or closed full ι-leaf segment. An *ι-leaf segment* is either an open or closed ι-leaf segment. The *endpoints* of a full ι-leaf segment are the points $\kappa_{\iota,x}(u)$ and $\kappa_{\iota,x}(v)$ where u and v are the endpoints of I_1. The *endpoints* of an ι-leaf segment I are the points of the minimal closed full ι-leaf segment containing I. The *interior* of an ι-leaf segment I is the complement of its boundary. In particular, an ι-leaf segment I has empty interior if, and only if, it is an ι-leaf gap. A map $c : I \to \mathbb{R}$ is an *ι-leaf chart* of an ι-leaf segment I if has an extension $c_E : I_E \to \mathbb{R}$ to a full ι-leaf segment I_E with the following properties: $I \subset I_E$ and c_E is a homeomorphism onto its image.

1.2 Marking

If Λ is a hyperbolic invariant set of a diffeomorphism $f : M \to M$, for $0 < \varepsilon < \varepsilon_0$ there is $\delta = \delta(\varepsilon) > 0$ such that, for all points $w, z \in \Lambda$ with $d(w, z) < \delta$, $W^u(w, \varepsilon)$ and $W^s(z, \varepsilon)$ intersect in a unique point that we denote by $[w, z]$. The hyperbolic set Λ has a *local product structure*, if $[w, z] \in \Lambda$. Furthermore, the following properties are satisfied: (i) $[w, z]$ varies continuously with $w, z \in \Lambda$; (ii) the bracket map is continuous on a δ-uniform neighbourhood of the diagonal in $\Lambda \times \Lambda$; and (iii) whenever both sides are defined $f([w, z]) = [f(w), f(z)]$. Note that the bracket map does not really depend on δ provided it is sufficiently small.

Let us underline that it is a standing hypothesis that all the hyperbolic sets considered here have such a local product structure.

A *rectangle* R is a subset of Λ which is (i) closed under the bracket i.e $x, y \in R$ implies $[x, y] \in R$, and (ii) proper i.e. is the closure of its interior in Λ. This definition imposes that a rectangle has always to be proper which is more restrictive than the usual one which only insists on the closure condition.

If ℓ^s and ℓ^u are respectively stable and unstable leaf segments intersecting in a single point, then we denote by $[\ell^s, \ell^u]$ the set consisting of all points of the form $[w, z]$ with $w \in \ell^s$ and $z \in \ell^u$. We note that if the stable and unstable leaf segments ℓ and ℓ' are closed, then the set $[\ell, \ell']$ is a rectangle. Conversely in this 2-dimensional situations, any rectangle R has a product structure in the following sense: for each $x \in R$ there are closed stable and unstable leaf segments of Λ, $\ell^s(x, R) \subset W^s(x)$ and $\ell^u(x, R) \subset W^u(x)$ such that $R = [\ell^s(x, R), \ell^u(x, R)]$. The leaf segments $\ell^s(x, R)$ and $\ell^u(x, R)$ are called *stable and unstable spanning leaf segments* for R (see Figure 1.1). For $\iota \in \{s, u\}$, we denote by $\partial\ell^\iota(x, R)$ the set consisting of the endpoints of $\ell^\iota(x, R)$, and we denote by $\mathrm{int}\ell^\iota(x, R)$ the set $\ell^\iota(x, R) \setminus \partial\ell^\iota(x, R)$. The *interior of* R is given by $\mathrm{int}R = [\mathrm{int}\ell^s(x, R), \mathrm{int}\ell^u(x, R)]$, and the *boundary of* R is given by $\partial R = [\partial\ell^s(x, R), \ell^u(x, R)] \bigcup [\ell^s(x, R), \partial\ell^u(x, R)]$.

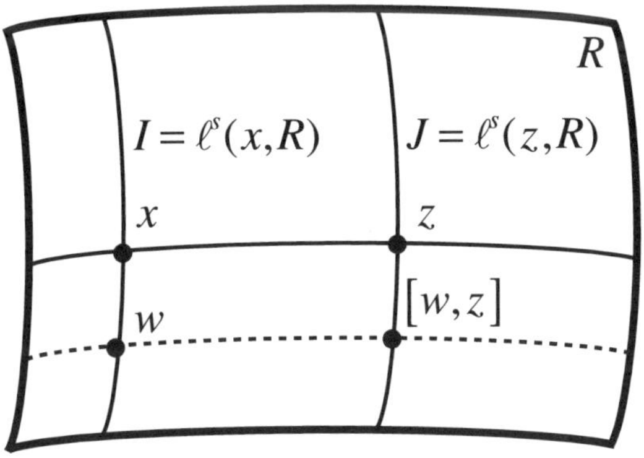

Fig. 1.1. A rectangle.

A *Markov partition of* f is a collection $\mathcal{R} = \{R_1, \ldots, R_k\}$ of rectangles such that (i) $\Lambda \subset \bigcup_{i=1}^{k} R_i$; (ii) $R_i \bigcap R_j = \partial R_i \bigcap \partial R_j$ for all i and j; (iii) if $x \in int\, R_i$ and $fx \in int\, R_j$, then

(a) $f(\ell^s(x, R_i)) \subset \ell^s(fx, R_j)$ and $f^{-1}(\ell^u(fx, R_j)) \subset \ell^u(x, R_i)$
(b) $f(\ell^u(x, R_i)) \bigcap R_j = \ell^u(fx, R_j)$ and $f^{-1}(\ell^s(fx, R_j)) \bigcap R_i = \ell^s(x, R_i)$.

The last condition means that $f(R_i)$ goes across R_j just once. In fact, it follows from condition (a) providing the rectangles R_j are chosen sufficiently small (see Mañé [73]). The rectangles making up the Markov partition are called *Markov rectangles.*

A Markov partition $\mathcal{R}$ of f satisfies the *disjointness property*, if:

(i) if $0 < \delta_{f,s} < 1$ and $0 < \delta_{f,u} < 1$, then the stable and unstable leaf boundaries of any two Markov rectangles do not intersect.
(ii) if $0 < \delta_{f,\iota} < 1$ and $\delta_{f,\iota'} = 1$, then the ι'-leaf boundaries of any two Markov rectangles do not intersect except, possibly, at their endpoints.

For simplicity of our exposition, we will just consider Markov partitions satisfying the disjointness property.

Let us give the definition of an infinite two-sided subshift of finite type Θ. The elements of $\Theta = \Theta_A$ are all infinite two-sided words $w = \ldots w_{-1} w_0 w_1 \ldots$ in the symbols $1, \ldots, k$ such that $A_{w_i w_{i+1}} = 1$, for all $i \in \mathbb{Z}$. Here $A = (A_{ij})$ is any matrix with entries 0 and 1 such that A^n has all entries positive for some $n \geq 1$. We write $w \overset{n_1,n_2}{\sim} w'$ if $w_j = w'_j$ for every $j = -n_1, \ldots, n_2$. The metric d on Θ is given by $d(w, w') = 2^{-n}$ if $n \geq 0$ is the largest such that $w \overset{n,n}{\sim} w'$. Together with this metric Θ is a compact metric space. The two-sided shift map $\tau : \Theta \to \Theta$ is the mapping which sends $w = \ldots w_{-1} w_0 w_1 \ldots$ to $v = \ldots v_{-1} v_0 v_1 \ldots$ where $v_j = w_{j+1}$ for every $j \in \mathbb{Z}$.

The properties of the Markov partition $\mathcal{R} = \{R_1, \ldots, R_k\}$ of f imply the existence of a unique two-sided subshift τ of finite type $\Theta = \Theta_A$ and a continuous surjection $i : \Theta \to \Lambda$ such that (a) $f \circ i = i \circ \tau$ and (b) $i(\Theta_j) = R_j$ for every $j = 1, \ldots, k$. We call such a map $i : \Theta \to \Lambda$ *a marking of a* C^{1+} *hyperbolic diffeomorphism* (f, Λ).

Hence, a C^{1+} hyperbolic diffeomorphism (f, Λ) admits always a marking which is not necessarily unique. For a proof, see Bowen [17], Newhouse and Palis [104] and Sinai [200].

1.3 Metric

For $\iota = s$ or u, an *ι-leaf primary cylinder of a Markov rectangle R* is a spanning ι-leaf segment of R. For $n \geq 1$, an *ι-leaf n-cylinder of R* is an ι-leaf segment I such that

(i) $f_\iota^n I$ is an ι-leaf primary cylinder of a Markov rectangle M;
(ii) $f_\iota^n \left(\ell^{\iota'}(x, R)\right) \subset M$ for every $x \in I$.

For $n \geq 2$, an *ι-leaf n-gap G of R* is an ι-leaf gap $\{x, y\}$ in a Markov rectangle R such that n is the smallest integer such that both leaves $f_\iota^{n-1}\ell^{\iota'}(x, R)$ and $f_\iota^{n-1}\ell^{\iota'}(y, R)$ are contained in ι'-boundaries of Markov rectangles; An *ι-leaf primary gap* G is the image $f_\iota G'$ by f_ι of an ι-leaf 2-gap G'.

We note that an ι-leaf segment I of a Markov rectangle R can be simultaneously an n_1-cylinder, $(n_1 + 1)$-cylinder, ..., n_2-cylinder of R if $f^{n_1}(I)$, $f^{n_1+1}(I)$, ..., $f^{n_2}(I)$ are all spanning ι-leaf segments. Furthermore, if I is an ι-leaf segment contained in the common boundary of two Markov rectangles R_i and R_j, then I can be an n_1-cylinder of R_i and an n_2-cylinder of R_j with n_1 distinct of n_2. If $G = \{x, y\}$ is an ι-gap of R contained in the interior of R, then there is a unique n such that G is an n-gap. However, if $G = \{x, y\}$ is contained in the common boundary of two Markov rectangles R_i and R_j, then G can be an n_1-gap of R_i and an n_2-gap of R_j with n_1 distinct of n_2. Since the number of Markov rectangles $R_1, \ldots, R_k$ is finite, there is $C \geq 1$ such that, in all the above cases for cylinders and gaps we have $|n_2 - n_1| \leq C$.

We say that a leaf segment K is the i-th *mother* of an n-cylinder or an n-gap J of R if $J \subset K$ and K is a leaf $(n-i)$-cylinder of R. We denote K by $m^i J$.

By the properties of a Markov partition, the smallest full ι-leaf $\hat{K}$ containing a leaf n-cylinder K of a Markov rectangle R is equal to the union of all smallest full ι-leaves containing either a leaf $(n+j)$-cylinder or a leaf $(n+i)$-gap of R, with $i \in \{1, \ldots, j\}$, contained in K.

We say that a rectangle R is an *(n_s, n_u)-rectangle* if there is $x \in R$ such that, for $\iota = s$ and u, the spanning leaf segments $\ell^\iota(x, R)$ are either an ι-leaf n_ι-cylinder or the union of two such cylinders with a common endpoint.

The reason for allowing the possibility of the spanning leaf segments being inside two touching cylinders is to allow us to regard geometrically very small rectangles intersecting a common boundary of two Markov rectangles to be small in the sense of having n_s and n_u large.

If $x, y \in \Lambda$ and $x \neq y$, then $d_\Lambda(x, y) = 2^{-n}$ where n is the biggest integer such that both x and y are contained in an (n_s, n_u)-rectangle with $n_s \geq n$ and $n_u \geq n$. Similarly, if I and J are ι-leaf segments, then $d_\Lambda(I, J) = 2^{-n_{\iota'}}$ where $n_\iota = 1$ and $n_{\iota'}$ is the biggest integer such that both I and J are contained in an (n_s, n_u)-rectangle.

1.4 Interval notation

We also use the notation of interval arithmetic for some inequalities where:

(i) if I and J are intervals, then $I + J$, $I.J$ and I/J have the obvious meaning as intervals,
(ii) if $I = \{x\}$, then we often denote I by x, and
(iii) $I \pm \varepsilon$ denotes the interval consisting of those x such that $|x - y| < \varepsilon$ for all $y \in I$.

By $\phi(n) \in 1 \pm \mathcal{O}(\nu^n)$ we mean that there exists a constant $c > 0$ depending only upon explicitly mentioned quantities such that for all $n \geq 0$, $1 - c\nu^n < \phi(n) < 1 + c\nu^n$. By $\phi(n) = \mathcal{O}(\nu^n)$ we mean that there exists a constant $c \geq 1$ depending only upon explicitly mentioned quantities such that for all $n \geq 0$, $c^{-1}\nu^n \leq \phi(n) \leq c\nu^n$.

1.5 Basic holonomies

Suppose that x and y are two points inside any rectangle R of Λ. Let $\ell(x, R)$ and $\ell(y, R)$ be two stable leaf segments respectively containing x and y and inside R. We define the *basic stable holonomy* $\theta : \ell(x, R) \to \ell(y, R)$ by $\theta(w) = [w, y]$ (see Figure 1.2). The basic stable holonomies generate the pseudo-group of all stable holonomies. Similarly we define the basic unstable holonomies. We say that a basic holonomy $\theta : \ell(x, R) \to \ell(y, R)$ is C^r, if it has a C^r diffeomorphic extension $\hat{\theta} : \hat{\ell}(x, R) \to \hat{\ell}(y, R)$ from the full leaf segment $\hat{\ell}(x, R)$ containing $\ell(x, R)$ to the full leaf segment $\hat{\ell}(y, R)$ containing $\ell(y, R)$.

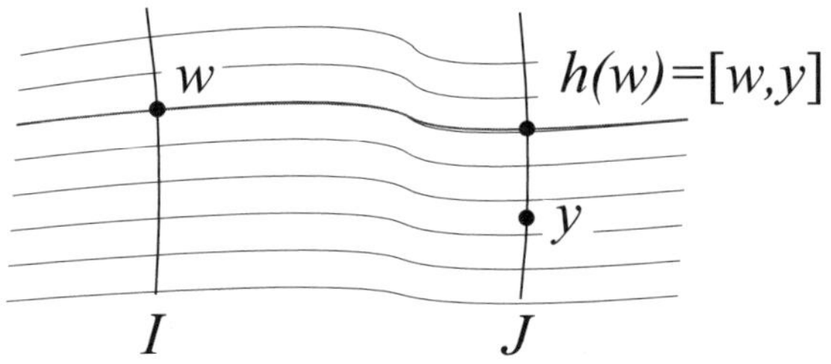

Fig. 1.2. A basic stable holonomy from I to J.

1.6 Foliated atlas

In this section when we refer to a C^r object r is allowed to take the values $k + \alpha$ where k is a positive integer and $0 < \alpha \leq 1$. Two ι-leaf charts i and j are C^r compatible if whenever U is an open subset of an ι-leaf segment contained in the domains of i and j, then $j \circ i^{-1} : i(U) \to j(U)$ extends to a C^r diffeomorphism of the real line. Such maps are called *chart overlap maps.* A *bounded C^r ι-lamination atlas* $\mathcal{A}^\iota$ is a set of such charts which (a) cover Λ, (b) are pairwise C^r compatible, and (c) the chart overlap maps are uniformly bounded in the C^r norm.

Let $\mathcal{A}^\iota$ be a bounded $C^{1+\alpha}$ ι-lamination atlas, with $0 < \alpha \leq 1$. If $i : I \to \mathbb{R}$ is a chart in $\mathcal{A}^\iota$ defined on the leaf segment I and K is a leaf segment in I, then we define $|K|_i$ to be the length of the minimal closed interval containing

$i(K)$. Since the atlas is bounded, if $j : J \to \mathbb{R}$ is another chart in $\mathcal{A}^\iota$ defined on the leaf segment J which contains K, then the ratio between the lengths $|K|_i$ and $|K|_j$ is universally bounded away from 0 and ∞. If $K' \subset I \bigcap J$ is another such segment, then we can define the ratio $r_i(K : K') = |K|_i/|K'|_i$. Although this ratio depends upon i, the ratio is exponentially determined in the sense that if T is the smallest segment containing both K and K', then

$$r_j(K : K') \in (1 \pm \mathcal{O}(|T|_i^\alpha))\, r_i(K : K') \ .$$

This follows from the $C^{1+\alpha}$ smoothness of the overlap maps and Taylor's Theorem.

A C^r lamination atlas $\mathcal{A}^\iota$ has *bounded geometry* (i) if f is a C^r diffeomorphism with C^r norm uniformly bounded in this atlas; (ii) if for all pairs I_1, I_2 of ι-leaf n-cylinders or ι-leaf n-gaps with a common point, we have that $r_i(I_1 : I_2)$ is uniformly bounded away from 0 and ∞ with the bounds being independent of i, I_1, I_2 and n; and (iii) for all endpoints x and y of an ι-leaf n-cylinder or ι-leaf n-gap I, we have that $|I|_i \le \mathcal{O}\left((d_\Lambda(x,y))^\beta\right)$ and $d_\Lambda(x,y) \le \mathcal{O}\left(|I|_i^\beta\right)$, for some $0 < \beta < 1$, independent of i, I and n.

Definition 2 *A C^r bounded lamination atlas $\mathcal{A}^\iota$ is C^r* foliated *(i) if $\mathcal{A}^\iota$ has bounded geometry; and (ii) if the basic holonomies are C^r and have a C^r norm uniformly bounded in this atlas, except possibly for the dependence upon the rectangles defining the basic holonomy. A bounded lamination atlas $\mathcal{A}^\iota$ is C^{1+}* foliated *if $\mathcal{A}^\iota$ is C^rfoliated for some $r > 1$.*

The following result relates smoothness of the holonomy with ratio distortion and will be used several times. It follows directly from Theorem B.28 (see also Theorem 3 in Pinto and Rand [159]).

Lemma 1.1. *Suppose that $\theta : I \to J$ is a basic ι-holonomy for the rectangle R and $i : I \to \mathbb{R}$ and $j : J \to \mathbb{R}$ are in $\mathcal{A}^\iota$. The holonomy $\theta : I \to J$ is $C^{1+\beta}$, for every $0 < \beta < \alpha$, with respect to the charts of the lamination atlas $\mathcal{A}^\iota$ if, and only if, for every $0 < \beta < \alpha$ and for all $I_1, I_2 \subset I$ with I_1 a leaf n-cylinder and I_2 a leaf n-cylinder or a leaf n-gap, we have*

$$\left|\log \frac{r_j(\theta(I_1) : \theta(I_2))}{r_i(I_2 : I_1)}\right| \le \mathcal{O}\left(|i(K)|^\beta\right), \tag{1.1}$$

whenever K is an ι-leaf segment containing I_1 and I_2, and where the constant of proportionality in the $\mathcal{O}$ term depends only upon the choice of i, j and the rectangle R. Moreover, there exist some constants $0 < \beta, \eta < \alpha$ and some affine map $a : \mathbb{R} \to \mathbb{R}$ such that

$$\|j \circ \theta \circ i^{-1} - a\| \le \mathcal{O}\left((d_\Lambda(I : J))^\beta\right) \tag{1.2}$$

if, and only if, there exist some constants $0 < \beta, \nu < 1$ such that, for all I_1 and I_2 as above, we have

$$\left|\log \frac{r_j(\theta(I_1):\theta(I_2))}{r_i(I_2:I_1)}\right| \le \mathcal{O}\left((d_\Lambda(I,J))^\beta \nu^n\right). \tag{1.3}$$

For $L \subset \mathbb{R}$, by $|L|$ we mean the Euclidean length of the minimal interval in $\mathbb{R}$ containing L.

1.7 Foliated atlas $\mathcal{A}^\iota(g,\rho)$

Let $g \in \mathcal{T}(f,\Lambda)$ and $\rho = \rho_g$ be a C^{1+} Riemannian metric on the manifold containing Λ. The *ι-lamination atlas $\mathcal{A}^\iota(g,\rho)$ determined by ρ* is the set of all maps $e: I \to \mathbb{R}$ where $I = \Lambda \cap \hat{I}$ with $\hat{I}$ a full ι-leaf segment, such that e extends to an isometry between the induced Riemannian metric on $\hat{I}$ and the Euclidean metric on the reals. We call the maps $e \in \mathcal{A}^\iota(g,\rho)$ the *ι-lamination charts.* If I is an ι-leaf segment (or a full ι-leaf segment), then by $|I| = |I|_\rho$ we mean the length in the Riemannian metric ρ of the minimal full ι-leaf containing I.

Fix a bounded atlas for the $C^{1+\gamma}$ structure on M. Suppose that I, J and K are full ι-leaf segments with $I, J \subset K$ and that in some chart i of the atlas, K has the form $y = u(x)$ with $x \in (x_0, x_1)$ and $u'(x) = 0$, for some $x \in (x_0, x_1)$. Let $I' = \{(x,0) : x'_0 < x < x'_1\}$ and $\{(x,0) : x''_0 < x < x''_1\}$ be, respectively, the projection of $i(I)$ and $i(J)$ onto the x-axis, and let $I'' = i^{-1}(I')$ (see Figure 1.3). Let $||i(I)||$ and $||i(J)||$ be, respectively, the Euclidean distances between the endpoints of $i(I)$ and $i(J)$.

Lemma 1.2. *There exists $0 < \alpha < 1$ such that*

$$\frac{|I|}{|I''|} \in 1 \pm \mathcal{O}(|K|^\alpha),$$

$$\frac{|I|}{|J|} \in (1 \pm \mathcal{O}(|K|^\alpha)) \frac{|x'_1 - x'_0|}{|x''_1 - x''_0|} \tag{1.4}$$

$$\frac{|I|}{|J|} \in (1 \pm \mathcal{O}(|K|^\alpha)) \frac{||i(I)||}{||i(J)||}. \tag{1.5}$$

The constants of proportionality depend only upon the atlas, ρ and the $C^{1+\gamma}$ norm of u, and α depends only upon the atlas.

Proof. Since ρ is $C^{1+\gamma}$, we can assume that in each chart of the atlas it can be written in the form $g_{11}dx^2 + g_{12}dxdy + g_{22}dy^2$, where the g_{ij} are C^γ with uniformly bounded C^γ norm. Then, integrating ρ along $y = u(x)$ and $y = 0$, and using that $|u'|$ is uniformly bounded, we get

$$|I|, |I^{''}| \in (1 \pm O(|K|^\alpha))\sqrt{g_{11}(x_0)}\,|x'_1 - x'_0|.$$

Similarly for J. Hence, (1.4) follows from combining these results. □

It follows from Lemma 1.1 and Lemma 1.2 (4) that the charts of the stable manifold are C^{1+} compatible with the charts in $\mathcal{A}^\iota(g,\rho)$.

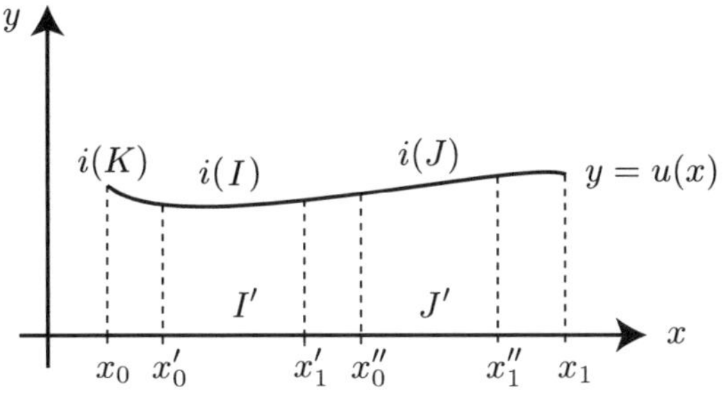

Fig. 1.3. The images of I and J by i and their projections in the horizontal axis.

Lemma 1.3. *Fix a bounded atlas for the $C^{1+\gamma}$ structure on M. Suppose that I, J and K are full ι-leaf segments with $I, J \subset K$. Then,*

$$\frac{|I|}{|J|} \in (1 \pm O(|K|_\rho^\gamma))\frac{||i(I)||}{||i(J)||},$$

where i is any chart in the atlas that contains K in its domain. The constants of proportionality depend only upon the atlas ρ and the bounded atlas considered.

Proof. Consider a chart i whose domain contains K. After composing i with a rotation and a translation, if necessary, we obtain that if K is sufficiently small, then $i(K)$ is of the form $y = u(x)$ with $x \in (x_0, x_1)$ and $u(x_0) = 0 = u(x_1)$, where the $C^{1+\gamma}$ norm of u is uniformly bounded. The result then follows directly from Lemma 1.2. □

We present a version of the naive distortion lemma that we shall use. We shall consider the case where g is $C^{1+\gamma}$ and the full u-leaf segments are 1-dimensional. The case where the full s-leaf segments are 1-dimensional is analogous.

Lemma 1.4. *For all u-leaf segments I and J with a common endpoint and for all $n \geq 0$, we have*

$$\left|\log\frac{|g^{-n}(I)|}{|g^{-n}(J)|}\frac{|J|}{|I|}\right| \leq \mathcal{O}\left(|I \cup J|^\gamma\right), \tag{1.6}$$

where the constant of proportionality in the $\mathcal{O}$ term depends only upon the choice of the Riemannian metric ρ.

Proof. Let $\hat{I}$ and $\hat{J}$ be the minimal full u-leaf segments such that $I = \hat{I} \cap \Lambda$ and $J = \hat{J} \cap \Lambda$. Also, let $k_n : g^{-n}(\hat{I} \cup \hat{J}) \to \mathbb{R}$ be an isometry between the Riemannian metric on the full u-leaf segments and the Euclidean metric on the reals.

The maps $\hat{g}_n : k_n \circ g^{-n}(\hat{I} \cup \hat{J}) \to k_{n+1} \circ g^{-(n+1)}(\hat{I} \cup \hat{J})$ defined by $\hat{g}_n = k_{n+1} \circ g^{-1} \circ k_n$ are $C^{1+\gamma}$ and have $C^{1+\gamma}$ norm uniformly bounded for all $n \geq 0$.

Hence, by the Mean Value Theorem and by the hyperbolicity of Λ for g, we get

$$\left|\log \frac{|g^{-n}(I)|}{|g^{-n}(J)|}\frac{|J|}{|I|}\right| \leq \sum_{i=0}^{n-1} |\log \hat{g}_i'(x_i) - \log \hat{g}_i'(y_i)|$$
$$\leq \mathcal{O}\left(|I \cup J|^{\gamma}\right),$$

where $x_i \in k_i \circ g^{-i}(\hat{I})$ and $y_i \in k_i \circ g^{-i}(\hat{J})$. □

We also need the following geometrical result.

Lemma 1.5. *The lamination atlas $\mathcal{A}^u(g, \rho)$ has* bounded geometry *in the sense that*

(i) for all pairs I_1, I_2 of u-leaf n-cylinders or u-leaf n-gaps with a common point, we have $|I_1|/|I_2|$ uniformly bounded away from 0 and ∞, with the bounds being independent of i, I_1, I_2 and n;
(ii) for all endpoints x and y of an u-leaf n-cylinder or u-leaf n-gap I, we have $|I| \leq \mathcal{O}\left((d_\Lambda(x,y))^{\beta}\right)$ and $d_\Lambda(x,y) \leq \mathcal{O}\left(|I|^{\beta}\right)$, for some $0 < \beta < 1$ which is independent of i, I and n.

Proof. By the continuity of the stable and unstable bundles (see Section 6 in Hirsch and Pugh [48]) the length $|I|$ of the leaf segments varies continuously with the endpoints. Thus, by the compactness of Λ, the results follow for all pairs I_1, I_2 of u-leaf 1-cylinders or u-leaf 1-gaps with a common point. Hence, by Lemma 1.4, we obtain the result for all pairs I_1, I_2 of u-leaf n-cylinders or u-leaf n-gaps with a common point and for all $n > 1$. □

1.8 Straightened graph-like charts

A chart $i : U \to \mathbb{R}^2$ in the smooth structure on M is called *graph-like*, if each full u-leaf segment and each full s-leaf segment in U are, respectively, the graph of a C^{1+} function over the x-axis and over the y-axis. Let $u(x)$ and $v(y)$ be, respectively, C^{1+} functions whose graphs are the images by i of the stable and unstable leaves passing through the point $i^{-1}(0,0)$. Given such a chart and $x \in U$, by changing the coordinates by the local diffeomorphism of the form $(x, y) \mapsto (x - u(y), y - v(x))$, we obtain a new chart $j : U \to \mathbb{R}^2$ for which the images of the stable and unstable leaves through x are respectively contained in the y and x axes. We call such charts *straightened graph-like charts*. Hence, for simplicity, one can choose an atlas of the smooth structure on M consisting only of straightened graph-like charts.

Consider a basic holonomy $\theta : I \to J$ between the u-leaf segments I and J. Suppose that the domains of the lamination charts $i, j \in \mathcal{A}^u(g, \rho)$, respectively, contain I and J, and suppose moreover that there is $x \in I$ such that $i(x) = j \circ \theta(x)$. Let $d_\Lambda(I, J)$ be as in §1.3.

Theorem 1.6. *There exists $0 < \alpha \leq 1$ such that all the ι-basic holonomies are $C^{1+\alpha}$. Furthermore, there are $0 < \alpha, \beta < 1$ such that, for all θ as above, there is a diffeomorphic extension $\tilde{\theta}$ of $j \circ \theta \circ i^{-1}$ to $\mathbb{R}$ such that*

$$||\tilde{\theta} - \mathrm{id}||_{C^{1+\alpha}} \leq \mathcal{O}\left((d_\Lambda(I, J))^\beta\right), \tag{1.7}$$

where the constant of proportionality in the $\mathcal{O}$ term depends only upon the choice of i, j and the rectangle R.

From Lemma 1.5 and Theorem 1.6, we obtain the following result.

Corollary 1.7. *The lamination atlas $\mathcal{A}^\iota(g, \rho)$ is C^{1+} foliated.*

Proof of Theorem 1.6. Fix a $C^{1+\gamma}$ Riemannian metric ρ and a finite atlas $\mathcal{G}$ for M consisting of straightened graph-like charts. For a leaf-segment I, by $|I|$ we mean the length $|I|_\rho$ in the Riemannian metric as defined above.

Let $I_1, I_2 \subset I_\theta$ be u-leaf n-cylinders or u-leaf n-gaps with a common point and $I = I_1 \cup I_2$. By Lemma 1.5, there are constants $0 < \psi \leq \alpha < 1$ such that, for $0 \leq i \leq n$,

$$\mathcal{O}(\psi^{n-i}) \leq |f^i(I_1)|, |f^i(I_2)| \leq \mathcal{O}(\alpha^{n-i}). \tag{1.8}$$

Therefore, $|f^i(I)| \leq \mathcal{O}(\alpha^{n-i})$, for $0 \leq i \leq n$.

Let $[x]$ denote the integer part of $x \in \mathbb{R}$, and let $0 < \varepsilon < 1$. By Lemma 1.4, we have

$$\left| \log \frac{|I_1|}{|I_2|} \frac{\left|f^{[n(1-\varepsilon)]}(I_2)\right|}{\left|f^{[n(1-\varepsilon)]}(I_1)\right|} \right| \leq \mathcal{O}\left(\left|f^{[n(1-\varepsilon)]}(I)\right|^\gamma\right) \leq \mathcal{O}(\alpha^{\varepsilon\gamma n}). \tag{1.9}$$

Inequality (1.9) is also satisfied if we replace the leaf segment I_j by the leaf segment $\theta(I_j)$. Thus,

$$\frac{|I_1|}{|I_2|} \cdot \frac{|\theta(I_2)|}{|\theta(I_1)|} \in (1 \pm \mathcal{O}(\alpha^{\varepsilon\gamma n})) \frac{\left|f^{[n(1-\varepsilon)]}(I_1)\right|}{\left|f^{[n(1-\varepsilon)]}(I_2)\right|} \cdot \frac{\left|f^{[n(1-\varepsilon)]}(\theta(I_2))\right|}{\left|f^{[n(1-\varepsilon)]}(\theta(I_1))\right|}. \tag{1.10}$$

For $j \in \{1, 2\}$, $f^{[n(1-\varepsilon)]}(I_j)$ and $f^{[n(1-\varepsilon)]}(\theta(I_j))$ are $[\varepsilon n]$-cylinders contained in a rectangle R' whose spanning s-leaf segments are contained in either an $[n(1-\varepsilon)]$-cylinder or the union of two of them with a common endpoint.

Let us consider a straightened graph-like chart $i : U \to \mathbb{R}^2$ whose domain contains the rectangle R'. Let $u : (a, b) \to \mathbb{R}$ be the map whose graph contains the image under i of the full unstable leaf segment containing $f^{[n(1-\varepsilon)]}(I_j)$, and let $(a_j, u(a_j))$ and $(b_j, u(b_j))$ be the images under i of the endpoints of $f^{[n(1-\varepsilon)]}(I_j)$. By changing the coordinates by a local diffeomorphism of the form $(x, y) \mapsto (x, y - u(x))$, we obtain a partially straightened graph-like chart $k : U \to \mathbb{R}^2$ for which the image of $f^{[n(1-\varepsilon)]}(I_j)$ under k is contained in the horizontal axes. Let $v : (c, d) \to \mathbb{R}$ be the map for which the graph is the image under k of the stable or unstable manifold containing $f^{[n(1-\varepsilon)]}(\theta(I_j))$,

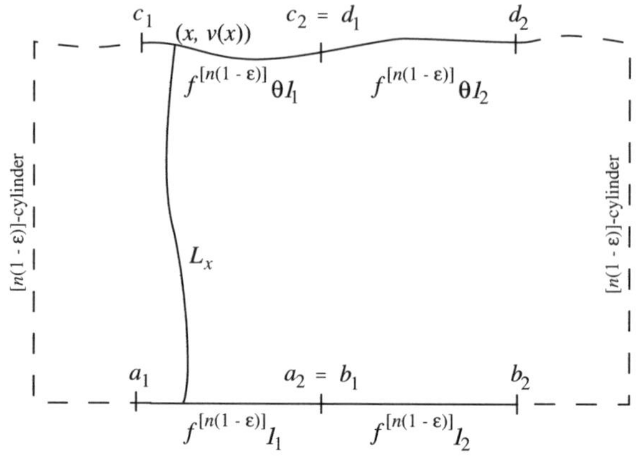

Fig. 1.4. This figure shows the various u leaf-segments in R'.

and let $(c_j, v(c_j))$ and $(d_j, v(d_j))$ be the images under i of the endpoints of $f^{[n(1-\varepsilon)]}(\theta(I_j))$ (see Figure 1.4). If in this chart the Riemannian metric is given by $ds^2 = g_{11}dx^2 + 2g_{12}dxdy + g_{11}dy^2$, then

$$\left|f^{[n(1-\varepsilon)]}(I_j)\right| = \int_{a_j}^{b_j} (g_{11}(x,0))^{1/2}\,dx,$$

$$\left|f^{[n(1-\varepsilon)]}(\theta(I_j))\right| = \int_{c_j}^{d_j} \big(g_{11}(x,v(x)) + 2g_{12}(x,v(x))v'(x) + g_{22}(x,v(x))v'(x)^2\big)^{1/2}\,dx.$$

By $C^{1+\gamma}$ smoothness of the Riemannian metric, we obtain

$$|g_{11}(x,0) - g_{11}(x,v(x))| \le \mathcal{O}\left(|v(x)|^{\gamma}\right).$$

By the Hölder continuity of the stable and unstable bundles (see Section 6 in Hirsch and Pugh [48]), there exists $0 < \eta < \gamma$ such that $|v'(x)| \le \mathcal{O}(|v(x)|^{\eta})$. Let L_x be the 1-dimensional submanifold with endpoints contained in the leaf segments $f^{[n(1-\varepsilon)]}(I)$ and $f^{[n(1-\varepsilon)]}(\theta(I))$, such that the image under k of one of its endpoints is $(x, v(x))$, and such that L_x is contained in a full s-leaf segment (see Figure 1.5).

By hyperbolicity of Λ for f, there exists $0 < \lambda < 1$ such that

$$|a_j - c_j| \le \|(a_j,0) - (c_j, v(c_j))\| \le \mathcal{O}(|L_{c_j}|) \le \mathcal{O}\left(\lambda^{n(1-\varepsilon)}\right),$$

$$|b_j - d_j| \le \|(b_j,0) - (d_j, v(d_j))\| \le \mathcal{O}(|L_{d_j}|) \le \mathcal{O}\left(\lambda^{n(1-\varepsilon)}\right), \quad \text{and}$$

$$|v(x)|^{\eta} \le \mathcal{O}\left(\lambda^{n\eta(1-\varepsilon)}\right). \tag{1.11}$$

Thus, for $j \in \{1,2\}$ and taking $\omega = \lambda^{\eta} < 1$, we have

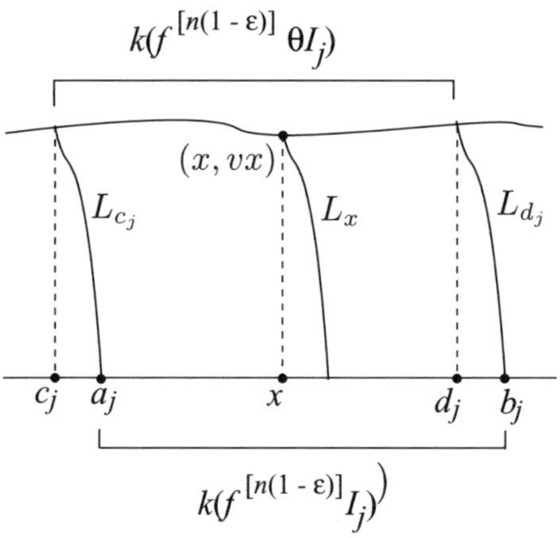

Fig. 1.5. The leaves $f^{[n(1-\varepsilon)]}(I)$ and $f^{[n(1-\varepsilon)]}(\theta(I))$.

$$\left|\left|f^{[n(1-\varepsilon)]}(\theta(I_j))\right| - \left|f^{[n(1-\varepsilon)]}(I_j)\right|\right| \leq \mathcal{O}\left(\omega^{n(1-\varepsilon)}\right). \tag{1.12}$$

Let $\nu \geq 0$ be such that $\omega^\nu = \psi$. By inequality (1.8), $\left|f^{[n(1-\varepsilon)]}(I_j)\right| \geq \mathcal{O}(\omega^{n\nu\varepsilon})$. Therefore,

$$\left|\log \frac{\left|f^{[n(1-\varepsilon)]}(\theta(I_j))\right|}{\left|f^{[n(1-\varepsilon)]}(I_j)\right|}\right| \leq \mathcal{O}\left(\omega^{n(1-\varepsilon(1+\nu))}\right). \tag{1.13}$$

Choose $0 < \varepsilon < 1$ such that $0 < \mu = \max\{\alpha^{\varepsilon\gamma}, \omega^{1-\varepsilon(1+\nu)}\} < 1$. By inequalities (1.11) and (1.13), we obtain

$$\left|\log \frac{|I_1|}{|I_2|}\frac{|\theta(I_2)|}{|\theta(I_1)|}\right| \leq \mathcal{O}(\mu^n). \tag{1.14}$$

Since this is true for all $n > 0$, and for every I_1 that is an u-leaf n-cylinder and every I_2 that is either an u-leaf n-cylinder or an u-leaf n-gap and has one common endpoint with I_1, it follows by Proposition 1.1 that the holonomy $\theta : I_\theta \to J_\theta$ is $C^{1+\beta}$, for some $\beta = \beta(\mu) > 0$ that depends only upon μ.

Now we prove that the holonomy $\theta : I_\theta \to J_\theta$ varies Hölder continuously with respect to I_θ, J_θ. As for our proof of inequality (1.12), we deduce that there exists $0 < \varepsilon_1 < 1$ such that $||I_j| - |\theta(I_j)|| \leq \mathcal{O}\left((d_\Lambda(I_\theta, J_\theta))^{\varepsilon_1}\right)$, for $j \in \{1, 2\}$. Now, we choose η small enough so that $0 < \rho = \eta^{\frac{\varepsilon_1}{2}}\psi^{-1} < 1$. If $d_\Lambda(I_\theta, J_\theta) \leq \mathcal{O}(\eta^n)$, then, as in inequality (1.13),

$$\left|\log \frac{|\theta(I_j)|}{|I_j|}\right| \leq \mathcal{O}\left((d_\Lambda(I_\theta, J_\theta))^{\frac{\varepsilon_1}{2}}\eta^{\frac{n\varepsilon_1}{2}}\psi^{-n}\right) \leq \mathcal{O}\left((d_\Lambda(I_\theta, J_\theta))^{\frac{\varepsilon_1}{2}}\rho^n\right). \tag{1.15}$$

Therefore,

$$\left|\log \frac{|I_1|}{|\theta(I_1)|}\frac{|\theta(I_2)|}{|I_2|}\right| \leq \mathcal{O}\left((d_\Lambda(I_\theta, J_\theta))^{\frac{\varepsilon_1}{2}}\rho^n\right).$$

Let $\varepsilon_2 > 0$ be such that $\mu = \eta^{2\varepsilon_2}$. If $d_\Lambda(I_\theta, J_\theta) \geq \mathcal{O}(\eta^n)$, then, by inequality (1.14),

$$\left|\log \frac{|I_1|}{|I_2|}\frac{|\theta(I_2)|}{|\theta(I_1)|}\right| \leq \mathcal{O}\left((d_\Lambda(I_\theta, J_\theta))^{\varepsilon_2}\, \mu^{\frac{n}{2}}\right).$$

Therefore, by Proposition 1.1, there is an affine map $a : \mathbb{R} \to \mathbb{R}$ such that

$$||j \circ \theta \circ i^{-1} - a||_{C^{1+\alpha}} \leq \mathcal{O}((d_\Lambda(I_\theta, J_\theta))^{\varepsilon_2}).$$

By inequality (1.15) and since there is a point x such that $j \circ \theta \circ i^{-1}(x) = x$, we get from last inequality that a is $\mathcal{O}((d_\Lambda(I_\theta, J_\theta))^{\varepsilon_3})$-close to the identity in the $C^{1+\alpha}$-norm, for some $\varepsilon_3 > 0$, and so inequality (1.7) follows. □

Consider a straightened graph-like chart $i : U \to \mathbb{R}^2$ and a rectangle R contained in U and containing $i^{-1}(0,0)$. For $y \in \mathbb{R}$ with $(0, y)$ in the image of R under i, let $I_y = \ell(i^{-1}(0, y), R)$. Let $\pi : \mathbb{R}^2 \to \mathbb{R}$ be the projection into the first coordinate.

Lemma 1.8. *Let $j : I_y \to \mathbb{R}$ be in $\mathcal{A}^u(g, \rho)$. There exists $0 < \alpha < 1$ such that the function $\pi \circ i \circ j^{-1}$ has a $C^{1+\alpha}$ diffeomorphic extension to $\mathbb{R}$. The $C^{1+\alpha}$ norm of the extension is bounded above by a quantity that depends only upon i, R and ρ.*

Proof. Let $I_1, I_2 \subset I_y$ be u-leaf n-cylinders or u-leaf n-gaps with a common point and $I = I_1 \cup I_2$. Let $I_\pi = \pi \circ i(I)$ and let $I_{\pi,k} = \pi \circ i(I_k)$ for $k \in \{1, 2\}$. Since $|I_\pi| = \mathcal{O}(|I|)$, we obtain by (1.8) that there exist $0 < \psi \leq \alpha < 1$ such that

$$\mathcal{O}(\psi^n) \leq |I_\pi| \leq \mathcal{O}(\alpha^n). \qquad (1.16)$$

The image of the full u-leaf segment $\hat{I}_y$ with $\hat{I}_y \cap \Lambda = I_y$ under i is a graph of the form $(x, v_y(x))$, where v_y is $C^{1+\gamma}$. Letting a_k and b_k be the endpoints of $I_{\pi,k}$, we find that

$$|I_k| = \int_{a_k}^{b_k} \left(g_{11}(x, v(x)) + 2g_{12}(x, v(x))v'(x) + g_{22}(x, v(x))v'(x)^2\right)^{1/2} dx.$$

Since v_y is $C^{1+\gamma}$, we obtain

$$|v_y'(w) - v_y'(z)| \leq \mathcal{O}(|I_\pi|^\gamma) \leq \mathcal{O}(\alpha^{n\gamma}), \qquad (1.17)$$

for all $w, z \in I_\pi$. By the Hölder continuity of the Riemannian metric, there exists $0 < \eta \leq 1$ such that

$$|g_{j,l}(w) - g_{j,l}(z)| \leq \mathcal{O}(|I_\pi|^\eta) \leq \mathcal{O}(\alpha^{n\eta}), \qquad (1.18)$$

for all $w, z \in I_\pi$. Let $\nu = \max\{\alpha^\gamma, \alpha^\eta\}$. By (1.17) and (1.18), and taking t such that $|I_1| = t\,|I_{\pi,1}|$, we obtain

$$|I_2| = t|I_{\pi,2}|(1 \pm \mathcal{O}(\nu^n)).$$

Hence,

$$\left|\log \frac{|I_2|}{|I_1|}\frac{|I_{\pi,1}|}{|I_{\pi,2}|}\right| \leq \mathcal{O}\left(\nu^n\right), \tag{1.19}$$

and so, by Proposition 1.1, the overlap map $\pi \circ i \circ j^{-1}$ has a $C^{1+\alpha}$ diffeomorphic extension to $\mathbb{R}$ with $C^{1+\alpha}$ norm bounded above by a quantity that depends only upon i, R and ρ. □

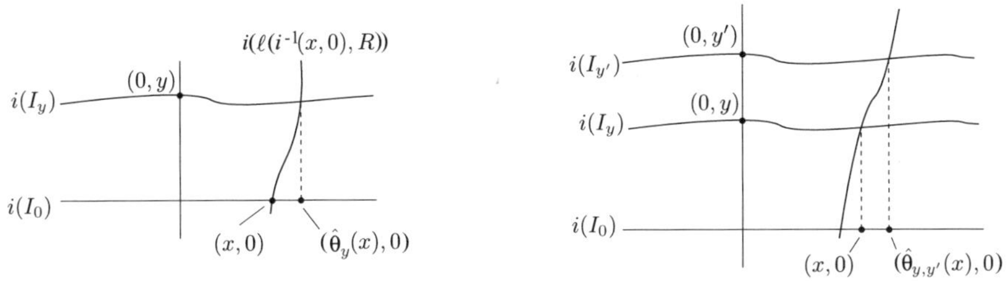

Fig. 1.6. The map $\hat{\theta}_{y,y'}$.

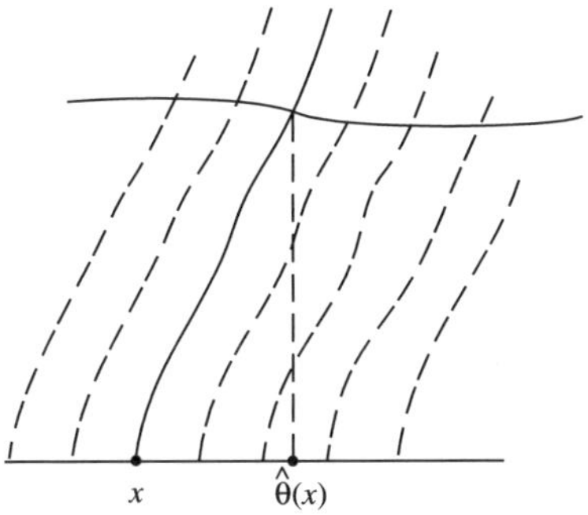

Fig. 1.7. The construction of the map $\hat{\theta}_y$.

Let $i : I_y \to \mathbb{R}^2$ be given by $i(\xi) = (x(\xi), z(x(\xi)))$, where $z : \mathbb{R} \to \mathbb{R}$ is a function, and consider a basic holonomy $\theta_{y,y'} : I_y \to I_{y'}$ in R, where $I_y = \ell(i^{-1}(0,y), R)$ and $I_{y'} = \ell(i^{-1}(0,y'), R)$. Let $\hat{\theta}_{y,y'} : \pi \circ i\,(I_y) \subset \mathbb{R} \to \mathbb{R}$ be given by $\hat{\theta}_{y,y'}(x) = \pi \circ i \circ \theta_{y,y'} \circ i^{-1}(x, z(x))$ (see Figure 1.6). Let $\theta_y : I_0 \to I_y$ in R be given by $\theta_y = \theta_{y',y}$ and $\hat{\theta}_y = \hat{\theta}_{y',y}$, with $y' = 0$ (see Figure 1.7).

Lemma 1.9. *There are $0 < \alpha, \beta < 1$ such that the maps $\hat{\theta}_y$ and $\hat{\theta}_{y,y'}$ have $C^{1+\alpha}$ diffeomorphic extensions $\tilde{\theta}_y$ and $\tilde{\theta}_{y,y'}$, respectively, to $\mathbb{R}$. Furthermore,*

$$\left|\left|\tilde{\theta}_y - \tilde{\theta}_{y'}\right|\right|_{C^{1+\alpha}} \leq \mathcal{O}\left(|y - y'|^{\beta}\right) \tag{1.20}$$

and

$$\left|\left|\tilde{\theta}_{y,y'} - \mathrm{id}\right|\right|_{C^{1+\alpha}} \leq \mathcal{O}\left(|y-y'|^{\beta}\right), \tag{1.21}$$

where the constant of proportionality in the $\mathcal{O}$ term depends only upon the choice of i and upon the rectangle R.

Proof. Let $I_1, I_2 \subset I_y$ be ι-leaf n-cylinders or u-leaf n-gaps with a common point and $I = I_1 \cup I_2$. Let $I_\pi = \pi \circ i(I)$, $J = \theta(I)$ and $J_\pi = \pi \circ i(J)$. For $k \in \{1, 2\}$, let $I_{\pi,k} = \pi \circ i(I_k)$, $J_k = \theta(I_k)$ and $J_{\pi,k} = \pi \circ i(J_k)$. By (1.14), there exists $0 < \nu < 1$ such that

$$\left|\log \frac{|I_1|}{|I_2|}\frac{|J_2|}{|J_1|}\right| \leq \mathcal{O}\left(\nu^n\right).$$

Thus, by (1.19), we obtain

$$\left|\log \frac{|I_{\pi,1}|}{|I_{\pi,2}|}\frac{|J_{\pi,2}|}{|J_{\pi,1}|}\right| \leq \mathcal{O}\left(\nu^n\right). \tag{1.22}$$

Therefore, by Proposition 1.1, the map $\hat{\theta}_{y,y'}$ has a $C^{1+\alpha}$ diffeomorphic extension to $\mathbb{R}$.

Let L_z be the 1-dimensional submanifold contained in a full s-leaf segment with minimal length and with endpoints $z \in I_y$ and $\theta_{y,y'}(z) \in I_{y'}$. By the hyperbolicity of Λ for f, there exists $0 < \varepsilon_1 \leq 1$ such that

$$|\pi \circ i(x) - \pi \circ i \circ \theta_{y,y'}(x)| \leq \mathcal{O}(|L_z|) \leq \mathcal{O}\left(|y-y'|^{\varepsilon_1}\right).$$

Thus, for $k \in \{1, 2\}$,

$$||J_{\pi,k}| - |I_{\pi,k}|| \leq \mathcal{O}\left(|y-y'|^{\varepsilon_1}\right). \tag{1.23}$$

Let ψ be as in (1.16). Choose η small enough such that $0 < \tau = \eta^{\varepsilon_1/2}\psi^{-1} < 1$. If $|y-y'| \leq \mathcal{O}(\eta^n)$, then, by (1.16) and (1.23), we obtain

$$\left|\log \frac{|J_{\pi,k}|}{|I_{\pi,k}|}\right| \leq \mathcal{O}\left(|y-y'|^{\varepsilon_1/2}\eta^{n\varepsilon_1/2}\psi^{-n}\right) \leq \mathcal{O}\left(|y-y'|^{\varepsilon_1/2}\tau^n\right). \tag{1.24}$$

Therefore,

$$\left|\log \frac{|J_{\pi,1}|}{|I_{\pi,1}|}\frac{|I_{\pi,2}|}{|J_{\pi,2}|}\right| \leq \mathcal{O}\left(|y-y'|^{\varepsilon_1/2}\tau^n\right).$$

Let $\varepsilon_2 > 0$ be such that $\nu = \eta^{2\varepsilon_2}$. If $|y-y'| \geq \mathcal{O}(\eta^n)$, then, by inequality (1.22), we obtain

$$\left|\log \frac{|J_{\pi,1}|}{|J_{\pi,2}|}\frac{|I_{\pi,2}|}{|I_{\pi,1}|}\right| \leq \mathcal{O}\left(|y-y'|^{\varepsilon_2}\nu^{n/2}\right).$$

Therefore, by Proposition 1.1, there is an affine map $a : \mathbb{R} \to \mathbb{R}$ and there exists a constant $0 < \alpha \leq 1$ such that

$$\left\|\tilde{\theta}_{y,y'} - a\right\|_{C^{1+\alpha}} \leq \mathcal{O}\left(|y - y'|^{\varepsilon_2}\right).$$

Since $\hat{\theta}_{y,y'}(0) = 0$ and by (1.24), there exists $\varepsilon_3 > 0$ such that a is $\mathcal{O}\left(|y - y'|^{\varepsilon_3}\right)$-close to the identity in the $C^{1+\alpha}$ norm. Therefore,

$$\left\|\tilde{\theta}_{y,y'} - \mathrm{id}\right\|_{C^{1+\alpha}} \leq \mathcal{O}\left(|y - y'|^{\beta}\right),$$

and so inequality (1.21) holds. Since $\tilde{\theta}_{y,y'} = \tilde{\theta}_{y'} \circ \tilde{\theta}_y^{-1}$ and the $C^{1+\alpha}$ norm of $\tilde{\theta}_y$ is uniformly bounded, we have that

$$\begin{aligned}
\left\|\tilde{\theta}_{y'} - \tilde{\theta}_y\right\|_{C^{1+\alpha}} &= \left\|\left(\tilde{\theta}_{y,y'} - \mathrm{id}\right) \circ \tilde{\theta}_y\right\|_{C^{1+\alpha}} \\
&\leq \mathcal{O}\left(\left\|\tilde{\theta}_{y,y'} - \mathrm{id}\right\|_{C^{1+\alpha}} \left\|\tilde{\theta}_y\right\|_{C^{1+\alpha}}\right) \\
&\leq \mathcal{O}\left(\left\|\tilde{\theta}_{y,y'} - \mathrm{id}\right\|_{C^{1+\alpha}}\right) \\
&\leq \mathcal{O}\left(|y - y'|^{\beta}\right).
\end{aligned}$$

□

1.9 Orthogonal atlas

An *orthogonal chart* (i, U) on Λ is an embedding $i : U \to \mathbb{R}^2$ of an open subset U of Λ that embeds every leaf segment in U into a horizontal or vertical arc of $\mathbb{R}$ (say stable leaf segments into horizontals and unstable leaf segments into verticals). Two such charts (i_1, U_1) and (i_2, U_2) on Λ are C^r *compatible* if the *chart overlap map* $i_2 \circ i_1^{-1} : i_1(U_1 \bigcap U_2) \to i_2(U_1 \bigcap U_2)$ is C^r in the sense that it extends to a C^r diffeomorphism of a neighbourhood of $i_1(U_1 \bigcap U_2)$ in $\mathbb{R}^2$ onto a neighbourhood of $i_2(U_1 \bigcap U_2)$ in $\mathbb{R}^2$.

Definition 3 *A* C^r orthogonal atlas $\mathcal{O}$ *on* Λ *is a set of orthogonal charts that cover* Λ *and are* C^r *compatible with each other. Such an atlas is said to be* bounded, *if its overlap maps have a uniformly bounded* C^r *norm, with the bound depending only upon the atlas* $\mathcal{O}$.

Let (f, Λ) be a C^{1+} hyperbolic diffeomorphism. Since Λ is compact, any atlas contains a bounded atlas. Let $i : R \to \mathbb{R}^2$ be defined by $i(w) = (i_s([w, z]), i_u([z, w]))$, where $i_s : \ell^s(z, R) \to \mathbb{R}$ and $i_u : \ell^s(z, R) \to \mathbb{R}$ are C^{1+H} charts given by the Stable Manifold Theorem.

Proposition 1.10. *The orthogonal chart* $i : R \to \mathbb{R}^2$ *is* C^{1+H} *compatible with* $\mathcal{S}_f$, *i.e, for every chart* $j \in \mathcal{S}_f$, *the overlap map* $j \circ i^{-1}$ *has a* C^{1+H} *diffeomorphic extension to an open neighbourhood of* $\mathbb{R}^2$.

Corollary 1.11. *Every C^{1+} hyperbolic diffeomorphism (f, Λ) has a finite C^{1+} orthogonal atlas $\mathcal{O}_f$ that is C^{1+} compatible with the C^{1+H} structure $\mathcal{S}_f$.*

Proof of Proposition 1.10. Take a straightened graph-like chart $(j, V') \in \mathcal{S}_f$ such that (i) $j(z) = 0$; and (ii) $j \circ i^{-1}$ is the identity along the leaf segments $\ell^s(z, R)$ and $\ell^u(z, R)$. Thus, $j \circ i^{-1}(0) = 0$. Let $K = i(R)$, and the map $u : K \to \mathbb{R}^2$ be defined by $u = j \circ i^{-1}$. We are going to prove that u has a C^{1+} extension $\tilde{u} : \mathbb{R}^2 \to \mathbb{R}^2$ and that the derivative $d\tilde{u}(0)$ of $\tilde{u}$ at 0 is an isomorphism. Thus, there is small open set $V \subset V'$ containing z such that $V \bigcap \Lambda = V \bigcap R$ and such that $\tilde{u}|j(V)$ is a C^{1+} diffeomorphism onto its image. Hence, $(v = \tilde{u}^{-1} \circ j, V)$ is a chart C^{1+} compatible with the structure $\mathcal{S}_f$ and $v|(V \bigcap \Lambda) = i|(V \bigcap R)$. To prove that u has a C^{1+} extension $\tilde{u} : \mathbb{R}^2 \to \mathbb{R}^2$ we start by finding the natural candidates $\partial_x u(x, y)$ and $\partial_y u(x, y)$ to be the derivatives $\partial_x \tilde{u}(x, y)$ and $\partial_y \tilde{u}(x, y)$ of the extension $\tilde{u}$ at the points $(x, y) \in K$.

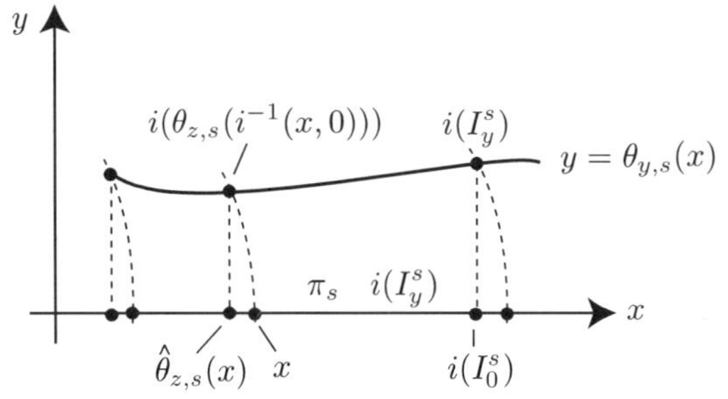

Fig. 1.8. The map $\hat{\theta}_{z,\iota}$.

Let $\pi_s : \mathbb{R}^2 \to \mathbb{R}$ and $\pi_u : \mathbb{R}^2 \to \mathbb{R}$ be the projections onto the x- and y-axis, respectively. For every $(0, y) \in K$, consider the s-spanning leaf segments I^s_y in R of the form $(x, v_{y,s}(x))$ for $x \in \pi_s \circ i(I^s_y)$ in this chart, and, for every $(x, 0) \in K$, consider the u-spanning leaf segments I^u_y in R of the form $(v_{x,u}(y), y)$ for $y \in \pi_u \circ i(I^u_s)$ in this chart, where $v_{y,s}$ and $v_{x,u}$ are C^{1+} functions. Consider the basic holonomies $\theta_{z,\iota} : I^\iota_0 \to I^\iota_z$ in R, and let $\hat{\theta}_{z,\iota} : \pi_\iota \circ i(I^\iota_0) \subset \mathbb{R} \to \mathbb{R}$ be defined by $\hat{\theta}_{z,\iota}(x) = \pi_\iota \circ i \circ \theta_{z,\iota} \circ i^{-1}(x, 0)$ (see Figure 1.8). Hence,

$$\begin{aligned} u(x, y) &= \left(\hat{\theta}_{y,s}(x), v_{y,s}\left(\hat{\theta}_{y,s}(x)\right)\right) \\ &= \left(v_{x,u}\left(\hat{\theta}_{x,u}(y)\right), \hat{\theta}_{x,u}(y)\right). \end{aligned}$$

By Lemma 1.9, the maps $\hat{\theta}_{y,s}$ and $\hat{\theta}_{x,u}$ have $C^{1+\alpha_1}$ extensions $\tilde{\theta}_{y,s}$ and $\tilde{\theta}_{x,u}$ that vary Hölder continuously with y and x, respectively, for some $0 < \alpha_1 \leq 1$. Thus, we define

$$\begin{aligned} \partial_x u(x, y) &= \left(\tilde{\theta}'_{y,s}(x), v'_{y,s}\left(\tilde{\theta}_{y,s}(x)\right)\tilde{\theta}'_{y,s}(x)\right), \\ \partial_y u(x, y) &= \left(v'_{x,u}\left(\tilde{\theta}_{x,u}(y)\right)\tilde{\theta}'_{x,u}(y), \tilde{\theta}'_{x,u}(y)\right). \end{aligned}$$

Since $\tilde{\theta}_{y,s}$ and $v'_{y,s}$ are C^{1+}, for every $y \in \pi_u \circ i\,(\ell^u(z,R))$, $\partial_x u(x,y)$ varies Hölder continuously with $x \in \pi_s \circ i\,(\ell^s(z,R))$. Since the $C^{1+\alpha_1}$ extensions $\tilde{\theta}_{y,s}$ and $\tilde{\theta}_{x,u}$ vary Hölder continuously with y and x, and by the Hölder continuity of the stable and unstable bundles (see §6 in Hirsch and Pugh [48]), for every $x \in \pi_s \circ i\,(\ell^s(z,R))$, $\partial_x u(x,y)$ varies Hölder continuously with $y \in \pi_u \circ i\,(\ell^u(z,R))$. Therefore, $\partial_x u(x,y)$ varies Hölder continuously with $(x,y) \in K$. Similarly, we obtain that $\partial_y u(x,y)$ varies Hölder continuously with $(x,y) \in K$.

By the Whitney Extension Lemma (see Abraham and Robbin [1]), the map u has a C^{1+} extension $\tilde{u}$ with $\partial_x \tilde{u}(x,y) = \partial_x u(x,y)$ and $\partial_y \tilde{u}(x,y) = \partial_y u(x,y)$, if

$$||U((x,y),(x+h_x,y+h_y))|| \leq \mathcal{O}\left(||(h_x,h_y)||^{1+\alpha}\right),$$

for some $\alpha > 0$, where

$$U((x,y),(x',y')) = u(x',y') - u(x,y) - \partial_x u(x,y)(x'-x) - \partial_y u(x,y)(y'-y).$$

Since $\tilde{\theta}_{y,s}$ and $v_{y,s}$ are C^{1+}, for all $y \in \pi_u \circ i\,(\ell^u(z,R))$, we have that the maps $u_y : \pi_s \circ i\left(\ell^s_g(z,R)\right) \to \mathbb{R}^2$ defined by $u_y(x) = \left(\tilde{\theta}_{y,s}(x), v_{y,s}\left(\tilde{\theta}_{y,s}(x)\right)\right)$ are $C^{1+\alpha_1}$, for some $\alpha_1 > 0$. Since $\tilde{\theta}_{x,u}$ and $v_{x,u}$ are C^{1+}, for all $x \in \pi_s \circ i\,(\ell^u(z,R))$, we have that the maps $u_x : \pi_u \circ i\left(\ell^u_g(z,R)\right) \to \mathbb{R}^2$ defined by $u_x(y) = \left(v_{x,u}\left(\tilde{\theta}_{x,u}(y)\right), \tilde{\theta}_{x,u}(y)\right)$ are $C^{1+\alpha_1}$, for some $\alpha_1 > 0$. Therefore,

$$\begin{aligned} u(x+h_x,y+h_y) - u(x,y) &= u_{y+h_y}(x+h_x) - u_{y+h_y}(x) + u_x(y+h_y) - u_x(y) \\ &\in \partial_x u(x,y+h_y)h_x + \partial_y u(x,y)h_y \\ &\quad \pm \mathcal{O}\left(||(h_x,h_y)||^{1+\alpha_1}\right). \end{aligned}$$

Since $\partial_x u(x,y)$ varies Hölder continuously with $(x,y) \in K$, there exists $0 < \alpha \leq \alpha_1$ such that

$$\begin{aligned} U((x,y),(x+h_x,y+h_y)) &\in \partial_x u(x,y+h_y)h_x - \partial_x u(x,y)h_x \\ &\quad \pm \mathcal{O}\left(||(h_x,h_y)||^{1+\alpha_1}\right) \\ &\subset \pm\mathcal{O}\left(||(h_x,h_y)||^{1+\alpha}\right). \end{aligned}$$

□

1.10 Further literature

There are a number of results about smoothness of the holonomies of Anosov diffeomorphisms. Anosov used the fact that the holonomies of C^2 Anosov diffeomorphisms have a Hölder continuous Jacobian to show that, when such maps preserve Lebesgue measure, they are ergodic. In the case of codimension 1 Anosov systems, one can use this Jacobian to show that the holonomies are $C^{1+\alpha}$, for some $\alpha > 0$ (see Exercise 3.1 of Chapter III of Mañé [73]). For more general hyperbolic sets, a number of papers address the question of the regularity of the invariant foliations via the regularity of their tangent distributions. As explained in Pugh, Shub and Wilkinson [177], this is not the same as regularity of holonomies. In Schmeling and Siegmund-Schultze [197] it is proved that the holonomies associated with hyperbolic sets are Hölder continuous. The paper Pugh, Shub and Wilkinson [177] contains a very interesting discussion of different notions of smooth foliation, and gives necessary and sufficient conditions for a $C^{1+\alpha}$ foliation in terms of the smoothness of both the leaves and holonomies plus the variation in the holonomies from leaf to leaf. This chapter is based on Pinto and Rand [164].

2

HR structures

We study the flexibility of smooth hyperbolic dynamics on surfaces. By the flexibility of a given topological model of hyperbolic dynamics we mean the extent of different smooth realizations of this model. We construct moduli spaces for hyperbolic sets of diffeomorphisms on surfaces which will be used in other chapters, for instance, to study the rigidity of diffeomorphisms on surfaces, and also to construct all smooth hyperbolic systems with an invariant Hausdorff measure.

2.1 Conjugacies

Let (f, Λ) be a C^{1+} hyperbolic diffeomorphism. Somewhat unusually we also desire to highlight the C^{1+} structure on M in which f is a diffeomorphism. By a C^{1+} *structure on* M we mean a maximal set of charts with open domains in M such that the union of their domains cover M and whenever U is an open subset contained in the domains of any two of these charts i and j, then the overlap map $j \circ i^{-1} : i(U) \to j(U)$ is $C^{1+\alpha}$, where $\alpha > 0$ depends on i, j and U. We note that by compactness of M, given such a C^{1+} structure on M, there is an atlas consisting of a finite set of these charts which cover M and for which the overlap maps are $C^{1+\alpha}$ compatible and uniformly bounded in the $C^{1+\alpha}$ norm, where $\alpha > 0$ just depends upon the atlas. We denote by $\mathcal{C}_f$ the C^{1+} structure on M in which f is a diffeomorphism. Usually one is not concerned with this as, given two such structures, there is a homeomorphism of M sending one onto the other and thus, from this point of view, all such structures can be identified. For our discussion it will be important to maintain the identity of the different smooth structures on M.

We say that a map $h : \Lambda_f \to \Lambda_g$ is a *topological conjugacy* between two C^{1+} hyperbolic diffeomorphisms (f, Λ_f) and (g, Λ_g) if there is a homeomorphism $h : \Lambda_f \to \Lambda_g$ with the following properties:

(i) $g \circ h(x) = h \circ f(x)$ for every $x \in \Lambda_f$.

(ii) The pull-back of the ι-leaf segments of g by h are ι-leaf segments of f.

Definition 4 *Let $\mathcal{T}(f,\Lambda)$ be the set of all C^{1+} hyperbolic diffeomorphisms (g,Λ_g) such that (g,Λ_g) and (f,Λ) are topologically conjugate by h.*

Hence, if $i:\Theta\to\Lambda_f$ is a marking for $(f,\Lambda_f),(g,\Lambda_g)\in\mathcal{T}(f,\Lambda)$, the map $h\circ i:\Theta\to\Lambda_g$ is a marking for (g,Λ_g), where $h:\Lambda_f\to\Lambda_g$ is the topological conjygacy between (f,Λ_f) and (g,Λ_g).

We say that a topological conjugacy $h:\Lambda_f\to\Lambda_g$ is a *Lipschitz conjugacy* if h has a bi-Lipschitz homeomorphic extension to an open neighbourhood of Λ_f in the surface M (with respect to the C^{1+} structures $\mathcal{C}_f$ and $\mathcal{C}_g$, respectively).

Similarly, we say that a topological conjugacy $h:\Lambda_f\to\Lambda_g$ is a *C^{1+} conjugacy* if h has a $C^{1+\alpha}$ diffeomorphic extension to an open neighbourhood of Λ_f in the surface M, for some $\alpha>0$.

Our approach is to fix a C^{1+} hyperbolic diffeomorphism (f,Λ) and consider C^{1+} hyperbolic diffeomorphism (g_1,Λ_{g_1}) topologically conjugate to (f,Λ). The topological conjugacy $h:\Lambda\to\Lambda_{g_1}$ between f and g_1 extends to a homeomorphism H defined on a neighbourhood of Λ. Then, we obtain the new C^{1+}-realization (g_2,Λ_{g_2}) of f defined as follows: (i) the map $g_2=H^{-1}\circ g_1\circ H$; (ii) the basic set is $\Lambda_{g_2}=H^{-1}|\Lambda_{g_1}$; (iii) the C^{1+} structure $\mathcal{C}_{g_2}$ is given by the pull-back $(H)_*\mathcal{C}_{g_1}$ of the C^{1+} structure $\mathcal{C}_{g_1}$. From (i) and (ii), we get that $\Lambda_{g_2}=\Lambda$ and $g_2|\Lambda=f$. From (iii), we get that g_2 is C^{1+} conjugated to g_1. Hence, to study the conjugacy classes of C^{1+} hyperbolic diffeomorphisms (f,Λ) of f, we can just consider the C^{1+} hyperbolic diffeomorphisms (g,Λ_g) with $\Lambda_g=\Lambda$ and $g|\Lambda=f|\Lambda$.

2.2 HR - Hölder ratios

A *HR structure* associates an affine structure to each stable and unstable leaf segment in such a way that these vary Hölder continuously with the leaf and are invariant under f. (The abbreviation HR stands for Hölder ratios).

An affine structure on a stable or unstable leaf is equivalent to a *ratio function* $r(I:J)$ which can be thought of as prescribing the ratio of the size of two leaf segments I and J in the same stable or unstable leaf. A *ratio function* $r(I:J)$ is positive (we recall that each leaf segment has at least two distinct points) and continuous in the endpoints of I and J. Moreover,

$$r(I:J)=r(J:I)^{-1} \quad\text{and}\quad r(I_1\cup I_2:K)=r(I_1:K)+r(I_2:K), \tag{2.1}$$

provided I_1 and I_2 intersect in at most one of their endpoints.

Definition 5 *We say that r is an* ι-ratio function *if (i) for all ι-leaf segments K, $r(I:J)$ defines a ratio function on K, where I and J are leaf segments contained in K; (ii) r is invariant under f, that is $r(I:J)=r(f(I):f(J))$,* *for all ι-leaf segments; and (iii) for every basic ι-holonomy $\theta : I \to J$ between the leaf segment I and the leaf segment J defined with respect to a rectangle R and for every ι-leaf segment $I_0 \subset I$ and every ι-leaf segment or gap $I_1 \subset I$,*

$$\left| \log \frac{r(\theta(I_0) : \theta(I_1))}{r(I_0 : I_1)} \right| \le \mathcal{O}\left((d_\Lambda(I, J))^\varepsilon\right), \tag{2.2}$$

where $\varepsilon \in (0, 1)$ depends upon r and the constant of proportionality also depends upon R, but not on the segments considered.

Definition 6 *A* HR structure *on Λ, invariant by f, is a pair (r^s, r^u) consisting of a stable and an unstable ratio function.*

2.3 Foliated atlas $\mathcal{A}(r)$

Given an ι-ratio function r, we define the embeddings $e : I \to \mathbb{R}$ by

$$e(x) = r(\ell(\xi, x), \ell(\xi, R)), \tag{2.3}$$

where ξ is an endpoint of the ι-leaf segment I, R is a Markov rectangle containing ξ (not necessarily containing I) and $\ell(\xi, x)$ is the ι-leaf segment with endpoints x and ξ. We denote the set of all these embeddings e by $\mathcal{A}(r)$.

The embeddings e in $\mathcal{A}(r)$ have overlap maps with affine extensions. Therefore, the atlas $\mathcal{A}(r)$ extends to a $C^{1+\alpha}$ lamination structure $\mathcal{L}(r)$. In Proposition 2.1, it is proved that the atlas $\mathcal{A}(r)$ has a bounded geometry, and, in Proposition 2.3, it is proved that in this the basic holonomies are $C^{1+\beta}$, for some $0 < \beta \le 1$. Thus, this lamination structure is C^{1+}-foliated. Moreover, it is a unique structure compatible with r in the sense that it and r induce the same C^{1+} structures on leaf segments.

Proposition 2.1. *If r is an ι-ratio function, then $\mathcal{A}(r)$ is a C^{1+} bounded atlas with bounded geometry.*

Proof. Suppose that I and J are either both ι-leaf n-cylinders or else that one of them is and the other is an ι-leaf n-gap. In addition, suppose that they have a common endpoint. Consider the set of ratios $r(I : J)$. By compactness and continuity, when we restrict n to be 1, the set S of such ratios is bounded away from 0 and ∞. However, since r is f-invariant, all other such ratios $r(I : J)$ are in this set S. This also implies that, for all endpoints x and y of an ι-leaf n-cylinder or ι-leaf n-gap I, we have that $|I|_i \le \mathcal{O}\left((d_\Lambda(x, y))^\beta\right)$ and $d_\Lambda(x, y) \le \mathcal{O}\left(|I|_i^\beta\right)$, for some $0 < \beta < 1$ independent of i, I and R. □

Lemma 2.2. *Let r be an ι-ratio function. There exists $0 < \alpha \le 1$ such that, for every basic holonomy $\theta : I \to J$ defined with respect to the rectangle R,*

$$\left|\log \frac{r(\theta(I_1):\theta(I_2))}{r(I_1:I_2)}\right| \le \mathcal{O}\left((d_\Lambda(I,J)|K|)^\alpha\right), \tag{2.4}$$

for all ι-leaf segments $I_1, I_2 \subset K$ in I. Here, for $|K|$ one takes $r(K:\ell(\xi,R))$ which is its length measured in a chart of the bounded atlas $\mathcal{A}(r)$, where $\xi \in K$. The constant α depends only upon r and the constant of proportionality depends only upon r and R.

Proof. Take the largest n such that the ι-leaf segments I_1 and I_2 are contained in the union of two n-cylinders with a common endpoint. By inequality (2.2) and since the ratio functions are f-invariant, we have

$$\begin{aligned}\left|\log \frac{r(\theta(I_1):\theta(I_2))}{r(I_1:I_2)}\right| &= \left|\log \frac{r(f_\iota^{-n}(\theta(I_1)):f_\iota^{-n}(\theta(I_2)))}{r(f_\iota^{-n}(I_1):f_\iota^{-n}(I_2))}\right| \\ &\le \mathcal{O}\left((d_\Lambda(f_\iota^{-n}(I),f_\iota^{-n}(J)))^\alpha\right).\end{aligned}$$

By bounded geometry, there exist $0<\nu<1$ and $0<\beta\le 1$ such that

$$\begin{aligned}d_\Lambda(f_\iota^{-n}(I),f_\iota^{-n}(J)) &\le \mathcal{O}\left(d_\Lambda(I,J)\nu^n\right) \\ &\le \mathcal{O}\left(d_\Lambda(I,J)|K|^\beta\right).\end{aligned}$$

□

Proposition 2.3. *The lamination atlas $\mathcal{A}(r)$ is $C^{1+\alpha}$-foliated, for some $0<\alpha\le 1$. Moreover, there exists $0<\beta<1$ such that if $\theta: I\to J$ is an ι-basic holonomy defined with respect to the rectangle R, then, for all segments $I_1, I_2 \subset K$ in I,*

$$\left|\log \frac{j(\theta(I_1))}{j(\theta(I_2))}\frac{i(I_2)}{i(I_1)}\right| \le \mathcal{O}\left((d_\Lambda(I,J))^\beta\,|K|_i^\beta\right), \tag{2.5}$$

where $i: I\to\mathbb{R}$ and $j: J\to\mathbb{R}$ are in $\mathcal{A}(r)$. The constant of proportionality in the $\mathcal{O}$ term depends only upon the choice of $\mathcal{A}(r)$ and upon the rectangle R.

Proof. By Proposition 2.1, $\mathcal{A}(r)$ is a $C^{1+\alpha}$ bounded atlas. Inequality (2.5) follows from Lemma 2.2, and so, by Proposition 1.1, the holonomies are $C^{1+\alpha}$ smooth, for some $0<\alpha\le 1$. Therefore, $\mathcal{L}(r)$ is a $C^{1+\alpha}$-foliated lamination structure. □

Combining Proposition 1.1 and Proposition 2.3, we get the following result.

Proposition 2.4. *Let $\theta: I\to J$ be a basic holonomy between ι-leaf segments in a rectangle R. There is $0<\eta<1$ such that the holonomy θ is $C^{1+\eta}$ with respect to the charts in $\mathcal{A}(r^\iota)$. Furthermore, there is $0<\beta<1$ with the property that for all charts $i: I\to\mathbb{R}$ and $j: J\to\mathbb{R}$ in $\mathcal{A}(r^\iota)$ there is an affine map $a:\mathbb{R}\to\mathbb{R}$ such that $j\circ\theta\circ i^{-1}$ has a $C^{1+\eta}$ diffeomorphic extension $\tilde{\theta}$ and*

$$||\tilde{\theta}-a||_{C^{1+\eta}} \le \mathcal{O}\left((d_\Lambda(I,J))^\beta\right),$$

where η and β depend upon r^ι and the constant of proportionality also depends upon R.

2.4 Invariants

For every $g \in \mathcal{T}(f, \Lambda)$ we will determine a unique HR structure associated to g as follows. Let $\mathcal{A}^{\iota}(g, \rho)$ be the $C^{1+\alpha}$ foliated lamination atlases associated with g and with a $C^{1+\gamma}$ Riemannian metric ρ on M (see §1.6 and §1.7). Recall that for an ι-leaf segment I, by $|I| = |I|_\rho$ we mean the length in the Riemannian metric ρ of the minimal full ι-leaf segment containing I.

Lemma 2.5. *For all ι-leaf segments I and J with a common endpoint and for all $n \geq 0$, the following limit exists:*

$$r^{\iota}_{\rho}(I : J) = \lim_{n\to\infty} \frac{|f_{\iota}^{-n}(I)|}{|f_{\iota}^{-n}(J)|} \in \frac{|I|}{|J|}\left(1 \pm \mathcal{O}\left(|I \cup J|^{\gamma}\right)\right), \tag{2.6}$$

where the constant of proportionality in the $\mathcal{O}$ term depends only upon the choice of the Riemannian metric ρ. Furthermore, (r^s_ρ, r^u_ρ) is a HR structure associated to g.

Lemma 2.6. *Suppose that instead of using equation (2.6) to define the ratios $r(I : J)$ we use the Euclidean distances so that*

$$r^{\iota}_{e}(I : J) = \lim_{n\to\infty} \frac{||i\left(f_{\iota}^{n}(I)\right)||}{||i\left(f_{\iota}^{n}(J)\right)||},$$

where $||i\left(f_{\iota}^{n}(I)\right)||$ and $||i\left(f_{\iota}^{n}(J)\right)||$ are as in Lemma 1.2. Then, $(r^s_e, r^u_e) = (r^s_\rho, r^u_\rho)$.

Proof. Lemma 2.6 follows from putting together Lemmas 1.3 and 2.5. □

Combining Proposition 1.1 and Proposition 2.3, we get the following lemma.

Lemma 2.7. *The overlap map $e_1 \circ e_2^{-1}$ between a chart $e_1 \in \mathcal{A}(g, \rho)$ and a chart $e_2 \in \mathcal{A}(r^{\iota}_{\rho})$ has a C^{1+} diffeomorphic extension to the reals. Therefore, the atlases $\mathcal{A}(g, \rho)$ and $\mathcal{A}(r^{\iota}_{\rho})$ determine the same C^{1+} foliated ι-lamination. In particular, for all short leaf segments K and all leaf segments I and J contained in it, we obtain that*

$$r^{\iota}_{\rho}(I : J) = \lim_{n\to\infty} \frac{|g^n_{\iota'}(I)|_\rho}{|g^n_{\iota'}(J)|_\rho} = \lim_{n\to\infty} \frac{|g^n_{\iota'}(I)|_{i_n}}{|g^n_{\iota'}(J)|_{i_n}}, \tag{2.7}$$

where i_n is any chart in $\mathcal{A}(r^{\iota}_{g})$ containing the segment $g^n_{\iota'}(K)$ in its domain.

Lemma 2.8. *Let $g \in \mathcal{T}(f, \Lambda)$. There is a unique HR structure $HR_g = (r^s_g, r^u_g)$ on Λ such that the C^{1+} stable and unstable foliated lamination atlases $\mathcal{A}^s_g$ and $\mathcal{A}^u_g$ induced by g have the following property:*

(*) A map $i : I \to \mathbb{R}$ defined on an ι-leaf segment I is $C^{1+\alpha}$ compatible with all $j \in \mathcal{A}(r^{\iota}_{g})$ if, and only if, it is $C^{1+\alpha}$ compatible with all $j \in \mathcal{A}^{\iota}_{g}$.

Furthermore, $(r^s_g, r^u_g) = (r^s_\rho, r^u_\rho)$, *for any* $C^{1+\gamma}$ *Riemannian metric* ρ.

Proof of Lemma 2.5. Let us start proving that r^ι_ρ is an ι-ratio function. By construction (see (2.6)), we obtain that r^ι_ρ is continuous, satisfies (2.1) and is invariant under f. So, it is enough to prove that r^ι_ρ satisfies (2.2).

Let $\theta : I \to J$ be an ι-basic holonomy. Let n be the integer part of $(\log d_\Lambda(I,J))/(2\log 2)$. Let $\hat\theta : f_\iota^{-n}(I) \to f_\iota^{-n}(J)$ be the ι-basic holonomy given by $\hat\theta(x) = f_\iota^{-n} \circ \theta \circ f_\iota^{n}(x)$. By the f-invariance of r^ι_ρ, for all ι-leaf segments $I_1, I_2 \subset K$ in I, we have that

$$\left|\log \frac{r(\theta(I_1):\theta(I_2))}{r(I_1:I_2)}\right| = \left|\log \frac{r\left(\hat\theta(f_\iota^{-n}(I_1)) : \hat\theta(f_\iota^{-n}(I_2))\right)}{r\left(f_\iota^{-n}(I_1) : f_\iota^{-n}(I_2)\right)}\right| . \tag{2.8}$$

By (2.6) and bounded geometry, there exists $0 < \beta_1 \le 1$ such that

$$\begin{aligned}\left|\log r\left(\hat\theta\left(f_\iota^{-n}(I_1)\right) : \hat\theta\left(f_\iota^{-n}(I_2)\right)\right) \frac{\left|f_\iota^{-n}\left(\hat\theta(I_2)\right)\right|_\rho}{\left|f_\iota^{-n}\left(\hat\theta(I_1)\right)\right|_\rho}\right| &\le \mathcal{O}\left(\left|f_\iota^{-n}\left(\hat\theta(I)\right)\right|_\rho^\gamma\right)\\ &\le \mathcal{O}\left(2^{-n\gamma\beta_1}\right)\\ &\le \mathcal{O}\left(d_\Lambda(I:J)^{\gamma\beta_1/2}\right). \end{aligned}\tag{2.9}$$

Similarly, we have

$$\left|\log r\left(\hat\theta\left(f_\iota^{-n}(I_2)\right) : \hat\theta\left(f_\iota^{-n}(I_1)\right)\right) \frac{\left|f_\iota^{-n}(I_1)\right|_\rho}{\left|f_\iota^{-n}(I_2)\right|_\rho}\right| \le \mathcal{O}\left(d_\Lambda(I:J)^{\gamma\beta_1/2}\right). \tag{2.10}$$

By Theorem 1.6, the basic ι-holonomies satisfy (1.2) and so (1.3), with respect to the charts in the lamination atlas $\mathcal{A}^\iota(g,\rho)$. Hence, for some $0 < \beta_2 \le 1$, we have

$$\begin{aligned}\left|\log \frac{\left|f_\iota^{-n}\left(\hat\theta(I_1)\right)\right|_\rho}{\left|f_\iota^{-n}\left(\hat\theta(I_2)\right)\right|_\rho} \frac{\left|f_\iota^{-n}(I_2)\right|_\rho}{\left|f_\iota^{-n}(I_1)\right|_\rho}\right| &\le \mathcal{O}\left(d_\Lambda\left(f_\iota^{-n}(I) : f_\iota^{-n}(J)\right)^{\beta_2}\right)\\ &\le \mathcal{O}\left(d_\Lambda(I:J)^{\beta_2/2}\right), \end{aligned}\tag{2.11}$$

where the constant of proportionality in the $\mathcal{O}$ term depends upon the rectangle R.

Applying (2.9), (2.10) and (2.11) to (2.8), we obtain

$$\left|\log \frac{r\left(\theta(I_1):\theta(I_2)\right)}{r\left(I_1:I_2\right)}\right| \le \mathcal{O}\left(d_\Lambda(I,J)^{\beta_3}\right),$$

where $\beta_3 = \min\{\gamma\beta_1/2, \beta_2/2\}$. Thus, r^ι_ρ satisfies (2.2), and so is an ι-ratio function. □

Proof of Lemma 2.7: Let us prove that the overlap map between the charts $i : I \to \mathbb{R}$ in $\mathcal{A}(r^\iota_\rho)$ and the charts $j : I \to \mathbb{R}$ in $\mathcal{A}^\iota(g, \rho)$ are C^{1+} compatible. By (2.6), for all ι-leaf segments $I_1, I_2 \subset K$ in I, we have

$$\left|\log \frac{|I_1|_i}{|I_2|_i}\frac{|I_2|_j}{|I_1|_j}\right| = \left|\log r(I_1 : I_2)\frac{|I_2|_\rho}{|I_1|_\rho}\right| \le \mathcal{O}\left(|K|_i^\gamma\right).$$

Hence, the overlap map (or identity map) between the charts i and j satisfies (1.1), taking in (1.1) the holonomy θ equal to the identity map, and so the overlap map has a C^{1+} extension to $\mathbb{R}$. □

Proof of Lemma 2.8: Let us take $r^\iota_g = r^\iota_\rho$, for some chosen C^{1+} Riemannian metric ρ. As observed in §1.7, the charts in $\mathcal{A}^\iota_g$ are C^{1+} compatible with the charts in $\mathcal{A}^\iota(g, \rho)$. Hence, by Lemma 2.7, $\mathcal{A}(r^\iota_\rho)$ satisfies (*). Now, the uniqueness of the HR structure follows from the f-invariance of r^s_ρ and r^u_ρ, because two HR structures that are compatible with the lamination structures have arbitrarily close ratios on sufficiently small segments, and therefore, since the ratios are f-invariant, they must be the same. □

Lemma 2.9. *Let $g_1, g_2 \in \mathcal{T}(f, \Lambda)$. If g_1 is a C^{1+} conjugated to g_2, then $(r^s_{g_1}, r^u_{g_1}) = (r^s_{g_2}, r^u_{g_2})$.*

Proof. Suppose that g_1 and g_2 are $C^{1+\beta}$ conjugated. By conjugating g_2 with the conjugacy, we obtain a new diffeomorphism g_0 that has the same invariant set Λ as g_1 and for which $g_1|\Lambda = g_0|\Lambda$. Moreover, the HR structures of g_0 and g_2 are the same, since the conjugacy maps the full ι-leaf segments of g_2 to the full ι-leaf segments of g_0, i.e $(r^s_{g_2}, r^u_{g_2}) = (r^s_{g_0}, r^u_{g_0})$. Hence, it is enough to show that $(r^s_{g_0}, r^u_{g_0}) = (r^s_{g_1}, r^u_{g_1})$. In particular, this means that K is an ι-leaf segment for g_1 if, and only if, it is one for g_2. Since g_0 and g_1 are C^{1+} conjugated, they admit a common C^{1+} atlas A. (We note that the minimal full leaf segments K_0 and K_1 containing K for g_1 and g_2 do not have to coincide). However, by Lemma 2.6 and Lemma 2.8, $r^\iota_{g_1} = r^\iota_e = r^\iota_{g_2}$ where r^ι_e is the ratio obtained using the chart $e \in A$. Therefore, the ι-ratio functions are the same for g_1 and g_2, and hence that they induce the same HR structures.

□

2.5 HR Orthogonal atlas

Let (r^s, r^u) be a HR orthogonal structure on Λ. For every rectangle R and $x \in R$, we define a unique *HR rectangle chart* $i = i_{x,R} : R \to \mathbb{R}$ as follows. For every $y \in \ell^s(x, R)$, let $i_s(y) = \pm r^s(\ell^s(x, y) : \ell^s(x, R))$ where the plus sign is chosen if y is positively oriented with respect to x, and the minus sign otherwise. Define similarly i_u. The chart i is given by $i(z) = (i_s([z, x]), i_u([x, z])) \in \mathbb{R}^2$. The *HR atlas* associated to (r^s, r^u) is the set of all HR rectangle charts constructed as above and covering Λ.

Lemma 2.10. *Let (r^s, r^u) be a HR orthogonal structure on Λ. The HR atlas associated to (r^s, r^u) is a C^{1+} orthogonal atlas with the following properties: (i) the image by $i_{x,R}$ of the ι-leaf segments passing through x determines the same affine structure on these leaf segments as the one given by the HR structure; and (ii) the map $i_{f(x),f(R)} \circ f \circ i_{x,R}^{-1}$ has an affine extension to $\mathbb{R}^2$.*

Proof. By construction, the HR rectangle charts satisfy property (i). Since the HR structure determines an affine structure along leaf segments that is kept invariant by f, for every $x \in \Lambda$, the map $i_{f(x),f(R)} \circ f \circ i_{x,R}^{-1}$ has an affine extension to $\mathbb{R}^2$. Since a HR structure determines a unique affine structure on all leaf segments and since, by Proposition 2.3, the basic holonomies for this are $C^{1+\alpha}$, for some $\alpha > 0$, the overlap map between any two canonical charts i_x and i_y has a C^{1+} extension (not necessarily unique). □

Proposition 2.11. *Let $g \in \mathcal{T}(f, \Lambda)$ with associated structure $\mathcal{S}_g$, and let $\mathcal{O}(r^s, r^u)$ denote the HR orthogonal atlas associated to (r^s, r^u). The atlas $\mathcal{O}(r^s, r^u)$ is C^{1+H} compatible with the structure $\mathcal{S}_g$, i.e for every charts $i \in \mathcal{O}(r^s, r^u)$ and $j \in \mathcal{S}_g$, the overlap map $j \circ i^{-1}$ has a C^{1+H} diffeomorphic extension to an open set of $\mathbb{R}^2$.*

Proof. Let $i_{x,R} : R \to \mathbb{R}^2$ be a chart in $\mathcal{O}(r^s, r^u)$. By Lemma 2.8, $i_{x,R}|\ell^s(x, R)$ and $i_{x,R}|\ell^s(x, R)$ have extensions to the minimal full leaf segments containing $\ell^s(x, R)$ and $\ell^u(x, R)$, respectively, C^{1+H} compatible with the C^{1+H} charts given by the Stable Manifold Theorem. Hence, by Proposition 1.10, for every chart $j \in \mathcal{S}_g$, the overlap map $j \circ i^{-1}$ has a C^{1+H} diffeomorphic extension to an open neighbourhood of $\mathbb{R}^2$. □

2.6 Complete invariant

By Lemma 2.9, if g_1 and g_2 are C^{1+} conjugated, then they determine the same HR structure on Λ. We are going to prove that if g_1 and g_2 determine the same HR structure on Λ, then g_1 and g_2 are C^{1+} conjugated.

Lemma 2.12. *Let $g \in \mathcal{T}(f, \Lambda)$ and let $h : \Lambda \to \Lambda$ be a homeomorphism preserving the order along the leaf segments. Let $\mathcal{S}$ and $\mathcal{S}'$ be C^{1+} structures on M such that there are charts $(u_1, U_1), \ldots, (u_p, U_p) \in \mathcal{S}$ and $(v_1, V_1), \ldots, (v_p, V_p) \in \mathcal{S}'$ with the following properties:*

(i) $\Lambda \subset \bigcup_{q=1}^{p} U_q$;
(ii) For every $q = 1, \ldots, p$, there is a C^{1+} diffeomorphism $h_q : U_q \to V_q$ between $\mathcal{S}$ and $\mathcal{S}'$ that extends $h|(\Lambda \bigcap U_q)$.

Then, $h : \Lambda \to \Lambda$ extends to a C^{1+} diffeomorphism defined on an open set of M.

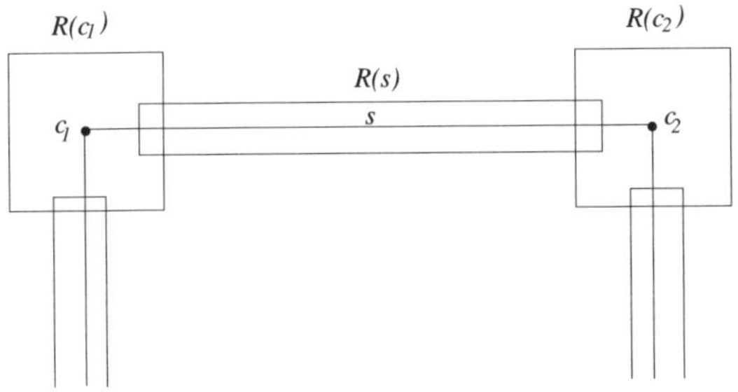

Fig. 2.1. The corner and side rectangles.

Proof. Let us just introduce some useful notions for the proof of this lemma. Recall that a rectangle R_n is an (N_s, N_u)- *Markov rectangle* if, for all $x \in R_n$, the spanning ι-leaf segments $\ell^\iota(x, R_n)$ are ι-leaf N_ι-cylinders. Let us consider the set of all (N, N)-Markov rectangles R_n, for some fixed $N > 1$. A *corner* c is an endpoint of a spanning stable leaf segment and of a spanning unstable leaf segment contained in the boundary of an (N, N)-Markov rectangle R_n. An *ι-partial side* s is a closed ι-leaf segment whose endpoints are corners and such that $\mathrm{int} s$ does not contain any corner. Let $\mathcal{C}_N$ be the set of all such corners and $\mathcal{S}_N$ be the set of all such s-partial sides and u-partial sides. For all corners $c \in \mathcal{C}_N$ and for all partial sides $s \in \mathcal{S}_N$, there are *corner rectangles* $R(c)$ and *side rectangles* $R(s)$ with the following properties (see Figure 2.1):

(i) $c \in R(c)$;
(ii) If c_1 and c_2 are corners of the ι-partial side s, then $s \subset R(c_1) \bigcup R(s) \bigcup R(c_2)$ and the ι'-boundary of $R(s)$ is contained in $R(c_1) \bigcup R(c_2)$;
(iii) The rectangles $R(c)$ are pairwise disjoint, for all $c \in \mathcal{C}_N$;
(iv) The rectangles $R(s)$ are pairwise disjoint, for all $s \in \mathcal{S}_N$.

We will consider separately the cases where (i) both the stable and unstable leaf segments are one-dimensional topological manifolds (the Anosov case); (ii) both the stable and unstable leaf segments are Cantor sets (e.g. Smale horseshoes); (iii) the stable leaf segments are Cantor sets and the unstable leaf segments are one-dimensional topological manifolds (attractors); and (iv) the stable leaf segments are one-dimensional topological manifolds and the unstable leaf segments are Cantor sets (repellers).

Case (i). In this case $\Lambda = M$, and so, by the hypotheses of this lemma, $h : M \to M$ is a C^{1+} diffeomorphism.

Case (ii). Since Λ is compact and a Cantor set, there is a finite set $\{R_n : 1 \leq n \leq m\}$ of pairwise disjoint rectangles with the following properties: (i) $\bigcup_{n=1}^m R_n \supset \Lambda$; (ii) for each rectangle R_n, there are charts $(u_n, U_n) \in \mathcal{S}$ and $(v_n, V_n) \in \mathcal{S}'$ such that $U_n \supset R_n$ and h has a C^{1+} diffeomorphic extension $h_n : U_n \to V_n$. Take pairwise disjoint open sets $U'_n \subset U_n$ such that $R_n \subset U'_n$ and the sets $V'_n = h_n(U'_n)$ are also pairwise disjoint. The map

$$\hat{h} : \bigcup_{n=1}^{m} U'_n \to \bigcup_{n=1}^{m} V'_n$$

defined by $\hat{h}|U'_n = h_n$ is a C^{1+} diffeomorphic extension of the conjugacy $h : \Lambda \to \Lambda$.

Case (iii). Since Λ is compact, there exists N large enough such that, for every (N, N)-Markov rectangle R_n, there are charts $(u_n, U_n) \in \mathcal{S}$ and $(v_n, V_n) \in \mathcal{S}'$ such that

$$U_n \supset \left(R_n \bigcup \left(\bigcup_{s \in \mathcal{S}_N \cap R_n} R(s) \right) \bigcup \left(\bigcup_{c \in \mathcal{C}_N \cap R_n} R(c) \right) \right)$$

and h has a C^{1+} diffeomorphic extension $h_n : U_n \to V_n$. For every corner $c \in \mathcal{C}_N$, we choose an (N, N)-Markov rectangle $R_{n(c)}$ containing c, and an open set $U(c) \supset R(c)$ with the following properties:

(i) For every (N, N)-Markov rectangle R_m containing c,

$$U(c) \subset U_m \quad \text{and} \quad V(c) = h_{n(c)}(U(c)) \subset V_m;$$

(ii) The sets $U(c)$ are pairwise disjoint, for all $c \in \mathcal{C}_N$; and
(iii) The sets $V(c) = h_{n(c)}(U(c))$ are also pairwise disjoint, for all $c \in \mathcal{C}_N$.

We define the C^{1+} diffeomorphic extension

$$h_{\mathcal{C}} : \bigcup_{c \in \mathcal{C}_N} U(c) \to \bigcup_{c \in \mathcal{C}_N} V(c)$$

of $h| \left(\Lambda \bigcap \left(\bigcup_{c \in \mathcal{C}_N} U(c) \right) \right)$ by $h_{\mathcal{C}}|U(c) = h_{n(c)}|U(c)$. Similarly, for every partial side $s \in \mathcal{S}_N$, we choose an (N, N)-Markov rectangle $R_{n(s)}$ containing s, and an open set $U(s) \supset R(s)$ with the following properties:

(i) For every (N, N)-Markov rectangle R_m containing s,

$$U(s) \subset U_m \quad \text{and} \quad V(s) = h_{n(s)}(U(s)) \subset V_m;$$

(ii) The sets $U(s)$ are pairwise disjoint, for all $s \in \mathcal{S}_N$;
(iii) The sets $V(s) = h_{n(s)}(U(s))$ are also pairwise disjoint, for all $s \in \mathcal{S}_N$.

We define the C^{1+} diffeomorphic extension

$$h_{\mathcal{S}} : \bigcup_{s \in \mathcal{S}_N} U(s) \to \bigcup_{s \in \mathcal{S}_N} V(s)$$

of $h| \left(\Lambda \bigcap \left(\bigcup_{s \in \mathcal{S}} U(s) \right) \right)$ by $h_{\mathcal{S}}|U(s) = h_{n(s)}|U(s)$. Let $s \in \mathcal{S}_N$ be a partial side with endpoints c_1 and c_2. We define

$$H_s : u_{n(s)}(U(s)) \to v_{n(s)}(V(s)) \quad \text{and} \quad H_{c_k} : u_{n(s)}(U(c_k)) \to v_{n(s)}(V(c_k))$$

by

$$H_s = v_{n(s)} \circ h_S \circ u^{-1}_{n(s)} \quad \text{and} \quad H_{c_k} = v_{n(s)} h_C \circ u^{-1}_{n(s)},$$

for $k \in \{1,2\}$. We choose open sets $U'(s)$, $U'(c_1)$, $U'(c_2)$, $U''(s)$ and sets $U''(c_1)$ and $U''(c_2)$ with the following properties:

(i) $U'(s) = U'(c_1) \bigcup U''(c_1) \bigcup U''(s) \bigcup U''(c_2) \bigcup U'(c_2)$;
(ii) $s \bigcap U'(s) = s \bigcap U(s)$;
(iii) $\overline{U'(c_1) \bigcup U''(c_1)} \subset U(c_1)$ and $\overline{U'(c_2) \bigcup U''(c_2)} \subset U(c_2)$;
(iv) $\overline{U''(c_1)} \subset U(s)$ and $\overline{U''(c_2)} \subset U(s)$; and
(v) $\overline{U'(c_1)} \bigcap \overline{U''(s)} = \emptyset$ and $\overline{U'(c_2)} \bigcap \overline{U''(s)} = \emptyset$.

Now, using bump functions, there is a $C^{1+H\ddot{o}lder}$ map $\tilde{H}_s : u_{n(s)}(U'(s)) \subset \mathbb{R}^2 \to \mathbb{R}^2$ with the following properties:

(i) $\tilde{H}_s | u_{n(s)}(U''(s)) = H_s$;
(ii) $\tilde{H}_s | u_{n(s)}(U'(c_k)) = H_{c_k}$, for all $k \in \{1,2\}$; and
(iii) $\tilde{H}_s(z) = v_{n(s)} \circ h \circ u^{-1}_{n(s)}(z)$, for all $z \in u_{n(s)} (U'(s) \bigcap \Lambda)$.

Using the facts that H_s and H_{c_k} coincide on $u_{n_s}(U'(s) \bigcap U(c_k) \bigcap \Lambda)$ and that $R(s)$ is compact, there is an open set $\tilde{U}(s) \subset U'(s)$ such that $s \bigcap \tilde{U}(s) = s \bigcap U'(s)$ and such that $\tilde{H}_s$ restricted to $u_{n(s)}(\tilde{U}(s))$ is injective. Set $\tilde{V}(s) = v^{-1}_{n(s)} \circ \tilde{H}_s \circ u(\tilde{U}(s))$. Letting, for every $c \in \mathcal{C}_N$, $\tilde{U}(c)$ and $\tilde{V}(c)$ be the open sets defined by

$$\tilde{U}(c) = U(c) \setminus \left(U(c) \bigcap \left(\bigcup_{s \in \mathcal{S}_N} U(s) \right) \right) \text{ and } \tilde{V}(c) = h_C(\tilde{U}(c)) ,$$

we obtain that the map

$$\tilde{h} : \left(\bigcup_{c \in \mathcal{C}_N} \tilde{U}(c) \right) \bigcup \left(\bigcup_{s \in \mathcal{S}_N} \tilde{U}(s) \right) \to \left(\bigcup_{c \in \mathcal{C}_N} \tilde{V}(c) \right) \bigcup \left(\bigcup_{s \in \mathcal{S}_N} \tilde{V}(s) \right)$$

defined by

$$\tilde{h}(z) = \begin{cases} v^{-1}_{n(s)} \circ \tilde{H}_s \circ u_{n(s)}(z), & \text{for all } z \in \bigcup_{s \in \mathcal{S}_N} \tilde{U}(s) \\ h_C(z), & \text{for all } z \in \bigcup_{c \in \mathcal{C}_N} \tilde{U}(c) \end{cases}$$

is a C^{1+} diffeomorphic extension of

$$h \left| \left(\Lambda \bigcap \left(\left(\bigcup_{c \in \mathcal{C}_N} \tilde{U}(c) \right) \bigcup \left(\bigcup_{s \in \mathcal{S}_N} \tilde{U}(s) \right) \right) \right) \right. .$$

For any (N,N)-Markov rectangle R_n, letting

$$\tilde{U}_n = \left(\bigcup_{c \in R_n \cap \mathcal{C}_N} \tilde{U}(c^n_k) \right) \bigcup \left(\bigcup_{s \in R_n \cap \mathcal{S}_N} \tilde{U}(s^n_k) \right),$$

we have that $\partial R_n \subset \tilde{U}_n$. We take open sets $\tilde{\tilde{U}}_n$ with pairwise disjoint closures and such that $\tilde{U}_n \bigcup \tilde{\tilde{U}}_n \supset R_n$. Using bump functions, there is a C^{1+} injective map

$$\hat{H}_n : u_n\left(\tilde{U}_n \bigcup \tilde{\tilde{U}}_n\right) \subset \mathbb{R}^2 \to \mathbb{R}^2$$

with the following properties:

(i) $\hat{H}_n(z) = v_n \circ \tilde{h} \circ u_n^{-1}(z)$, for all $z \in u_n\left(\tilde{U}_n \setminus \left(\tilde{\tilde{U}}_n \bigcap \tilde{U}_n\right)\right)$;
(ii) $\hat{H}_n(z) = v_n \circ h_n \circ u_n^{-1}(z)$, for all $z \in u_n\left(\tilde{\tilde{U}}_n \setminus \left(\tilde{\tilde{U}}_n \bigcap \tilde{U}_n\right)\right)$; and
(iii) $\hat{H}_n(z) = v_n \circ h \circ u_n^{-1}(z)$, for all $z \in u_n\left(\Lambda \bigcap \left(\tilde{\tilde{U}}_n \bigcup \tilde{U}_n\right)\right)$.

Using the fact that $v_n \circ \tilde{h} \circ u_n^{-1}$ and $v_n \circ h_n \circ u_n^{-1}$ coincide on $u_n\left(\Lambda \bigcap \tilde{U}_n\right)$, there is an open set $\hat{U}_n \subset \tilde{\tilde{U}}_n \bigcup \tilde{U}_n$ containing R_n such that $\hat{H}_n$ restricted to $u_n\left(\hat{U}_n\right)$ is injective. Set $\hat{V}_n = v_n^{-1} \circ \hat{H}_n \circ u_n\left(\hat{U}_n\right)$. Therefore, the map

$$\hat{h} : \bigcup_{R_n} \hat{U}_n \to \bigcup_{R_n} \hat{V}_n$$

defined by

$$\hat{h}(z) = v_n^{-1} \circ \hat{H}_n \circ u_n(z), \text{ for all } z \in \hat{U}_n,$$

is a C^{1+} diffeomorphism with $\hat{h}|\Lambda = h$, which ends the proof of this case.

Case (iv) The proof follows in a similar way to the case (iii). □

Lemma 2.13. *Let $g_1, g_2 \in \mathcal{T}(f, \Lambda)$. The maps g_1 and g_2 are C^{1+} conjugated if, and only if, they determine the same HR structures on Λ.*

Proof. Since the HR structures induced by g_1 and g_2 are the same, for every $z \in \Lambda$ and every rectangle R containing z, we have that the orthogonal charts $i : R \to \mathbb{R}^2$ defined by the HR structure are also the same for g_1 and g_2. By Proposition 2.11, for every $z \in \Lambda$, there is an open set W of M and there is an orthogonal chart $i : R \to \mathbb{R}^2$ with the following properties:

(i) $W \bigcap R = W \bigcap \Lambda$;
(ii) $i|(W \bigcap \Lambda)$ extends to a chart (u, W) that is C^{1+} compatible with the structure $\mathcal{S}_{g_1}$; and
(iii) $i|\left(W \bigcap \Lambda\right)$ extends to a chart (v, W) that is C^{1+} compatible with the structure $\mathcal{S}_{g_2}$.

Hence, the map $v \circ u^{-1} : u(W) \to v(W)$ is a C^{1+} diffeomorphism that extends the topological conjugacy between g_1 and g_2 restricted to R, given by the identity map id : $R \to R$. Hence, taking a finite set of rectangles that cover Λ, by Lemma 2.12, the topological conjugacy between g_1 and g_2 has a C^{1+} diffeomorphic extension to an open set of M. □

2.7 Moduli space

Given a HR structure (r^s, r^u), we are going to construct a corresponding C^{1+} structure $\mathcal{S}(r^s, r^u)$. Let $\{R_1, \ldots, R_n\}$ be a Markov partition for f. For every Markov rectangle R_m, we take a rectangle $\tilde{R}_m \supset R_m$ that contains a small neighbourhood of R_m with respect to the distance d_Λ. We construct an orthogonal chart $i_m : \tilde{R}_m \to \mathbb{R}^2$ as in Lemma 2.10. Let $h_{m,k} : i_m\left(\tilde{R}_m \bigcap \tilde{R}_k\right) \to i_k\left(\tilde{R}_m \bigcap \tilde{R}_k\right)$ be the map defined by $h_{m,k}(x) = i_m \circ i_k^{-1}(x)$. By Lemma 2.10, there is a C^{1+} diffeomorphic extension $H_{m,k} : U_{m,k} \to U_{k,m}$ of $h_{m,k}$ that sends vertical lines into vertical lines and horizontal lines into horizontal lines. Let us denote by S_m the rectangle in $\mathbb{R}^2$ whose boundary contains the image under i_m of the boundary of R_m. For every pair of Markov rectangles R_m and R_k that intersect in a partial side $I_{m,k} = R_m \bigcap R_k$, let $J_{m,k}$ and $J_{k,m}$ be the smallest line segments containing, respectively, the sets $i_m(I_{m,k})$ and $i_k(I_{m,k})$. Hence, $J_{k,m} = H_{m,k}(J_{m,k})$. Let $\tilde{M} = \bigsqcup_{m=1}^n S_m/\{H_{m,k}\}$ be the disjoint union of the squares S_m where we identify two points $x \in J_{m,k}$ and $y \in J_{k,m}$ if $H_{k,m}(x) = y$. Hence, $\tilde{M}$ is a topological surface possibly with boundary. By taking appropriate extensions E_m of the rectangles S_m and using the maps $H_{m,k}$ to determine the identifications along the boundaries, we get a surface $\hat{M} = \bigsqcup_{m=1}^n E_m/\{H_{m,k}\}$ without boundary. The surface $\hat{M}$ has a natural smooth structure $\mathcal{S}_{\mathrm{HR}}$ that we now describe: if a point z is contained in the interior of E_m, then we take a small open neighbourhood U_z of z contained in E_m and we define a chart $u_z : U_z \to \mathbb{R}^2$ as being the inclusion of $U_z \bigcap E_m$ into $\mathbb{R}^2$. Otherwise z is contained in a boundary of two or three or four sets $E_{m_1}, \ldots, E_{m_n}$ that we order such that the maps $I_{m_1,m_2}, \ldots, I_{m_n,m_1}$ are well-defined. In this case we take a small open neighbourhood U_z of z and we define the chart $u_z : U_z \to \mathbb{R}^2$ as follows:

(i) $u_n | (U_z \bigcap E_n)$ is the inclusion of $U_z \bigcap E_n$ into $\mathbb{R}^2$; and
(ii) $u_n | (U_z \bigcap E_j) = H_{m_{n-1},m_n} \circ \ldots \circ H_{m_j,m_{j+1}}$, for $j \in \{1, \ldots, n-1\}$.

Since the maps $H_{m_1,m_2}, \ldots, H_{m_{n-1},m_n}$ and H_{m_n,m_1} are smooth, we obtain that the set of all these charts form a C^{1+} structure $\mathcal{S}(r^s, r^u)$ on $\hat{M}$.

We will also denote by Λ its embedding into $\hat{M}$. A rectangle is also the embedding of a rectangle into $\hat{M}$. By Lemma 2.10 and by construction of the maps $H_{m,k}$, for every $z \in \Lambda$ and for every rectangle R_z, the orthogonal chart $i_z : R_z \to \mathbb{R}^2$ has an extension v_z whose restriction to an open set V_z of z is a chart C^{1+} compatible with the structure $\mathcal{S}(r^s, r^u)$.

We state a proposition due to Journé [64] that we will use in the proof of the Theorem 5.7.

Proposition 2.14. *If f is a continuous function in an open set $V \subset \mathbb{R}^2$ that is C^r along the leaves of two transverse foliations with uniformly smooth leaves, then f is C^r.*

Lemma 2.15. *Given an HR structure (r^s, r^u) on Λ, there is $g \in \mathcal{T}(f,\Lambda)$ such that $(r^s_g, r^U_g) = (r^s, r^u)$.*

Proof. For every $z \in \Lambda$, there is a rectangle R_z and a chart $(u_z, U_z) \in \mathcal{S}(r^s, r^u)$ with the following properties:

(i) $z \in R_z \bigcap U_z$ and $R_z \bigcap U_z = \Lambda \bigcap U_z$; and
(ii) $u_z | (\Lambda \bigcap U_z) = i_z | (R_z \bigcap U_z)$, where $i_z : U_z \to \mathbb{R}^2$ is an orthogonal chart as constructed in §2.5.

Hence, the map $u_{f(z)} \circ f \circ u_z^{-1} | \left(R_z \bigcap f^{-1}(R_{f(z)})\right)$ has an affine diffeomorphic extension $F_z : \mathbb{R}^2 \to \mathbb{R}^2$. Taking $U'_z = u_z(U_z) \bigcap F_z^{-1}\left(u_{f(z)}\left(U_{f(z)}\right)\right)$ and $V'_z = F_z\left(U'_z\right)$, we obtain that the map

$$f_z : u_z^{-1}\left(U'_z\right) \to u_{f(z)}^{-1}\left(V'_z\right)$$

defined by $u_z^{-1} \circ F_z \circ u_{f(z)}$ is a C^{1+} diffeomorphic extension of $f| \left(\Lambda \bigcap u_z^{-1}\left(U'_z\right)\right)$ with respect to the C^{1+} structure $\mathcal{S}(r^s, r^u)$. Thus, by compactness of Λ and by Lemma 2.12, the map $f : \Lambda \to \Lambda$ has a C^{1+} diffeomorphic extension g to an open set U_M of M with respect to the structure $\mathcal{S}(r^s, r^u)$. Let X^s be the horizontal axis in $\mathbb{R}^2$ and X^u be the vertical axis in $\mathbb{R}^2$. For every $z \in \Lambda$, we have that $T_z M = E^s_z \bigoplus E^u_z$, where $E^\iota_z = du_z(z)^{-1}(X^\iota)$. Since $u_z\left(\ell^\iota\left(z, R_z\right) \bigcap U_z\right) \subset X^\iota$, we obtain that $d\phi(z)\left(E^\iota_z\right) = E^\iota_{\phi(z)}$. Now, we take a C^{1+} Riemannian metric ρ on M compatible with the C^{1+} structure $\mathcal{S}(r^s, r^u)$. By bounded geometry of the atlases $\mathcal{A}(r^s)$ and $\mathcal{A}(r^u)$ associated with the HR structure (r^s, r^u), there exist constants $C > 0$ and $\lambda > 1$ with the following properties:

(i) $\left|d\phi^{-n}(z)v^s\right|_\rho \geq C\lambda^n \left|v^s\right|_\rho$, for all $v^s \in E^s_z$; and
(ii) $\left|d\phi^{n}(z)v^u\right|_\rho \geq C\lambda^n \left|v^u\right|_\rho$, for all $v^u \in E^u_z$.

Therefore, ϕ is a C^{1+} hyperbolic diffeomorphism in $\mathcal{T}(f,\Lambda)$. By the Stable Manifold Theorem, for every $z \in \Lambda$, there are stable and unstable C^{1+} full leaf segments passing through z. For every triple (y, z, w) of points in $\ell^\iota(z, R_z)$, let $\ell^\iota(y, z)$ be the ι-leaf segment with endpoints y and z and $\ell^\iota(z, w)$ be the ι-leaf segment with endpoints z and w. Applying Lemma 1.2, we obtain that

$$\left|\log r^\iota(\ell^\iota(y,z), \ell^\iota(z,w)) \frac{|\ell^\iota(z,w)|_\rho}{|\ell^\iota(y,z)|_\rho}\right| \leq \mathcal{O}\left(|\ell^\iota(y,w)|_\rho^\alpha\right),$$

where $0 < \alpha \leq 1$ and the constant of proportionality are uniform on $z \in \Lambda$. Therefore, the HR structure determined by ϕ is equal to the initial HR structure (r^s, r^u). □

Putting together Lemmas 2.8, 2.13 and 2.15, we obtain the following theorem.

Theorem 2.16. *The map $g \mapsto (r^s_g, r^u_g)$ determines a one-to-one correspondence between C^{1+} conjugacy classes of $g \in \mathcal{T}(f,\Lambda)$ and HR structures*

$$[g]_{C^{1+}} \quad \longleftrightarrow \quad \left(r^s_g, r^u_g\right).$$

A structure $\mathcal{S}_g$ of a C^r hyperbolic diffeomorphism $g \in \mathcal{T}(f, \Lambda)$ is *holonomically optimal*, if it maximizes the smoothness of the holonomies amongst the systems in the C^{1+} conjugacy class of g.

Theorem 2.17. *(i) The C^{1+} structure $\mathcal{S}(r^s, r^u)$ is the holonomically optimal representative of its C^{1+} conjugacy class.*

(ii) If g_1 and g_2 are C^r Anosov diffeomorphisms, with $r > 1$, determining the same HR structure, then they are C^r conjugated.

Proof. Let $g \in \mathcal{T}(f, \Lambda)$. Let $c_n : I \to [0,1]$ be defined as $d_{2,n} \circ d_{1,n}$, where $d_{1,n} : I \to f_\iota^{-n}(I)$ is given by f_ι^{-n}, and $d_{2,n} = \lambda \circ i_n$, where $i_n : f_\iota^{-n}(I) \to \mathbb{R}$ is contained in a bounded C^r lamination atlas with bounded geometry $\mathcal{A}^\iota_\phi$ induced by ϕ, and λ is the affine map of $\mathbb{R}$ that sends the endpoints of $i_n(f_\iota^{-n} I)$ into 0 and 1. Using (2.6), we obtain that $c = \lim c_n$ is a chart of the form given in (2.3) with respect to r^ι (up to scale) and c is C^{1+} compatible with the charts in $\mathcal{A}^\iota_\phi$. Since the atlas $\mathcal{A}^\iota_\phi$ has bounded geometry, the function $d_{2,n} \circ d_{1,n} \circ i_0^{-1}$ is the composition of an exponential contraction $i_n \circ d_{1,n} \circ i_0^{-1}$ in the C^r norm followed by a linear map λ. Hence, there exists $C > 0$ such that the C^r norm of $d_{2,n} \circ d_{1,n} \circ i_0^{-1}$ is bounded by C, for all $n \geq 0$. Thus, by Arzelá-Ascoli's Theorem, we obtain that the sequence $d_{2,n} \circ d_{1,n} \circ i_0^{-1}$ converges in the $C^{r-\varepsilon}$ norm to a C^r map d with C^r norm also bounded by C. Hence, $c = d \circ i_0^{-1}$ is C^r compatible with the charts in $\mathcal{A}^\iota_\phi$, and so $\mathcal{A}(r^\iota)$ is a C^r atlas. Thus, if the ι-basic holonomies $\theta : I \to J$ are C^s, for some $1 < s \leq r$, with respect to the charts in $\mathcal{A}^\iota_\phi$, then the basic holonomies are also C^s with respect to the charts in $\mathcal{A}(r^\iota)$. Since by Lemma 2.8 the charts in $\mathcal{A}(r^\iota)$ do not depend upon the $C^{r'}$ hyperbolic realizations ψ that are C^{1+} compatible with ϕ, we obtain that the basic holonomies attain, with respect to the atlas $\mathcal{A}(r^\iota)$, at least the maximum smoothness of the basic holonomies with respect to any atlas $\mathcal{A}^\iota_\psi$ induced by these realizations ψ.

By construction of the structure $\mathcal{S}_{\mathrm{HR}}$ in Lemma 2.15, the smoothness of the hyperbolic representative in this structure and the smoothness of the basic holonomies in this structure are equal to the smoothness of the basic holonomies with respect to the atlases $\mathcal{A}(r^s)$ and $\mathcal{A}(r^u)$, which ends the proof of part (i) of this theorem.

Let ϕ and ψ be two C^r Anosov diffeomorphisms that are C^{1+} conjugated, and let $\mathcal{A}^\iota_\phi$ and $\mathcal{A}^\iota_\psi$ be, respectively, C^r atlases induced by ϕ and ψ. By Lemma 2.8, ϕ and ψ determine the same pair of ratio functions (r^s, r^u). As before, the charts in $\mathcal{A}(r^\iota)$ are C^r compatible with the charts in $\mathcal{A}^\iota_\phi$ and $\mathcal{A}^\iota_\psi$, and the overlap maps have C^r uniformly bounded norm. Therefore, the conjugacy between ϕ and ψ is C^r along the stable and unstable leaves of the transverse stable and unstable foliations with uniformly smooth leaves. Hence, by Proposition 2.14 due to Journé [64], the conjugacy is C^r, which ends the proof of part (ii) of this theorem. □

2.8 Further literature

Sullivan and Ghys [44] defined affine structures on leaves for Anosov diffeomorphisms. Pinto, Rand and Sullivan developed a similar notion to HR structures for 1-dimensional expanding dynamics (see [158, 175, 230]). For Anosov diffeomorphisms of the torus that are C^2, the eigenvalue spectrum is also known to be a complete invariant of smooth conjugacy (see De la Llave [70], De la Llave, Marco and Moriyon [71], and Marco and Moriyon [75, 76]), but for hyperbolic systems on surface other than Anosov systems the eigenvalue spectra is only a complete invariant of Lipschitz conjugacy. A moduli space for Anosov diffeomorphisms of tori has been constructed by Cawley [21]. This is in terms of cohomology classes of Hölder cocycles defined on the torus. Its effectiveness for Anosov systems relies on the fact that the Lipshitz and C^{1+} theories coincide. This chapter is based on Pinto and Rand [163].

3

Solenoid functions

We present the definition of *stable* and *unstable solenoid functions*, and introduce the set $\mathcal{PS}(f)$ of all pairs of solenoid functions. To each HR structure we associate a pair (σ^s, σ^u) of solenoid functions corresponding to the stable and unstable laminations of Λ, where the solenoid functions σ^s and σ^u are the restrictions of the ratio functions r^s and r^u, respectively, to sets determined by a Markov partition of f. Since these solenoid function pairs form a nice space with a simply characterized completion, they provide a good moduli space for C^{1+} conjugacy classes of hyperbolic diffeomorphisms. For example, in the classical case of Smale horseshoes, the moduli space is the set of all pairs of positive Hölder continuous functions with the domain $\{0,1\}^{\mathbb{N}}$.

3.1 Realized solenoid functions

We are going to give an explicit construction of the solenoid functions for each hyperbolic diffeomorphism $g \in \mathcal{T}(f, \Lambda)$.

Definition 7 *Let* sol^ι *denote the set of all ordered pairs* (I, J) *of* ι*-leaf segments with the following properties:*

(i) The intersection of I *and* J *consists of a single endpoint.*
(ii) If $\delta_{\iota,f} = 1$*, then* I *and* J *are primary* ι*-leaf cylinders.*
(iii) If $0 < \delta_{\iota,f} < 1$*, then* $f_{\iota'}(I)$ *is an* ι*-leaf* 2*-cylinder of a Markov rectangle* R *and* $f_{\iota'}(J)$ *is an* ι*-leaf* 2*-gap also contained in the same Markov rectangle* R*.*

(See section 1.2 for the definitions of leaf cylinders and gaps). Pairs (I, J) where both are primary cylinders are called *leaf-leaf pairs.* Pairs (I, J) where J is a gap are called *leaf-gap pairs* and in this case we refer to J as a *primary gap.* The set sol^ι has a very nice topological structure. If $\delta_{\iota',f} = 1$ then the set sol^{ι} is isomorphic to a finite union of intervals, and if $\delta_{\iota',f} < 1$ then the set sol^{ι} is isomorphic to an embedded Cantor set on the real line.

We define a pseudo-metric $d_{\mathrm{sol}^{\iota}} : \mathrm{sol}^{\iota} \times \mathrm{sol}^{\iota} \to \mathbb{R}^{+}$ on the set sol^{ι} by

$$d_{\mathrm{sol}^{\iota}}\left((I,J),(I',J')\right) = \max\left\{d_{\Lambda}\left(I,I'\right), d_{\Lambda}\left(J,J'\right)\right\}.$$

Definition 8 *Let $g \in \mathcal{T}(f,\Lambda)$. We call the restriction of an ι-ratio function r_g^{ι} to* sol^{ι} *a* realized solenoid function σ_g^{ι}, *i.e for every $(I,J) \in \mathrm{sol}^{\iota}$,*

$$\sigma_g^{\iota}(I:J) = \lim_{n\to\infty} \frac{|g_{\iota'}^{n}(I)|_{\rho}}{|g_{\iota'}^{n}(J)|_{\rho}} = \lim_{n\to\infty} \frac{|g_{\iota'}^{n}(I)|_{i_n}}{|g_{\iota'}^{n}(J)|_{i_n}}, \tag{3.1}$$

Equality (3.1) follows from equality (2.7). By construction, the restriction of an ι-ratio function to sol^{ι} gives an Hölder continuous function satisfying the matching condition, the boundary condition and the cylinder-gap condition as we now describe.

3.2 Hölder continuity

The Hölder continuity of realized solenoid functions means that for $t = (I,J)$ and $t' = (I',J')$ in sol^{ι}, $\left|\sigma_g^{\iota}(t) - \sigma_g^{\iota}(t')\right| \le \mathcal{O}\left(\left(d_{\mathrm{sol}^{\iota}}\left(t,t'\right)\right)^{\alpha}\right)$, for some $\alpha > 0$. The Hölder continuity of σ_g^{ι} and the compactness of its domain imply that σ_g^{ι} is bounded away from zero and infinity.

3.3 Matching condition

Let $(I,J) \in \mathrm{sol}^{\iota}$ be leaf-leaf pair and suppose that we have leaf-leaf pairs

$$(I_0,I_1),(I_1,I_2),\ldots,(I_{n-2},I_{n-1}) \in \mathrm{sol}^{\iota}$$

such that $f_{\iota}(I) = \bigcup_{j=0}^{k-1} I_j$ and $f_{\iota}(J) = \bigcup_{j=k}^{n-1} I_j$. Then,

$$\frac{|f_{\iota}(I)|}{|f_{\iota}(J)|} = \frac{\sum_{j=0}^{k-1}|I_j|}{\sum_{j=k}^{n-1}|I_j|} = \frac{1+\sum_{j=1}^{k-1}\prod_{i=1}^{j}|I_i|/|I_{i-1}|}{\sum_{j=k}^{n-1}\prod_{i=1}^{j}|I_i|/|I_{i-1}|}.$$

Hence, noting that $g|\Lambda = f|\Lambda$, the realized solenoid function σ_g^{ι} must satisfy the following *matching condition* (see Figure 3.1), for all such leaf segments:

$$\sigma_g^{\iota}(I:J) = \frac{1+\sum_{j=1}^{k-1}\prod_{i=1}^{j}\sigma_g^{\iota}(I_i:I_{i-1})}{\sum_{j=k}^{n-1}\prod_{i=1}^{j}\sigma_g^{\iota}(I_i:I_{i-1})}. \tag{3.2}$$

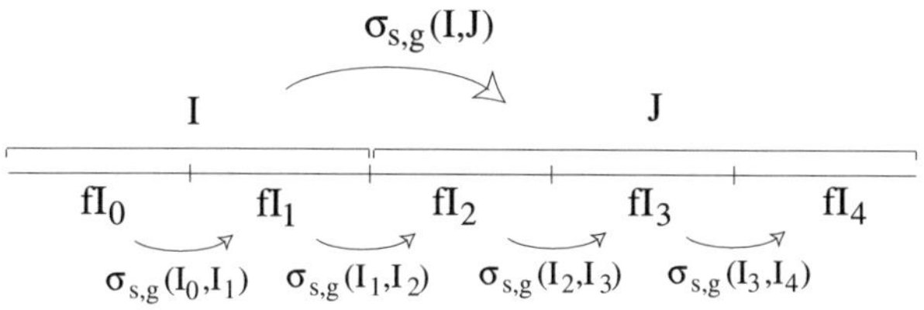

Fig. 3.1. The f-matching condition for stable leaf segments.

3.4 Boundary condition

If the stable and unstable leaf segments have Hausdorff dimension equal to 1, then leaf segments I in the boundaries of Markov rectangles can sometimes be written as the union of primary cylinders in more than one way. This gives rise to the existence of a boundary condition that the realized solenoid functions have to satisfy as we now explain.

If J is another leaf segment adjacent to the leaf segment I, then the value of $|I|/|J|$ must be the same whichever decomposition we use. If we write $J = I_0 = K_0$ and I as $\bigcup_{i=1}^{m} I_i$ and $\bigcup_{j=1}^{n} K_j$ where the I_i and K_j are primary cylinders with $I_i \neq K_j$, for all i and j, then the above two ratios are

$$\sum_{i=1}^{m} \prod_{j=1}^{i} \frac{|I_j|}{|I_{j-1}|} = \frac{|I|}{|J|} = \sum_{i=1}^{n} \prod_{j=1}^{i} \frac{|K_j|}{|K_{j-1}|} .$$

Thus, noting that $g|\Lambda = f|\Lambda$, a realized solenoid function σ_g^ι must satisfy the following *boundary condition* (see Figure 3.2), for all such leaf segments:

$$\sum_{i=1}^{m} \prod_{j=1}^{i} \sigma_g^\iota (I_j : I_{j-1}) = \sum_{i=1}^{n} \prod_{j=1}^{i} \sigma_g^\iota (K_j : K_{j-1}) . \tag{3.3}$$

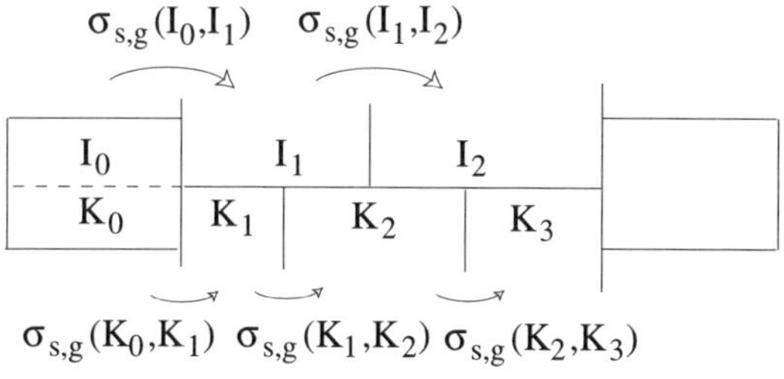

Fig. 3.2. The boundary condition for stable leaf segments.

3.5 Scaling function

If the ι-leaf segments have Hausdorff dimension less than one and the ι'-leaf segments have Hausdorff dimension equal to 1, then a primary cylinder I in the ι-boundary of a Markov rectangle can also be written as the union of gaps and cylinders of other Markov rectangles. This gives rise to the existence of a cylinder-gap condition that the ι-realized solenoid functions have to satisfy.

Before defining the cylinder-gap condition, we will introduce the scaling function that will be useful to express the cylinder-gap condition, and also the bounded equivalence classes of solenoid functions (see Definitions 10) and the (δ, P)-bounded solenoid equivalence classes of a Gibbs measure (see Definition 27).

Let scl^ι be the set of all pairs (K, J) of ι-leaf segments with the following properties:

(i) K is a leaf n_1-cylinder or an n_1-gap segment for some $n_1 > 1$;
(ii) J is a leaf n_2-cylinder or an n_2-gap segment for some $n_2 > 1$;
(iii) $m^{n_1-1}K$ and $m^{n_2-1}J$ are the same primary cylinder.

Lemma 3.1. *For every function $\sigma^\iota : \mathrm{sol}^\iota \to \mathbb{R}^+$, we present a unique extension s^ι of σ^ι to scl^ι. Furthermore, if σ^ι is the restriction of a ratio function $r^\iota|\mathrm{sol}^\iota$ to sol^ι, then $s^\iota = r^\iota|\mathrm{scl}^\iota$.*

Remark 3.2. We call $s^\iota : \mathrm{scl}^\iota \to \mathbb{R}^+$ the *scaling function* determined by the solenoid function $\sigma^\iota : \mathrm{sol}^\iota \to \mathbb{R}^+$.

Proof of Lemma 3.1. Let us construct the ι-scaling function $s : \mathrm{scl}^\iota \to \mathbb{R}^+$ from an ι-solenoid function σ. Let us proceed to construct the ι-scaling function $s : \mathrm{scl}^\iota \to \mathbb{R}^+$ from an ι-solenoid function σ. Suppose that J is an ι-leaf n-cylinder or n-gap. Then, there are pairs

$$(I_0, I_1), (I_1, I_2), \ldots, (I_{l-1}, I_l) \in \mathrm{sol}^\iota$$

such that $mJ = \bigcup_{j=0}^{l} f_{\iota'}^{n-1}(I_j)$ and $J = f_{\iota'}^{n-1}(I_s)$, for some $0 \le s \le l$. Let us denote $f_{\iota'}^{n-1}(I_j)$ by I'_j, for every $0 \le s \le l$. Then,

$$\frac{|mJ|}{|J|} = \sum_{j=0}^{l} \frac{|I'_j|}{|I'_s|} = 1 + \sum_{j=0}^{s-1} \prod_{i=s}^{j+1} \frac{|I'_{i-1}|}{|I'_i|} + \sum_{j=s+1}^{l} \prod_{i=s}^{j-1} \frac{|I'_{i+1}|}{|I'_i|},$$

where the first sum above is empty if $s = 0$, and the second sum above is empty if $s = 1$. Therefore, we define the extension s_g from σ_g to the pairs (mJ, J) by

$$s_g(mJ, J) = 1 + \sum_{j=0}^{s-1} \prod_{i=s}^{j+1} \sigma_g(I_{i-1}, I_i) + \sum_{j=s+1}^{l} \prod_{i=s}^{j-1} \sigma_g(I_{i+1}, I_i),$$

where the first sum above is empty if $s = 0$, and the second sum above is empty if $s = 1$. For every $(K, J) \in \mathrm{scl}^\iota$, there is a primary leaf segment I such that $m^{m_1} K = I = m^{m_2} J$, for some $m_1 \geq 1$ and $m_2 \geq 1$. Then,

$$\frac{|K|}{|J|} = \prod_{j=1}^{m_1} \frac{|m^j J|}{|m^{j-1} J|} \prod_{j=1}^{m_2} \frac{|m^{j-1} K|}{|m^j K|}.$$

Therefore, we define the extension s to (K, J) by

$$s(K, J) = \prod_{j=1}^{m_1} s(m^j J, m^{j-1} J) \prod_{j=1}^{m_2} s(m^{j-1} K, m^j K).$$

Hence, we have constructed a scaling function s from σ on the set scl^ι such that if σ is the restriction of a ratio function $r_g^\iota|\mathrm{sol}^\iota$ to sol^ι, then $s = r_g^\iota|\mathrm{scl}^\iota$. □

3.6 Cylinder-gap condition

Let (I, K) be a leaf-gap pair such that the primary cylinder I is the ι-boundary of a Markov rectangle R_1. Then, the primary cylinder I intersects another Markov rectangle R_2 giving rise to the existence of a cylinder-gap condition that the realized solenoid functions have to satisfy as we proceed to explain. Take the smallest $l \geq 0$ such that $f_{\iota'}^l(I) \cup f_{\iota'}^l(K)$ is contained in the intersection of the boundaries of two Markov rectangles M_1 and M_2. Let M_1 be the Markov rectangle with the property that $M_1 \cap f_{\iota'}^l(R_1)$ is a rectangle with non-empty interior (and so $M_2 \cap f_{\iota'}^l(R_2)$ also has non-empty interior). Then, for some positive n, there are distinct n-cylinder and leaf gap segments $J_1, \ldots, J_m$ contained in a primary cylinder of M_2 such that $f_{\iota'}^l(K) = J_m$ and the smallest full ι-leaf segment containing $f_{\iota'}^l(I)$ is equal to the union $\cup_{i=1}^{m-1} \hat{J}_i$, where $\hat{J}_i$ is the smallest full ι-leaf segment containing J_i. Therefore, we have that

$$\frac{|f_{\iota'}^l(I)|}{|f_{\iota'}^l(K)|} = \sum_{i=1}^{m-1} \frac{|J_i|}{|J_m|}.$$

Hence, noting that $g|\Lambda = f|\Lambda$, a realized solenoid function σ_g^ι must satisfy the *cylinder-gap condition* (see Figure 3.3), for all such leaf segments:

$$\sigma_g^\iota(I, K) = \sum_{i=1}^{m-1} s_g^\iota(J_i, J_m),$$

where s_g^ι is the scaling function determined by the solenoid function σ_g^ι.

3.7 Solenoid functions

Now, we are ready to present the definition of an ι-solenoid function.

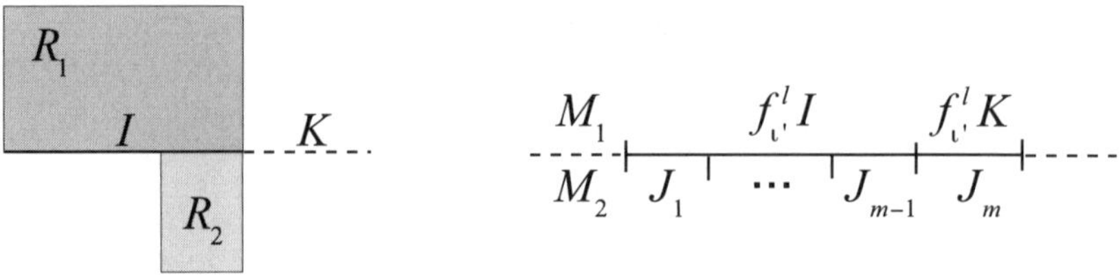

Fig. 3.3. The cylinder-gap condition for ι-leaf segments.

Definition 9 *An Hölder continuous function* $\sigma^\iota : \mathrm{sol}^\iota \to \mathbb{R}^+$ *is an* ι-solenoid function, *if* σ^ι *satisfies the following conditions:*

(i) If $\delta_{\iota,f} = 1$ *the matching condition. Furthermore, if* $\delta_{s,f} = \delta_{u,f} = 1$, *the boundary condition;*
(ii) If $\delta_{\iota,f} < 1$ *and* $\delta_{\iota',f} = 1$, *the cylinder-gap condition.*

We denote by $\mathcal{PS}(f)$ the set of pairs (σ^s, σ^u) of stable and unstable solenoid functions.

Lemma 3.3. *The map* $r^\iota \to r^\iota|\mathrm{sol}^\iota$ *gives a one-to-one correspondence between* ι*-ratio functions and* ι*-solenoid functions.*

Proof. Every ι-ratio function restricted to the set sol^ι determines an ι-solenoid function $r^\iota|\mathrm{sol}^\iota$. Now we prove the converse. Since the solenoid functions are continuous and their domains are compact, they are bounded away from 0 and ∞. By this boundedness and the f-matching condition of the solenoid functions and by iterating the domains sol^s and sol^u of the solenoid functions backward and forward by f, we determine the ratio functions r^s and r^u at very small (and large) scales, such that f leaves the ratios invariant. Then, using the boundedness again, we extend the ratio functions to all pairs of small adjacent leaf segments by continuity. By the boundary condition and the cylinder-gap condition of the solenoid functions, the ratio functions are well determined at the boundaries of the Markov rectangles. Using the Hölder continuity of the solenoid function, we deduce inequality (2.2). □

The set $\mathcal{PS}(f)$ of all pairs (σ^s, σ^u) has a natural metric given by the supremo. Combining Theorem 2.16 with Lemma 3.3, we obtain that the set $\mathcal{PS}(f)$ forms a moduli space for the C^{1+} conjugacy classes of C^{1+} hyperbolic diffeomorphisms $g \in \mathcal{T}(f, \Lambda)$:

Theorem 3.4. *The map* $g \to (r_g^s|\mathrm{sol}^s, r_g^u|\mathrm{sol}^u)$ *determines a one-to-one correspondence between* C^{1+} *conjugacy classes of* $g \in \mathcal{T}(f, \Lambda)$ *and pairs of solenoid functions in* $\mathcal{PS}(f)$.

Definition 10 *We say that any two* ι*-solenoid functions* $\sigma_1 : \mathrm{sol}^\iota \to \mathbb{R}^+$ *and* $\sigma_2 : \mathrm{sol}^\iota \to \mathbb{R}^+$ *are in the same* bounded equivalence class, *if the corresponding scaling functions* $s_1 : \mathrm{scl}^\iota \to \mathbb{R}^+$ *and* $s_2 : \mathrm{scl}^\iota \to \mathbb{R}^+$ *satisfy the following* *property: There exists a constant $C > 0$ such that, for every ι-leaf $(i+1)$-cylinder or $(i+1)$-gap J,*

$$\left|\log s_1(J, m^i J) - \log s_2(J, m^i J)\right| < C. \tag{3.4}$$

In Lemma 10.9, we prove that two C^{1+} hyperbolic diffeomorphisms g_1 and g_2 are lippeomorphic conjugate if, and only if, the solenoid functions $\sigma^{\iota}_{g_1}$ and $\sigma^{\iota}_{g_2}$ are in the same bounded equivalence class.

3.8 Further literature

The solenoid functions were first introduced in Pinto and Rand [158, 163] inspired in the scaling functions introduced by E. Faria [28], Feigenbaum [31, 32], Sullivan [230], Y. Jiang et al. [59] and Y. Jiang et al. [60]. The completion of the image of c is the set of pairs of continuous solenoid functions which is a closed subset of a Banach space. They correspond to f-invariant affine structures on the stable and unstable laminations for which the holonomies are uniformly asymptotically affine (uaa) (see definition of (uaa) in Ferreira [35], Ferreira and Pinto [36] and Sullivan [231]). This chapter is based on Pinto and Rand [163].

4

Self-renormalizable structures

We present a construction of C^{1+} stable and unstable self-renormalizable structures living in 1-dimensional spaces called *train-tracks*. The train-tracks are a form of optimal local leaf-quotient space of the stable and unstable laminations of Λ. Locally, these train-tracks are just the quotient space of stable or unstable leaves within a Markov rectangle, but globally the identification of leaves common to two more than one rectangle gives a non-trivial structure and introduces junctions. They are characterised by being the compact quotient on which the Markov map induced by the action of f is continuous with the minimal number of identifications. A smooth structure on the stable or unstable leaves of Λ induces a smooth structure on the corresponding train-tracks and *vice-versa*. Then we use the fact that the holonomies of codimension one hyperbolic systems are C^{1+} to see that the holonomies induce C^{1+} mappings of train-tracks. Together with the Markov maps, give rise to what we call C^{1+} self-renormalizable structures. We prove then the existence of a one-to-one correspondence between stable and unstable pairs of C^{1+} self-renormalizable structures and C^{1+} conjugacy classes of hyperbolic diffeomorphisms. We use this result to prove that given C^{1+H} hyperbolic diffeomorphisms f and g that are topologically conjugate, if the topological conjugacy is differentiable at a point $x \in \Lambda_f$ and the derivative at x has non-zero determinant, then h admits a C^{1+H} extension to an open neighbourhood of Λ_f.

4.1 Train-tracks

Roughly speaking, train-tracks are the optimal leaf-quotient spaces on which the stable and unstable Markov maps induced by the action of f on leaf segments are local homeomorphisms.

For each Markov rectangle R, let t_R^ι be the set of ι'-segments of R. Thus by the local product structure one can identify t_R^ι with any spanning ι-leaf segment $\ell^\iota(x, R)$ of R. We form the space $\mathbf{B}^\iota$ by taking the disjoint union $\bigsqcup_{R \in \mathcal{R}} t_R^\iota$ (union over all Markov rectangles R of the Markov partition $\mathcal{R}$) and identifying two points $I \in t^{\iota}_R$ and $J \in t^{\iota}_{R'}$ if either (i) the ι'-leaf segments I and J are ι'-boundaries of Markov rectangles and their intersection contains at leat a point which is not an endpoint of I or J, or (ii) there is a sequence $I = I_1, \dots, I_n = J$ such that all I_i, I_{i+1} are both identified in the sense of (i). This space is called the *ι-train-track* and is denoted $\mathbf{B}^{\iota}$.

Let $\pi_{\mathbf{B}^{\iota}} : \bigsqcup_{R \in \mathcal{R}} R \to \mathbf{B}^{\iota}$ be the natural projection sending $x \in R$ to the point in $\mathbf{B}^{\iota}$ represented by $\ell^{\iota'}(x, R)$. A *topologically regular point* I in $\mathbf{B}^{\iota}$ is a point with a unique preimage under $\pi_{\mathbf{B}^{\iota}}$ (that is the pre-image of I is not a union of distinct ι'-boundaries of Markov rectangles). If a point has more than one preimage by $\pi_{\mathbf{B}^{\iota}}$, then we call it a *junction*. Since there are only a finite number of ι'-boundaries of Markov rectangles, there are only finitely many junctions (see Figure 4.1).

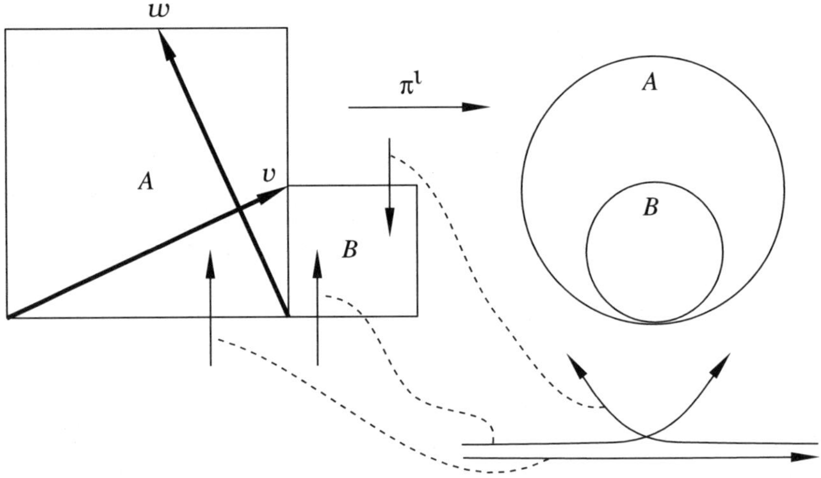

Fig. 4.1. This figure illustrates a (unstable) train-track for the Anosov map $g : \mathbb{R}^2 \setminus (\mathbb{Z}v \times \mathbb{Z}w) \to \mathbb{R}^2 \setminus (\mathbb{Z}v \times \mathbb{Z}w)$ defined by $g(x, y) = (x + y, y)$. The rectangles A and B are the Markov rectangles and the vertical arrows show paths along unstable manifolds from A to A and from B to A. The train-track is represented by the pair of circles and the curves below it show the smooth paths through the junction of the two circles which arise from the smooth paths between the rectangles A and B along unstable manifolds. Note that there is no smooth path from B to B even though in this representation of the train-track it looks as though there ought to be. This is because there is no unstable manifold running directly from the rectangle B to itself.

Let $d_{\mathbf{B}^{\iota}}$ be the metric on $\mathbf{B}^{\iota}$ defined as follows: if $\xi, \eta \in \mathbf{B}^{\iota}$, $d_{\mathbf{B}^{\iota}}(\xi, \eta) = d_{\Lambda}(\xi, \eta)$.

4.2 Charts

We say that I_T is a *train-track segment*, if there is an ι-leaf segment I, not intersecting ι-boundaries of Markov rectangles, such that $\pi_{\mathbf{B}^\iota}|I$ is an injection and $\pi_{\mathbf{B}^\iota}(I) = I_T$. Let $\mathcal{A}$ be an ι-lamination atlas (take for instance $\mathcal{A}$ equal to $\mathcal{A}^\iota(f,\rho)$ or to $\mathcal{A}(r_f^\iota)$). The chart $i : I \to \mathbb{R}$ in $\mathcal{A}$ determines a *train-track chart* $i_T : I_T \to \mathbb{R}$ for I_T given by $i_T = i \circ \pi_{\mathbf{B}^\iota}^{-1}$. We denote by $\mathcal{B}$ the set of all train-track charts for all train-track segments determined by $\mathcal{A}$.

Two train-track charts (i_T, I_T) and (j_T, J_T) on the train-track $\mathbf{B}^\iota$ are C^{1+} *compatible*, if the *overlap map* $j_T \circ i_T^{-1} : i_T(I_T \cap J_T) \to j_T(I_T \cap J_T)$ has a C^{1+} extension. A C^{1+} atlas $\mathcal{B}$ is a set of C^{1+} compatible charts with the following property: For every short train-track segment K_T there is a chart $(i_T, I_T) \in \mathcal{B}$ such that $K_T \subset I_T$. A C^{1+} *structure* $\mathcal{S}$ *on* $\mathbf{B}^\iota$ is a maximal set of C^{1+} compatible charts with a given atlas $\mathcal{B}$ on $\mathbf{B}^\iota$. We say that two C^{1+} structures $\mathcal{S}$ and $\mathcal{S}'$ are in the same *Lipschitz equivalence class*, if, for every chart e_1 in $\mathcal{S}$ and every chart e_2 in $\mathcal{S}'$, the overlap map $e_1 \circ e_2^{-1}$ has a bi-Lipschitz extension.

Given any train-track charts $i_T : I_T \to \mathbb{R}$ and $j_T : J_T \to \mathbb{R}$ in $\mathcal{B}$, the overlap map $j_T \circ i_T^{-1} : i_T(I_T \cap J_T) \to j_T(I_T \cap J_T)$ is equal to $j_T \circ i_T^{-1} = j \circ \theta \circ i^{-1}$, where $i = i_T \circ \pi_{\mathbf{B}^\iota} : I \to \mathbb{R}$ and $j = j_T \circ \pi_{\mathbf{B}^\iota} : J \to \mathbb{R}$ are charts in $\mathcal{A}$, and

$$\theta : i^{-1}(i_T(I_T \cap J_T)) \to j^{-1}(j_T(I_T \cap J_T))$$

is a basic ι-holonomy. Let us denote by $\mathcal{B}^\iota(g,\rho)$ and $\mathcal{B}(r_g^\iota)$ the train-track atlases determined, respectively, by $\mathcal{A}^\iota(g,\rho)$ and $\mathcal{A}(r_g^\iota)$ with $g \in \mathcal{T}(f,\Lambda)$.

Lemma 4.1. *The atlases* $\mathcal{B}^\iota(g,\rho)$ *and* $\mathcal{B}(r_g^\iota)$ *are* C^{1+}.

Proof. Since $\mathcal{A}^\iota(g,\rho)$ and $\mathcal{A}(r_g^\iota)$ are C^{1+} foliated atlases, there exists $\eta > 0$ such that, for all train-track charts i_T and j_T in $\mathcal{B}^\iota(g,\rho)$ (or in $\mathcal{B}(r_g^\iota)$), the overlap maps $j_T \circ i_T^{-1} = j \circ \theta \circ i^{-1}$ have $C^{1+\eta}$ diffeomorphic extensions with a uniformly bound for their $C^{1+\eta}$ norm. Hence, $\mathcal{B}^\iota(g,\rho)$ and $\mathcal{B}(r_g^\iota)$ are $C^{1+\eta}$ atlases. □

4.3 Markov maps

The *Markov map* $m_\iota : \mathbf{B}^\iota \to \mathbf{B}^\iota$ is the mapping induced by the action of f on leaf segments, that it is defined as follows: If $I \in \mathbf{B}^\iota$, $m_\iota(I) = \pi_{\mathbf{B}^\iota}(f_\iota(I))$ is the ι'-leaf segment containing the f_ι-image of the ι'-leaf segment I. This map m_ι is a local homeomorphism because f_ι sends a short ι-leaf segment homeomorphically onto a short ι-leaf segment.

Consider the Markov map m_ι on $\mathbf{B}^\iota$ induced by the action of f on ι'-leaves and described above. For $n \geq 1$, an n-cylinder is the projection into $\mathbf{B}^\iota$ of an ι-leaf n-cylinder segment in Λ. Thus, each Markov rectangle in Λ

projects in a unique primary ι-leaf segment in $\mathbf{B}^\iota$. For $n \geq 1$, an n-gap of m_ι is the projection into $\mathbf{B}^\iota$ of a ι-leaf n-gap in Λ. We say that $\mathbf{B}^\iota$ is a *no-gap train-track* if $\mathbf{B}^\iota$ does not have gaps. Otherwise, we call $\mathbf{B}^\iota$ a *gap train-track.*

Given a topological chart (e, U) on the train-track $\mathbf{B}^\iota$ and a train-track segment $C \subset U$, we denote by $|C|_e$ the length of the smallest interval containing $e(C)$. We say that m_ι has *bounded geometry* in a C^{1+} atlas $\mathcal{B}$ if there is $\kappa_1 > 0$ such that, for every n-cylinder C_1 and n-cylinder or n-gap C_2 with a common endpoint with C_1, we have $\kappa_1^{-1} < |C_1|_e/|C_2|_e < \kappa_1$, where the lengths are measured in any chart (e, U) of the atlas such that $C_1 \cup C_2 \subset U$.

We note that if m_ι has bounded geometry in a C^{1+} atlas $\mathcal{B}$, then there are $\kappa_2 > 0$ and $0 < \nu < 1$ such that $|C|_e \leq \kappa_2 \nu^n$ for every n-cylinder or n-gap C and every $e \in \mathcal{B}$. We say that the Markov map m_ι is expanding with respect to an atlas $\mathcal{B}$ if there are $c \geq 0$ and $\lambda > 1$ such that, for every $x \in \mathbf{B}^\iota$ and every $n \geq 0$,

$$\left(j \circ m_\iota^n \circ i^{-1}\right)'(x) > c\lambda^n,$$

where $i : I \to \mathbb{R}$ and $j_n : J \to \mathbb{R}$ are any charts in $\mathcal{B}$ such that $x \in I$ and $f^n(x) \in J_n$. We note that m_ι has bounded geometry in $\mathcal{B}$ if, and only if, m_ι is expanding with respect to $\mathcal{B}$.

Lemma 4.2. *The Markov map m_ι is a C^{1+} local diffeomorphism with bounded geometry with respect to the atlases $\mathcal{B}(r^\iota)$ and $\mathcal{B}^\iota(g, \rho_g)$.*

Proof. Since f on Λ along leaves has affine extensions with respect to the charts in $\mathcal{A}(r^\iota)$ and the basic ι-bolonomies have $C^{1+\eta}$ extensions we get that the Markov maps m_ι also have $C^{1+\eta}$ extensions with respect to the charts in $\mathcal{B}(r^\iota)$ for some $\eta > 0$. Since $\mathcal{A}(r^\iota)$ has bounded geometry, we obtain that m_ι also has *bounded geometry* in $\mathcal{B}(r^\iota)$. Since, for every $g \in \mathcal{T}(f, \Lambda)$, the C^{1+} lamination atlas $\mathcal{A}^\iota(g, \rho_g)$ has bounded geometry we obtain that the Markov map m_ι has $C^{1+\eta}$ extensions with respect to the charts in $\mathcal{B}^\iota(g, \rho_g)$, for some $\eta > 0$, and has bounded geometry. □

4.4 Exchange pseudo-groups

The elements $\tilde{\theta}_\iota = \tilde{\theta}_{f,\iota}$ of the *holonomy pseudo-group on* $\mathbf{B}^\iota$ are the mappings defined as follows. Suppose that I and J are ι-leaf segments and $\theta : I \to J$ a holonomy. Then, it follows from the definition of the train-track $\mathbf{B}^\iota$ that the map $\tilde{\theta} : \pi_{\mathbf{B}^\iota}(I) \to \pi_{\mathbf{B}^\iota}(J)$ given by $\tilde{\theta}(\pi_{\mathbf{B}^\iota}(x)) = \pi_{\mathbf{B}^\iota}(\theta(x))$ is well-defined. The collection of all such local mappings forms the *basic holonomy pseudo-group of* $\mathbf{B}^\iota$. Note that if x is a junction of $\mathbf{B}^\iota$, then there may be segments I and J containing x such that $I \cap J = \{x\}$. The image of I and J under the holonomies will not agree in that they will map x differently.

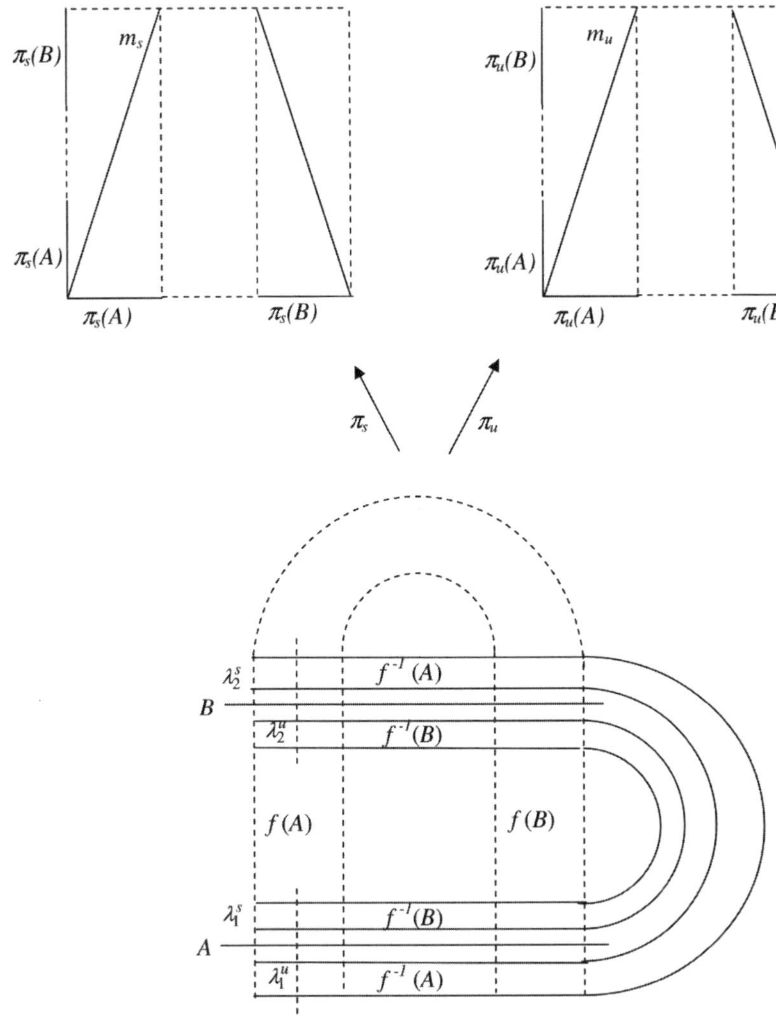

Fig. 4.2. A Markov partition for the Smale horseshoe f into two rectangles A and B. A representation of the Markov maps $m_s : \Theta^s \to \Theta^s$ and $m_u : \Theta^u \to \Theta^u$ for Smale horseshoes.

4.5 Markings

Recall, from §1.2, the definition of the two-sided shift $\tau : \Theta \to \Theta$ on the two sided symbol space Θ and of the marking $i : \Theta \to \Lambda$.

Let Θ^u be the set of all words $w_0 w_1 \dots$ which extend to words $\dots w_0 w_1 \dots$ in Θ, and, similarly, let Θ^s be the set of all words $\dots w_{-1} w_0$ which extend to words $\dots w_{-1} w_0 \dots$ in Θ. Then, $\pi_u : \Theta \to \Theta^u$ and $\pi_s : \Theta \to \Theta^s$ are the natural projection given, respectively, by

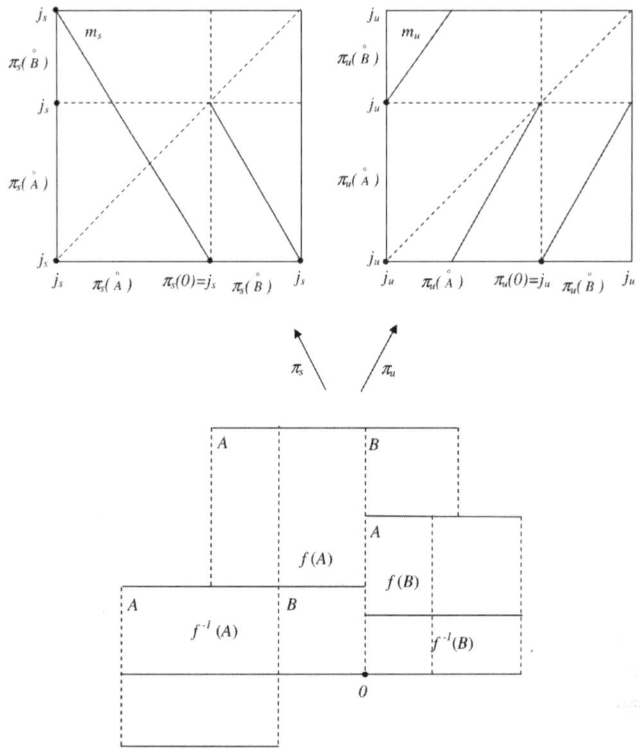

Fig. 4.3. A representation of the Markov maps $m_s : \Theta^s \to \Theta^s$ and $m_u : \Theta^u \to \Theta^u$ as maps of the interval for Anosov diffeomorphisms.

$$\pi_u(\ldots w_{-1}w_0w_1\ldots) = w_0w_1\ldots \quad \text{and} \quad \pi_s(\ldots w_{-1}w_0w_1\ldots) = \ldots w_{-1}w_0 \; .$$

An n-cylinder $\Theta^u_{w_0\cdots w_{n-1}}$ is equal to $\pi_u(\Theta_{w_0\cdots w_{n-1}})$ where $\Theta_{w_0\cdots w_{n-1}}$ is a $(0, n-1)$-cylinder of Θ, and an n-cylinder $\Theta^s_{w_{-(n-1)}\cdots w_0}$ is equal to $\pi_s(\Theta_{w_{-(n-1)}\cdots w_0})$ where $\Theta_{w_{-(n-1)}\cdots w_0}$ is a $(n-1,0)$-cylinder of Θ. Let $\tau_u : \Theta^u \to \Theta^u$ and $\tau_s : \Theta^s \to \Theta^s$ be the corresponding one-sided shifts.

The Markov partition $\mathcal{R} = \{R_1, \ldots, R_m\}$ for (f, Λ) induces a Markov partition $\mathcal{R}^\iota = \{R_1^\iota, \ldots, R_m^\iota\}$ for the Markov map m_ι on the train-track $\mathbf{B}^\iota$. The marking $i : \Theta \to \Lambda$ determines unique *markings* $i_u : \Theta^u \to \mathbf{B}^u$ and $i_s : \Theta^s \to \mathbf{B}^s$ such that $i_u(w_0w_1\ldots) = \cap_{i\geq 0} R^u_{w_i}$ and $i_s(\ldots w_{-1}w_0) = \cap_{i\geq 0} R^u_{w_i}$.

We note that $\pi_{\mathbf{B}^\iota} \circ i = i_\iota \circ \pi_\iota$. The map i_ι is continuous, onto $\mathbf{B}^\iota$ and semiconjugates the shift map on Θ^ι to the Markov map on $\mathbf{B}^\iota$. Defining $\varepsilon, \varepsilon' \in \Theta^\iota$ to be equivalent ($\varepsilon \sim \varepsilon'$) if $i^\iota(\varepsilon) = i^\iota(\varepsilon')$, we get that the space $\Theta^\iota / \sim$ is homeomorphic to the train-track $\mathbf{B}^\iota$.

4.6 Self-renormalizable structures

The C^{1+} structure $\mathcal{S}_\iota$ on $\mathbf{B}^\iota$ is an ι *self-renormalizable structure*, if it has the following properties:

(i) In this structure the Markov mapping m_ι is a local diffeomorphism and has bounded geometry in some C^{1+} atlas of this structure.
(ii) The elements of the basic holonomy pseudo-group are local diffeomorphisms in $\mathcal{S}_\iota$.

We say that $\mathcal{B}$ is a C^{1+} *self-renormalizable atlas*, if $\mathcal{B}$ has bounded geometry and extends to a C^{1+} self-renormalizable structure. By definition, a C^{1+} self-renormalizable structure contains a C^{1+} self-renormalizable atlas.

Lemma 4.3. *A C^{1+} foliated ι-lamination atlas $\mathcal{A}$ induces a C^{1+} ι self-renormalizable atlas $\mathcal{B}$ on $\mathbf{B}^\iota$ (and vice-versa).*

Since $\mathcal{A}(r^\iota)$ and $\mathcal{A}^\iota(\rho)$ are C^{1+} foliated ι-lamination atlases, we obtain that the atlases $\mathcal{B}(r^\iota)$ and $\mathcal{B}^\iota(g, \rho)$ determine, respectively, C^{1+} self-renormalizable structures $\mathcal{S}(r^\iota)$ and $\mathcal{S}(g, \iota)$ (see also lemmas 4.1 and 4.2).

Proof of lemma 4.3. The holonomies are C^{1+} with respect to the atlas $\mathcal{A}$, and so the charts in $\mathcal{B}$ are C^{1+} compatible and the basic holonomy pseudo-group of $\mathbf{B}^\iota$ are local diffeomorphisms. Since $\mathcal{A}$ has bounded geometry, the Markov mapping m_ι is a local diffeomorphism and also has bounded geometry in $\mathcal{B}$. Therefore, $\mathcal{B}$ is a C^{1+} self-renormalizable atlas and extends to a C^{1+} self-renormalizable structure $\mathcal{S}(\mathcal{B})$ on $\mathbf{B}^\iota$. □

Lemma 4.4. *The map $r^\iota \to \mathcal{S}(r^\iota)$ determines a one-to-one correspondence between ι-ratio functions (or, equivalently, ι-solenoid functions $r^\iota|\mathrm{sol}^\iota$) and C^{1+} self-renormalizable structures on $\mathbf{B}^\iota$.*

Proof. Every ratio function r^ι determines a unique C^{1+} self-renormalizable $\mathcal{S}$. Conversely, let us prove that a given C^{1+} self-renormalizable structure $\mathcal{S}$ on $\mathbf{B}^\iota$ also determines a unique ratio function $r^\iota_\mathcal{S}$. Let $\mathcal{B}$ be a bounded atlas for $\mathcal{S}$. Consider a small leaf segment K and two leaf segments I and J contained in K. Since the elements of the basic holonomy pseudo-group on $\mathbf{B}^\iota$ are C^{1+} and the Markov map is also C^{1+} and has bounded geometry, we obtain by Taylor's Theorem that the following limit exists

$$
\begin{aligned}
r^{\iota}_{\mathcal{S}}(I:J) &= \lim_{n\to\infty} \frac{|\pi_\iota(f^n_{\iota'}(I))|_{i_n}}{|\pi_\iota(f^n_{\iota'}(J))|_{i_n}} \\
&\in \frac{|\pi_\iota(I)|_{i_0}}{|\pi_\iota(J)|_{i_0}} \left(1 \pm \mathcal{O}(|\pi_\iota(K)|^{\gamma}_{i_0})\right),
\end{aligned} \tag{4.1}
$$

where the size of the leaf segments are measured in charts of the bounded atlas $\mathcal{B}$. Furthermore, by §2 and (4.1), the charts in $\mathcal{B}(r^\iota)$ and the charts in $\mathcal{B}$ are C^{1+} equivalent, and so determine the same C^{1+} self-renormalizable structure. □

4.7 Hyperbolic diffeomorphisms

Let $g \in \mathcal{T}(f,\Lambda)$ and $\mathcal{A}(g,\rho)$ be the C^{1+}foliated ι-lamination atlas determined by the Riemannian metric ρ. As shown in §4.6, the atlas $\mathcal{A}(g,\rho)$ induces a C^{1+} self-renormalizable atlas $\mathcal{B}(g,\rho)$ on $\mathbf{B}^\iota$ which generates a C^{1+} self-renormalizable structure $\mathcal{S}(g,\iota)$.

Lemma 4.5. *The mapping $g \to (\mathcal{S}(g,s),\mathcal{S}(g,u))$ gives a 1-1 correspondence between C^{1+} conjugacy classes in $\mathcal{T}(f,\Lambda)$ and pairs $(\mathcal{S}(g,s),\mathcal{S}(g,u))$ of C^{1+} self-renormalizable structures. Furthermore, $r^s_g = r^s_{\mathcal{S}(g,s)}$ and $r^u_g = r^u_{\mathcal{S}(g,u)}$.*

Proof. By Lemma 4.4, the pair $(\mathcal{S}_s,\mathcal{S}_u)$ determines a pair $(r^s_{\mathcal{S}}|\mathrm{sol}^s, r^u_{\mathcal{S}}|\mathrm{sol}^u)$ of solenoid functions and vice-versa. By Theorem 3.4, the pair $(r^s_{\mathcal{S}}|\mathrm{sol}^s, r^u_{\mathcal{S}}|\mathrm{sol}^u)$ determines a unique C^{1+} conjugacy class of diffeomorphisms $g \in \mathcal{T}(f,\Lambda)$ which realize the pair $(r^s_{\mathcal{S}}|\mathrm{sol}^s, r^u_{\mathcal{S}}|\mathrm{sol}^u)$ and vice-versa (and so $(\mathcal{S}(g,s), \mathcal{S}(g,u)) = (\mathcal{S}_s,\mathcal{S}_u)$). Furthermore, by Lemma 3.3, we get $r^s_g = r^s_{\mathcal{S}(g,s)}$ and $r^u_g = r^u_{\mathcal{S}(g,u)}$. □

4.8 Explosion of smoothness

The following result for C^{1+} hyperbolic diffeomorphisms f and g topologically conjugate by h shows that the smoothness of the conjugacy extends from a point to a neighbourhood of the invariant set Λ_f.

Theorem 4.6. *Let f and g be $C^{1+\text{Hölder}}$ hyperbolic diffeomorphisms that are topologically conjugate on their basic sets Λ_f and Λ_g. If the conjugacy is differentiable at a point $x \in \Lambda_f$, then f and g are $C^{1+\text{Hölder}}$ conjugate with non-zero determinant.*

Proof. Given a Markov partition $\mathcal{M}_f = \{R_1,\ldots,R_m\}$ of f, we consider the Markov partition of g given by $\mathcal{M}_g = \{h(R_1),\ldots,h(R_m)\}$. The conjugacy $h:\Lambda_f \to \Lambda_g$ determines the conjugacy $\psi_s : \mathbf{B}^s_f \to \mathbf{B}^s_g$ between the Markov maps $m_{f,s}$ and $m_{g,s}$, and the conjugacy $\psi_u : \mathbf{B}^u_f \to \mathbf{B}^u_g$ between the Markov maps $m_{f,u}$ and $m_{g,u}$ such that the following diagrams commute:

$$\begin{array}{ccc} \Lambda_f & \xrightarrow{h} & \Lambda_g \\ \downarrow{\pi_{f,s}} & & \downarrow{\pi_{f,s}} \\ \mathbf{B}^s_f & \xrightarrow{\psi_s} & \mathbf{B}^s_g \end{array} \quad \text{and} \quad \begin{array}{ccc} \Lambda_f & \xrightarrow{h} & \Lambda_g \\ \downarrow{\pi_{f,u}} & & \downarrow{\pi_{f,u}} \\ \mathbf{B}^u_f & \xrightarrow{\psi_u} & \mathbf{B}^u_g \end{array}$$

□

Since the conjugacy h is differentiable at a point $x \in \Lambda$, the conjugacies ψ_s and ψ_u are differentiable at the points $\pi_{f,s}(x)$ and $\pi_{f,u}(x)$ with respect to the atlases $\mathcal{B}^s(f, \rho_f)$, $\mathcal{B}^s(g, \rho_g)$, $\mathcal{B}^u(f, \rho_f)$ and $\mathcal{B}^u(g, \rho_g)$ compatible with the C^{1+} structure of the full leaf segments determined by the Stable Manifold Theorem. By Alves et al. [6], the Markov maps $m_{f,s}$ and $m_{g,s}$ are C^{1+} conjugate, and the Markov maps $m_{f,u}$ and $m_{g,u}$ are C^{1+} conjugate. Hence, in particular, the charts in the atlas $\mathcal{B}^s(f, \rho_f)$ are C^{1+} compatible with the charts in $\mathcal{B}^s(g, \rho_g)$, and the charts in the atlas $\mathcal{B}^u(f, \rho_f)$ are C^{1+} compatible with the charts in $\mathcal{B}^u(g, \rho_g)$. Therefore, by Lemma 4.5, the conjugacy $h : \Lambda_f \to \Lambda_g$ has a C^{1+} extension to an open set of Λ_f.

4.9 Further literature

Sullivan [231] stated the following rigidity theorem for a topological conjugacy between two expanding circle maps: if the conjugacy is differentiable at a point, then the conjugacy is smooth everywhere. De Faria [28] proved a stronger version of D. Sullivan's result, showing that it is sufficient the conjugacy to be uniformly asymptotically affine (uaa) at a point to imply that the conjugacy is smooth everywhere. In Ferreira and Pinto [38], a generalization of these results to a larger class of one-dimensional expanding maps is presented. In Ferreira et al. [37], these results are extended to C^{1+} hyperbolic diffeomorphisms. In Alves et al. [6], these results are extended to non-uniformly one-dimensional expanding maps. This chapter is based on Ferreira and Pinto [37], Pinto, Rand and Ferreira [173] and Pinto and Rand [168].

5

Rigidity

In dynamics, rigidity occurs when simple topological and analytical conditions on the model system imply that there is no flexibility and so there is a unique smooth realization. One can paraphrase this by saying that the moduli space for such systems is a singleton. For example, a famous result of this type due to Arnol'd, Herman and Yoccoz is that a sufficiently smooth diffeomorphism of the circle with an irrational rotation number satisfying the usual Diophantine condition is C^{1+} conjugate to a rigid rotation. The rigidity depends upon both the analytical hypothesis concerning the smoothness and the topological condition given by the rotation number, and if either are relaxed, then it fails. The analytical part of the rigidity hypotheses for hyperbolic surface dynamics will be a condition on the smoothness of the holonomies along stable and unstable manifolds. Given a diffeomorphism f on a surface with a hyperbolic invariant set Λ (with local product structure and with a dense orbit on Λ), we show that if the holonomies are sufficiently smooth, then the diffeomorphism f is rigid; i.e., there is a conjugacy on Λ between f and a hyperbolic affine model which has a C^{1+} extension to the surface.

5.1 Complete sets of holonomies

Before introducing the notion of a $C^{1,HD^{\iota}}$ complete set of holonomies, we define the $C^{1,\alpha}$ regularities for diffeomorphisms, with $0 < \alpha \leq 1$.

Definition 11 *Let $h : I \subset \mathbb{R} \to J \subset \mathbb{R}$ be a homeomorphism. For $0 < \alpha < 1$, the homeomorphism h is $C^{1,\alpha}$ if it is differentiable and, for all points $x, y \in I$,*

$$|h'(y) - h'(x)| \leq \chi_h(|y - x|), \tag{5.1}$$

where the positive function $\chi_h(t)$ is $o(t^\alpha)$, that is $\lim_{t \to 0} \chi_h(t)/t^\alpha = 0$.

The map $h : I \to J$ is $C^{1,1}$ if, for all points $x, y \in I$,

$$\left| \log h'(x) + \log h'(y) - 2 \log h' \left(\frac{x+y}{2} \right) \right| \leq \chi_h(|y - x|), \tag{5.2}$$

where the positive function $\chi_h(t)$ *is* $o(t)$*, that is* $\lim_{t\to 0}\chi_\theta(t)/t = 0$*. The functions* χ_h *are called the* modulus of continuity of h.

In particular, for every $\beta > \alpha > 0$, a $C^{1+\beta}$ diffeomorphism is $C^{1,\alpha}$, and, for every $\gamma > 0$, a $C^{2+\gamma}$ diffeomorphism is $C^{1,1}$. We note that the regularity $C^{1,1}$ (also denoted by $C^{1+zigmund}$) of a diffeomorphism h used here is stronger than the regularity $C^{1+Zigmund}$ (see de Melo and van Strien [99] and Pinto and Sullivan [175]). The importance of these $C^{1,\alpha}$ smoothness classes for a homeomorphism $h : I \to J$ follows from the fact that if $0 < \alpha < 1$, then the map h will distort ratios of lengths of short intervals in an interval $K \subset I$ by an amount that is $o(|I|^\alpha)$, and if $\alpha = 1$, the map h will distort the cross-ratios of quadruples of points in an interval $K \subset I$ by an amount that is $o(|I|)$.

Let $\mathcal{M}$ be a Markov partition for f satisfying the disjointness property (see §1.2). Suppose that M and N are Markov rectangles, and $x \in M$ and $y \in N$. We say that x and y are ι- *holonomically related* if (i) there is an ι'-leaf segment $\ell^{\iota'}(x, y)$ such that $\partial\ell^{\iota'}(x, y) = \{x, y\}$, and (ii) $\ell^{\iota'}(x, y) \subset \ell^{\iota'}(x, M) \cup \ell^{\iota'}(y, N)$. Let $P^\iota = P^\iota_{\mathcal{M}}$ be the set of all pairs (M, N) such that there are points $x \in M$ and $y \in N$ ι-holonomically related.

For every Markov rectangle $M \in \mathcal{M}$, choose an ι-spanning leaf segment ℓ^ι_M in M. Let $\mathcal{I}^\iota = \{\ell^\iota_M : M \in \mathcal{M}\}$. For every pair $(M, N) \in P^\iota$, there are maximal leaf segments $\ell^D_{(M,N)} \subset \ell^\iota_M$, $\ell^C_{(M,N)} \subset \ell^\iota_N$ such that there is a well-defined ι-holonomy $h^\iota_{(M,N)} : \ell^D_{(M,N)} \to \ell^C_{(M,N)}$. We call such holonomies $h^\iota_{(M,N)} : \ell^D_{(M,N)} \to \ell^C_{(M,N)}$ the ι-primitive holonomies associated to the Markov partition $\mathcal{M}$. The set $\mathcal{H}^\iota = \{h^\iota_{(M,N)} : \ell^D_{(M,N)} \to \ell^C_{(M,N)}; (M, N) \in P^\iota\}$ is a *complete set of* ι*-holonomies* (see Figures 5.1 and 5.2).

For every leaf segment $\ell^\iota_M \in \mathcal{I}^\iota$, let $\hat{\ell}^\iota_M$ be the smallest full ι-leaf segment containing ℓ^ι_M (see definition in §1.1). By the Stable Manifold Theorem, there are $C^{1+\alpha}$ diffeomorphisms $u^\iota_M : \hat{\ell}^\iota_M \to K^\iota_M \subset \mathbb{R}$.

Definition 12 *A complete set of holonomies* $\mathcal{H}^\iota$ *is* C^{1,HD^ι} *if for every holonomy* $h^\iota_{(M,N)} : \ell^D_{(M,N)} \to \ell^C_{(M,N)}$ *in* $\mathcal{H}^\iota$*, the map* $u^\iota_N \circ h^\iota_{(M,N)} \circ (u^\iota_M)^{-1}$ *and its inverse have a* C^{1,HD^ι} *diffeomorphic extension to* $\mathbb{R}$ *such that the modulus of continuity does not depend upon* $h^\iota_{(M,N)} \in \mathcal{H}^\iota$.

For many systems such as Anosov diffeomorphisms and codimension 1 attractors, there is only a finite number of holonomies in a complete set. In this case the uniformity hypothesis in the modulus of continuity of Definition 12 is redundant. However, for Smale horseshoes, this is not the case (see Figure 5.2).

Definition 13 *An* hyperbolic affine model for f on Λ *is an atlas* $\mathcal{A}$ *with the following properties (see Figure 5.3):*

(i) the union of the domains U *of the charts* $i : U \to \mathbb{R}^2$ *of* $\mathcal{A}$ *(which are open sets of* M*) cover* Λ*;*

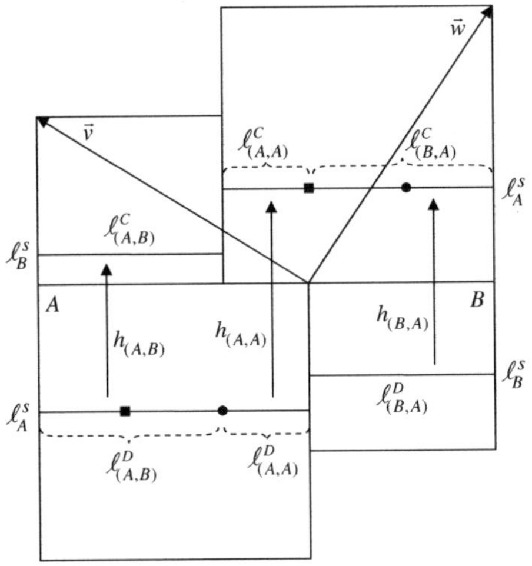

Fig. 5.1. The complete set of holonomies $\mathcal{H} = \{h_{(A,A)}, h_{(A,B)}, h_{(B,A)}, h^{-1}_{(A,A)}, h^{-1}_{(A,B)}, h^{-1}_{(B,A)}\}$ for the Anosov map $f : \mathbf{R}^2 \setminus (\mathbf{Zv} \times \mathbf{Zw}) \to \mathbf{R}^2 \setminus (\mathbf{Zv} \times \mathbf{Zw})$ defined by $f(x,y) = (x+y,y)$ and with Markov partition $\mathcal{M} = \{A, B\}$.

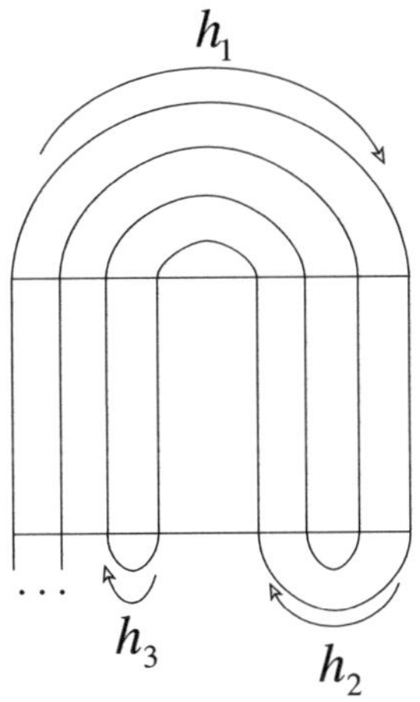

Fig. 5.2. The cardinality of the complete set of holonomies $\mathcal{H} = \{h_1, h_2, h_3, \ldots\}$ is not finite.

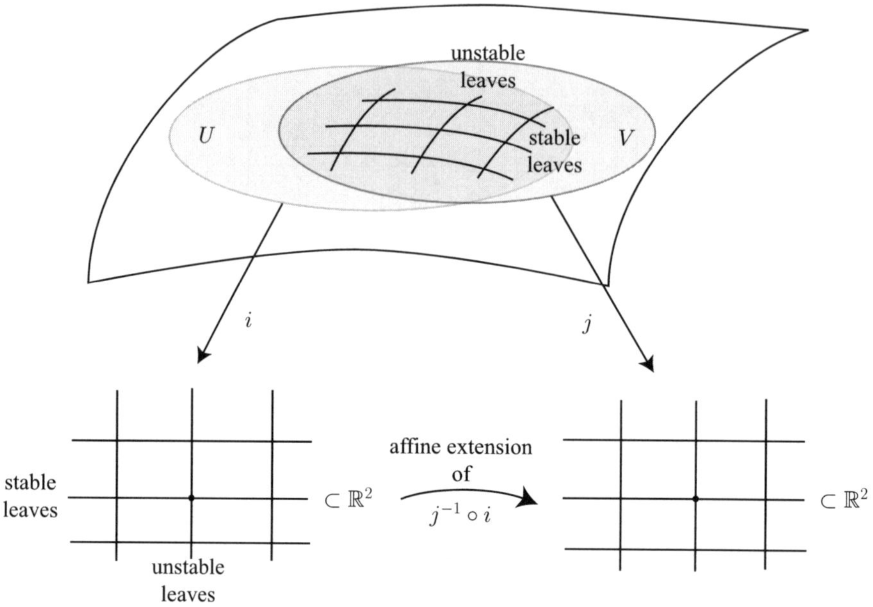

Fig. 5.3. Affine model for f.

(ii) any two charts $i : U \to \mathbb{R}^2$ and $j : V \to \mathbb{R}^2$ in $\mathcal{A}$ have overlap maps $j \circ i^{-1} : i(U \cap V) \to \mathbb{R}^2$ with affine extensions to $\mathbb{R}^2$;
(iii) f is affine with respect to the charts in $\mathcal{A}$;
(iv) Λ is a basic hyperbolic set;
(v) the images of the stable and unstable local leaves under the charts in $\mathcal{A}$ are contained in horizontal and vertical lines; and
(vi) the basic holonomies have affine extensions to the stable and unstable leaves with respect to the charts in $\mathcal{A}$.

5.2 $C^{1,1}$ diffeomorphisms

In Lemma 5.1 below, we will relate distinct regularities of smoothness of the holonomies and of the diffeomorphism f with ratio and cross-ratio distortions determined by the atlas $\mathcal{A}^\iota(f, \rho)$. For a complete discussion on the relations between smoothness of diffeomorphisms and cross-ratio distortions see de Melo and van Strien [99] and Pinto and Sullivan [175].

Let $h : J \to K$ be either a holonomy θ or f_ι, and let J and K be ι-leaf segments. Let $I_0, I_1, I_2 \subset J$ be leaf n-cylinders such that I_0 is adjacent to I_1, I_1 is adjacent to I_2 and $I = I_0 \cup I_1 \cup I_2$. Let $\mathcal{A}^\iota(\{, \rho)$ be an ι-lamination atlas induced by a Riemannian metric ρ on the surface, and let $|I'| = |I'|_\rho$, for every ι-leaf segment I'. We define $B(I_0, I_1, I_2)$ and $B_h(I_0, I_1, I_2)$ as follows:

$$B(I_0, I_1, I_2) = \frac{|I_1||I|}{|I_0||I_2|}$$
$$B_h(I_0, I_1, I_2) = \frac{|h(I_1)||h(I)|}{|h(I_0)||h(I_2)|}.$$

We define the cross-ratio distortion $\mathrm{crd}_{h,\rho}(I_0, I_1, I_2)$ of h with respect to $\mathcal{A}^\iota(\{,\rho)$ by

$$\mathrm{crd}_{h,\rho}(I_0, I_1, I_2) = \log(1 + B_h(I_0, I_1, I_2)) - \log(1 + B(I_0, I_1, I_2)).$$

We note that, for every $\varepsilon > 0$, a $C^{2+\varepsilon}$ diffeomorphism h is a $C^{1,1}$ diffeomorphism (see de Melo and van Strien [99]).

Lemma 5.1. *Let $h : J \subset \mathbb{R} \to K \subset \mathbb{R}$ be a $C^{1,1}$ diffeomorphism with respect to the atlas $\mathcal{A}(\rho)$. Then,*

$$\mathrm{crd}_{h,\rho}(I_0, I_1, I_2) \le o(|I|),$$

for all $n \ge 1$ and for all n-cylinders $I_0, I_1, I_2 \subset J$ such that I_0 is adjacent to I_1, I_1 is adjacent to I_2 and $I = I_0 \cup I_1 \cup I_2$.

Proof. By the theorem on page 294 of de Melo and van Strien [99], we get

$$|B_h(I_0, I_1, I_2) - B(I_0, I_1, I_2)| \le o(|I|B(I_0, I_1, I_2)). \tag{5.3}$$

Therefore,

$$|\mathrm{crd}_{h,\rho}(I_0, I_1, I_2)| = \left|\log\left(1 + \frac{B_h(I_0, I_1, I_2) - B(I_0, I_1, I_2)}{1 + B(I_0, I_1, I_2)}\right)\right|$$
$$\le o\left(\frac{|I|B(I_0, I_1, I_2)}{1 + B(I_0, I_1, I_2)}\right) \le o(|I|).$$

□

5.3 C^{1,HD^ι} and cross-ratio distortions for ratio functions

Consider an ι-ratio function r^ι and let $\theta : K \to K'$ be a basic ι-holonomy. We will consider two distinct cases, (*i*) (presence of gaps) when the ι-leaf segments have gaps, and (*ii*) (absence of gaps) when the ι-leaf segments do not have gaps.

Case (i) (presence of gaps): The *ratio distortion of θ in $I \subset K$* with respect to a ratio function r^ι is defined by

$$\mathrm{rd}(\theta, I) = \sup_{I_0, I_1} \log \frac{r^\iota(\theta(I_0) : \theta(I_1))}{r^\iota(I_0 : I_1)},$$

where the supremum is over all pairs $I_0, I_1 \subset I$ such that I_0 is a leaf n-cylinder and I_1 is either a leaf n-cylinder or a leaf n-gap that has a unique common endpoint with I_0 and $n \geq 1$.
Case (ii) (absence of gaps): Suppose that J_0, J_1 and J_2 are distinct leaf n-cylinders such that J_0 and J_1 have a common endpoint, and J_1 and J_2 also have a common endpoint. Let J be the union of J_0, J_1 and J_2. Then, the *Poincaré length* with respect to a ratio function r^ι is defined by

$$P_{r^\iota}(J_1 : J) = \log\left(1 + \frac{r^\iota(J_1 : J_0)}{r^\iota(J_2 : J)}\right).$$

The *cross-ratio distortion of θ in $I \subset K$* with respect to a ratio function r^ι is defined by

$$\mathrm{crd}(\theta, I) = \sup_{J_0, J_1, J_2} P_{r^\iota}(\theta(J_1) : \theta(J)) - P_{r^\iota}(J_1 : J),$$

where the supremum is taken over all such triples J_0, J_1, J_2 with the property that $J \subset I$.

We observe that if $\mathrm{rd}(\theta, I) = 0$, then θ is affine on I, and if $\mathrm{crd}(\theta, I) = 0$, then θ is Möbius with respect to the atlas $\mathcal{A}(r^\iota)$ determined by r^ι. Here, for simplicity of exposition, we give a slightly distinct definition of cross-ratio distortion from the usual one (see de Melo and van Strien [99]); however, this is equivalent for our purposes.

Definition 14 *The ratio function r^ι has $C^{1,\alpha}$* distortion *with respect to a complete set of holonomies $\mathcal{H}^\iota$, if there is a modulus of continuity χ with the following properties:*

(i) $\lim_{t \to 0} \chi(t)/t^\alpha = 0$, *that is $\chi(t)$ is $o(t^\alpha)$;*
(ii) For every $\theta : K \to K'$ contained in $\mathcal{H}^\iota$ and for every ι-leaf segment $I \subset K$, let ξ be an endpoint of K and R be a Markov rectangle containing ξ.
   (a) If $\alpha < 1$, then the ι-leaf segments have gaps and $|\mathrm{rd}(\theta, I)| \leq \chi\left(r^\iota(I, \ell(\xi, R))\right)$.
   (b) If $\alpha = 1$, then the ι-leaf segments do not have gaps and $|\mathrm{crd}(\theta, I)| \leq \chi\left(r^\iota(I, \ell(\xi, R))\right)$.

The following lemma gives the essential link between a $C^{1,\alpha}$ complete set of holonomies $\mathcal{H}^\iota$ and $C^{1,\alpha}$ distortion of r^ι with respect to $\mathcal{H}^\iota$.

Lemma 5.2. *Suppose that $0 < \alpha, \alpha' \leq 1$. Let (r^s_f, r^u_f) be the HR structure determined by f on Λ. If $r - 1 > \max\{\alpha, \alpha'\}$ and there is a complete set of holonomies $\mathcal{H}^\iota$ for f in which the stable holonomies are $C^{1,\alpha}$ and the unstable holonomies are $C^{1,\alpha'}$, then r^s_f has $C^{1,\alpha}$ distortion and r^u_f has $C^{1,\alpha'}$ distortion with respect to $\mathcal{H}^\iota$.*

Proof. Let $\theta : K \to K'$ be a $C^{1,\alpha}$ holonomy in the ι-complete set of holonomies. Let ξ be an endpoint of K and R be a Markov rectangle containing ξ. We will prove seperately the cases where (i) $0 < \alpha < 1$ and (ii)

$\alpha = 1$. For simplicity of notation, we will denote r_f^{ι} by r^{ι}. Let $I \subset K$ be an ι-leaf segment. Using inequality (2.2), we obtain that

$$|\theta(I)|_\rho < \mathcal{O}\left(r^{\iota}(I, \ell(\xi, R))\right) \quad \text{and} \quad |I|_\rho < \mathcal{O}\left(r^{\iota}(I, \ell(\xi, R))\right). \tag{5.4}$$

Case (i). Let I_1, I_2 be disjoint ι-leaf segments contained in $I \subset K$ such that I_1 is a leaf n-cylinder and I_2 is either a leaf n-cylinder or a leaf n-gap that has a common endpoint with I_1. From inequality (5.4), we get

$$\frac{r^{\iota}(\theta(I_1) : \theta(I_2))}{r^{\iota}(I_1 : I_2)} \in \frac{|\theta(I_1)|_\rho}{|\theta(I_2)|_\rho} \frac{|I_2|_\rho}{|I_1|_\rho} \left(1 \pm \mathcal{O}\left(\left(r^{\iota}(I, \ell(\xi, R))\right)^{\beta}\right)\right), \tag{5.5}$$

where $\beta = \min\{1, r-1\}$. Since θ is $C^{1,\alpha}$, using the Mean Value Theorem we get

$$\frac{|\theta(I_1)|_\rho}{|\theta(I_2)|_\rho} \frac{|I_2|_\rho}{|I_1|_\rho} \in \left(1 \pm o\left(\left(r^{\iota}(I, \ell(\xi, R))\right)^{\alpha}\right)\right). \tag{5.6}$$

Noting that $\alpha < \beta$ and putting (5.5) together with (5.6), we obtain

$$\frac{r^{\iota}(\theta(I_1) : \theta(I_2))}{r^{\iota}(I_1 : I_2)} \in \left(1 \pm o\left(\left(r^{\iota}(I, \ell(\xi, R))\right)^{\alpha}\right)\right).$$

Therefore, for every ι-lef segment $I \subset K$, we have $|\mathrm{rd}(\theta, I)| \le o\left(r^{\iota}(I, \ell(\xi, R))^{\alpha}\right)$.
Case (ii). Let J_0, J_1 and J_2 be leaf n-cylinders contained in an ι-leaf segment $I \subset K$ such that J_0 and J_1 have a common endpoint and J_1 and J_2 have also a common endpoint. Let J be the union of J_0, J_1 and J_2. Let

$$P_\rho(J_1 : J) = \log\left(1 + \frac{|J_1|_\rho |J|_\rho}{|J_0|_\rho |J_2|_\rho}\right). \tag{5.7}$$

Since f_ι is C^r with $r > 2$, from Lemma 5.1 and (5.4), we get

$$\begin{aligned} P_\rho(f_\iota^{-(n+1)}(J_1) : f_\iota^{-(n+1)}(J)) - P_\rho\left(f_\iota^{-n}(J_1) : f_\iota^{-n}(J)\right) &\in \pm o\left(\nu^n |J|_\rho\right) \\ &\subset \pm o\left(\nu^n r^{\iota}(J, \ell(\xi, R))\right). \end{aligned}$$

Therefore,

$$\begin{aligned} P_{r^{\iota}}(J_1 : J) &= \lim_{n \to \infty} P_\rho(f_\iota^{-n}(J_1) : f_\iota^{-n}(J)) \\ &= P_\rho(f_\iota^{-m}(J_1) : f_\iota^{-m}(J)) + \\ &\quad + \sum_{n=m}^{\infty} \left(P_\rho(f_\iota^{-(n+1)}(J_1) : f_\iota^{-(n+1)}(J)) - P_\rho(f_\iota^{-n}(J_1) : f_\iota^{-n}(J))\right) \\ &\in P_\rho\left(f_\iota^{-m}(J_1) : f_\iota^{-m}(J)\right) \pm o\left(\nu^m \left(r^{\iota}(J, \ell(\xi, R)\right)\right). \end{aligned}$$

Thus, since θ is $C^{1,1}$, and from Lemma 5.1, we get

$$\begin{aligned}P_{r^\iota}(\theta(J_1):\theta(J)) - P_{r^\iota}(J_1:J) &= \lim_{n\to\infty}\left(P_\rho(f_\iota^{-n}(\theta(J_1)):f_\iota^{-n}(\theta(J)))-\right.\\ &\qquad \left.-P_\rho(f_\iota^{-n}(J_1):f_\iota^{-n}(J))\right)\\ &\in P_\rho(\theta(J_1):\theta(J)) - P_\rho(J_1:J) \pm\\ &\qquad \pm o\left((r^\iota(J,\ell(\xi,R))\right))\\ &\subset \pm o\left(r^\iota(J,\ell(\xi,R))\right).\end{aligned}$$

Therefore, for every ι-leaf segment $I \subset K$, we have

$$|\mathrm{crd}(\theta,I)| \le o(r^\iota(I,\ell(\xi,R))).$$

□

5.4 Fundamental Rigidity Lemma

We use the following proposition in the proof of the Fundamental Rigidity Lemma. It can be deduced from standard results about Gibbs states such as those in Bowen [17], and it also follows from the results proved in §6.4 (see also Pinto and Rand [162]).

Proposition 5.3. *Let $m_\iota : \mathbf{B}^\iota \to \mathbf{B}^\iota$ be a Markov map on a train-track $\mathbf{B}^\iota$, as defined in §4.3. There is a unique m_ι-invariant probability measure μ on $\mathbf{B}^\iota$ such that, if δ is the Hausdorff dimension of $\mathbf{B}^\iota$, then there exists a constant $C \ge 1$ satisfying*

$$C^{-1} \le \mu(I)/|I|_i^\delta \le C,$$

for all n-cylinders I, for all $n \ge 1$ and for all train-track charts $i \in \mathcal{B}(r^\iota)$. It follows from this that the Hausdorff δ-measure $\mathcal{H}^\delta$ is finite and positive on $\mathbf{B}^\iota$, and μ is absolutely continuous (equivalent) with respect to $\mathcal{H}^\delta$.

Theorem 5.4 (Fundamental Rigidity Lemma). *If the ι-ratio function r^ι has C^{1,HD^ι} distortion, then all basic holonomies are affine with respect to the atlas $\mathcal{A}(r^\iota)$, that is they leave r^ι invariant.*

Proof. We shall prove Theorem 5.4 for the stable holonomies. The unstable result is proved in the same way by replacing f by f^{-1}.

Let $\theta : I \to I'$ be a basic stable holonomy in the rectangle R, where I and I' are spanning stable leaves of R and R has the property that every spanning stable and unstable leaf segment of R is either contained inside a single primary cylinder or inside the union of two touching primary cylinders. We shall prove that, since there is a complete set of holonomies with C^{1,HD^s} distortion, θ has an affine extension with respect to the charts in $\mathcal{A}(r^s)$.

For every $n \ge 1$, the rectangle $f^n(R)$ is equal to $\cup_{j=0}^{m(n)} M_j^n$, where the rectangles $M_j^n = [J_j^n, U_j^n]$ have the following properties (see Figure 5.4):

(i) For j equal to 0 and $m(n)$, we have the following:

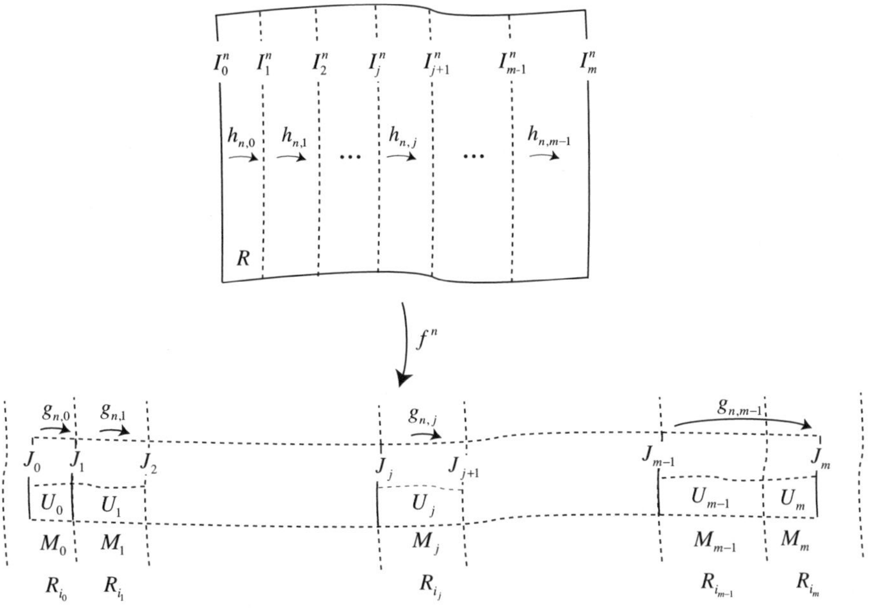

Fig. 5.4. The rectangles R and $f^n(R)$.

(a) $f^n(I) = J_0^n$ and $f^n(I') = J_{m(n)}^n$.

(b) If J_j^n is contained in a single Markov rectangle, then U_j^n is an unstable spanning leaf of this Markov rectangle intersected with $f^n(R)$.

(c) If J_j^n is not contained in a Markov rectangle, then U_j^n is the biggest possible unstable leaf segment in $f^n(R)$ contained in the union of the unstable boundaries of Markov rectangles and intersecting J_j^n.

(ii) For $j = 1, \ldots, m(n) - 1$, one of the following holds.

(a) J_j^n is a spanning stable leaf segment of M_j^n contained in a leaf segment of the domain $\mathcal{I}^\iota$ of the complete set of holonomies $\mathcal{H}^\iota$, and U_j^n is a spanning unstable leaf segment of the Markov rectangle containing J_j^n;

(b) J_j^n is a stable leaf segment not contained in a single Markov rectangle, and U_j^n is the biggest possible unstable leaf segment contained in the union of the unstable boundaries of Markov rectangles and intersecting J_j^n.

(iii) M_j^n intersects M_{j+1}^n only along a common stable boundary, and $M_i^n \cap M_j^n = \emptyset$ if $|j - i| \geq 1$.

Let Θ_n^a be the set of $j \in \{1, \ldots, m(n) - 1\}$ such that J_j^n and J_{j+1}^n are contained in the domain $\mathcal{I}^\iota$, and let Θ_n^b be equal to $\{0, \ldots, m(n) - 1\} \setminus \Theta_n^a$. Since the number of Markov rectangles is finite, the cardinality of the set Θ_n^b is uniformly bounded independent of n.

Set $I_j^n = f^{-n}(J_j^n)$. Then, we can decompose θ as the composition $\theta_{n,m-1} \circ \cdots \circ \theta_{n,0}$, where $\theta_{n,j}$ is the basic holonomy between I_j and I_{j+1} defined by R. Now, consider the holonomies $\tilde{\theta}_{n,j} = f^n \circ \theta_{n,j} \circ f^{-n} : J_j^n \to J_{j+1}^n$ and observe that, since f is affine with respect to the HR structure, $\mathrm{rd}\left(\theta_{n,j}, I_j^n\right) = \mathrm{rd}\left(\tilde{\theta}_{n,j}, J_j^n\right)$ and $\mathrm{crd}\left(\theta_{n,j}, I_j^n\right) = \mathrm{crd}\left(\tilde{\theta}_{n,j}, J_j^n\right)$. Furthermore, if $j \in \Theta_n^a$, then $\tilde{\theta}_{n,j}$ belongs to the complete set of holonomies. Let us first consider the case where $HD^s < 1$. By hypothesis, for every $j \in \Theta_n^a$, we have

$$\sum_{j \in \Theta_n^a} \left|\mathrm{rd}\left(\tilde{\theta}_{n,j}, J_j^n\right)\right| \leq \sum_{j \in \Theta_n^a} \chi\left(r\left(J_j^n, \ell(x_j^n, R_j^n)\right)\right),$$

where x_j^n is an endpoint of J_j^n, R_j^n is a Markov rectangle containing x_j^n, the positive function χ is independent of θ and $\chi(t) = o\left(t^{HD^s}\right)$. From inequality (2.2), for every $j \in \Theta_n^b$, we get

$$\sum_{j \in \Theta_n^b} \left|\mathrm{rd}\left(\theta_{n,j}, I_j^n\right)\right| \leq \sum_{j \in \Theta_n^b} \mathcal{O}\left(d_\Lambda\left(I_j^n, I_{j+1}^n\right)^\alpha\right).$$

Therefore,

$$\begin{aligned}
|\mathrm{rd}(\theta, I)| &\leq \sum_{j=0}^{m-1} \left|\mathrm{rd}\left(\theta_{n,j}, I_j^n\right)\right| \\
&\leq \sum_{j \in \Theta_n^b} \left|\mathrm{rd}\left(\theta_{n,j}, I_j^n\right)\right| + \sum_{j \in \Theta_n^a} \left|\mathrm{rd}\left(\tilde{\theta}_{n,j}, J_j^n\right)\right| \\
&\leq \sum_{j \in \Theta_n^b} \mathcal{O}\left(d_\Lambda\left(I_j^n, I_{j+1}^n\right)^\alpha\right) + \sum_{j \in \Theta_n^a} \chi\left(r\left(J_j^n, \ell\left(x_j^n, R_j^n\right)\right)\right).
\end{aligned}$$

Now, we note that

$$r\left(J_j^n, \ell\left(x_j^n, R_j^n\right)\right) \leq \mathcal{O}\left(\left|K_j^n\right|\right),$$

where $K_j^n = \pi_{\mathbf{B}^s}(J_j^n)$ is the projection of J_j^n into the train-track $\mathbf{B}^s$ under $\pi_{\mathbf{B}^s}$ and the size $|K_j^n|$ of K_j^n is measured in any chart of the bounded atlas $\mathcal{B}(r^s)$ of $\mathbf{B}^s$. Therefore,

$$|\mathrm{rd}(\theta, I)| \leq \sum_{j \in \Theta_n^b} \mathcal{O}\left(d_\Lambda\left(I_j^n, I_{j+1}^n\right)^\alpha\right) + \sum_{j \in \Theta_n^a} \hat{\chi}\left(\left|K_j^n\right|\right), \tag{5.8}$$

where $\hat{\chi}$ is a positive function independent of θ and $\hat{\chi}(t) = o\left(t^{HD^s}\right)$. In the case where $HD^s = 1$, a similar argument gives

$$|\mathrm{crd}(\theta, I)| \leq \sum_{j \in \Theta_n^b} \mathcal{O}\left(d_\Lambda\left(I_j^n, I_{j+1}^n\right)^\alpha\right) + \sum_{j \in \Theta_n^a} C_1 \hat{\chi}\left(\left|K_j^n\right|\right), \tag{5.9}$$

where $\hat{\chi}$ is a positive function independent of θ and $\hat{\chi}(t) = o(t)$. We now show that the right-hand sides of (5.8) and (5.9) tend to zero as n tends to infinity and thus that the left-hand sides are zero. For every $j \in \Theta_n^b$, the distance $d_\Lambda\left(I_j^n, I_{j+1}^n\right)$ converges to zero when n tends to infinity, and, since the cardinal of Θ_n^b is uniformly bounded independently of n, we get

$$\sum_{j\in\Theta_n^b} \mathcal{O}\left(d_\Lambda\left(I_j^n, I_{j+1}^n\right)^\alpha\right) \to 0 \tag{5.10}$$

when n tends to infinity. Now, we are going to prove that $\sum_{j\in\Theta_n^a} \chi\left(\left|K_j^n\right|\right)$ also converges to zero when n tends to infinity. Since R has the property that every spanning stable leaf segment of R is either contained inside a single primary cylinder or inside the union of two touching primary cylinders, we obtain that the train-track segments K_j^n can only intersect in endpoints, and moreover each of them is either contained in an n-cylinder or two adjacent n-cylinders of the Markov map m_s on $\mathbf{B}^s$. Hence, there is a continuous positive function η with $\eta(0) = 0$ such that

$$\sum_{j\in\Theta_n^a} \chi\left(\left|K_j^n\right|\right) \le \eta(\nu^n) \sum_{n-\text{cyls}} |C^n|^{HD^s}, \tag{5.11}$$

where the sum on the right-hand side is over all n-cylinders. By Proposition 5.3, there is an m_ι-invariant probability measure μ and a positive constant C_1 such that

$$\sum_{n-\text{cyls}} |C^n|^{HD^s} \le C_1 \sum_{n-\text{cyls}} \mu(C^n) \le C_1. \tag{5.12}$$

Putting together (5.11) and (5.12), we get

$$\sum_{j\in\Theta_n^a} \chi\left(\left|K_j^n\right|\right) \to 0 \tag{5.13}$$

when n tends to infinity. If $HD^s < 1$, applying (5.10) and (5.13) to (5.8), we get that $rd(\theta, I) = 0$. Therefore, θ is affine on I, which completes the proof for this case. If $HD^s = 1$, applying (5.10) and (5.11) to (5.9), we get that $crd(\theta, I) = 0$. Therefore, θ is Möbius on I and extends to a Möbius homeomorphism of the global leaf, where the affine structures of the global leaves are determined by the invariance of the affine structures under iteration by f. Since a Möbius homeomorphism of $\mathbb{R}$ is an affine map, the holonomies θ are affine. □

5.5 Existence of affine models

In Lemma 5.5, (r^s, r^u) is any HR structure and not necessarily the HR structure determined by f.

Lemma 5.5 (Existence of affine models). *If r^s has C^{1,HD^s} distortion and r^u has C^{1,HD^u} distortion, then there is a hyperbolic affine model for g on $\hat{\Lambda}$ such that g on $\hat{\Lambda}$ is topological conjugated to f on Λ and such that the HR structures are the same (i.e. $r^\iota(I:J) = r^\iota_g(\psi(I):\psi(J))$, where $\psi : \Lambda \to \hat{\Lambda}$ is the conjugacy between f and g).*

Proof. Let $\{R_1, \ldots, R_k\}$ be a Markov partition for f. For every Markov rectangle R_m, we take a rectangle $M_m \supset R_m$ that contains a small neigbourhood of R_m with respect to the distance d_Λ. We construct an ortogonal chart $i_m : M_m \to \mathbb{R}^2$ as follows. Choose an $x \in M_m$ and let $e_s : \ell^s(x, M_m) \to \mathbb{R}$ be in $\mathcal{A}(r^s)$ and $e_u : \ell^u(x, M_m) \to \mathbb{R}$ be in $\mathcal{A}(r^u)$. The ortogonal chart i_m on M_m is now given by $i_m(z) = (e_s([z,x]), e_u([x,z])) \in \mathbb{R}^2$.

Let $\phi_{m,n} : i_m(M_m \bigcap M_n) \to i_k(M_m \bigcap M_n)$ be the map defined by $\phi_{m,n}(x) = i_m \circ i_n^{-1}(x)$. By Theorem 5.4, the stable and unstable holonomies have affine extensions with respect to the charts in $\mathcal{A}(r^s)$ and $\mathcal{A}(r^u)$. Hence, there is a unique affine extension $\Phi_{m,n} : \mathbb{R}^2 \to \mathbb{R}^2$ of $\phi_{m,n}$. This extension sends vertical lines into vertical lines and horizontal lines into horizontal lines.

Let us denote by S_m the rectangle in $\mathbb{R}^2$ whose boundary contains the image under i_m of the boundary of R_m. For every pair of Markov rectangles R_m and R_n that intersect in a partial side $I_{m,n} = R_m \bigcap R_n$, let $J_{m,n}$ and $J_{n,m}$ be the smallest line segments containing respectively the sets $i_m(I_{m,n})$ and $i_n(I_{m,n})$. We call $J_{m,n}$ and $J_{n,m}$ *partial sides.* Hence, $J_{m,n} = \Phi(J_{n,m})$. Let $\tilde{M} = \bigsqcup_{m=1}^k S_m/\{\Phi_{m,n}\}$ be the disjoint union of the squares S_m where we identify two points $x \in J_{m,n}$ and $y \in J_{n,m}$ if $\Phi_{n,m}(x) = y$. Hence, $\tilde{M}$ is a topological surface possibly with boundary. By taking appropriate extensions E_m of the rectangles S_m and using the maps $\Phi_{m,n}$ to determine the identifications along the boundaries, we get a surface $\hat{M} = \bigsqcup_{m=1}^k E_m/\{\Phi_{m,n}\}$ without boundary. The surface $\hat{M}$ has a natural affine atlas that we now describe: if a point z is contained in the interior of E_m, then we take a small open neighbourhood U_z of z contained in E_m and we define a chart $u_z : U_z \to \mathbb{R}^2$ as being the inclusion of $U_z \bigcap E_m$ into $\mathbb{R}^2$. Otherwise z is contained in a boundary of two, three or four sets $E_{m_1}, \ldots, E_{m_k}$ that we order such that the $J_{m_i,m_{i+1}}$ are partial sides. In this case, for a small open neighbourhood U_z of z we define the chart $u_z : U_z \to \mathbb{R}^2$ as follows:

(i) $u_z | (U_z \bigcap E_{m_k})$ is the inclusion of $U_z \bigcap E_{m_k}$ into $\mathbb{R}^2$;
(ii) $u_z | (U_z \bigcap E_j) = \Phi_{m_{k-1},m_k} \circ \ldots \circ \Phi_{m_j,m_{j+1}}$, for $j \in \{1, \ldots, k-1\}$.

Since the maps $\Phi_{m_1,m_2}, \ldots, \Phi_{m_{k-1},m_k}$ and Φ_{m_k,m_1} are affine, we deduce that the set of all these charts form an affine atlas $\mathcal{S}$ on $\hat{M}$.

Let $\psi : \Lambda \to \hat{\Lambda}$ be the natural embedding of Λ into $\hat{M}$, and $\hat{f} : \hat{\Lambda} \to \hat{\Lambda}$ be the map $\hat{f} = \psi \circ f \circ \psi^{-1}$ conjugate to f.

For every $x \in \hat{\Lambda}$, we take charts $u : U \to \mathbb{R}^2$ and $v : V \to \mathbb{R}$ in the affine atlas $\mathcal{S}$ such that $x \in U$ and $\hat{f}(x) \in V$. Since $\hat{f}$ along leaves and also the holonomies have affine extensions with respect to the charts in $\mathcal{A}(r^s)$ and

$\mathcal{A}(r^u)$, the map $v \circ \hat{f} \circ u^{-1}$ has a unique affine extension g_x to $\mathbb{R}^2$. These affine extensions determine a unique affine extension g of $\hat{f}$ to an open set of $\hat{M}$.

The maps g_x send horizontal lines into horizontal lines and vertical lines into vertical lines. Furthermore, $g_{g^n(x)} \circ \ldots \circ g_x$ contracts horizontal lines exponentially fast and expands vertical lines exponentially fast with respect to any fixed finite set of charts in $\mathcal{S}$ covering $\hat{M}$. Hence, g is hyperbolic on $\hat{\Lambda}$ and the image under these charts of the stable and unstable leaves are contained, respectively, in horizontal and vertical lines.

Since the holonomies have affine extensions with respect to the charts in $\mathcal{A}(r^s)$ and $\mathcal{A}(r^u)$, they also have affine extensions along leaves with respect to the charts in this affine atlas. By construction of the affine model for g on $\hat{\Lambda}$, we get that $r^\iota(I : J) = r^\iota_g(\psi(I) : \psi(J))$. □

5.6 Proof of the hyperbolic and Anosov rigidity

Here we show how to use the Fundamental Rigidity Lemma and the existence of affine models (Lemma 5.5) to prove the Hyperbolic Rigidity Theorem.

Theorem 5.6 (Hyperbolic rigidity). *Let HD^s and HD^u be, respectively, the Hausdorff dimension of the intersection with Λ of the stable and unstable leaves of f. If f is C^r, with $r-1 > \max\{HD^s, HD^u\}$, and there is a complete set of holonomies for f in which the stable holonomies are C^{1,HD^s} and the unstable holonomies are C^{1,HD^u}, then the map f on Λ is $C^{1+\gamma}$ conjugate to a hyperbolic affine model, for some $0 < \gamma < 1$.*

In assuming that f is C^r with $r - 1 > \max\{HD^s, HD^u\}$ in the previous theorem, we actually only use the fact that f is C^{1,HD^ι} along ι-leaves.

Proof of Theorem 5.6. By Lemma 2.8, f determines on Λ an HR structure (r^s, r^u). By Lemma 5.2, r^ι has C^{1,HD^ι} distortion. By Theorem 5.4, all the basic ι-holonomies are affine with respect to the atlas $\mathcal{A}(r^\iota)$. Hence, by Lemma 5.5, there is a diffeomorphism g with a hyperbolic basic set $\hat{\Lambda}$ and a hyperbolic affine model for g on $\hat{\Lambda}$ such that there is a conjugacy between f and g such that $r^\iota(I : J) = r^\iota_g(\psi(I) : \psi(J))$. By Lemma 2.13, we get that f is C^{1+} conjugated to g. □

We use Theorem 5.6 to prove the following theorem which partially extends the previous result mentioned above of Ghys [44].

Theorem 5.7 (Anosov rigidity). *If f is a C^r Anosov diffeomorphism on a surface, with $r > 2$, and there is a complete set of holonomies for f in which the stable and unstable holonomies are $C^{1,1}$, then f is C^r conjugate to an affine model.*

We note that Theorem 5.7 also follows from the fact that the holonomies and f are affine with respect to the atlases $\mathcal{A}(r^s)$ and $\mathcal{A}(r^u)$ (see the proof of Theorem 5.6) and Corollary 3.3 in Ghys [44].

Proof of Theorem 5.7. If $f : M \to M$ is a C^r surface Anosov diffeomorphism, then $\Lambda = M$. By Franks [40, 41], Manning [74] and Newhouse [103], there is a unique hyperbolic toral automorphism $\hat{f} : \hat{M} \to \hat{M}$ topologically conjugate to f. By Theorem 5.6, there is a C^{1+} conjugacy $\psi : M \to \hat{M}$ between f and $\hat{f}$. By Lemma 2.8, we have that $r^\iota(I : J) = r^\iota_{\hat{f}}(\psi(I) : \psi(J))$. By a somewhat standard blow-down-blow-up argument, we get that ψ is C^r along stable and unstable leaves (see de Melo and van Strien [99] and Pinto and Sullivan [175]). Hence, by Proposition 2.14 due to Journé [64], ψ is C^r. □

5.7 Twin leaves for codimension 1 attractors

We introduce the notion of a twinned pair of leaves for a diffeomorphism f of a surface with a basic set Λ. We prove that every proper codimension 1 attractor Λ contains a twinned pair of leaves.

Definition 15 *A* twinned pair of u-leaves (I, J) *in a basic set Λ consists of a pair of u-leaf segments I and J with the following properties (see Figure 5.5):*

(i) an endpoint p of I and an endpoint q of J are periodic points under f;
(ii) $(I \setminus \{p\}) \cap (J \setminus \{q\}) = \emptyset$;
(iii) for all $z \in I \setminus \{p\}$ there is a full s-leaf segment γ_z in the stable manifold through z which has endpoints z and z' such that $z' \in J \setminus \{q\}$ and $\gamma_z \cap \Lambda = \{z, z'\}$.

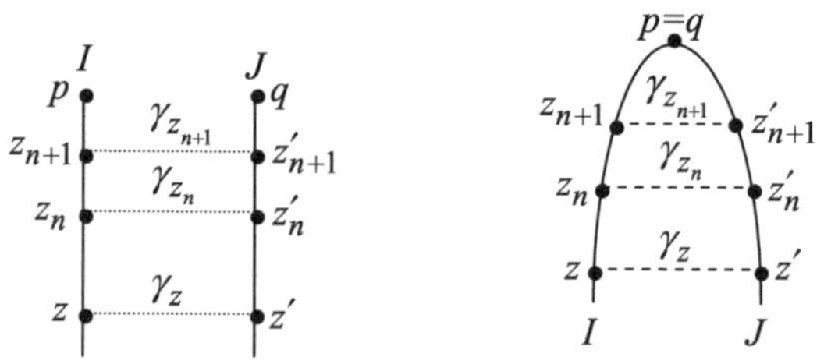

Fig. 5.5. An illustration of twinned pair of u-leaves.

It follows from this that if a sequence $z_n \in I \setminus \{p\}$ converges to p, then the corresponding sequence $z'_n \in J \cap \gamma_{z_n}$ converges to q. Also, it follows that the periodic points p and q must have the same period. A twinned pair of s-leaves in a basic set Λ is similarly defined.

Remark 5.8. In the previous definition we allow the points p and q to coincide. However, if p is different from q, then there is no stable leaf containing both p and q (otherwise they would converge under iteration by f which is absurd).

The set $\Lambda \subset M$ is an *attractor* for f if there is an open set $U \subset M$ such that $\Lambda = \cap_{i=0}^{\infty} f^i(U)$. We say that Λ is a *proper codimension* 1 *attractor* if Λ is an attractor basic set, the Hausdorff dimension of the unstable leaf segments is one, and the Hausdorff dimension of the stable leaf segments is strictly less than one.

Theorem 5.9. *If Λ is a proper codimension* 1 *attractor, then Λ contains a twinned pair of u-leaves.*

We call an unstable leaf ℓ an *unstable free-leaf* if there is a full s-leaf segment I transversal to the leaf ℓ which is the union $I_1 \cup \{p\} \cup I_2$ of two disjoint (non-empty) full s-leaf segments I_1 and I_2 such that I_1 and I_2 have a common endpoint $p \in \ell \cap \Lambda$ and I_2 does not intersect Λ.

By Kollmer [66], the set $\mathcal{L}$ of all unstable free-leaves is non-empty and finite. Since the free-leaves are permuted by f, each one of these leaves ℓ contains a single periodic point P_ℓ. Furthermore, $\mathcal{L}$ is equal to the union of pairwise disjoint subsets $\mathcal{L}_1, \ldots, \mathcal{L}_j$ which are characterized by the following property: the leaves of each set $\mathcal{L}_m$ form the boundary of an open connected set $\mathcal{O}_m$ in M which does not intersect the basic set Λ.

Remark 5.10. We observe that, by Ruas [43], $f|\Lambda$ is topologically conjugate to an Anosov or pseudo-Anosov map that has been unzipped along a finite set of leaves. It is these unzipped leaves which form $\mathcal{L}$. Each set $\mathcal{L}_m \subset \mathcal{L}$ corresponds to the unzipping a k-prong singularity where k is the number of leaves contained in $\mathcal{L}_m$ (see Figure 5.6). The sets $\mathcal{L}_m$ of cardinality one and two correspond respectively to umbilic singularities and regular points.

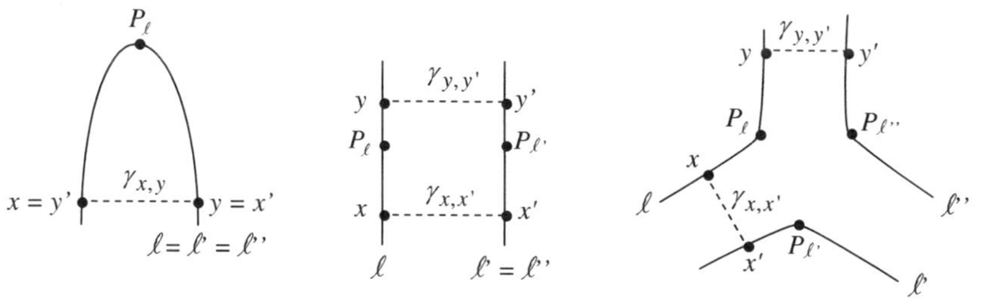

Fig. 5.6. Examples of sets $\mathcal{L}_m$ with cardinality 1, 2 and 3.

Proof of Theorem 5.9. We claim that for each leaf $\ell \in \mathcal{L}_m$ there are two leaves $\ell', \ell'' \in \mathcal{L}_m$, two points $x \in \ell$ and $y \in \ell$ on different sides of the periodic point P_ℓ in ℓ and two points $x' \in \ell'$ and $y' \in \ell''$ such that x and x', and y and y'

are the endpoints of two full s-leaf segments $\gamma_{x,x'}$ and $\gamma_{y,y'}$ whose interiors meet no unstable leaves of Λ. If the cardinality of $\mathcal{L}_m$ is greater or equal to three, then ℓ, ℓ' and ℓ'' are distinct leaves. If the cardinality of $\mathcal{L}_m$ is one, then $\ell = \ell' = \ell''$ and the claim just says that there are x and y in ℓ on either side of P_ℓ with x and y joined by a full s-leaf segment $\gamma_{x,y}$ whose interior meets no unstable leaves. If the cardinality of $\mathcal{L}_m$ is two, then $\ell' = \ell'' \neq \ell$ and $x', y' \in \ell'$ are on either side of the periodic point in ℓ'. This claim follows from the density of the unstable manifold in Λ and the local product structure as we now describe. If $x \in \ell$, then, for some $n > 0$, $f^n(x)$ lies inside of a small full s-leaf segment γ and, in γ, is contained between two points contained in Λ. We can then find a non-trivial full s-leaf segment γ' inside γ which also contains $f^n(x)$ so that to one side of $f^n(x)$ there is only a single point $w \neq f^n(x)$ in $\gamma' \cap \Lambda$. Let γ'' denote the part of γ between $f^n(x)$ and w. Then, $f^{-n}(\gamma'')$ is a full s-leaf segment through x such that $x' = f^n(w)$ is the other endpoint of $f^{-n}(\gamma'')$. Since by construction $f^{-n}(\gamma'') \setminus \{x, x'\}$ meets no unstable leaves of Λ, $f^{-n}(\gamma'')$ is the required full s-leaf segment $\gamma_{x,x'}$, and ℓ' is the stable leaf passing through x'. One finds y' and ℓ'' by taking y on the other side of P_ℓ in ℓ and proceeding in a similar fashion which ends the proof of the claim. Let $\ell(x)$ be an unstable leaf segment containing x and having P_ℓ as one of its endpoints. Let $\ell'(x')$ be the unstable leaf containing x' such that there is a local holonomy $h : \ell(x) \to \ell'(x')$ with $h(x) = x'$ (and so $h(\ell(x)) = \ell'(x')$). Then, the pair $(\ell(x), \ell'(x'))$ form a twinned pair of leaves. □

5.8 Non-existence of affine models

The relevance of the existence of a twinned pair of leaves is that these basic sets do not have affine models. Hence, if Λ is a proper codimension 1 attractor, then there are no affine models for f on Λ.

Definition 16 *A ι-ratio function r is* transversely affine, *if r is invariant under f, i.e $r(I : J) = r(f(I) : f(J))$, and r is invariant under holonomies h, i.e. $r(I : J) = r(h(I) : h(J))$.*

Lemma 5.11. *If Λ contains a twinned pair of ι-leaves, then there is not a transversely affine ι'-ratio function r.*

Proof. For simplicity of exposition we will consider the case $\iota = u$ and $\iota' = s$. The other case is similar by replacing f by f^{-1} and stable by unstable, and vice-versa. Let us suppose by contradiction that there is an affine model for f. For arguments sake assume that the twinned pair leaves are unstable. Let the full u-leaf segments I and J and the periodic points $p \in I \cap \Lambda$ and $q \in J \cap \Lambda$ be as in the definition of a twinned pair leaves. Let m be the common period of the periodic points. Fix $z \in I \cap \Lambda$ and $z' \in J \cap \Lambda$ such that z and z' are the endpoints of a full s-leaf segment which does not intersect Λ. Choose a ful u-leaf segment K such that there is a holonomy $h : J \cap \Lambda \to K \cap \Lambda$. For every

$n \neq 1$, let $y_n \in I \cap \Lambda$, $y'_n \in J \cap \Lambda$ and $y''_n \in K \cap \Lambda$ be such that $f^{mp}(y_n) = z$, $f^{mp}(y'_n) = z'$ and $h(y'_n) = y''_n$ (see Figure 5.7). The ratio $r(y_n, y'_n, y''_n)$ between

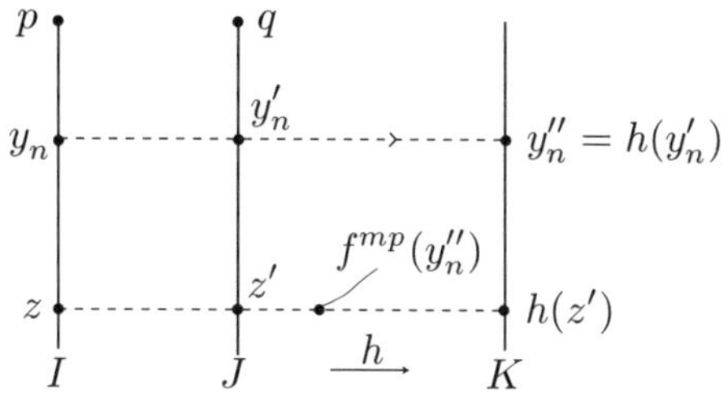

Fig. 5.7. The nonexistence of transversely affine ratio function.

the length of the full u-leaf segment with endpoints y''_n and y'_n and the length of the full u-leaf segment with endpoints y'_n and y_n, when measured in a chart of the affine atlas, is well-defined and does not depend upon the chart considerd.

Since the holonomy is affine, the value of the ratio $r(y_n, y'_n, y''_n)$ does not depend upon $n \neq 1$. Since f is also affine, $r(y_n, y'_n, y''_n)$ is equal to $r(z, z', f^{mp}(y''_n))$. Therefore, the value of the ratio $r(z, z', f^{mp}(y''_n))$ does not depend upon $n \neq 1$. But, by construction the sequence $f^{mp}(y''_n)$ converges to z' which implies that the ratio $r(z, z', f^{mp}(y''_n))$ converges to zero, which is absurd. □

Theorem 5.12. *If a basic set Λ contains a twinned pair of ι-leaves, then there are no affine models for f on Λ.*

Proof. If there is an affine model for f on Λ, then r is a transversely affine ι-ratio function, which contradicts Lemma 5.11. □

5.9 Non-existence of uniformly $C^{1,HD^{\iota}}$ complete sets of holonomies for codimension 1 attractors

For a Smale horseshoe there is an infinite number of holonomies in a complete set. However, if there is only a finite number of holonomies in a complete set, then the uniformity hypothesis on the modulus of continuity of $h^{\iota}_{(M,N)} \in \mathcal{H}^{\iota}$ is redundant.

Lemma 5.13. *For a proper codimension 1 attractor the stable complete set of holonomies consists of a finite set of holonomies.*

However, there are cases where the complete set of holonomies is forced to be infinite. This is the case for systems like the Smale horseshoe (see Figure 5.2).

Proof of Lemma 5.13. Since the u-leaf segments are manifolds, the number of holonomies in the complete sets of s-holonomies is two times the minimal number $\mathcal{N}$ of stable leaves which cover the s-boundaries of the rectangles contained in the Markov partition with the property that the interior of each one of these leaves is contained in at most two s-boundaries of Markov rectangles. □

Theorem 5.14. *Let Λ be a basic set for a $C^{1+\gamma}$ diffeomorphism f of a surface with $\gamma > HD^{\iota}$. If Λ contains a twinned pair of ι'-leaves, then the complete set of ι-holonomies $\mathcal{H}^{\iota}$ is not $C^{1,HD^{\iota}}$.*

Proof. By Lemma 5.2 and Theorem 5.4, if f is $C^{1+\gamma}$, with $\gamma > HD^{\iota}$, and the complete set of ι-holonomies is $C^{1,HD^{\iota}}$, then r is a transversely affine ι-ratio function. This contradicts Lemma 5.11. □

5.10 Further literature

For Anosov diffeomorphisms of the torus, the hyperbolic affine model is a hyperbolic toral automorphism (see Franks [41], Manning [74] and Newhouse [103]). In general, the topological conjugacy between such a system and the corresponding hyperbolic affine model is only Hölder continuous and need not be any smoother. This is the case if there is a periodic orbit of f whose eigenvalues differ from those of the hyperbolic affine model. For Anosov diffeomorphisms f of the torus, there are the following results, all of which have the form that if a C^k f has C^r foliations, then f is C^s-rigid, that is f is C^s conjugate to the corresponding hyperbolic affine model:

(i) Area-preserving Anosov diffeomorphisms f with $r = \infty$ are C^{∞}-rigid (Avez [11]).
(ii) C^k area-preserving Anosov diffeomorphisms f with $r = 1 + o(t|\log t|)$ are C^{k-3}-rigid (Hurder and Katok [50]).
(iii) C^1 area-preserving Anosov diffeomorphisms f with $r \geq 2$ are C^r-rigid (Flaminio and Katok [39]).
(iv) C^k Anosov diffeomorphisms f $(k \geq 2)$ with $r \geq 1 +$ Lipshitz are C^k-rigid (Ghys [44]).

Coelho et al. [22] have also proved a rigidity result for comuting pairs of the circle. This chapter is based on Pinto and Rand [165] and Pinto, Rand and Ferreira [170].

6

Gibbs measures

We give a novel and elementary proof of existence and uniqueness of Gibbs states for Hölder weight systems. A bonus of this approach is that it leads directly to a decomposition of the measure as an integral of an explicitly given canonical ratio function with respect to a measure dual to the Gibbs state. The ratio decomposition is particularly useful in certain situations and it is used to link certain Gibbs states with Hausdorff measures on basic sets of C^{1+} hyperbolic diffeomorphisms (cf. chapter 7).

6.1 Dual symbolic sets

Let us recall the definition of a one-sided subshift of finite type $\Sigma = \Theta_A^u$ from §4.5. The elements of Σ are all the infinite right-handed words $w = w_0 w_1 \ldots$ in the symbols $1, \ldots, k$ such that for all $i \geq 0$, $A_{w_i w_{i+1}} = 1$. Here, $A = (A_{ij})$ is any matrix with entries o and 1 such that A^n has all entries positive for some $n \geq 1$. We write $w \overset{n_1,n_2}{\sim} w'$ if the two words $w, w' \in \Sigma$ agree on their first n entries. The metric d on Σ is given by $d(w, w') = 2^{-n}$ if $n \geq 0$ is the largest such that $w \overset{n,n}{\sim} w'$. Together with this metric Σ is a compact metric space. The shift $\tau : \Sigma \to \Sigma$ is the mapping that sends $w_0 w_1 \ldots$ to $w_1 w_2 \ldots$. It is a local homeomorphism.

An n-cylinder Σ_w, $w \in \Sigma_n$, consists of all those words $w' \in \Sigma$ such that $w \overset{n}{\sim} w'$ If C is an n-cylinder, then we define mC to be the $(n-1)$-cylinder containing C and denote by $n(C)$ the depth n of C. A 1-cylinder is also called a *primary cylinder*.

Together with Σ we will consider the augmented space Δ that consists of both the infinite right-handed words in Σ and their finite subwords. Let Δ_{fin} denote the subset of all finite words. Then, we can identity Δ_{fin} with the set of all cylinders in Σ via the association $w \leftrightarrow \Sigma_w$. This set has two natural oriented tree structures:

(a) Δ_{fin}^m in which all the oriented edges connect a cylinder C to mC; and

(b) $\Delta_{\text{fin}}^{\tau}$ in which all the oriented edges are from the cylinder C to τC.

An admissible backward path in either of these trees is a finite or infinite sequence $\{C_j\}$ of cylinders indexed by either $j = 0, \ldots, n$ or $j = 0, 1, \ldots$ and such that C_0 is a primary cylinder and such that there is an oriented edge from C_j to C_{j-1} for all $j > 0$. The infinite paths in Δ_{fin}^{m} correspond to points of Σ^u.

Definition 17 *The dual* $(\Sigma^u)^*$ *of* Σ^u *is the set of all infinite admissible backward paths in* $\Delta_{\text{fin}}^{\tau}$ *together with the metric defined as follows:* $d^*(\{C_j\}, \{C'_j\}) = 2^{-n}$ *if* $C_j = C'_j$, *for* $0 < j < n$ *and* $C_n \neq C'_n$.

Note that one can identify the elements of $(\Sigma^u)^*$ with those infinite left-handed words $\ldots w_1 w_0$ in the symbols $1, \ldots, k$ such that $A_{w_i w_{i-1}} = 1$, which leads to the following remark:

Remark 6.1. Via the association $w \leftrightarrow \Sigma_w^u$, there is a homeomorphism $\psi : (\Sigma^u)^* \to \Sigma^s$ such that $\psi\tau^* = \tau_s\psi$ and $\psi m^* = m_s\psi$.

We note that for both Σ^u and $(\Sigma^u)^*$, a cylinder is given by prescribing a finite admissible backward path $\{C_j\}_{j=0}^{n-1}$ (respectively in Δ_{fin}^{m} and in $\Delta_{\text{fin}}^{\tau}$), and it is then equal to the set of all infinite admissible backward paths $\{D_j\}$ such that $D_j = C_j$ for $0 < j < n$. Since this finite path is determined by C_{n-1} there is a one-to-one correspondence between the cylinders of Σ^u and $(\Sigma^u)^*$. Specifically, this is given as follows: if C is an n-cylinder of Σ^u, then the cylinder C^* of $(\Sigma^u)^*$ consists of all infinite admissible backward paths $\{C_j\}_{j=0}^{\infty}$ in $\Delta_{\text{fin}}^{\tau}$ such that $C_{n-1} = C$. We also define duals to m and τ: if $C^* = \{C_j\}_{j=0}^{n-1}$ is an n-cylinder of $(\Sigma^u)^*$, then m^*C^* is the $(n-1)$-cylinder $\{\tau C_j\}_{j=1}^{n-1}$ of $(\Sigma^u)^*$ containing C^*, and τ^*C^* is the $(n-1)$-cylinder $\{mC_j\}_{j=1}^{n-1}$. Note how these translate under duality:

$$m^*C^* = (\tau C)^* \quad \text{and} \quad \tau^*C^* = (mC)^*. \tag{6.1}$$

The dual set $(\Theta^s)^*$ of Θ^s and the maps τ_s^* and m_s^* are constructed similarly to the above ones. The set $(\Theta^s)^*$ can be identified with Θ^u, and the maps τ_s^* and m_s^* with the maps τ_u and m_u, respectively.

6.2 Weighted scaling function and Jacobian

Now consider a function l defined on Δ_{fin} and with the following properties: there exists $0 < \omega < \omega' < 1$ such that if C is an n-cylinder, then

$$\mathcal{O}(\omega^n) < l(C) < \mathcal{O}(\omega'^n) \tag{6.2}$$

and there exists $0 < \nu < 1$ such that the following two equivalent conditions hold:

(i) If C is an n-cylinder with $n > 0$, then $\sigma_l(C) = l(C)/l(mC)$ converges exponentially along backward orbits i.e. $\sigma_l(C) \in (1 \pm \mathcal{O}(\nu^n))\, \sigma_l(\tau(C))$.
(ii) If C is an n-cylinder with $n > 0$, then $J_l(C) = l(\tau C)/l(C)$ converges exponentially along nested sequences, i.e. $J_l(C) \in (1 \pm \mathcal{O}(\nu^n))\, J_l(mC)$.

We leave the proof of the equivalence to the reader, but note that it comes from the relation

$$\frac{\sigma_l(\tau C)}{\sigma_l(C)} = \frac{J_l(C)}{J_l(mC)}.$$

It also follows from these conditions that the limits defining the following functions σ_l and J_l are reached exponentially fast and that consequently these functions are Hölder continuous: if $\xi = \{C_n\}_{n=0}^{\infty} \in \Sigma^*$, where C_n is an n-cylinder and $\tau C_{n+1} = C_n$, and if $x = \bigcap_{n \geq 0} D_n$, where D_n is an n-cylinder with $mD_{n+1} = D_n$, then

$$\sigma_l(\xi) = \lim_{n \to \infty} \sigma_l(C_n) \text{ and } J_l(x) = \lim_{n \to \infty} J_l(D_n).$$

Definition 18 *Such a system of weights l is called a* Hölder weight system . *We call σ_l the* weighted scaling function *of l and J_l the* weighted Jacobian. *The Hölder weighted scaling function is said to satisfy the* matching condition *or to* match, *if, for all $\xi \in (\Sigma^u)^*$,*

$$\sum_{\tau^*(\xi')=\xi} \sigma_l(\xi') = 1. \tag{6.3}$$

The matching condition is equivalent to the following: There exists $0 < \theta < 1$ such that $\sum \sigma_l(C') = 1 \pm \mathcal{O}(\theta^n)$ (sum over $(n+1)$-cylinders C' contained in C), for all $n \geq 0$ and all n-cylinders C.

Consider the sums $Z_s^n = \sum_C l(C)e^{-sn}$, where the sum is over all n-cylinders C. From (6.2), for $s > 0$ sufficiently large, Z_s^n is bounded away from infinity uniformly in $n \geq 0$. On the other hand, if s is sufficiently negative, then Z_s^n diverges to ∞ as $n \to \infty$. Since if this divergence occurs for a particular value of s then it occurs for all smaller values, there is a critical value P given by $P = \inf\{s : Z_s^n \text{ uniformly bounded in } n\}$. This is called the *pressure* of l. It corresponds to the usual definition (see Bowen [17]).

6.3 Weighted ratio structure

Before proceeding we need to introduce some notation. Consider a cylinder C in Σ^u and let C_1 denote the primary cylinder containing C. If C_n is an n-cylinder such that $\tau^{n-1}(C_n) = C_1$, then by $C(C_n)$ we denote $(\tau^{n-1}|_{C_n})^{-1}(C)$.

From (6.9), (6.10) and (6.11), we also get bounds for $r_l(C : D)$ as presented in the following remark.

Remark 6.2. Suppose that C is an m-cylinder contained in the n-cylinder D. Then,

$$r_l(C:D) = \mathcal{O}\left(\mathrm{e}^{-(m-n)P} l(C)/l(D)\right).$$

If l satifies the matching condition, then $P = 0$ and, for some $0 < \theta < 1$,

$$r_l(C:D) \in (1 \pm \mathcal{O}(\theta^n)) \frac{l(C)}{l(D)}, \tag{6.4}$$

whenever C and D are contained in a common n-cylinder. Therefore, for all $\xi = \{\xi_n\}_{n=0}^{\infty} \in (\Sigma^u)^*$,

$$\sigma(\xi) = \sigma_l(\xi) = \lim_{n\to\infty} l(\xi_n)/l(m\xi_n) \quad \text{and} \quad r_{l,\xi}(C) = \lim_{n\to\infty} l(C(\xi_n))/l(\xi_n).$$

In these cases the limits are reached exponentially fast and $\sigma_l(\xi)$ and $r_{l,\xi}(C)$ are Hölder in ξ.

6.4 Gibbs measure and its dual

Consider a Hölder weight system l with pressure P. We omit the proof of the following lemma because it closely follows the proof of Lemma 3.1 of Paterson [137].

Lemma 6.3. *There is a positive decreasing continuous function k on $[0,\infty]$ with the following properties:*

(i) The sums $Z_s = \sum_C k(l(C))l(C)\mathrm{e}^{-n(C)s}$ (sum over all cylinders C) converge for $s > P$ and diverge for $s = P$; and
(ii) For all $\varepsilon > 0$, there exists $y_0(\varepsilon) > 0$ such that $\lambda^{-\varepsilon} \le k(\lambda y)/k(y) \le 1$, whenever $\lambda > 1$ and $0 < \lambda y < y_0(\varepsilon)$.

Definition 19 *Suppose that μ is a τ-invariant probability measure on Σ^u and ν a τ_*-invariant probability measure on $(\Sigma^u)^*$. Then, the* dual measures *μ^* and ν^*, respectively, to μ and ν are the probability measures defined on $(\Sigma^u)^*$ and Σ^u by $\mu^*(C^*) = \mu(C)$ and $\nu^*(C^*) = \nu(C)$.*

In the above definition, we use the fact that μ^* is a probability measure (respectively, τ_*-invariant) if, and only if, μ is τ-invariant (respectively, a probability measure). Similarly for ν. This is because $\tau C = D$ (respectively, $\tau_* C^* = D^*$) if, and only if, $m_* C^* = D^*$ (respectively, $mC = D$).

Theorem 6.4. *There exist a unique pair of Borel probability measures ν on Σ^u and ν^* on $(\Sigma^u)^*$ with the following property, for some $0 < \theta < 1$: If C is an n-cylinder of Σ^u,*

$$\frac{\nu(\tau C)}{\nu(C)} \in (1 \pm \mathcal{O}(\nu^n))\, J_l(C)\mathrm{e}^P, \qquad \frac{\nu^*(\tau_* C^*)}{\nu^*(C^*)} \in (1 \pm \mathcal{O}(\theta^n))\, \sigma_l^{-1}(C^*)\mathrm{e}^P$$

and if C and D are two cylinders, then $\nu(C)/\nu(D) = r_l(C : D)$. Moreover, the weights $l_\nu(C) = \nu(C)$ form a matching Hölder weight system and $\sigma_{l_\nu} = \sigma$.

If the weight function l satisfies the matching condition, then ν^* is τ_*-invariant and its dual measure μ satisfies the following equivalent conditions:

(i) If C and D are two cylinders contained in the same n-cylinder, then $\mu(D)/\mu(C) \in (1 \pm \mathcal{O}(\theta^n))\, l(D)/l(C)$;
(ii) If C is an n-cylinder and $\xi = (\xi_i) \in (\Sigma^u)^*$ has $\xi_n = C$, then

$$\mu(C)/\mu(mC) \in (1 \pm \mathcal{O}(\theta^n))\, \sigma_l(\xi);$$

(iii) (Ratio decomposition) *If C is an n-cylinder and C_0 is the primary cylinder containing C, then*

$$\mu(C) = \int_{C_0^*} r_{l,\xi}(C)\mu^*(d\xi).$$

Here, μ^* is the dual measure to μ.

Moreover, for each of the conditions (i), (ii) and (iii), μ is the unique measure with the given property.

If J_μ is the Jacobian $d(\mu \circ \tau)/d\mu$ and $x = \bigcap_{n\geq 0} C_n \in \Sigma^u$, where C_n is an n-cylinder with $mC_{n+1} = C_n$, then $J_\mu(x) = \lim_{n\to\infty} \nu^*(m_* C_n^*)/\nu^*(C_n^*)$. The Jacobian $J_{\nu^*}(\xi) = d(\nu^* \circ \tau)/d\nu^*(\xi)$ is $\sigma_l^{-1}(\xi)$.

Remark 6.5. As part of the proof of the theorem, we will prove that if the Hölder weight system l matches and if μ is any τ-invariant probability measure satisfying the ratio decomposition (iii), then, for all cylinders C of Σ^u,

$$\sum r_{l,\xi_D}(C)\, \mu^*(D^*) \in (1 \pm \mathcal{O}(\nu^n))\, \mu(C),$$

where the sum is over all n-cylinders D such that $C \subset \tau^{n-1}(D)$, and, for each D, $\xi_D = \{\xi_j\}_{j=0}^\infty$ is an infinite backward path with the property that $\xi_n = D$.

Proof of Theorem 6.4. Firstly, consider the sum $Z_s = \sum_C k(l(C))l(C)e^{-sn(C)}$, where the sum is over all cylinders C and k is the function given by Lemma 6.3. As we have seen above, $Z_s < \infty$ for $s > P$, and Z_s diverges if $s = P$. We denote $k(l(C))l(C)$ by $\tilde{l}(C)$ and $\tilde{l}(C)e^{-sn(C)}$ by $\tilde{l}_s(C)$.

Note that the condition (ii) of Lemma 6.3 on k and the fact that $l(\tau C) = J_l(C) \cdot l(C)$ implies that, for all $\varepsilon > 0$, if $J_l(C) \geq 1$ then $J_l(C)^{-\varepsilon} \leq k(l(\tau C))/k(l(C)) \leq 1$, and if $J_l(C) < 1$ then $1 \leq k(l(\tau C))/k(l(C)) \leq J_l(C)^{-\varepsilon}$, provided $\max\{l(C), l(\tau C)\} < y_0(\varepsilon)$. Since $J_l(C)$ is bounded away from 0 and ∞ uniformly in C, we deduce that, for all $\varepsilon > 0$,

$$\frac{\tilde{l}(\tau C)}{\tilde{l}(C)} \in (1 \pm \varepsilon) J_l(C), \tag{6.5}$$

provided $l(C)$ is sufficiently small. Similarly, we deduce that

$$\frac{\tilde{l}(mC)}{\tilde{l}(C)} \in (1 \pm \varepsilon)\sigma_l(C)^{-1}, \tag{6.6}$$

provided $l(C)$ is sufficiently small.

For $s > P$, let ν_s and ν_s^* be the probability measures on Δ and Δ^* defined by $\nu_s = Z_s^{-1} \sum_{x \in \Delta_{\mathrm{fin}}} \tilde{l}_s(C)\delta_x$ and $\nu_s^* = Z_s^{-1} \sum_{\xi \in \Delta_{\mathrm{fin}}^*} \tilde{l}_s(C)\delta_\xi$, where δ_x and δ_ξ are, respectively, the Dirac measures at x and ξ.

Let Δ be the set of all infinite right-handed words in Σ^u and their finite subwords. Similarly to the dual $(\Sigma^u)^*$ of Σ^u, let the dual Δ^* of Δ be the set of all finite and infinite admissible backward paths in $\Delta_{\mathrm{fin}}^\tau$.

Since Δ and Δ^* are compact metric spaces, there exist sequences $s_i > 0$ and $s_i^* > 0$ converging to P as $i \to \infty$ so that the sequence ν_{s_i} (respectively, $\nu_{s_i^*}^*$) converges weakly to a Borel probability measure ν on Δ (respectively, ν^* on Δ^*). Since Z_{s_i} and $Z_{s_i^*}$ diverge as $i \to \infty$, ν and ν^* are, respectively, concentrated on Σ and Σ^*. Thus, ν and ν^*, respectively, define measures on Σ^u and $(\Sigma^u)^*$, which we also denote by ν and ν^*.

If w is a finite word, consider the cylinder Σ_w in Σ^u and also the subset Δ_w in Δ^* consisting of all finite and infinite right-handed words agreeing with w. We have

$$\nu(\Sigma_w) = \nu(\Delta_w) \approx \nu_{s_i}(\Delta_w) = Z_{s_i}^{-1} \sum_{C \subset \Sigma_w} \tilde{l}_{s_i}(C),$$

where the sum is over all cylinders C contained in Σ_w and with the approximation converging as $i \to \infty$. Therefore, by (6.6), for $\varepsilon > 0$,

$$\frac{\nu(\tau\Sigma_w)}{\nu(\Sigma_w)} = \lim_{i\to\infty} \frac{\sum_{D \subset \tau\Sigma_w} \tilde{l}_{s_i}(D)}{\sum_{C \subset \S_w} \tilde{l}_{s_i}(C)} = \lim_{i\to\infty} \frac{\sum_{C \subset \S_w} \tilde{l}_{s_i}(\tau C)}{\sum_{C \subset \Sigma_w} \tilde{l}_{s_i}(C)} \in (1 \pm \varepsilon) J_l(\Sigma_w)\mathrm{e}^P,$$

providing $l(\Sigma_w)$ is sufficiently small. This implies that the Jacobian of ν at $x \in \bigcap_{j=0}^\infty C_n$ is $J_\nu(x) = d(\nu \circ f)/d\nu = \lim_{n\to\infty} J_l(C_n)\mathrm{e}^P$. Since this is Hölder continuous, we obtain that if Σ_w is an n-cylinder, then

$$\frac{\nu(\tau\Sigma_w)}{\nu(\Sigma_w)} \in (1 \pm \mathcal{O}(\theta^n))\, J_l(\Sigma_w)\mathrm{e}^P, \tag{6.7}$$

for some $0 < \theta < 1$. Thus, the weights $l_\nu(\Sigma_w) = \nu(\Sigma_w)$ form a Hölder weight system.

If w is a word, consider the cylinder Σ_w^* in $(\Sigma^u)^*$ and also the subset Δ_w^* in Δ^* consisting of all admissible backward finite and infinite paths agreeing with w. We have

$$\nu^*(\Sigma_w^*) = \nu^*(\Delta_w^*) \approx \nu_{s_i^*}^*(\Delta_w^*) = Z_{s_i^*}^{-1} \sum_{C^* \subseteq \Sigma_w^*} \tilde{l}_{s_i^*}(C) = Z_{s_i^*}^{-1} \sum_{C \to \Sigma_w} \tilde{l}_{s_i^*}(C),$$

where $C \to \Sigma_w$ means that $\tau^k(C) = \Sigma_w$, for some $k \geq 0$, with the approximation marked $\approx$ converging as $i \to \infty$. The first sum in this equation is over all cylinders C^* contained in or equal to Σ_w^* and the second equals this because by duality (6.1), $C^* \subseteq \Sigma_w^*$ if, and only if, $\tau^k(C) = \Sigma_w$. Therefore, by construction of σ_l, we have that, for all $\varepsilon > 0$,

$$\frac{\nu^*(\tau_* \Sigma_w^*)}{\nu^*(\Sigma_w^*)} = \lim_{i\to\infty} \frac{\sum_{\tau^k(C)=m\Sigma_w} \tilde{l}_{s_i^*}(C)}{\sum_{\tau^k(C)=\Sigma_w} \tilde{l}_{s_i^*}(C)}$$

$$= \lim_{i\to\infty} \frac{\sum_{\tau^k(C)=\Sigma_w} \tilde{l}_{s_i^*}(mC)}{\sum_{\tau^k(C)=\Sigma_w} \tilde{l}_{s_i^*}(C)} \in (1 \pm \varepsilon)\sigma_l(\Sigma_w)^{-1}\mathrm{e}^P,$$

providing $l(\Sigma_w)$ is sufficiently small. This implies that the Jacobian of ν^* is $J_{\nu^*}(\xi) = d(\nu^* \circ \tau_*)/d\nu^* = \sigma_l(\xi)^{-1}\mathrm{e}^P$. Since this is Hölder continuous, we obtain that the weights $l_*(\Sigma_w^*) = \nu^*(\Sigma_w^*)$ form a Hölder weight system and, indeed, if Σ_w is an n-cylinder,

$$\frac{\nu^*(\tau_*(\Sigma_w^*))}{\nu^*(\Sigma_w^*)} \in (1 \pm \mathcal{O}(\theta^n))\, \sigma_l(\Sigma_w)^{-1}\mathrm{e}^P, \tag{6.8}$$

for some $0 < \theta < 1$.

Now we consider the uniqueness of ν and ν^*. Suppose that ν' is another measure satisfying (6.7). Then, if C is an n-cylinder,

$$\frac{\nu'(C)}{\nu(C)} = \frac{\nu'(C)}{\nu'(\tau C)} \cdot \frac{\nu'(\tau C)}{\nu(\tau C)} \cdot \frac{\nu(\tau C)}{\nu(C)} \in (1 \pm \mathcal{O}(\theta^n))\, \frac{\nu'(\tau C)}{\nu(\tau C)},$$

because $\nu'(\tau C)/\nu'(C) = (1 \pm \mathcal{O}(\theta^n))\,(\nu(\tau C)/\nu(C))$ by (6.7). Thus, if $\xi = (\xi_n) \in (\Sigma^u)^*$, where ξ_n is an n-cylinder and $J_{\nu',\nu}(\xi) = \lim_{n\to\infty} \nu'(\xi_n)/\nu(\xi_n)$, the limit is achieved exponentially fast, and $J_{\nu',\nu}$ is Hölder continuous on $(\Sigma^u)^*$. Also, since

$$\frac{J_{\nu',\nu}(\tau_*\xi)}{J_{\nu',\nu}(\xi)} \in (1 \pm \mathcal{O}b\theta^n) \cdot \frac{\nu'(\tau\xi_n)}{\nu'(\xi_n)} \cdot \frac{\nu(\xi_n)}{\nu(\tau\xi_n)} \in 1 \pm \mathcal{O}b\theta^n,$$

$J_{\nu',\nu}(\tau_*\xi) = J_{\nu',\nu}(\xi)$, i.e. $J_{\nu',\nu}$ is τ_*-invariant. Therefore, it is constant on a dense set of $(\Sigma^u)^*$, for example the full backward orbit of a single point. Since it is Hölder continuous, it must be constant everywhere and, therefore, equal to 1 everywhere. Thus, $\nu = \nu'$ and ν is the unique measure satisfying (6.7). It follows that $\nu = \lim_{s\searrow P} \nu_s$. A similar argument shows that ν^* is the unique measure satisfying (6.8) and $\nu^* = \lim_{s\searrow P} \nu_s^*$.

By the properties of the weight function l and by (6.7), for all n-cylinders C, we get

$$\frac{\nu(C)}{l(C)e^{-nP}} = \frac{\nu(\tau^n C)}{l(\tau^n C)} \cdot \prod_{j=0}^{n-1} \frac{\nu(\tau^j C)}{\nu(\tau^{j+1}C)\mathrm{e}^{-P}} \cdot \frac{l(\tau^{j+1}C)}{l(\tau^j C)} \in \frac{\nu(\tau^n C)}{l(\tau^n C)} \prod_{j=0}^{n-1} (1 \pm \mathcal{O}(\theta^j)). \tag{6.9}$$

Thus, the ratios $\nu(C)/l(C)e^{-nP}$ are uniformly bounded away from 0 and ∞. Similarly as above, using (6.8) instead of (6.7), we obtain that the ratios $\nu^*(C^*)/l(C)\mathrm{e}^{-nP}$ are uniformly bounded away from 0 and ∞. Therefore,

$$\lim_{s\searrow P}\sum_C l(C)\mathrm{e}^{-n(C)s} \geq c_1 \lim_{s\searrow P}\sum_C \nu(C)\mathrm{e}^{-n(C)(p-s)} \geq c_2 \lim_{s\searrow P}\sum_{n=1}^{\infty} \mathrm{e}^{-n(p-s)}$$

diverges at $s = P$. The first sum is over all cylinders C.

Therefore, since ν and $\nu*$ are the unique probability measures satisfying, respectively, (6.7) and (6.8), we deduce that $\nu = \lim_{s\searrow P}\rho_s$ and $\nu^* = \lim_{s\searrow P}\rho_s^*$, where ρ_s and ρ_s^* are defined as ν_s and ν_s^* above, but with $k \equiv 1$. For all cylinders C and D, it follows that

$$\frac{\nu(C)}{\nu(D)} = \lim_{s\searrow P}\frac{\sum_{C'\subset C} l(C')\mathrm{e}^{-n(C')s}}{\sum_{D'\subset D} l(D')\mathrm{e}^{-n(D')s}} = r_l(C:D), \tag{6.10}$$

which ends the proof of the first assertion of this theorem.

From now on in this proof we assume that the weight function l matches. In this case, $\sum_{C_n\subset C} l(C_n)/\sum_{D_{n-1}\subset C} l(D_{n-1}) \in (1 \pm \mathcal{O}(\theta^n))$, if the first and second sums are, respectively, over all n-cylinders and all $(n-1)$-cylinders contained in C. Thus, $\sum_{C_n} l(C_n) = \mathcal{O}(1)$, and consequently $\sum_C l(C)e^{-n(C)s} = \mathcal{O}(\sum_{n=0}^{\infty}\mathrm{e}^{-ns})$ converges for every $s > 0$ and diverges at $s = 0$. This implies that $P = 0$. Furthermore, we obtain that

$$r_l(C:D) \in (1 \pm \mathcal{O}(\theta^n))\,\frac{l(C)}{l(D)}, \tag{6.11}$$

where C and D are contained in a common n-cylinder. This implies (6.4).

For all cylinder Σ_w, we have that

$$\frac{\nu^*(\tau_*^{-1}\Sigma_w^*)}{\nu^*(\Sigma_w^*)} \approx \frac{\rho_s^*(\tau_*^{-1}\Sigma_w^*)}{\rho_s^*(\Sigma_w^*)} = \frac{\sum_{mD=C:\tau^k C=\Sigma_w} l(D)\mathrm{e}^{-n(D)s}}{\sum_{\tau^k C=\Sigma_w} l(C)e^{-n(C)s}}$$

with the approximation converging as $s \searrow 0$. Since the ratios $l(C)/l(mC)$ converge exponentially fast along backward orbits there are continuous functions $\tau_1(s)$ and $\tau_2(s)$ which converge to 1 as $s \searrow 0$ such that, for all cylinders Σ_w,

$$\tau_1(s) < \frac{\sum_{mD=C:\tau^k C=\Sigma_w} l(D)\mathrm{e}^{-n(D)s}}{\sum_{\tau^k C=\Sigma_w} l(C)\mathrm{e}^{-n(C)s}} < \tau_2(s).$$

Thus, we deduce that $\nu^*(\tau_*^{-1}C^*) = \nu^*(C^*)$, for all cylinders, and hence that ν^* is τ_*-invariant. It follows from this that if we define μ on Σ by $\mu(C) = \nu^*(C^*)$, for all cylinders C of Σ^u, then μ is a τ-invariant probability measure on Σ^u. The fact that it is a measure follows from the τ_*-invariance of ν^*, and the fact that it is τ-invariant follows from the fact that ν^* is a probability measure.

Now we consider the ratios $\mu(C_1)/\mu(C_2) = \nu^*(C_1^*)/\nu^*(C_2^*)$, where C_1 and C_2 are cylinders and C_1 is contained in C_2. Then, there exists $r \geq 0$ such that $m^r C_1 = C_2$. In this case, $\tau_*^r C_1^* = C_2^*$. Thus, the ratio is approximated by

$$\frac{\sum_{m_*^k C^* = C_1^*} l(C) \mathrm{e}^{-n(C)s}}{\sum_{m_*^k C^* = \tau_*^r C_1^*} l(C) \mathrm{e}^{-n(C)s}} = \frac{\sum_{\tau^k C = C_1} l(C) \mathrm{e}^{-n(C)s}}{\sum_{\tau^k C = m^r C_1} l(C) \mathrm{e}^{-n(C)s}} \tag{6.12}$$

with convergence as $s \searrow 0$. To each summand $l(C)$ of the top sum there corresponds a summand $l(C')$ of the bottom sum such that $m^r C = C'$, and the pair (C, C') is mapped by some power of τ onto the pair (C_1, C_2). It follows that $l(C)/l(C') \in (1 \pm \mathcal{O}(\theta^{n(C_2)}))l(C_1)/l(C_2)$, where the constant of proportionality in the $\mathcal{O}$ term is independent of C, C', C_1 and C_2. Thus, we deduce that the last term for $s = 0$ of (6.12) is in the interval $(1 \pm \mathcal{O}(\theta^n))l(C_1)/l(C_2)$. We have proved that if C_2 is an n-cylinder, then

$$\frac{\mu(C_1)}{\mu(C_2)} \in (1 \pm \mathcal{O}(\theta^n)) \frac{l(C_1)}{l(C_2)}. \tag{6.13}$$

Parts (i) and (ii) of Theorem 6.4 follow from this.

It remains to prove part (iii), the ratio decomposition. To do this, recall the meaning of $C(C_n)$ given in §6.3. If C_p is a primary cylinder, let $\mathcal{C}_n(C_p)$ denote the set of n-cylinders C such that $\tau^{n-1}C = C_p$. Let C_p be the primary cylinder containing Σ_w. We have

$$\begin{aligned} \mu(\Sigma_w) &\stackrel{1}{=} \sum_{C_n \in \mathcal{C}_n(C_p)} \frac{\mu(\Sigma_w(C_n))}{\mu(C_n)} \nu^*(C_n^*) \\ &\stackrel{2}{\approx} \sum_{C_n \in \mathcal{C}_n(C_p)} \frac{l(\Sigma_w(C_n))}{l(C_n)} \nu^*(C_n^*) \to \int_{C_p^*} r_{l,\xi}(\Sigma_w) \nu^*(d\xi), \end{aligned}$$

as $n \to \infty$. The equality marked $\stackrel{1}{=}$ follows from the τ-invariance of μ and also by duality, that marked $\stackrel{2}{\approx}$ from (6.13) and the convergence from property (ii) of the potential, from the definition of $r_\xi(\Sigma_w)$ in §6.3 and the comments in Remark 6.2.

The final point is to check uniqueness of invariant measures satisfying either (i), (ii) or (iii). Since (i) implies (ii), it suffices to check (ii) to verify both. However, if ρ^* is another measure satisfying the condition in part (ii), then one can prove that $\rho = \mu$ in a similar fashion to the proof of the uniqueness of ν above, using ρ^* and ν^*, the fact that $\tau_* C^* = (mC)^*$, and condition (ii) of this theorem.

Suppose that ρ is a measure satisfying the ratio decomposition (iii) of the theorem and let ρ^* denote its dual. First, we note that if ξ_{n+1} is an $(n+1)$-cylinder and $\xi_n = \tau(\xi_{n+1})$, then $r_l(C(\xi_{n+1}) : \xi_{n+1}) \in (1 \pm \mathcal{O}(\theta^n))\, r_l(C(\xi_n) : \xi_n)$. Moreover, since ρ is τ-invariant, $\sum_{\xi_{n+1}^*} \rho^*(\xi_{n+1}^*) = \rho^*(\xi_n^*)$, where the sum is over all ξ_{n+1}^* contained in ξ_n^* or equivalently over all τ-preimages ξ_{n+1} of ξ_n. Thus,

$$\sum_{(n+1)-\text{cyls.}\xi_{n+1}} r_l(C(\xi_{n+1}) : \xi_{n+1})\rho^*(\xi_{n+1})$$
$$\in (1 \pm \mathcal{O}(\theta^n)) \sum_{n-\text{cyls.}\xi_n} r_l(C(\xi_n) : \xi_n)\, \rho^*(\xi_n).$$

This with condition (iii) proves Remark 6.5.

Therefore, if C and D are cylinders of Σ^u contained in the cylinder E and $\xi \in (\Sigma^u)^*$ has $E \subset \xi_0$, then, denoting by $\mathcal{C}_n(E)$ the set of n-cylinders C' such that $\tau^{n-1}(C')$ contains E,

$$\begin{aligned}
\frac{\rho(C)}{\rho(D)} &\approx \frac{\sum_{\xi_n \in \mathcal{C}_n(E)} r_l(C(\xi_n) : \xi_n)\, \rho^*(\xi_n)}{\sum_{\xi_n \in \mathcal{C}_n(E)} r_l(D(\xi_n) : \xi_n)\, \rho^*(\xi_n)} \\
&= \frac{\sum_{\xi_n \in \mathcal{C}_n(E)} r_l(C(\xi_n) : D(\xi_n)) r_l(D(\xi_n) : \xi_n)\, \rho^*(\xi_n)}{\sum_{\xi_n \in \mathcal{C}_n(E)} r_l(D(\xi_n) : \xi_n)\, \rho^*(\xi_n)} \\
&\in (1 \pm \mathcal{O}(\theta^n))\, \frac{l(C)}{l(D)}
\end{aligned} \tag{6.14}$$

because

$$r_l(C(\xi_n) : D(\xi_n)) = \frac{l(C(\xi_n))}{l(D(\xi_n))} \in (1 \pm \mathcal{O}(\theta^n))\, \frac{l(C)}{l(D)},$$

since C and D are in the n-cylinder E. Thus, if $\xi = (\xi_n)_{n=0}^{\infty} \in (\Sigma^u)^*$, then

$$\frac{\rho(\xi_n)}{\rho(m\xi_n)} \in (1 \pm \mathcal{O}(\theta^n))\, \sigma_l(\xi)$$

by (6.14) and, consequently, ρ, like μ, satisfies condition (ii) of the theorem. But we have already shown that there is only one measure satisfying this. Hence $\rho = \mu$. □

Lemma 6.6. *Let l be a Hölder weight system.*

(i) We define the ratio $r_l(C : D)$ between two cylinders C and D by

$$r_l(C : D) = \lim_{s \searrow P} \frac{\sum_{C' \subset C} l(C')\mathrm{e}^{-n(C')s}}{\sum_{D' \subset D} l(D')\mathrm{e}^{-n(D')s}},$$

where the sums are, respectively, over all cylinders contained in or equal to C and D. For $s > P$, both numerator and denominator are finite and positive. As part of the proof of the following theorem we will show that the limit as $s \searrow P$ is finite and positive.

(ii) For $\xi = (\xi_n) \in (\Sigma^u)^*$, let $\sigma(\xi) = \lim_{n\to\infty} r_l(\xi_n : m\xi_n)$.

(iii) For $\xi \in (\Sigma^u)^*$ and C contained in the primary cylinder ξ_0, define $r_{l,\xi}(C) = \lim_{n\to\infty} r_l(C(\xi_n) : \xi_n)$.

Proof. The limits in (i), (ii) and (iii) exist and are finite and positive (use (6.9) and (6.10) to deduce (i), and use that $l_\nu(C) = \nu(C)$ form a matching Hölder weight system to deduce (ii) and (iii) where ν is the probability measure constructed in Theorem 6.4). □

Theorem 6.7. *(Existence and uniqueness of Gibbs states) There exist a unique pair of Borel probability measures μ on Σ^u and μ^* on $(\Sigma^u)^*$ with the following properties, for some $0 < \theta < 1$:*

(i) μ and μ^* are dual to each other and, respectively, τ-invariant and τ_*-invariant;

(ii) If C and D are two cylinders contained in the same n-cylinder, then

$$\mu(C)/\mu(D) \in (1 \pm \mathcal{O}(\theta^n))\, r_l(C : D);$$

(iii) (Ratio decomposition) *If C is an n-cylinder and C_0 is the primary cylinder containing C, then*

$$\mu(C) = \int_{C_0^*} r_{l,\xi}(C)\mu^*(dx).$$

Either of the conditions (ii) and (iii) characterise the measure μ, i.e. it is the unique measure with the given property.

If J_μ is the Jacobian $d(\mu \circ \tau)/d\mu$ and $x = \bigcap_{n \geq 0} C_n \in \Sigma^u$, where C_n is an n-cylinder with $mC_{n+1} = C_n$, then $J_\mu(x) = \lim_{n\to\infty} \mu^*(m_* C_n^*)/\mu^*(C_n^*)$. Finally, $d(\mu^* \circ \tau_*)/d\mu^* = \sigma^{-1}$.

The measure μ is the Gibbs state for the potential J_l in the sense of Bowen [17], i.e. it is the unique τ-invariant probability measure which for all cylinders C the ratios $\mu(C)/l(C)\mathrm{e}^{-n(C)P}$ are uniformly bounded away from 0 and ∞.

Note that the ratios $r_{l,\xi}$ and r_ξ can be different, if the weights do not match and the logaritmic scaling functions $\log \sigma_l$ and $\log \sigma$ differ at most by a coboundary, i.e. there is a Hölder continuous function $u : (\Sigma^u)^* \to \mathbb{R}$ such that $\log(\sigma_l(\xi)) = \log(\sigma(\xi)) + u(\tau_*\xi) - u(\xi)$. However, if the weight system l matches, then $r_{l,\xi} = r_\xi$ and $\sigma_l = \sigma$.

Corollary 6.8. *(Moduli space for Gibbs states) The correspondence between σ and μ given in Theorem 6.7 gives a natural one-to-one correspondence between Hölder Gibbs states and Hölder measured scaling functions on the dual space $(\Sigma^u)^*$ which satisfy the matching condition (6.3).*

Proof of Theorem 6.7. First, we apply Theorem 6.4 to the weight system l to obtain the measure ν. Then we consider the new weight system $l_\nu(C) = \nu(C)$. By Theorem 6.4, this is Hölder and, since ν is a probability measure, it satisfies the matching condition. Now, apply Theorem 6.4 to this to obtain measures ν_1, ν_1^* and $\mu = \mu_1$ (corresponding to ν, ν^* and μ of the theorem). It follows immediately from Theorem 6.4 that μ is the required Gibbs state. As is well-known, since μ has a Hölder jacobian, it is ergodic. Therefore, it is the unique invariant measure in its measure class and, hence, the unique invariant measure for which the ratios $\mu(C)/\ell(C)e^{-n(C)P}$ are uniformly bounded away from 0 and ∞. □

6.5 Further literature

The novelty of the approach presented is to use the notion of duality and combined with the approach to construct measures pioneered by Paterson [137] in the context of the limit sets of Fuchsian groups and used by Sullivan [229] to construct conformal measures for Julia sets. This chapter is based on Bowen [17] and Pinto and Rand [162].

7

Measure scaling functions

We present some basic facts on Gibbs measures and measure scaling functions, linking them with two dimensional hyperbolic dynamics.

7.1 Gibbs measures

Let us give the definition of an infinite two-sided subshift of finite type Θ. The elements of $\Theta = \Theta_A$ are all infinite two-sided words $w = \ldots w_{-1}w_0w_1\ldots$ in the symbols $1, \ldots, k$ such that $A_{w_i w_{i+1}} = 1$, for all $i \in \mathbb{Z}$. Here $A = (A_{ij})$ is any matrix with entries 0 and 1 such that A^n has all entries positive for some $n \geq 1$. We write $w \overset{n_1,n_2}{\sim} w'$ if $w_j = w'_j$ for every $j = -n_1, \ldots, n_2$. The metric d on Θ is given by $d(w, w') = 2^{-n}$ if $n \geq 0$ is the largest such that $w \overset{n,n}{\sim} w'$. Together with this metric Θ is a compact metric space. The two-sided shift map $\tau : \Theta \to \Theta$ is the mapping which sends $w = \ldots w_{-1}w_0w_1\ldots$ to $v = \ldots v_{-1}v_0v_1\ldots$ where $v_j = w_{j+1}$ for every $j \in \mathbb{Z}$. We will denote τ by τ_u and τ^{-1} by τ_s. An (n_1, n_2)-*rectangle* $\Theta_{w_{-n_1}\ldots w_{n_2}}$, where $w \in \Theta$, consists of all those words w' in Θ such that $w \overset{n_1,n_2}{\sim} w'$. Let Θ^u be the set of all right-handed words $w_0w_1\ldots$ which extend to words $\ldots w_0w_1\ldots$ in Θ, and, similarly, let Θ^s be the set of all left-handed words $\ldots w_{-1}w_0$ which extend to words $\ldots w_{-1}w_0\ldots$ in Θ. Then, $\pi_u : \Theta \to \Theta^u$ and $\pi_s : \Theta \to \Theta^s$ are the natural projection given, respectively, by

$$\pi_u(\ldots w_{-1}w_0w_1\ldots) = w_0w_1\ldots \qquad \text{and} \qquad \pi_s(\ldots w_{-1}w_0w_1\ldots) = \ldots w_{-1}w_0\ .$$

The metric d determines, naturally, a metric d_u in Θ^u and d_s in Θ^s. An n-rectangle $\Theta^u_{w_0\ldots w_{n-1}}$ is equal to $\pi_u(\Theta_{w_0\ldots w_{n-1}})$ and an n-rectangle $\Theta^s_{w_{-(n-1)}\ldots w_0}$ is equal to $\pi_s(\Theta_{w_{-(n-1)}\ldots w_0})$. Let $\tilde{\tau}_u : \Theta^u \to \Theta^u$ and $\tilde{\tau}_s : \Theta^s \to \Theta^s$ be the corresponding one-sided shifts. Noting that $\pi_u \circ \tau_u = \tilde{\tau}_u \circ \pi_u$ and $\pi_s \circ \tau_s^{-1} = \tilde{\tau}_s \circ \pi_s$, we will also denote $\tilde{\tau}_u$ by τ_u and $\tilde{\tau}_s$ by τ_s.

Definition 7.1. *For $\iota = s$ and u, we say that $s_{\iota'} : \Theta^\iota \to \mathbb{R}^+$ is an ι-*measure scaling function *if s_ι is a Hölder continuous function, and for every $\xi \in \Theta^\iota$*

$$\sum_{\tau_\iota \eta = \xi} s_\iota(\eta) = 1 ,$$

where the sum is upon all $\xi \in \Theta^\iota$ such that $\tau_\iota \eta = \xi$.

For $\iota \in \{s, u\}$, a τ-invariant measure ν on Θ determines a unique τ_ι-invariant measure $\nu_\iota = (\pi_\iota)_* \nu$ on Θ^ι. We note that a τ_ι-invariant measure ν_ι on Θ^ι has a unique τ-invariant natural extension to an invariant measure ν on Θ such that $\nu(\Theta_{w_0 \dots w_{n_2}}) = \nu_\iota(\Theta^\iota_{w_0 \dots w_{n_2}})$.

Definition 7.2. *A τ-invariant measure ν on Θ is a* Gibbs measure*:*

(i) if the function $s_{\nu,s} : \Theta^u \to \mathbb{R}^+$ given by

$$s_{\nu,s}(w_0 w_1 \dots) = \lim_{n \to \infty} \frac{\nu(\Theta_{w_0 \dots w_n})}{\nu(\Theta_{w_1 \dots w_n})} ,$$

is well-defined and it is an s-measure scaling function; or

(ii) if the function $s_{\nu,u} : \Theta^s \to \mathbb{R}^+$ given by

$$s_{\nu,u}(\dots w_1 w_0) = \lim_{n \to \infty} \frac{\nu(\Theta_{w_n \dots w_0})}{\nu(\Theta_{w_n \dots w_1})} ,$$

is well-defined and it is a u-measure scaling function.

By Theorem 7.7, condition (i) is equivalent to condition (ii). By Corollary 6.8, an ι-measure scaling function s_ι determines a Gibbs measure ν_{s_ι}.

7.2 Extended measure scaling function

We will construct the ι-measure scaling set $\text{msc}^{\iota'}$ that contains $\Theta^{\iota'}$. We will construct a natural extension of any scaling function to the domain msc_ι that we call an extended measure scaling function or measure ratio function. The extended measure scaling function plays a key role in this subject.

Throughout the chapter, if $\xi \in \Theta^{\iota'}$, we denote by ξ_Λ the leaf primary cylinder segment $i(\pi_{\iota'}^{-1}\xi) \subset \Lambda$. Similarly, if C is an n_ι-rectangle of Θ^ι, then we denote by C_Λ the $(1, n_\iota)$-rectangle $i(\pi_\iota^{-1} C) \subset \Lambda$.

We say that $I \subset \Theta$ is an ι-symbolic leaf n-cylinder, if $i(I)$ is an ι-leaf n-cylinder. Every ι-symbolic leaf n-cylinder can be expressed as

$$\xi.C = \pi_\iota^{-1} C \cap \pi_{\iota'}^{-1} \xi,$$

where $\xi \in \Theta^{\iota'}$ and C is an n-cylinder of Θ^ι (see Figure 7.1). We call that such pairs $\xi.C$ *ι'-admissible.* The set of all ι-admissible pairs is the *ι-measure scaling set* msc^ι.

Let C be an n-cylinder of Θ^ι. For all $0 < l < n$, we say that $m^l C$ is the *l-th mother* of C, if $m^l C$ is an $(n - l)$-cylinder and $m^l C \supset C$.

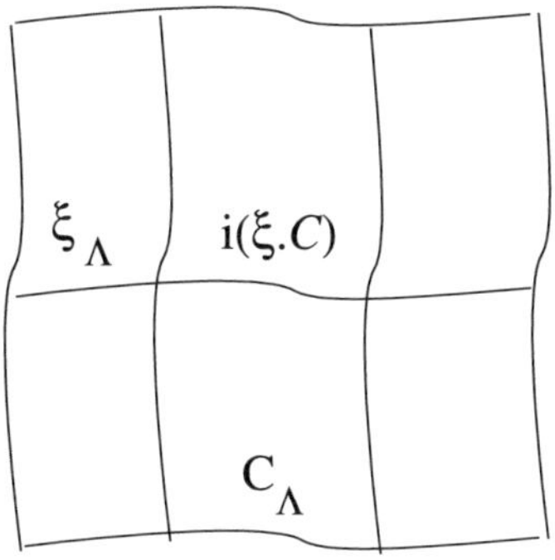

Fig. 7.1. An ι-admissible pair (ξ, C) where $\xi_\Lambda = i(\pi_{\iota'}^{-1}\xi)$, $C_\Lambda = i(\pi_\iota^{-1}C)$ and $i(\xi.C)$ is a leaf n-cylinder.

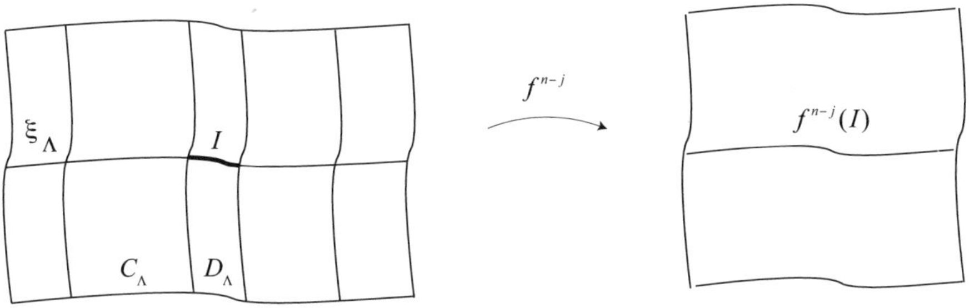

Fig. 7.2. The $(n-j+1)$-cylinder leaf segment $I = \xi_\Lambda \cap D_\Lambda$ and the primary leaf segment $f^{n-j}(I) = i(\pi_{\iota'}\tau_\iota^{n-j}(\xi.D))$, where $D = m_\iota^{j-1}C$.

Given an ι-measure scaling function s_ι, we construct the ι-extended measure scaling function $\rho_\iota : \mathrm{msc}^\iota \to \mathbb{R}^+$ induced by the ι-scaling function s_ι as follows: If C is an 1-rectangle on Θ^ι, then we define $\rho_\iota(\xi.C) = \rho_{\iota,\xi}(C) = 1$. If C is an n-rectangle on Θ^ι, with $n \geq 2$, then we define

$$\rho_\iota(\xi.C) = \prod_{j=1}^{n-1} s_\iota(\pi_{\iota'}\tau_\iota^{n-j}(\xi.m_\iota^{j-1}C))$$

(see Figure 7.2). We will denote, from now on, $\rho_\iota(\xi.C)$ by $\rho_{\iota,\xi}(C)$. By Lemma 7.6, $\rho_\iota(\xi.C)$ is the conditional measure $\nu(\xi.C|\xi)$ of $\xi.C$ in ξ.

Let $\xi \in \Theta^{\iota'}$ be such that $i(\xi)$ is an ι-leaf segment spanning of a Markov rectangle $i(M)$. Let $i(R)$ be a rectangle inside $i(M)$. There are pairwise disjoint rectangles $C_j \in \Theta^\iota$ such that $\pi_\iota R$ is the countable (or finite) union $\cup_{j\in\mathrm{Ind}}C_j$ of rectangles. If $\xi \cap R \neq \emptyset$, we define the ratio $\rho_{\iota,\xi}(R : M)$ by

$$\rho_{\iota,\xi}(R : M) = \sum_{j\in\mathrm{Ind}} \rho_{\iota,\xi}(C_j) \ . \tag{7.1}$$

If $\xi \cap R = \emptyset$, we define $\rho_{\iota,\xi}(R : M) = 0$. More generally, suppose that R_0 and R_1 are ι'-spanning rectangles contained in R. We define the ratio $\rho_{\iota,\xi}(R_0 : R_1)$

by

$$\rho_{\iota,\xi}(R_0 : R_1) = \rho_{\iota,\xi}(R_0 : M)\rho_{\iota,\xi}(R_1 : M)^{-1} . \tag{7.2}$$

Remark 7.3. The ratios determined by ρ_ι are f-invariant, i.e $\rho_{\iota',\xi}(R_0 : R_1) = \rho_{\iota',\xi}(fR_0 : fR_1)$. Furthermore, $\rho_{\iota'}$ determines affine structures on ι-symbolic leaves, i.e $\rho_{\iota'}(R_1 : R_2) = \rho_{\iota'}(R_2 : R_1)^{-1}$ and

$$\rho_{\iota'}(R_1 \cup R_2 : R_3) = \rho_{\iota'}(R_1 : R_3) + \rho_{\iota'}(R_2 : R_3),$$

where R_1, R_2 and R_3 are pairwise disjoint rectangles.

Lemma 7.4. *Let $\rho_{\iota'} : \mathrm{msc}_{\iota'} \to \mathbb{R}^+$ be an extended ι'-measure scaling function. There is $\gamma = \gamma(\rho_i) > 0$ such that*

$$\frac{\rho_{\iota',\xi}(C)}{\rho_{\iota',\eta}(C)} = 1 \pm \mathcal{O}\left(d_\iota(\xi,\eta)^\gamma\right),$$

for every n-rectangle C in Θ^ι and for all $\xi, \eta \in \Theta^{\iota'}$.

Proof. Let $d_\iota(\xi,\eta) = 2^{-m}$. We obtain that

$$d_\iota\left(\pi_{\iota'}\xi.m_\iota^{j-1}C, \pi_{\iota'}\eta.m_\iota^{j-1}C\right) = 2^{-(m+n-j)}.$$

Since, for some $\alpha > 0$ the scaling function $s_\iota : \Theta^{\iota'} \to \mathbb{R}^+$ is α-Hölder continuous, we get

$$\left|s_\iota(\pi_{\iota'}\eta.m_\iota^{j-1}C) - s_\iota(\pi_{\iota'}\xi.m_\iota^{j-1}C)\right| \le K_1 2^{-\alpha(m+n-j)},$$

for some $K_1 \ge 1$. Therefore,

$$\begin{aligned}
|\log \rho_\xi(C) - \log \rho_\eta(C)| &\le \sum_{j=1}^{n-1} \left|s_\iota(\pi_{\iota'}\eta.m_\iota^{j-1}C) - s_\iota(\pi_{\iota'}\xi.m_\iota^{j-1}C)\right| \\
&\le \sum_{j=1}^{n-1} K_1 2^{-\alpha(m+n-j)} \\
&\le K_2 2^{-\alpha m}.
\end{aligned}$$

□

Recall that a τ-invariant measure ν on Θ determines a unique τ_u-invariant measure $\nu_u = (\pi_u)_*\nu$ on Θ^u and a unique τ_s-invariant measure $\nu_s = (\pi_s)_*\nu$ on Θ^s.

Lemma 7.5. *(Ratio decomposition) Let ν be a Gibbs measure with ι'-extended scaling function $\rho_{\iota'}$. If $i(R)$ is a rectangle contained in a Markov rectangle $i(M)$, then*

$$\nu(R) = \int_{\pi_{\Theta^{\iota'}}(R)} \rho_{\iota,\xi}(R : M)\nu_{\iota'}(d\xi) . \tag{7.3}$$

Since any rectangle can be written as the union of rectangles R with the property hypothesised in the theorem for some Markov rectangle, the above theorem gives an explicit formula for the measure of any rectangle in terms of a ratio decomposition.

Proof of Lemma 7.5. Suppose that $i(R)$ is a rectangle contained in a Markov rectangle $i(M)$. There is $0 < \nu < 1$ such that for all $n > 0$ we can write $R = R_0 \cup \ldots \cup R_{N(n)}$ where

(i) $R_0, \ldots, R_{N(n)}$ are pairwise disjoint rectangles and the spanning ι-leaf segments of $i(R_i)$ are also $i(R)$-spanning ι-leaf segments, for every $0 \leq i \leq N(n)$;
(ii) $\pi_\iota(R_i)$ is an R-rectangle of Θ^ι, for every $0 < i < N(n)$;
(iii) R_0 and $R_{N(n)}$ are empty sets, or $\pi_{\iota'}(R_0)$ and $\pi_{\iota'}(R_{N(n)})$ are strictly contained in n-rectangles.

By property (iii), there is a sequence α_n tending to 0, such that $\mu(R_0) < \alpha_n$ and $\mu(R_{N(n)}) < \alpha_n$. Let $P_i = \pi_{\iota'}^{-1} \circ \pi_{\iota'}(R_i)$, for every $0 < i < N(n)$. Let $S_i = \tau_{\iota'}^n R_i$ and $Q_i = \tau_{\iota'}^n P_i$ for $0 < i < N(n)$. The rectangles $i(Q_i)$ are ι-spanning $(1, n)$-rectangles of some Markov rectangle $i(M_i)$. We note that $\pi_{\iota'}(S_i) = \pi_{\iota'}(Q_i)$. By Lemma 7.4, for all $\xi, \eta \in \pi_{\iota'}(Q_i)$,

$$\frac{\rho_\xi(S_i : Q_i)}{\rho_\eta(S_i : Q_i)} \in 1 \pm \mathcal{O}(\varepsilon^n),$$

for some $0 < \varepsilon < 1$. By Lemma 7.4, for every $\xi, \eta \in \pi_{\iota'}(R_i)$,

$$\frac{\rho_\xi(R_i : P_i)}{\rho_\eta(R_i : P_i)} \in 1 \pm \mathcal{O}(\varepsilon^n). \tag{7.4}$$

By invariance of the measure scaling function ρ under τ_ι, we get

$$\rho_\xi(R_i : P_i) = \rho_{\xi'}(S_i : Q_i) \quad \text{and} \quad \rho_\eta(R_i : P_i) = \rho_{\eta'}(S_i : Q_i), \tag{7.5}$$

where $\xi' = \pi_{\iota'}(\tau_{\iota'}^n(\xi))$ and $\eta' = \pi_{\iota'}(\tau_{\iota'}^n(\eta))$. Putting together (7.4) and (7.5), we get

$$\frac{\rho_{\xi'}(S_i : Q_i)}{\rho_{\eta'}(S_i : Q_i)} \in 1 \pm \mathcal{O}(\varepsilon^n). \tag{7.6}$$

By Theorem 6.7,

$$\begin{aligned} \nu(S_i) &= \int_{\pi_{\iota'}(M_i)} \rho_{\xi'}(S_i : M_i)(d\xi') \\ &= \int_{\pi_{\iota'}(M_i)} \rho_{\xi'}(S_i : Q_i)\rho_{\xi'}(Q_i : M_i)(d\xi'). \end{aligned} \tag{7.7}$$

By (7.6), we get that

$$\int_{\pi_{\iota'}(M_i)} \rho_{\xi'}(S_i : Q_i)\rho_{\xi'}(Q_i : M_i)(d\xi') \in$$

$$(1 \pm \mathcal{O}(\varepsilon^n))\, \rho_{\eta'}(S_i : Q_i) \int_{\pi_{\iota'}(M_i)} \rho_{\xi'}(Q_i : M_i)(d\xi'), \quad (7.8)$$

for any fixed $\eta' \in \pi_{\iota'}(S_i)$. By Theorem 6.7, we obtain that

$$\nu(Q_i) = \int_{\pi_{\iota'}(M_i)} \rho_{\xi'}(Q_i : M_i)(d\xi'). \quad (7.9)$$

Putting together (7.7), (7.8) and (7.9), we get that

$$\nu(S_i) \in (1 \pm \mathcal{O}(\varepsilon^n))\, \rho_{\eta'}(S_i : Q_i)\nu(Q_i). \quad (7.10)$$

By invariance of ν under τ, $\mu(S_i) = \mu(R_i)$ and $\mu(Q_i) = \mu(P_i)$. Therefore, putting together (7.5) and (7.10), we obtain that

$$\nu(R_i) \in (1 \pm \mathcal{O}(\varepsilon^n))\, \rho_{\eta}(R_i : P_i)\nu(P_i).$$

Hence,

$$\nu(R) \in \sum_{i=1}^{N(n)-1} \nu(R_i) \pm 2\alpha_n \subset (1 \pm \mathcal{O}(\varepsilon^n)) \sum_{i=1}^{N(n)-1} \rho_{\eta_i}(R : M)\nu_{\iota'}(P_i) \pm 2\alpha_n,$$

where $\eta_i \in \pi_{\iota'}(R_i)$. Hence, equation (7.3) follows on taking the limit $n \to \infty$. □

Lemma 7.6. *Let ν be a Gibbs measure with ι'-extended scaling function $\rho_{\iota'}$. Let R be contained in an (n_s, n_u)-rectangle such that $i(R)$ is contained in a Markov rectangle. Let R_1 and R_2 be rectangles in R such that the ι'-spanning leaves of $i(R_1)$ and $i(R_2)$ are also ι'-spanning leaves of $i(R)$. For all ι-leaf segments $\xi \in \pi_{\iota'}(R)$, we have that*

$$\frac{\nu(R_1)}{\nu(R_2)} \in \left(1 \pm \mathcal{O}(\varepsilon^{n_s+n_u})\right) \rho_{\xi}(R_1 : R_2), \quad (7.11)$$

for some constant $0 < \varepsilon < 1$ independent of R, R_1, R_2, n_s and n_u.

Proof. By invariance of ν and of the measure scaling function ρ under τ, we get

$$\rho_{\xi}(R_i : R) = \rho_{\xi'}(R'_i : R'), \quad (7.12)$$

where $R'_i = \tau_{\iota}^{n_\iota}(R_i)$, $R' = \tau_{\iota}^{n_\iota}(R)$, $\xi' \in \pi_{\iota'}(R')$ and $\xi = \pi_{\iota'}\tau_{\iota}^{-n_\iota}(\xi')$. By Hölder continuity of the measure scaling function, we get that

$$\frac{\rho_{\xi'}(R'_i : R')}{\rho_{\eta'}(R'_i : R')} \in 1 \pm \mathcal{O}(\varepsilon^{n_s+n_u}), \quad (7.13)$$

for every $\xi', \eta' \in \pi_{\iota'}(R')$. Putting together (7.12) and (7.13), we get that

$$\frac{\rho_\xi(R_i : R)}{\rho_\eta(R_i : R)} \in 1 \pm \mathcal{O}(\varepsilon^{n_s + n_u}), \tag{7.14}$$

for every $\xi, \eta \in \pi_{\iota'}(R)$. By Lemma 7.5 and (7.14), we get that

$$\begin{aligned}
\frac{\nu(R_i)}{\nu(R)} &= \frac{\int_{\pi_{\iota'}(R)} \rho_\xi(R_i : R)\rho_\xi(R : M)(d\xi)}{\int_{\pi_{\iota'}(R)} \rho_\xi(R : M)(d\xi)} \\
&= \left(1 \pm \mathcal{O}(\varepsilon^{n_s+n_u})\right) \rho_\eta(R_i : R) \frac{\int_{\pi_{\iota'}(R)} \rho_\xi(R : M)(d\xi)}{\int_{\pi_{\iota'}(R)} \rho_\xi(R : M)(d\xi)} \\
&= \left(1 \pm \mathcal{O}(\varepsilon^{n_s+n_u})\right) \rho_\eta(R_i : R),
\end{aligned}$$

for every $\eta \in \pi_{\iota'}(R)$. Hence,

$$\begin{aligned}
\frac{\nu(R_1)}{\nu(R_2)} &\in \left(1 \pm \mathcal{O}(\varepsilon^{n_s+n_u})\right) \frac{\rho_\eta(R_1 : R)}{\rho_\eta(R_2 : R)} \\
&\subset \left(1 \pm \mathcal{O}(\varepsilon^{n_s+n_u})\right) \rho_\eta(R_1 : R)\rho_\eta(R : R_2) \\
&\subset \left(1 \pm \mathcal{O}(\varepsilon^{n_s+n_u})\right) \rho_\eta(R_1 : R_2).
\end{aligned}$$

□

Theorem 7.7. *If $\sigma_{\nu,\iota'} : \Theta^{\iota'} \to \mathbb{R}^+$ is a scaling function, then the (dual) scaling function $\sigma_{\nu,\iota} : \Theta^\iota \to \mathbb{R}^+$ is well-defined.*

Recall from Corollary 6.8 that if $s_\iota : \Theta^\iota \to \mathbb{R}^+$ is an ι-measure scaling function for $\iota = s$ or u, then there is a unique τ-invariant Gibbs measure ν such that $s_{\nu,\iota} = s_\iota$.

Proof of Theorem 7.7. The dual ρ_ι of $\rho_{\iota'}$ is constructed as follows: Let I and K be two ι'-symbolic leaf segments contained in a common n-cylinder ι'-symbolic leaf ξ. Choose $p \in I$ and $p' \in K$. Let a_m be the ι-leaf N-cylinders containing p, and b_m the ι-leaf containing p' and holonomic to a_m. Let $A_m = [I, a_m]$ and $B_m = [K, b_m]$ (see Figure 7.3). By Lemma 7.6, there is $0 < \varepsilon < 1$ such that

$$\nu(A_{m+1})/\nu(A_m) \in (1 \pm \mathcal{O}(\varepsilon^{n+m}))\rho_{\iota,a_m}(A_{m+1} : A_m) ,$$

and, similarly,

$$\nu(B_{m+1})/\nu(B_m) \in (1 \pm \mathcal{O}(\varepsilon^{n+m}))\rho_{\iota,b_m}(B_{m+1} : B_m) .$$

By Lemma 7.6, we get

$$\frac{\rho_{\iota,a_m}(A_{m+1} : A_m)}{\rho_{\iota,b_m}(B_{m+1} : B_m)} \in 1 \pm \mathcal{O}(\varepsilon^{n+m}) .$$

Hence,

$$\frac{\nu(A_{m+1})}{\nu(B_{m+1})} \in (1 \pm \mathcal{O}(\varepsilon^{n+m}))\frac{\nu(A_m)}{\nu(B_m)}, \tag{7.15}$$

for some $0 < \varepsilon < 1$. For every $l \geq 1$, $I = A_l \cap \xi$ and $K = B_l \cap \xi$, where I and K do not depend upon l. Therefore, the following ratio

$$\rho_{\iota',\xi}(A_l : B_l) = \lim_{m \to \infty} \frac{\nu(A_m)}{\nu(B_m)} \tag{7.16}$$

is well-defined. Furthermore, by (7.15), the corresponding scaling function is Hölder continuous. Therefore, $\rho_{\iota'}$ is an extended measure scaling function for the Gibbs measure ν. □

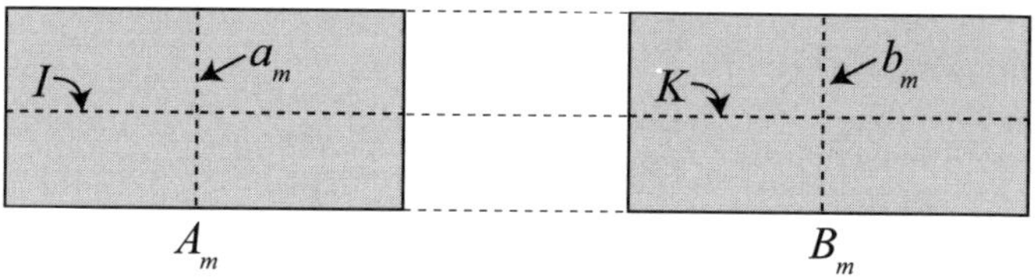

Fig. 7.3. The rectangles $A_m = [I, a_m]$ and $B_m = [K, b_m]$.

7.3 Further literature

This chapter is based on Bowen [17], Pinto and Rand [162] and Pinto and Rand [166].

8

Measure solenoid functions

We introduce the stable and unstable measure solenoid functions and stable and unstable measure ratio functions, which determine the Gibbs measures C^{1+}-realizable by C^{1+} hyperbolic diffeomorphisms and by C^{1+} self-renormalizable structures.

8.1 Measure solenoid functions

Let Msol^ι be the set of all pairs (I, J) with the following properties: (a) If $\delta_\iota = 1$, then $\mathrm{Msol}^\iota = \mathrm{sol}^\iota$. (b) If $\delta_\iota < 1$, then $f_{\iota'} I$ and $f_{\iota'} J$ are ι-leaf 2-cylinders of a Markov rectangle R such that $f_{\iota'} I \cup f_{\iota'} J$ is an ι-leaf segment, i.e. there is a unique ι-leaf 2-gap between them. Let msol^ι be the set of all pairs $(I, J) \in \mathrm{Msol}^\iota$ such that the leaf segments I and J are not contained in an ι-global leaf containing an ι-boundary of a Markov rectangle. By construction, the set msol^ι is dense in Msol^ι, and for every pair $(C, D) \subset \mathrm{msol}^\iota$ there is a unique $\psi \in \Theta^{\iota'}$ and a unique $\xi \in \Theta^{\iota'}$ such that $i(\pi_{\iota'}^{-1}(\psi)) = C$ and $i(\pi_{\iota'}^{-1}(\xi)) = D$. We will denote, in what follows, $i(\pi_{\iota'}^{-1}(\psi))$ by ψ_Λ and $i(\pi_{\iota'}^{-1}(\xi))$ by ξ_Λ.

Lemma 8.1. *Let ν be a Gibbs measure on Θ. The* s-measure pre-solenoid function $\sigma_{\nu,s} : \mathrm{msol}^s \to \mathbb{R}^+$ *of ν and the* u-measure pre-solenoid function $\sigma_{\nu,u} : \mathrm{msol}^u \to \mathbb{R}^+$ *of ν given by*

$$\sigma_{\nu,s}(\psi_\Lambda, \xi_\Lambda) = \lim_{n\to\infty} \frac{\nu(\Theta_{\psi_0 \dots \psi_n})}{\nu(\Theta_{\xi_0 \dots \xi_n})}$$

and

$$\sigma_{\nu,u}(\psi_\Lambda, \xi_\Lambda) = \lim_{n\to\infty} \frac{\nu(\Theta_{\psi_n \dots \psi_0})}{\nu(\Theta_{\xi_n \dots \xi_0})}$$

are both well-defined.

Proof. Let $(I, J) \in \text{msol}^\iota$. By Property (iii) of msol^ι, there is $k = k(I, J)$ such that $f_{\iota'}^k I$ and $f_{\iota'}^k J$ are cylinders with the same mother $mf_{\iota'}^k I = mf_{\iota'}^k J$. Let $(\xi : C)$ and $(\xi : D)$ be the admissible pairs in msc_ι such that $i(\xi.C) = f_{\iota'}^k I$ and $i(\xi.D) = f_{\iota'}^k J$. Since the measure ν is τ-invariant, we obtain that

$$\sigma_{\nu,\iota}(I, J) = \rho_\xi(C)\rho_\xi(D)^{-1} ,$$

where ρ is the extended scaling function determined by the Gibbs measure ν. Therefore, the ι-measure pre-solenoid function $\sigma_{\nu,\iota}$ is well-defined for $\iota \in \{s, u\}$. □

Lemma 8.2. *Suppose* $\delta_{f,\iota} = 1$. *If an ι-measure pre-solenoid function* $\sigma_{\nu,\iota} : \text{msol}^\iota \to \mathbb{R}^+$ *has a continuous extension to* sol^ι, *then its extension satisfies the matching condition.*

Proof. Let $(J_0, J_1) \in \text{sol}^\iota$ be a pair of primary cylinders and suppose that we have pairs

$$(I_0, I_1), (I_1, I_2), \ldots, (I_{n-2}, I_{n-1}) \in \text{sol}^\iota$$

of primary cylinders such that $f_\iota J_0 = \bigcup_{j=0}^{k-1} I_j$ and $f_\iota J_1 = \bigcup_{j=k}^{n-1} I_j$. Since the set msol^ι is dense in sol^ι there are pairs $(J_0^l, J_1^l) \in \text{msol}^\iota$ and pairs (I_j^l, I_{j+1}^l) with the following properties:

(i) $f_\iota J_0^l = \bigcup_{j=0}^{k-1} I_j^l$ and $f_\iota J_1^l = \bigcup_{j=k}^{n-1} I_j^i$.
(ii) The pair (J_0^l, J_1^l) converges to (J_0, J_1) when i tends to infinity.

Therefore, for every $j = 0, \ldots, n-2$ the pair (I_j^l, I_{j+1}^l) converges to (I_j, I_{j+1}) when i tends to infinity. Since ν is a τ-invariant measure, we get that the matching condition

$$\sigma_{\nu,\iota}(J_0^l : J_1^l) = \frac{1 + \sum_{j=1}^{k-1} \prod_{i=1}^{j} \sigma_{\nu,\iota}(I_j^l : I_{i-1}^l)}{\sum_{j=k}^{n-1} \prod_{i=1}^{j} \sigma_{\nu,\iota}(I_j^l : I_{i-1}^l)}$$

is satisfied for every $l \geq 1$. Since the extension of $\sigma_{\nu,\iota} : \text{msol}^\iota \to \mathbb{R}^+$ to the set sol^ι is continuous, we get that the matching condition also holds for the pairs (J_0, J_1) and $(I_0, I_1), \ldots, (I_{n-2}, I_{n-1})$. □

8.1.1 Cylinder-cylinder condition

Similarly to the cylinder-gap condition given in § 3.6 for a given solenoid function, we are going to construct the cylinder-cylinder condition for a given measure solenoid function $\sigma_{\nu,\iota}$. We will use the cylinder-cylinder condition to classify all Gibbs measures that are C^{1+}-Hausdorff realizable by codimension one attractors.

Let $\delta_\iota < 1$ and $\delta_{\iota'} = 1$. Let $(I, J) \in \text{Msol}^\iota$ be such that the ι-leaf segment $f_{\iota'} I \cup f_{\iota'} J$ is contained in an ι-boundary K of a Markov rectangle R_1. Then,

$f_{\iota'}I \cup f_{\iota'}J$ intersects another Markov rectangle R_2. Take the smallest $k \geq 0$ such that $f_{\iota'}^k I \cup f_{\iota'}^k J$ is contained in the intersection of the boundaries of two Markov rectangles M_1 and M_2. Let M_1 be the Markov rectangle with the property that $M_1 \cap f_{\iota'}^k R_1$ is a rectangle with non empty interior, and so $M_2 \cap f_{\iota'}^k R_2$ has also non-empty interior. Then, for some positive n, there are distinct ι-leaf n-cylinders $J_1, \ldots, J_m$ contained in a primary cylinder L of M_2 such that $f_{\iota'}^k I = \cup_{i=1}^{p-1} J_i$ and $f_{\iota'}^k J = \cup_{i=p}^{m} J_i$. Let $\eta \in \Theta^{\iota'}$ be such that $\eta_\Lambda = L$ and, for every $i = 1, \ldots, m$, let D_i be a cylinder of Θ^ι such that $i(\eta.D_i) = J_i$. Let $\xi \in \Theta^{\iota'}$ be such that $\xi_\Lambda = K$ and C_1 and C_2 cylinders of Θ^ι such that $i(\xi.C_1) = f_{\iota'}I$ and $i(\xi.C_2) = f_{\iota'}J$. We say that an ι-extended scaling function ρ satisfies the *cylinder-cylinder condition* (see Figure 8.1), if, for all such leaf segments,

$$\frac{\rho_\xi(C_2)}{\rho_\xi(C_1)} = \frac{\sum_{i=p}^{m} \rho_\eta(D_i)}{\sum_{i=1}^{p-1} \rho_\eta(D_i)} .$$

Fig. 8.1. The cylinder-cylinder condition for ι-leaf segments.

Remark 8.3. A function $\sigma : \text{msol}^\iota \to \mathbb{R}^+$ that has a Hölder continuous extension to Msol^ι determines a unique extended scaling function ρ, and so we say that σ satisfies the cylinder-cylinder condition, if the extended scaling function ρ satisfies the cylinder-cylinder condition.

Definition 8.4. *A Hölder continuous functions* $\sigma_\iota : \text{Msol}^\iota \to \mathbb{R}^+$ *is a* measure solenoid function, *if* σ_ι *satisfies the following properties:*

(i) If $\mathbf{B}^\iota$ *is a no-gap train-track, then* σ_ι *is an* ι*-solenoid function.*
(ii) If $\mathbf{B}^\iota$ *is a gap train-track and* $\mathbf{B}^{\iota'}$ *is a no-gap train-track, then* σ_ι *satisfies the cylinder-cylinder condition.*
(iii) If $\mathbf{B}^\iota$ *and* $\mathbf{B}^{\iota'}$ *are no-gap train-tracks, then* σ_ι *does not have to satisfy any extra property.*

8.2 Measure ratio functions

We say that ρ is a ι*-measure ratio function*, if

(i) $\rho(I : J)$ is well-defined for every pair of ι-leaf segments I and J such that (a) there is an ι-leaf segment K such that $I, J \subset K$, and (b) I or J has non-empty interior;

(ii) if I is an ι-leaf gap, then $\rho(I : J) = 0$ (and $\rho(J : I) = +\infty$);
(iii) if I and J have non-empty interiors, then $\rho(I : J)$ is strictly positive;
(iv) $\rho(I : J) = \rho(J : I)^{-1}$;
(v) if I_1 and I_2 intersect at most in one of their endpoints, then $\rho(I_1 \cup I_2 : K) = \rho(I_1 : K) + \rho(I_2 : K)$;
(vi) ρ is invariant under f, i.e. $\rho(I : J) = r(fI : fJ)$ for all ι-leaf segments;
(vii) for every basic ι-holonomy map $\theta : I \to J$ between the leaf segment I and the leaf segment J defined with respect to a rectangle R and for every ι-leaf segment $I_0 \subset I$ and every ι-leaf segment or gap $I_1 \subset I$,

$$\left| \log \frac{\rho(\theta I_0 : \theta I_1)}{\rho(I_0 : I_1)} \right| \leq \mathcal{O}\left((d_\Lambda(I, J))^\varepsilon\right), \tag{8.1}$$

where $\varepsilon \in (0,1)$ depends upon ρ and the constant of proportionality also depends upon R, but not on the segments considered.

We note that if $\mathbf{B}^\iota$ is a no-gap train-track, then an ι-measure ratio function is an ι-ratio function.

Let $\mathcal{SOL}^\iota$ be the space of all ι-solenoid functions.

Lemma 8.5. *The map $\rho \to \rho|\mathrm{Msol}^\iota$ determines a one-to-one correspondence between ι-measure ratio functions and solenoid functions in $\mathcal{SOL}^\iota$.*

Proof. The proof follows similarly to the proof of Lemma 3.3. □

Remark 8.6. (i) By Lemma 8.5, a Gibbs measure ν with an ι-measure pre-solenoid function with an extension $\hat{\sigma}$ to Msol^ι such that $\hat{\sigma} \in \mathcal{SOL}^\iota$ determines a unique ι-measure ratio function ρ_ν.
(ii) A measure ratio function ρ determines naturally a measure scaling function, and so, by Corollary 6.8, a Gibbs measure ν_ρ.
(iii) By Lemma 8.5, a function $\sigma : \mathrm{msol}^\iota \to \mathbb{R}^+$ with an extension $\hat{\sigma}$ to Msol^ι such that $\hat{\sigma} \in \mathcal{SOL}^\iota$ determines an ι-measure ratio function, and, by (ii), a unique Gibbs measure ν such that $\sigma = \sigma_\nu$.

8.3 Natural geometric measures

In this section, we define the natural geometric measures $\mu_{\mathcal{S},\delta}$ associated with a self-renormalizable structure $\mathcal{S}$ and $\delta > 0$. The natural geometric measures are measures determined by the length scaling structure of the cylinders. We will prove that every natural geometric measure is a pushforward of a Gibbs measure with the property that the measure solenoid function determines a measure ratio function. In § 10.2, we will show that a Gibbs measure with the property that its measure solenoid function determines a measure ratio function is C^{1+}-realizable by a self-renormalizable structure.

Definition 20 *Let $\mathcal{S}$ be a C^{1+} self-renormalizable structure on $\mathbf{B}^\iota$. If $\mathbf{B}^\iota$ is a gap train-track let $0 < \delta < 1$, and if $\mathbf{B}^\iota$ is a no-gap train-track let $\delta = 1$.*

(i) We say that $\mathcal{S}$ has a natural geometric measure $\mu_\iota = \mu_{\mathcal{S},\delta}$ with pressure $P = P(\mathcal{S}, \delta)$ *if (a) μ_ι is a f_ι-invariant measure; (b), there exists $\kappa > 1$ such that for all $n \geq 1$ and all n-cylinders I of $\mathbf{B}^\iota$, we have*

$$\kappa^{-1} < \frac{\mu_\iota(I)}{|I|_i^\delta e^{-nP}} < \kappa \ , \tag{8.2}$$

where i is a chart containing I of a bounded atlas $\mathcal{B}$ of $\mathcal{S}$;

(ii) We say that $\mathcal{S}$ is a C^{1+} realization of a Gibbs measure $\nu = \nu_{\mathcal{S},\delta}$ *if $\mu_\iota = (i_\iota)_* \nu_\iota$ where $\nu_\iota = (\pi_\iota)_* \nu$ and $\mu_\iota = \mu_{\mathcal{S},\delta}$ is a natural geometric measure of $\mathcal{S}$.*

Suppose that we have a C^{1+} self-renormalizable structure $\mathcal{S}$ on $\mathbf{B}^\iota$ and that $\mathcal{B}$ is a bounded atlas for it. Let $\delta > 0$. If I is a segment in $\mathbf{B}^\iota$, let $|I| = |I|_i$ be its length in any chart i of this atlas which contains it. If C is a m-cylinder, let us denote m by $n(C)$ and $i_\iota(C)$ by I_C. For $m_1 \geq 1$ and $m_2 \geq 1$, let C be an m_1-cylinder and D an m_2-cylinder contained in the same 1-cylinder. Let

$$L_{\delta,s}(C : D) = \frac{\sum_{C' \subset C} |I_{C'}|^\delta e^{-n(C')s}}{\sum_{D' \subset D} |I_{D'}|^\delta e^{-n(D')s}} \tag{8.3}$$

where the sums are respectively over all cylinders contained in C and D and the values $|I_{C'}|$ and $|I_{D'}|$ are determined using the same chart in $\mathcal{B}$. Let the *pressure* $P = P(\mathcal{S}, \delta)$ be the infimum value of s for which the numerator (and the denominator) are finite.

If $\xi \in \Theta^{\iota'}$, then the leaf 1-cylinder segment $\xi_\Lambda = i(\pi_{\iota'}^{-1}\xi) \subset \Lambda$ is also regarded, without ambiguity, as a point in the train-track $\mathbf{B}^{\iota'}$. Similarly, if C is an n-cylinder of Θ^ι, then the $(1, n)$-rectangle $C_\Lambda = i(\pi_\iota^{-1}\xi) \subset \Lambda$ is also regarded, without ambiguity, as an n-cylinder of the train-track $\mathbf{B}^\iota$.

The following theorem follows from the results proved in Pinto and Rand [162]. It can also be deduced from standard results about Gibbs states such as those in Chapter 6.

Lemma 8.7. *Let $\mathcal{S}$ be a C^{1+} self-renormalizable structure on $\mathbf{B}^\iota$. For every $\delta > 0$, there is a unique geometric natural measure $\mu_\iota = \mu_{\mathcal{S},\delta}$ with pressure $P = P(\mathcal{S}, \delta) \in \mathbb{R}$, and there is a unique τ-invariant Gibbs measure $\nu = \nu_{\mathcal{S},\delta}$ on Θ such that $\mu_\iota = (i_\iota)_* \nu_\iota$ where $\nu_\iota = (\pi_\iota)_* \nu$. Furthermore, the measure μ_ι has the following properties:*

(i) There is $0 < \alpha < 1$ such that if C and D are any two n-cylinders in Θ^ι such that I_C and I_D are contained in a common segment K, then

$$\frac{\mu_\iota(I_C)}{\mu_\iota(I_D)} \in (1 \pm \mathcal{O}(|K|^\alpha))\, L_{\delta,P}(C : D) \ .$$

(ii) If $\rho : \mathrm{msol}^\iota \to \mathbb{R}$ *is the extended measure scaling function of* ν_ι*, then*

$$\rho_\xi(C) = \lim_{m\to\infty} L_{\delta,P}(C_m : \xi_m) \ ,$$

where C_m *and* ξ_m *are the cylinders given by* $I_{C_m} = f_{\iota'}^m(C_\Lambda \cap \xi_\Lambda)$ *and* $I_{\xi_m} = f_{\iota'}^{m-1}\xi_\Lambda$.

(iii) (ratio decomposition) *if* C *is an* n*-cylinder in* Θ^ι *and* C_p *is the primary cylinder containing* C*, then*

$$\mu_\iota(I_C) = \int_{\xi\in\pi_{\iota'}(C)} \rho_\xi(C)\mu_{\iota'}(d\xi) \ . \tag{8.4}$$

Proof. It follows from putting together Lemma 6.6 and Theorem 6.7. □

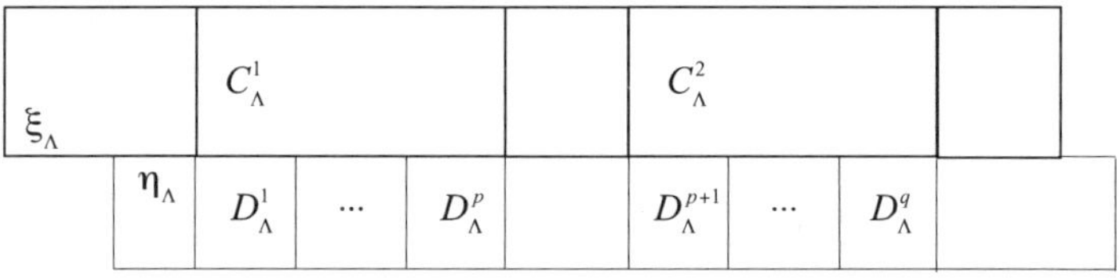

Fig. 8.2. The rectangles C^1_Λ, C^2_Λ and $D^1_\Lambda, \ldots, D^q_\Lambda$

Lemma 8.8. *Let* $\mathcal{S}$ *be a* C^{1+} *self-renormalizable structure on* $\mathbf{B}^\iota$ *and let* ρ *be the extended measure scaling function of the Gibbs measure* $\nu_{\mathcal{S},\delta}$.

(i) If C *and* D *are two cylinders contained in an* n*-cylinder* E *of* Θ^ι*, then, for all* ξ, η *contained in the* 1*-cylinder* $\pi_{\iota'}(\pi_\iota^{-1}E)$ *of* $\Theta^{\iota'}$*,*

$$\frac{\rho_\eta(C)}{\rho_\eta(D)} \in (1 \pm \mathcal{O}(\theta^n))\,\frac{\rho_\xi(C)}{\rho_\xi(D)}. \tag{8.5}$$

(ii) Let $\mathbf{B}^{\iota'}$ *be a no-gap train-track. Let* $\xi, \eta \in \Theta^{\iota'}$ *be such that the corresponding leaf segments in* Λ *have a common intersection* K *(or coincide). Let* $(\xi : C^1), (\xi : C^2), (\eta : D^1), \ldots, (\eta : D^q)$ *be admissible pairwise distinct pairs in* msc^ι *such that (a)* $\xi_\Lambda \cap C^1_\Lambda = \xi_\Lambda \cap (\cup_{i=1}^p D^i_\Lambda) \subset K$*, and (b)* $\xi_\Lambda \cap C^2_\Lambda = \xi_\Lambda \cap (\cup_{i=p+1}^q D^i_\Lambda) \subset K$ *(see Figure 8.2). Then,*

$$\frac{\rho_\xi(C^1)}{\rho_\xi(C^2)} = \frac{\sum_{i=1}^p \rho_\eta(D^i)}{\sum_{i=p+1}^q \rho_\eta(D^i)} \ . \tag{8.6}$$

(iii) Let $\mathbf{B}^\iota$ *be a no-gap train-track (and* $\delta = 1$*). Then, for every admissible pair* $(C : \xi) \in \mathrm{msc}^\iota$*, we get*

$$\rho_\xi(C) = r_\iota(C_\Lambda \cap \xi_\Lambda : \xi_\Lambda), \tag{8.7}$$

where r_ι *is the* ι*-ratio function determined by the* C^{1+} *self-renormalizable structure.*

Proof. Proof of (i) and (ii). Suppose that C and D are two cylinders contained in an n-cylinder E. Let E_1 be a $(n+1)$-cylinder whose image under the shift map τ is E and let C_1 and D_1 be the cylinders in E_1 such that $\tau C_1 = C$ and $\tau D_1 = D$. Then,

$$L_{\delta,P}(C_1 : D_1) \in (1 \pm \mathcal{O}(\theta^n))\, L_{\delta,P}(C : D),$$

where (i) $0 < \theta < 1$ is independent of C, D, E and E_1, and $P = P(\mathcal{S}, \delta)$ is the pressure. This follows directly from the definition of $L_{\delta,P}$ together with the fact that, for all cylinders C', D' in E_1,

$$\frac{|I_{D'}|}{|I_{C'}|} \in (1 \pm \mathcal{O}(\theta^n))\, \frac{|I_{\tau D'}|}{|I_{\tau C'}|}.$$

As a corollary of this we deduce (8.5). Then, equality (8.6) follows from using that the local holonomies are local diffeomorphisms in the self-renormalizable structure of $\mathbf{B}^\iota$.

Proof of (iii). In this case the self-renormalizable structure $\mathcal{S}$ is a local manifold structure as defined in § 4.6 (i.e. the charts are homeomorphisms onto a subinterval of $\mathbb{R}$), and $\delta = 1$. Using (8.3), we get $P(\mathcal{S}, \delta) = 0$ and so the ratios $\mu(I)/|I|$ are uniformly bounded away from 0 and ∞ for all segments I in $\mathbf{B}^\iota$. Moreover, in this case, the length system l matches in the sense that if C is an n-cylinder, then $\sum_{C'} |I_{C'}| = |I_C|$ where the sum is over all m-cylinders C' contained in C and $|I_C|$ and $|I_{C'}|$ are obtained using the same chart in $\mathcal{B}$. Thus, if C and D are n-cylinders and $I_C \cup I_D$ is a segment of $\mathbf{B}^\iota$, then

$$\frac{\mu_\iota(I_C)}{\mu_\iota(I_D)} \in (1 \pm \mathcal{O}(\theta^n))\frac{|I_C|}{|I_D|}.$$

Hence,

$$\rho_\xi(C) = \lim_{m \to \infty} \frac{|f_{\iota'}^m(C_\Lambda \cap \xi_\Lambda)|}{|f_{\iota'}^m \xi_\Lambda|},$$

which implies (8.7). □

Remark 8.9. If δ is the Hausdorff dimension of $\mathbf{B}^\iota$, then the ratios $\mu_\iota(I_C)/|I_C|^\delta$ are uniformly bounded away from 0 and ∞. It follows from this that the Hausdorff δ-measure $\mathcal{H}^\delta$ is finite and positive on $\mathbf{B}^\iota$ and such that μ_ι is absolutely continuous with respect to $\mathcal{H}^\delta$.

The above remark follows by using the orthogonal charts and the self-renormalizable structures.

8.4 Measure ratio functions and self-renormalizable structures

In this section, we prove that, for every $\delta > 0$, a given C^{1+} self-renormalizable structure $\mathcal{S}$ on $\mathbf{B}^{\iota'}$ determines an ι-measure ratio function $\rho_{\mathcal{S},\delta}$ such that the

Gibbs measure ν_ρ determined by $\rho_{\mathcal{S},\delta}$ (see Remark 8.6) is the same as the Gibbs measure $\nu_{\mathcal{S},\delta}$ that is C^{1+} realizable by the self-renormalizable structure $\mathcal{S}$.

Lemma 8.10. *Let R be an (n_s, n_u)-rectangle and R' and R'' be ι'-spanning rectangles contained in R. Let ξ be an ι-leaf segment of R. Let $\hat{R}'$, $\hat{R}''$ and $\hat{R}$ be rectangles in Θ such that $i(\hat{R}') = R'$, $i(\hat{R}'') = R''$ and $i(\hat{R}) = R$. Let $\hat{\xi} \in \pi_\iota^{-1}(\hat{R})$ be such that $i(\hat{\xi}) = \xi$. The values*

$$\rho_{\mathcal{S},\delta}(\xi \cap R' : \xi \cap R'') = \rho_{\iota,\hat{\xi}}(\hat{R}' : \hat{R}'')$$

are well-defined.

Proof. By (7.2), the ratios are well-defined for ι-spanning leaves ξ in the interior of R. By (7.2) and (7.6), the ratios are also well-defined for ι-spanning leaves in the boundary of R. □

From now on, for simplicity of notation, we will denote $\rho_{\mathcal{S},\delta}(\xi \cap R' : \xi \cap R'')$ by $\rho_{\iota,\xi}(R' : R'')$.

Lemma 8.11. *(2-dimensional ratio decomposition) Let $\mathcal{S}$ be a C^{1+} self-renormalizable structure and $\mu_\iota = \mu_{\mathcal{S},\delta}$ a natural geometric measure for some $\delta > 0$. Suppose that R is a rectangle contained in a Markov rectangle M. Then,*

$$\mu(R) = \int_{\pi_{\mathbf{B}^{\iota'}}(R)} \rho_{\iota,\xi}(R : M)\mu_{\iota'}(d\xi) \ . \tag{8.8}$$

We now consider the case where $\mathbf{B}^\iota$ is a no-gap train-track. Let $\mathcal{S}$ be a C^{1+} self-renormalizable structure and $\mu_\iota = \mu_{\mathcal{S},1}$ the natural measure (with pressure $P = 0$). Recall the definition of $t_R^{\iota'}$ as the set of spanning ι-leaf segments of the rectangle R (not necessarily a Markov rectangle). By the local product structure, one can identify $t_R^{\iota'}$ with any spanning ι'-leaf segment $l^{\iota'}(x, R)$ of R. Suppose that R is a rectangle and M is a Markov rectangle and that $\theta : l = l^{\iota'}(x, R) \to l' \subset l^{\iota'}(x', M)$ is a basic holonomy defined on the spanning ι'-leaf segment l. This defines an injection $t_\theta : t_R^{\iota'} \to t_M^{\iota'}$ which we call the *holonomy injection* induced by θ (see Figure 8.3). The measure $\mu_{\iota'}$ on $\mathbf{B}^{\iota'}$ induces a measure on $t_M^{\iota'}$ which we can pull back to $t_R^{\iota'}$ using t_θ to obtain a measure $\mu_{R,M}^\theta$ i.e. $\mu_{R,M}^\theta(E) = \mu_{\iota'}(\pi_{\mathbf{B}^{\iota'}}(t_\theta(E)))$.

Lemma 8.12. *(2-dimensional ratio decomposition for SRB measures) Let $\mathbf{B}^\iota$ be a no-gap train-track. Let $\mathcal{S}$ be a C^{1+} self-renormalizable structure and $\mu_\iota = \mu_{\mathcal{S},1}$ the natural measure (with pressure $P = 0$). If $t_\theta : t_R^{\iota'} \to t_M^{\iota'}$ is a holonomy injection as above with P a Markov rectangle, then*

$$\mu(R) = \int_{t_R^{\iota'}} r_\iota(\xi : t_\theta(\xi))\mu_{R,M}^\theta(d\xi) \ , \tag{8.9}$$

where r_ι is the ι-ratio function determined by $\mathcal{S}$.

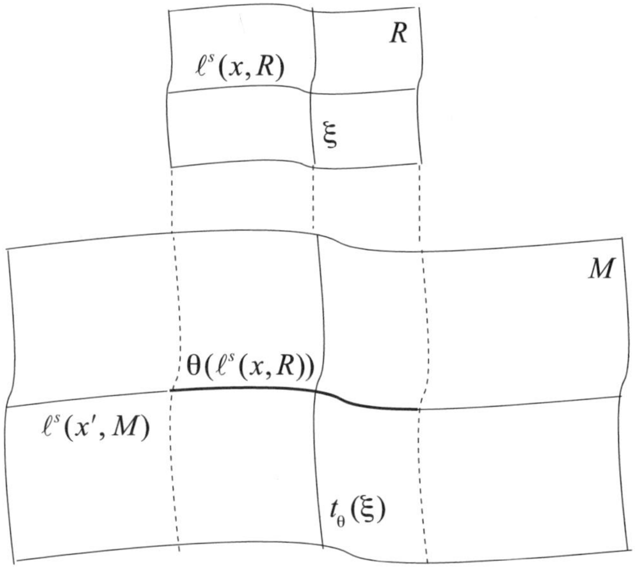

Fig. 8.3. The holonomy injection t_θ.

Remark 8.13. Note that if $R \subset M$, then $t_\theta(\xi)$ is just the M-spanning ι-leaf containing ξ and $\mu^\theta_{R,M} = \mu_{\iota'}$.

Since any rectangle can be written as the union of rectangles R with the property hypothesised in the theorem for some Markov rectangle, the above theorem gives an explicit formula for the measure of any rectangle in terms of a ratio decomposition using the ratio function which characterises the smooth structure of the train-track.

Proof of Lemmas 8.11 and 8.12. Suppose that R is any rectangle, M is a Markov rectangle and $t_\theta : t_R^{\iota'} \to t_M^{\iota'}$ is a holonomy injection as above (in the case of Lemma 8.11 t_θ is the identity map). Then, we note that there is $0 < \nu < 1$ such that for all $n > 0$ we can write $R = R_0 \cup \ldots \cup R_{N(n)}$ where

(i) $R_0, \ldots, R_{N(n)}$ are rectangles which intersect at most in their boundary leaves and their spanning ι-leaf segments are also R-spanning ι-leaf segments;
(ii) $P_i = t_\theta R_i$ and $\pi_{\iota'}(P_i)$ is an n-cylinder of $\mathbf{B}^{\iota'}$ for every $0 \le i \le N(n)$;
(iii) R_0 is the empty set, or $\pi_{\iota'}(P_0)$ is strictly contained in an n-cylinder of $\mathbf{B}^{\iota'}$, and so, using the bounded geometry of the Markov map (see § 4.3) and (8.2), $\mu(R_0) < \mathcal{O}(\varepsilon_0^n)$ for some $0 < \varepsilon_0 < 1$;

Let $S_i = f_{\iota'}^n R_i$ and $Q_i = f_{\iota'}^n P_i$ for $1 \le i \le N(n)$, and note that the rectangles Q_i are ι-spanning $(1,n)$-rectangles of some Markov rectangle M_i. We note that if t_θ is not the identity there might be a non-empty set V_n of values of i such that S_i is not be contained in the Markov rectangle M_i. However, since there are a finite number of Markov rectangles, the cardinality of the set V_n is bounded away from infinity, independently of $n \geq 0$. Hence, we desregard in what follows these values of $i \in V_n$, since the measure of the corresponding sets S_i converges to 0 when n tends to infinity. To prove the theorems we firstly note that by Lemma 8.8 and by (7.1) we obtain that, if Q_i, P_i and M_i are as above, for all $\xi, \eta \in M_i$,

$$\rho(\xi \cap S_i : \xi \cap Q_i) \in (1 \pm \mathcal{O}(\varepsilon^n))\, \rho(\eta \cap S_i : \eta \cap Q_i),$$

for some $0 < \varepsilon < 1$. Thus, since $\mu(S_i) = \mu(\pi_\iota(S_i))$ and $\mu(Q_i) = \mu(\pi_\iota(Q_i))$ and by (8.4), if $\xi \in t^{\iota'}_{M_i}$,

$$\mu(S_i) \in (1 \pm \mathcal{O}(\varepsilon^n))\, \rho(\xi \cap S_i : \xi \cap Q_i) \mu(Q_i).$$

Now consider the case of Lemma 8.11. Then, since R_i and P_i are contained in the same Markov rectangle, $\rho(\xi \cap S_i : \xi \cap Q_i)$ equals $\rho(\xi_i \cap R : \xi_i \cap M)$ for some $\xi_i \in t^{\iota'}_{R_i}$ and $\rho(Q_i) = \rho(P_i)$ which is equal to $\mu_{\iota'} P_i$ since P_i is an ι-spanning rectangle of the Markov rectangle M. Thus we have deduced that up to addition of a term that is $\mathcal{O}(\nu^n)$,

$$\mu(R) \in (1 \pm \mathcal{O}(\varepsilon^n)) \sum_{i=1}^{N(n)} \rho(\xi_i \cap R : \xi_i \cap M) \mu_{\iota'}(P_i).$$

Equation (8.8) follows on taking the limit $n \to \infty$.

Now consider the case of Lemma 8.12. Under its hypotheses we have that $\rho(\xi \cap S_i : \xi \cap Q_i) = r_\iota(\xi \cap S_i : \xi \cap Q_i)$ by (8.7) and (7.1). By the f-invariance of r_ι there is $\xi_i \in t^{\iota'}_{R_i}$ such that $r_\iota(\xi_i : t_\theta(\xi_i)) = r_\iota(\xi \cap S_i : \xi \cap Q_i)$. Thus, as above, we deduce that

$$\mu(R) \in (1 \pm \mathcal{O}(\varepsilon^n)) \sum_{i=1}^{N(n)} r_\iota(\xi_i : t_\theta(\xi_i)) \mu_{\iota'}(t_\theta R_i)\ .$$

Equation (8.9) follows on taking the limit $n \to \infty$. □

Lemma 8.14. *Let $\mathcal{S}$ be a C^{1+} self-renormalizable structure on $\mathbf{B}^\iota$ with natural measure $\mu_\iota = \mu_{\mathcal{S},\delta}$ for some $\delta > 0$. Suppose that R is contained in a (n_s, n_u)-rectangle and that R' and R'' are ι'-spanning rectangles contained in R. Suppose in addition that either (ii) R is contained in a Markov rectangle or (ii) $\mathbf{B}^\iota$ does not have gaps and there is a holonomy injection of R into a Markov rectangle (in this case $\delta = 1$ and $P = 0$). Then, for every ι-leaf segment $\xi \in t^{\iota'}_R$, we have that*

$$\frac{\mu(R')}{\mu(R'')} \in \left(1 \pm \mathcal{O}(\varepsilon^{n_s + n_u})\right) \rho(\xi \cap R' : \xi \cap R'') \tag{8.10}$$

for some constant $0 < \varepsilon < 1$ independent of R, R', R'', n_s and n_u, (and in case (ii) $\rho(\xi \cap R : \xi \cap R') = r_\iota(\xi \cap R : \xi \cap R')$).

Proof. We give the proof for the second case since that for the first is similar. By Lemma 8.12, we have that

$$\frac{\mu(R)}{\mu(R')} = \frac{\int_{t_R^{\iota'}} r_\iota(\xi : t_\theta(\xi))\mu^\theta_{R,M}(d\xi)}{\int_{t_{R'}^{\iota'}} r_\iota(\xi : t_\theta(\xi))\mu^\theta_{R',M}(d\xi)} = \frac{\int_{t_R^{\iota'}} r_{\iota,\xi}(R : R')r_\iota(\xi : t_\theta(\xi))\mu^\theta_{R,M}(d\xi)}{\int_{t_{R'}^{\iota'}} r_\iota(\xi : t_\theta(\xi))\mu^\theta_{R',M}(d\xi)} .$$

where $r_{\iota,\xi}(R : R') = r_\iota(R \cap \xi : R' \cap \xi)$. Let $F = f_{\iota'}^{n_s+n_u}$. By inequality (2.2) (or inequality (8.5) in case (i)), there is $0 < \varepsilon < 1$ such that

$$r_{\iota,F\eta}(FR : FR') \in (1 \pm \mathcal{O}(\varepsilon^{n_s+n_u}))r_{\iota,F\xi}(FR : FR')$$

for all $\xi, \eta \in t_R^{\iota'}$. Thus,

$$r_{\iota,\eta}(R : R') \in (1 \pm \mathcal{O}(\varepsilon^{n_s+n_u}))r_{\iota,\xi}(R : R')$$

and so

$$\frac{\mu(R)}{\mu(R')} \in (1 \pm \mathcal{O}(\varepsilon^{n_s+n_u}))r_{\iota,\xi}(R : R') .$$

Similarly,

$$\frac{\mu(R)}{\mu(R'')} \in (1 \pm \mathcal{O}(\varepsilon^{n_s+n_u}))r_{\iota,\xi}(R : R') .$$

Putting together the previous two equations we obtain (8.10). □

Theorem 8.15. *Let $\mathcal{S}$ be a C^{1+} self-renormalizable structure on $\mathbf{B}^\iota$ with natural measure $\mu_\iota = \mu_{\mathcal{S},\delta}$ for some $\delta > 0$. The values*

$$\rho_{\mathcal{S},\delta}(\xi \cap R' : \xi \cap R'')$$

(as in Lemma 8.10) determine an ι-measure ratio function $\rho_{\mathcal{S},\delta}$ with the following properties:

(i) The Gibbs measure ν_ρ determined by the ι-measure ratio function $\rho_{\mathcal{S},\delta}$ (see Remark 8.6) is the same as the Gibbs measure $\nu_{\mathcal{S},\delta}$ which is C^{1+} realizable by the self-renormalizable structure $\mathcal{S}$;
(ii) If $\mathbf{B}^\iota$ is a no-gap train-track, then $\rho_{\mathcal{S},1} = r$, where r is the ratio function determined by the C^{1+} self-renormalizable structure $\mathcal{S}$.

Putting together Theorem 8.15 and Lemma 8.5, we obtain the following corollary.

Corollary 8.16. *Let $\mathcal{S}$ be a C^{1+} self-renormalizable structure on $\mathbf{B}^\iota$ with natural measure $\mu_\iota = \mu_{\mathcal{S},\delta}$ for some $\delta > 0$. The measure pre-solenoid function $\sigma^\iota_{\nu_{\mathcal{S},\delta}} : \mathrm{msol}^\iota \to \mathbb{R}^+$ determines a solenoid function $\hat{\sigma}^\iota_{\nu_{\mathcal{S},\delta}} : \mathrm{Msol}^\iota \to \mathbb{R}^+$ and $\hat{\sigma}^\iota_{\nu_{\mathcal{S},\delta}} = \rho_{\mathcal{S},\delta}|\mathrm{Msol}^\iota$.*

Proof of Theorem 8.15. Let us prove this lemma first in the case where the train-track $\mathbf{B}^\iota$ does not have gaps and then in the case where the train-track $\mathbf{B}^\iota$ has gaps.
(i) $\mathbf{B}^\iota$ *does not have gaps.* Then, $\delta = 1$ and, by Lemma 8.8 (iii), we have $\rho(\xi \cap R' : \xi \cap R'') = r(\xi \cap R' : \xi \cap R'')$ where r is the ratio function determined by the C^{1+} self-renormalizable structure $\mathcal{S}$. Hence $\rho_{\mathcal{S},\delta} = r$ is an ι-measure ratio function. Using (8.10), we get that the Gibbs measure, that is a C^{1+} realization of the natural geometric measure $\mu_{\mathcal{S},\delta}$, determines an ι-measure solenoid function which induces the ι-measure ratio function $\rho_{\mathcal{S},\delta}$.
(ii) $\mathbf{B}^\iota$ *has gaps.* By f-invariance of μ and (8.10), we get that

$$\rho(\xi \cap R : \xi \cap R') = \rho(f_{\iota'}\xi \cap f_{\iota'}R : f_{\iota'}\xi \cap f_{\iota'}R') \tag{8.11}$$

is invariant under f. Let I and J be ι-leaf segments such that (a) there is an ι-leaf segment K such that $I, J \subset K$, and (b) I or J has non-empty interior. Then, there is $n > 0$, $\xi \in \mathbf{B}^{\iota'}$, R and R' such that $f_{\iota'}^n I = \xi \cap R$ and $f_{\iota'}^n J = \xi \cap R'$. Hence, using (8.11), the ratio

$$\rho_{\mathcal{S},\delta}(I : J) = \rho_{\mathcal{S},\delta}(f_{\iota'}^n I : f_{\iota'}^n J)$$

is well defined independently of n. Using (8.10), we get that (8.1) is satisfied and the Gibbs measure, that is a C^{1+} realization of the natural geometric measure $\mu_{\mathcal{S},\delta}$, determines an ι-measure solenoid function which induces the ι-measure ratio function $\rho_{\mathcal{S},\delta}$. □

8.5 Dual measure ratio function

We will show that an ι-measure ratio function ρ_ι determines a unique dual function $\rho_{\iota'}$ which is an ι'-measure ratio function.

Definition 21 *We say that the ι-measure ratio function ρ_ι and the ι'-measure ratio function $\rho_{\iota'}$ are* dual *if both determine the same Gibbs measure $\nu = \nu_{\rho_\iota} = \nu_{\rho_{\iota'}}$ on Θ (see Remark 8.6).*

Theorem 8.17. *Let $\mathcal{S}$ be a C^{1+} self-renormalizable structure on $\mathbf{B}^\iota$ with ι-measure ratio function $\rho_\iota = \rho_{\mathcal{S},\delta}$ corresponding to the Gibbs measure $\nu = \nu_{\mathcal{S},\delta}$. Then, there is an ι'-measure ratio function $\rho_{\iota'}$ dual to ρ_ι.*

Putting together Theorem 8.17 and Lemma 8.5, we obtain the following corollary.

Corollary 8.18. *The measure pre-solenoid function $\sigma^{\iota'}_{\nu_{\mathcal{S},\delta}} : \mathrm{msol}^{\iota'} \to \mathbb{R}^+$ determines a solenoid function $\hat{\sigma}^{\iota'}_{\nu_{\mathcal{S},\delta}} : \mathrm{Msol}^{\iota'} \to \mathbb{R}^+$ and $\hat{\sigma}^{\iota'}_{\nu_{\mathcal{S},\delta}} = \rho_{\iota'}|\mathrm{Msol}^{\iota'}$.*

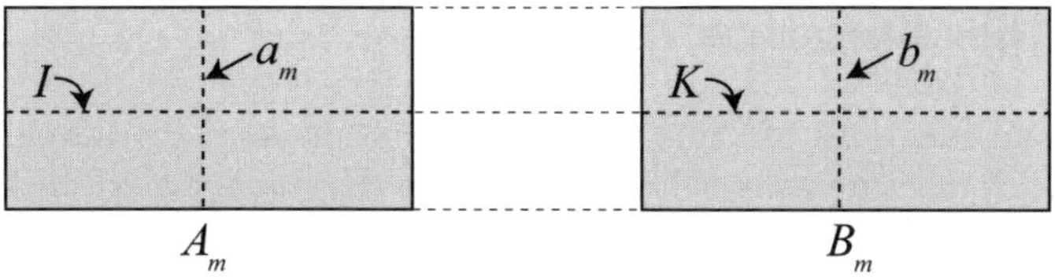

Fig. 8.4. The rectangles $A_m = [I, a_m]$ and $B_m = [K, b_m]$.

Proof of Theorem 8.17. Let $\mu = i_*\nu$. The dual $\rho_{\iota'}$ of ρ_ι is constructed as follows: Let I and K be (i) two ι'-leaf segments contained in a common n-cylinder ι'-leaf, or also (ii) two ι'-leaf segments contained in a union of two n-cylinders with a common endpoint in the case of a local manifold structure. Choose $p \in I$ and $p' \in K$. Let a_m be the ι-leaf N-cylinders containing p, and b_m the ι-leaf containing p' and holonomic to a_m. Let $A_m = [I, a_m]$ and $B_m = [K, b_m]$ (see Figure 8.4). Now, let us prove that

(i)
$$\frac{\mu(A_{m+1})}{\mu(B_{m+1})} \in (1 \pm \mathcal{O}(\varepsilon^{n+m}))\frac{\mu(A_m)}{\mu(B_m)} \tag{8.12}$$
for some $0 < \varepsilon < 1$;

(ii) the dual measure raio function is given by
$$\rho_{\iota'}(I : K) = \lim_{m \to \infty} \frac{\mu(A_m)}{\mu(B_m)} \ ; \tag{8.13}$$

By Lemma 8.14, there is $0 < \varepsilon < 1$ such that
$$\mu(A_{m+1})/\mu(A_m) \in (1 \pm \mathcal{O}(\varepsilon^{n+m}))\rho_\iota(a_{m+1} : a_m) \ ,$$
and, similarly,
$$\mu(B_{m+1})/\mu(B_m) \in (1 \pm \mathcal{O}(\varepsilon^{n+m}))\rho_\iota(b_{m+1} : b_m) \ .$$
Since ρ_ι is an ι-ratio function,
$$\rho_\iota(a_{m+1} : a_m) \in \mathcal{O}(\varepsilon^{n+m}))\rho_\iota(b_{m+1} : b_m) \ .$$
Therefore, (8.12) follows. Furthermore, (8.12) implies (8.13).

Using (8.12), we obtain that $\rho_{\iota'}$ is an ι'-measure ratio function: $\rho_{\iota'}$ is f-invariant, $\rho_{\iota'}(I : K) = \rho_{\iota'}(K : I)^{-1}$ and
$$\rho_{\iota'}(I : K) = \rho_{\iota'}(I_1 : K) + \rho_{\iota'}(I_2 : K)$$
for ι-leaf segments I_1 and I_2 with at most one common point and such that $I = I_1 \cup I_2$. Again using (8.12), $\rho_{\iota'}$ satisfies inequality (8.1). □

8.6 Further literature

The solenoid functions of Pinto and Rand [163] inspired the development of the notion of measure solenoid function. This chapter is based on Pinto and Rand [166].

9

Cocycle-gap pairs

We introduce the cocycle-gap pairs. If a Gibbs measure ν is C^{1+} realizable as a C^{1+} hyperbolic diffeomorphism, then the cocycle-gap pairs allow us to construct all C^{1+} hyperbolic diffeomorphisms that realize the Gibbs measure ν.

9.1 Measure-length ratio cocycle

Let $\mathbf{B}^\iota$ be a gap train-track. For each Markov rectangle R let $t_R^{\iota'}$ be the set of ι-segments of R. Let us denote by $\mathbf{B}_o^{\iota'}$ the disjoint union $\sqcup_{i=1}^m t_{R_i}^{\iota'}$ over all Markov rectangles $R_1, \ldots, R_m$ (without doing any extra-identification). In this section, for every $\xi \in \mathbf{B}_o^{\iota'}$ and $n \geq 1$, we denote by ξ_n the n-cylinder $\pi_\iota f_{\iota'}^{n-1} \xi$ of $\mathbf{B}^\iota$.

Definition 22 *Let $\mathbf{B}^\iota$ be a gap train-track and ρ be a ι-measure ratio function. We say that $J : \mathbf{B}_o^{\iota'} \to \mathbb{R}^+$ is a (ρ, δ, P) ι-measure-length ratio cocycle if $J = \kappa/(\kappa \circ f_{\iota'})$ where κ is a positive Hölder continuous function on $\mathbf{B}_o^{\iota'}$ and is bounded away from 0, and*

$$\sum_{f_{\iota'}\eta=\xi} J(\eta)\rho(f_{\iota'}\eta : m(f_{\iota'}\eta))^{1/\delta} e^{P/\delta} < 1 , \tag{9.1}$$

for every $\eta \in \mathbf{B}_o^{\iota'}$.

We note that in (9.1), the mother of η is not defined because η is a leaf primary cylinder segment, and so we used instead the mother of the leaf 2-cylinder $f_{\iota'}\eta$.

Let us consider a C^{1+} self-renormalizable structure $\mathcal{S}$ on $\mathbf{B}^\iota$, and fix a bounded atlas $\mathcal{B}$ for $\mathcal{S}$. Let $\delta > 0$. By Lemma 8.7, the C^{1+} self-renormalizable structure $\mathcal{S}$ C^{1+}-realizes a Gibbs measure $\nu = \nu_{\mathcal{S},\delta}$ as a natural invariant measure $\mu = \mu_{\mathcal{S},\delta} = i_*\nu$ with pressure $P = P(\mathcal{S}, \delta)$. Let $\rho = \rho_{\mathcal{S},\delta}$ be the corresponding ι-measure ratio function (see Theorem 8.15). Since μ is a natural geometric measure, for every $\xi \in \mathbf{B}_o^{\iota'}$, the ratios $|\xi_n|_i e^{-nP/\delta}/\mu(\xi_n)^{1/\delta}$ are uniformly bounded away from 0 and ∞, where the length $|\xi_n|_i$ is measured in any chart $i \in \mathcal{B}$ containing ξ_n in its domain. Therefore,

$$\kappa_i(\xi_n) = \frac{|\xi_n|_i e^{-nP/\delta}}{\mu(\xi_n)^{1/\delta}}$$

is well-defined. By Lemma 8.14, we get

$$\frac{\mu_\iota(\xi_n)}{\mu_\iota(m\xi_n)} \in (1 \pm \mathcal{O}(\varepsilon^n))\rho(f_{\iota'}\xi : m(f_{\iota'}\xi)),$$

for some $0 < \varepsilon < 1$. Hence, the ratios $\mu_\iota(\xi_n)/\mu_\iota(m\xi_n)$ converge exponentially fast along backward orbits ξ of cylinders. By (4.1), we get that $|\xi_n|/|m\xi_n|$ also converge exponentially fast along backward orbits ξ of cylinders. Therefore, it follows that there is a Hölder function $J_{\mathcal{S},\delta} : \mathbf{B}_o^{\iota'} \to \mathbb{R}$ such that

$$\frac{\kappa_i(\xi_n)}{\kappa_i(m\xi_n)} \in (1 \pm \mathcal{O}(\varepsilon^n))J_{\mathcal{S},\delta}(\xi), \tag{9.2}$$

for some $0 < \varepsilon < 1$.

Lemma 9.1. *Let $\mathbf{B}^\iota$ be a gap train-track. Let $\mathcal{S}$ be a C^{1+} self-renormalizable structure, and $\delta > 0$. Let $\mu_{\mathcal{S},\delta}$ be the natural geometric measure with pressure $P = P(\mathcal{S},\delta)$, and $\rho = \rho_{\mathcal{S},\delta}$ the corresponding ι-measure ratio function. The function $J_{\mathcal{S},\delta} : \mathbf{B}_o^{\iota'} \to \mathbb{R}^+$ given by (9.2) is a (ρ,δ,P) ι-measure-length ratio cocycle.*

Proof. If I is an n-cylinder in $\mathbf{B}^\iota$, then $\sum_{mI'=I} |I'| < |I|$, where the lengths are measured in the same chart. Thus, since $|I'| = \kappa_i(I')\mu_\iota(I')^{1/\delta}e^{(n+1)P/\delta}$ we deduce that

$$\sum_{mI'=I} \frac{\kappa_i(I')}{\kappa_i(I)} \left(\frac{\mu_\iota(I')}{\mu_\iota(I)}\right)^{1/\delta} e^{P/\delta} < 1.$$

For every $\xi \in \mathbf{B}_o^{\iota'}$, we have that $\tau_{\iota'}\eta = \xi$ if, and only if, $\eta_{n+1} \subset \xi_n$ for every $n \geq 1$. Hence, the Hölder continuous function $J = J_{\mathcal{S},\delta}$ satisfies (9.1).

Now, suppose that $\xi \in \mathbf{B}_o^{\iota'}$ is such that there exists $p \geq 1$ with the property that $\xi_{n_p} \subset \xi_n$ for every $n \geq 1$. By (9.2), we get

$$\begin{aligned}\frac{\kappa_{i_0}(\xi_{jp})}{\kappa_{i_0}(\xi_{(j-1)p})} &= \prod_{l=0}^{p-1} \frac{\kappa_{i_{l+1}}(\xi_{(j-1)p+l+1})}{\kappa_{i_l}(\xi_{(j-1)p+l})} \\ &\in (1 \pm \mathcal{O}(\nu^{(j-1)p})) \prod_{l=0}^{p-1} J(f_{\iota'}^l(\xi)),\end{aligned}$$

where $i_0, \ldots, i_{p-1}$ are charts contained in a bounded atlas of $\mathcal{S}$. Thus, for all $1 < m < M$, we have

$$\frac{\kappa_{i_0}(\xi_{Mp})}{\kappa_{i_0}(\xi_{mp})} = \prod_{n=0}^{M-m-1} \frac{\kappa_{i_0}(\xi_{(n+m+1)p})}{\kappa_{i_0}(\xi_{(n+m)p})}$$

$$\in (1 \pm \mathcal{O}(\nu^{mp})) \left[\prod_{l=0}^{p-1} J(f_{\iota'}^l(\xi))\right]^{M-m} .$$

Since the term on the left of this equation is uniformly bounded away from 0 and ∞, it follows that $\prod_{l=0}^{p-1} J(f_{\iota'}^l(\xi)) = 1$. From Livšic's theorem (e.g. see Katok and Hasselblatt [65]) we get that $J_{\mathcal{S},\delta} = \kappa/(\kappa \circ f_{\iota'})$ where κ is a positive Hölder continuous function on $\mathbf{B}_o^{\iota'}$ and is bounded away from 0. □

9.2 Gap ratio function

Let $\mathbf{B}^{\iota}$ be a gap train-track. Let $\mathcal{G}^{\iota'}$ be the set of all pairs $(\xi_1 : \xi_2) \in \mathbf{B}_o^{\iota'} \times \mathbf{B}_o^{\iota'}$ such that $mf_{\iota'}\xi_1 = mf_{\iota'}\xi_2$. The metric d_Λ induces a natural metric $d_{\mathcal{G}^{\iota'}}$ on $\mathcal{G}^{\iota'}$ given by

$$d_{\mathcal{G}^{\iota'}}((\xi_1 : \xi_2), (\eta_1 : \eta_2)) = \max\{d_\Lambda(\xi_1, \eta_1), d_\Lambda(\xi_2, \eta_2)\} .$$

Definition 23 *A function $\gamma : \mathcal{G}^{\iota'} \to \mathbb{R}^+$ is an* ι-gap ratio function *if it satisfies the following conditions:*

(i) $\gamma(\xi_1 : \xi_2)$ is uniformly bounded away from 0 and ∞;
(ii) $\gamma(\xi_1 : \xi_2) = \gamma(\xi_1 : \xi_3)\gamma(\xi_3 : \xi_2)$;
(iii) there are $0 < \theta < 1$ and $C > 1$ such that

$$|\gamma(\xi_1 : \xi_2) - \gamma(\eta_1 : \eta_2)| \le C \left(d_{\mathcal{G}^{\iota'}}((\xi_1 : \xi_2), \gamma(\eta_1 : \eta_2))\right)^\theta . \tag{9.3}$$

We note that part (ii) of this definition implies that $\gamma(\xi_1 : \xi_2) = \gamma(\xi_2 : \xi_1)^{-1}$.

Let $\mathcal{S}$ be a C^{1+} self-renormalizable structure on $\mathbf{B}^{\iota}$ and $\mathcal{B}$ a bounded atlas for $\mathcal{S}$. Then, the gap ratio function $\gamma_{\mathcal{S}}$ is well-defined by

$$\gamma_{\mathcal{S}}(\xi : \eta) = \lim_{n\to\infty} \frac{|\pi_{\mathbf{B}^{\iota}} f_{\iota}^n \xi|_{i_n}}{|\pi_{\mathbf{B}^{\iota}} f_{\iota}^n \eta|_{i_n}}, \tag{9.4}$$

where $i_n \in \mathcal{B}$ contains in its domain the n-cylinder $mf_{\iota}^n\xi$ (we note that $mf_{\iota}^n\xi = mf_{\iota}^n\eta$).

9.3 Ratio functions

We are going to construct the ratio function of a C^{1+} self-renormalizable structure from the gap ratio function and measure-length ratio cocycle.

Lemma 9.2. *Let $\mathbf{B}^\iota$ be a gap train-track. Let $\mathcal{S}$ be a C^{1+} self-renormalizable structure. Let $r_\mathcal{S}$ be the corresponding ι-ratio function. Let $\delta > 0$ and let $\mu = \mu_{\mathcal{S},\delta}$ be the natural geometric measure with pressure $P = P(\mathcal{S}, \delta)$ and $\rho_{\mathcal{S},\delta}$ the corresponding ι-measure ratio function. Let $J_{\mathcal{S},\delta}$ and $\gamma_\mathcal{S}$ be the corresponding ι-gap ratio function and ι-measure-length ratio cocycle. Then, the following equalities are satisfied:*

(i) Let I be an ι-leaf n-cylinder contained in the ι-leaf $(n-1)$-cylinder L. Then,

$$r_\mathcal{S}(I : L) = J_{\mathcal{S},\delta}(\xi_I)\, \rho_{\mathcal{S},\delta}(I : L)^{1/\delta}\, e^{P/\delta}, \tag{9.5}$$

where $\xi_I = f_\iota^{n-1} I \in \mathbf{B}_o^\iota$.

(ii) Let I be an n-cylinder and K an n-gap and both contained in a $(n-1)$-cylinder L. Then,

$$r_\mathcal{S}(I : K) = r_\mathcal{S}(I : L) \frac{\sum_{G \subset L} \gamma_\mathcal{S}(G : K)}{1 - \sum_{D \subset L} r_\mathcal{S}(D : L)}, \tag{9.6}$$

where the sum in the numerator is over all n-gaps $G \subset L$ and the sum in the denominator is over all n-cylinders $D \subset L$.

Proof. For every n-cylinder $I \subset \mathbf{B}^\iota$, define $\kappa_i(I) = |I|_i e^{-nP/\delta} / \mu_\iota(I)^{1/\delta}$ and let $J_{\mathcal{S},\delta}$ be the associate measure-length cocycle. Let I be an ι-leaf n-cylinder, L the ι-leaf $(n-1)$-cylinder containing I. Choose $p \in I$ and let U_m be the ι'-leaf m-cylinders containing p. Let A_m be the rectangle $[I, U_m]$ and B_m be the rectangle $[L, U_m]$. Then, $f_{\iota'}^{m-1} A_m$ and $f_{\iota'}^{n-1} B_m$ are ι'-spanning rectangles of some Markov rectangle. Let a_m and b_m be the projections of these into $\mathbf{B}^\iota$. Then, by the invariance of μ, $\mu(A_m)/\mu(B_m) = \mu_\iota(a_m)/\mu_\iota(b_m)$ and therefore

$$\begin{aligned}
\rho_{\mathcal{S},\delta}(I : L)^{1/\delta} &= \lim_{m \to \infty} \frac{\mu(A_m)^{1/\delta}}{\mu(B_m)^{1/\delta}} \\
&= \lim_{m \to \infty} \frac{\mu_\iota(a_m)^{1/\delta}}{\mu_\iota(b_m)^{1/\delta}} \\
&= \lim_{m \to \infty} \frac{\kappa(a_m)^{-1} |a_m|_{i_m} e^{-(n+m)P/\delta}}{\kappa(b_m)^{-1} |b_m|_{i_m} e^{-(n+m-1)P/\delta}} \\
&= J_{\mathcal{S},\delta}(\xi_I)^{-1} r_\mathcal{S}(I : L) e^{-P/\delta},
\end{aligned}$$

where $|a_m|_{i_m}$ and $|b_m|_{i_m}$ are measured in a chart i_m of the bounded atlas on $\mathbf{B}^\iota$, and ξ_I is the leaf primary cylinder segment $f_\iota^{n-1}(I)$. Thus, equation (9.5) is satisfied.

We note that the ratio of the size of K to the size l of the totality of gaps G in L is given by $\left(\sum_{G \subset L} \gamma_\mathcal{S}(G : K)\right)^{-1}$, where $\gamma_\mathcal{S}$ is the gap ratio function and the sum is over all n-gaps in L. But since the complement of the gaps in L is the union of n-cylinders we have that the ratio of l to the size of L is $1 - \sum_{D \subset L} r_\mathcal{S}(D : L)$ where the sum is over all n-cylinders D in L. Thus, we deduce that for $r_\mathcal{S}(I : K)$ we should take

$$r_S(I:K) = r_S(I:L)\frac{\sum_{G\subset L}\gamma_S(G:K)}{1-\sum_{D\subset L} r_S(D:L)}, \tag{9.7}$$

which proves (9.6). □

9.4 Cocycle-gap pairs

In this section, we are going to construct a *cocyle-gap* map b which reflects the cylinder-gap condition of an ι-solenoid function (see § 3.6), i.e the ratios are well-defined along the ι-boundaries of the Markov rectangles. Hence, r is an ι-ratio function.

Let $\mathbf{B}^{\iota}$ be a gap train-track and $\mathbf{B}^{\iota'}$ a no-gap train-track (as in the case of codimension one attractors or repellors). Let $\mathcal{Q}$ be the set of all periodic orbits O which are contained in the ι-boundaries of the Markov rectangles. For every periodic orbit $O \in \mathcal{Q}$, let us choose a point $x = x(O)$ belonging to the orbit O. Let us denote by $p(x)$ the smallest period of x. Let us denote by $M(1,x)$ and $M(2,x)$ the Markov rectangles containing the point x. Let us denote by $l_i(x)$ the ι-leaf i-cylinder segment of Markov rectangle $M(1,x)$ containing the point x. Let $A(f_\iota^i(x))$ be the smallest ι-leaf segment containing all the ι-boundary leaf segments of Markov rectangles intersecting the global leaf segment passing through the point $f_\iota^i(x)$. Let $q(x)$ be the smallest integer which is a multiple of $p(x)$, such that

$$A(f_\iota^i(x)) \subset f_\iota^{q(x)+i}(l_i(x)),$$

for every $0 \le i < p(x)$. Let us denote the ι-leaf segments $f_\iota^{q(x)+i}(l_i(x))$ by $L_i(x)$. We note that when using the notation $L_i(x)$, we will always consider i to be $i \bmod p(x)$. For every $j \in \{1,2\}$, let $J(j,x)$ be the primary ι'-leaf segment contained in $M(j,x)$ with x as an endpoint such that $R(j,i,x) = [f_\iota^{q(x)+i}(l_i(x)), f_\iota^{q(x)+i}(J(j,x))]$ is a rectangle for every $0 \le i < p(x)$. Let $\mathrm{Co}(j,i,x) \subset \mathbf{B}_o^{\iota'}$ be the set of all ι-primary leaves ξ of Markov rectangles M such that $f_{\iota'}\xi \subset L_i(x)$ and $f_{\iota'}M \cap R(j,i,x)$ has non-empty interior. Let $\mathrm{Gap}(j,i,x) \subset \mathcal{G}^{\iota'}$ be the set of all sister pairs (ξ_1,ξ_2) such that $mf_{\iota'}\xi_1 (= mf_{\iota'}\xi_2)$ is an ι-primary leaf of a Markov rectangle M with the property that $M \cap R(j,i,x)$ has non-empty interior. Let $\mathrm{Co}_j = \cup_{O\in\mathcal{Q}}\cup_{i=0}^{p(x(O))-1}\mathrm{Co}(j,i,x(O))$ and $\mathrm{Gap}_j = \cup_{O\in\mathcal{Q}}\cup_{i=0}^{p(x(O))-1}\mathrm{Gap}(j,i,x(O))$. Let ρ be an ι-measure ratio function with corresponding Gibbs measure ν. Let $\mathcal{D}_j(\rho,\delta,P)$ be the set of all pairs (γ_j, J_j) with the following properties:

(i) $\gamma_j : \mathrm{Gap}_j \to \mathbb{R}^+$ is a map;
(ii) $J_j : \mathrm{Co}_j \to \mathbb{R}^+$ is a map satisfying property (9.1), with respect to (ρ,δ,P), for every $\xi \in \mathrm{Co}_j$ such that $\xi \subset \cup_{O\in\mathcal{Q}}\cup_{i=0}^{p(x(O))-1} L_i(x(O))$.
(iii) For every $x(O) \in \mathcal{Q}$, letting $x = x(O)$, $\prod_{l=0}^{p(x)-1} J_j(f_{\iota'}^l I^j(i,x) = 1$, where $I^j(i,x) \subset \mathrm{Co}_j$ is a ι'-primary leaf segment containing the periodic point $f_{\iota'}^i(x)$.

For every $x(O) \in \mathcal{Q}$, let $x = x(O)$ and let $A(1,x)$ and $A(2,x)$ be the $p(x)$-cylinders of $M(1,x)$ and $M(2,x)$, respectively, containing the point x. The points

$$\pi_{\mathbf{B}_o^{\iota'}} A(j,x), \pi_{\mathbf{B}_o^{\iota'}} f_\iota A(j,x), \ldots, \pi_{\mathbf{B}_o^{\iota'}} f_\iota^{p(x)-1} A(j,x)$$

in $\mathbf{B}_o^{\iota'}$ form a periodic orbit, under f_ι, with period $p(x)$, where $\pi_{\mathbf{B}_o^{\iota'}} : \Lambda \to \mathbf{B}_o^{\iota'}$ is the natural projection. The primary cylinders contained in the sets $\mathrm{Co}(j,i,x)$ are pre-orbits of the points $\pi_{\mathbf{B}_o^{\iota'}} f_\iota^i A(j,x)$ in $\mathbf{B}_o^{\iota'}$, under f_ι. Hence, we note that, if $\prod_{i=0}^{p(x)-1} J(\pi_{\mathbf{B}_o^{\iota'}} f_\iota^i A(j,x)) = 1$, then, by Livšic's theorem (e.g. see Katok and Hasselblatt [65]), there is a map k such that, for every $\xi \in \mathrm{Co}(j,i,x)$, $J(\xi) = k(\xi)/(k \circ f_{\iota'})(\xi)$.

We say that C is an *out-gap segment of a rectangle* R if C is a gap segment of R and is not a leaf n-gap segment of any Markov rectangle M such that $M \cap R$ is a rectangle with non-empty interior.

We say that C is a *leaf* n*-cylinder segment of a rectangle* R, if C is a leaf n-cylinder segment of a Markov rectangle M such that $M \cap R$ is a rectangle with non-empty interior. We say that C is a *leaf* n*-gap segment of a rectangle* R, if C is a leaf n-gap segment of a Markov rectangle M such that $M \cap R$ is a rectangle with non-empty interior. We say that C is an n*-leaf segment of a rectangle* R, if C is a leaf n-cylinder segment of R or if C is a leaf n-gap segment of R.

Lemma 9.3. *Let* $(\gamma_1, J_1) \in \mathcal{D}_1(\rho, \delta, P)$. *Let* $x = x(O)$, *where* $O \in \mathcal{Q}$. *For every* $i \in \{0, 1, \ldots, p(x)-1\}$ *and for all* 2*-leaf segments* $C \subset L_i(x)$ *of* $R(1,i,x)$, *the ratios* $r(C : mC)$ *are uniquely determined such that they are invariant under* f, *satisfy the matching condition, and satisfy equalities (9.5) and (9.6).*

Proof. If $C \subset L_i(x)$ is a leaf 2-cylinder segment of $R(1,i,x)$, then we define the ratio $r(C : mC)$, using (9.5), by

$$r(C : mC) = J(\xi_C)\, \rho(C : mC)^{1/\delta}\, e^{P/\delta}, \tag{9.8}$$

where $\xi_C = f_\iota C \in \mathrm{Co}_1$. For every sister pair $(\xi_1 : \xi_2) \in \mathrm{Gap}_1$ we define the ratio $r(f_{\iota'}\xi_1 : f_{\iota'}\xi_2)$ equal to $\gamma(\xi_1 : \xi_2)$. If $C \subset L_i(x)$ is a leaf 2-gap segment of $R(1,i,x)$, then we define the ratio $r(C : mC)$ by

$$r(C : mC) = \frac{1 - \sum_{D \subset mC} r(D : mC)}{\sum_{G \subset mC} r(G : C)}, \tag{9.9}$$

where the sum, in the numerator, is over all 2-cylinders $D \subset mC$ of $R(1,i,x)$, and the sum, in the denominator, is over all 2-gaps $G \subset mC$ of $R(1,i,x)$. Hence,

$$\sum_{C \subset mC} r(C : mC) = 1,$$

where the sum is over all 2-leaf segments $C \subset mC$ of $R(1,i,x)$. □

Lemma 9.4. *Let $(\gamma_1, J_1) \in \mathcal{D}_1(\rho, \delta, P)$. Let $x = x(O)$, where $O \in \mathcal{Q}$, and let $i \in \{0, 1, \ldots, p(x) - 1\}$. For all $n \geq 0$, and for all out-gaps and all 2-leaf segments $C \subset f_\iota^{n+i}\ell_i(x)$ of $f_\iota^{n+i}M(1, x)$, the ratios $r(C : f_\iota^{n+i}\ell_i(x))$ are uniquely determined such that they are invariant under f, satisfy the matching condition, and satisfy equalities (9.5) and (9.6).*

Proof. Let us denote $f_\iota^n M(1, x)$ by M_n and $f_\iota^n \ell_i(x)$ by L_i^n. The proof follows by induction on $n \geq 0$. For the case $n = 0$, the ratios $r(C : L_i^{n+i})$ are uniquely determined by Lemma 9.3. Let us prove that the ratios $r(C : L_i^{n+1+i})$ are uniquely determined using the induction hypotheses with respect to n. For every out-gap and every primary cylinder segment $C \subset L_i^{n+1+i}$ of $f_\iota^{n+i}M(1, x)$, $f_{\iota'}C$ is a out-gap or a 2-leaf segment. Hence, by the induction hypotheses, the ratio $r(f_{\iota'}C : L_i^{n+i})$ is well-defined. Therefore, using the invariance of f, we define

$$r(C : L_i^{n+1+i}) = r(f_{\iota'}C : L_i^{n+i}) \ . \tag{9.10}$$

For every 2-leaf segment $C \subset L_i^{n+1+i}$ of $f_\iota^{n+i}M(1, x)$, the ratio $r(C : mC)$ is well-defined by Lemma 9.3. Hence, by (9.10), we define

$$r(C : L_i^{n+1+i}) = r(C : mC) r(mC : L_i^{n+1+i}),$$

which ends the proof of the induction. □

Lemma 9.5. *Let $(\gamma_1, J_1) \in \mathcal{D}_1(\rho, \delta, P)$. Let $x = x(O)$, where $O \in \mathcal{Q}$, and let $i \in \{0, 1, \ldots, p(x) - 1\}$. Let $n \geq 0$ and $j \in \{0, \ldots, n\}$. For all out-gaps and all $j+2$-leaf segments $C \subset f_{\iota'}^n L_i(x)$ of $f_{\iota'}^n R(1, i, x)$, the ratios $r(C : f_{\iota'}^n L_i(x))$ are uniquely determined such that they are invariant under f, satisfy the matching condition, and satisfy equalities (9.5) and (9.6).*

Proof. The proof follows by induction in $n \geq 0$. For the case $n = 0$, noting that $L_i(x) = f_\iota^{q(x)+i}\ell_i(x)$, the ratios $r(C : L_i(x))$ are well-defined by Lemma 9.4. Hence, using the matching condition, the ratio $r(f_{\iota'}^{n+1}L_{i+1}(x) : f_{\iota'}^n L_i(x))$ is well-defined. Let us prove that for all out-gaps and $j+2$-leaf segments $C \subset f_{\iota'}^{n+1}L_i(x)$ of $f_{\iota'}^{n+1}R(1, i, x)$, with $1 \leq j \leq n+1$, the ratios $r(C : f_{\iota'}^{n+1}L_i(x))$ are uniquely determined using the induction hypotheses with respect to n. By the induction hypotheses and by the matching condition, the ratio $r(f_\iota C : f_{\iota'}^n L_i(x))$ is well-defined. By invariance of f, we define $r(C : f_{\iota'}^{n+1}L_i(x)) = r(f_\iota C : f_{\iota'}^n L_i(x))$. which ends the proof of the induction. □

Let us attribute the ratios for the cylinders and gaps of $R(2, i, x)$ such that they agree with the ratios previously defined in $R(1, i, x)$.

Lemma 9.6. *Let $(\gamma_1, J_1) \in \mathcal{D}_1(\rho, \delta, P)$. Let $x = x(O)$, where $O \in \mathcal{Q}$, and let $i \in \{0, 1, \ldots, p(x) - 1\}$. Let $n \geq 0$ and $j \in \{1, \ldots, n\}$. For all out-gaps and all $j+2$-leaf segments $C \subset f_{\iota'}^n L_i(x) \setminus f_{\iota'}^{n+1}L_{i+1}(x)$ of $f_{\iota'}^n R(2, i, x)$, the ratios $r(C : f_{\iota'}^n L_i(x))$ are uniquely determined such that they are invariant under f, satisfy the matching condition, satisfy equalities (9.5) and (9.6), and are well-defined along the ι-boundaries of the Markov rectangles. Hence, r is an ι-ratio function.*

Proof. The proof follows by induction in $n \geq 0$. Let us prove the case $n = 0$. By construction, $L_i(x) \supset A(f_\iota^i(x))$, i.e $L_i(x)$ contains all the ι-boundary leaf segments of Markov rectangles intersecting the global leaf segment passing through the point $f_\iota^i(x)$. Hence, if $G_2 \subset L_i(x) \setminus f_{\iota'} L_{i+1}(x)$ is an out-gap of $R(2,i,x)$, then there is an out-gap or a leaf 2-gap segment G_1 of $R(1,i,x)$ such that $G_1 = G_2$. Therefore, we define $r(G_2 : L_i(x)) = r(G_1 : L_i(x))$. Since $L_i(x) \supset A(f_\iota^i(x))$, if $G_2 \subset L_i(x) \setminus f_{\iota'} L_{i+1}(x)$ is a leaf 2-gap segment of $R(2,i,x)$, then there is an out-gap or a leaf 2-gap segment G_1 of $R(1,i,x)$ such that $G_1 = G_2$. Hence, we define $r(G_2 : L_i(x)) = r(G_1 : L_i(x))$. If $C_2 \subset L_i(x) \setminus f_{\iota'} L_{i+1}(x)$ is a leaf 2-cylinder segment of $R(2,i,x)$, then there is a primary leaf segment or a leaf 2-cylinder segment C_1 of $R(1,i,x)$ such that $C_2 = C_1$. Therefore, we define $r(C_2 : L_i(x)) = r(C_1 : L_i(x))$. Let us prove that for all out-gaps and $j+2$-leaf segments $C \subset f_{\iota'}^{n+1} L_i(x) \setminus f_{\iota'}^{n+2} L_{i+1}(x)$ of $f_{\iota'}^{n+1} R(2,i,x)$, with $1 \leq j \leq n+1$, the ratios $r(C : f_{\iota'}^{n+1} L_i(x))$ are uniquely determined using the induction hypotheses with respect to n. By the induction hypotheses and by the matching condition, the ratio $r(f_\iota C : f_{\iota'}^n L_i(x))$ is well-defined. By invariance of f, we define $r(C : f_{\iota'}^{n+1} L_i(x)) = r(f_\iota C : f_{\iota'}^n L_i(x))$. which ends the proof of the induction. □

Lemma 9.7. *Let $(\gamma_1, J_1) \in \mathcal{D}_1(\rho, \delta, P)$. Let $x = x(O)$, where $O \in \mathcal{Q}$, and let $i \in \{0, 1, \ldots, p(x) - 1\}$. For all out-gaps and all 2-leaf segments $C \subset L_i(x)$ of $R(2,i,x)$, the ratios $r(C : L_i(x))$ are uniquely determined such that they are invariant under f, satisfy the matching condition, satisfy equalities (9.5) and (9.6), and are well-defined along the ι-boundaries of the Markov rectangles. Hence, r is an ι-ratio function.*

Proof. By construction of $L_i(x) \setminus f_{\iota'} L_{i+1}(x)$, there is $k = k(n,i,x)$ such that $L_i(x) \setminus f_{\iota'} L_{i+1}(x) = \cup_{l=1}^k D_l$, where D_l are out-gaps, primary leaf segments and 2-leaf segments of $R(1,i,x)$. Therefore, $f_{\iota'}^n L_i(x) \setminus f_{\iota'}^{n+1} L_{i+1}(x) = \cup_{l=1}^k f_{\iota'}^n D_l$ where $f_{\iota'}^n D_l$ are out-gaps and $j+2$-leaf segments of $R(1,i,f_{\iota'}^n(x))$ with $0 \leq j \leq n$. Hence, by Lemma 9.6 and using the matching condition, the ratio $r(f_{\iota'}^n L_i(x) \setminus f_{\iota'}^{n+1} L_{i+1}(x) : f_{\iota'}^n L_i(x))$ is well-defined. Hence, using the matching condition, we define

$$r(f_{\iota'}^{n+1} L_{i+1}(x) : f_{\iota'}^n L_i(x)) = 1 - r(f_{\iota'}^n L_i(x) \setminus f_{\iota'}^{n+1} L_{i+1}(x) : f_{\iota'}^n L_i(x)) \ .$$

Therefore, using again the matching condition, we define

$$r(f_{\iota'}^{n+1} L_{i+n+1}(x) : L_i(x)) = \prod_{j=0}^{n} r(f_{\iota'}^{j+1} L_{i+j+1}(x) : f_{\iota'}^j L_{i+j}(x)) \ . \tag{9.11}$$

Let $M(i,x)$ be the 2-cylinder of $R(2,i,x)$ containing the point x. Take $N > 0$, large enough, such that $f_{\iota'}^{N+1} L_{i+N+1}(x) \subset M(i,x)$. Hence, there is $m = m(N,i,x)$ such that $M(i,x) = (\cup_{l=0}^m D_l) \cup f_{\iota'}^{N+1} L_{i+N+1}(x)$ where D_l are out-gaps or $j+2$-leaf segments of $R(1,i,x)$ for some $0 \leq j \leq N$. Hence, by Lemma 9.6, (9.11) and using the matching condition, the ratio is well-defined by

$$r(M(i,x) : L_i(x)) = \sum_{l=0}^{m} r(D_l : L_i(x)) + r(f_{\iota'}^{N+1} L_{i+N+1}(x) : L_i(x)) .$$

If $C \subset L_i(x) \setminus M(i,x)$ is a out-gap or a 2-leaf segment of $R(2,i,x)$, then, by Lemma 9.6, the ratio is well-defined by $r(C : L_i(x))$. By construction of the ratios, in Lemmas 9.3-9.7, they are compatible with the cylinder-gap condition. □

Definition 24 *Let $(\gamma_1, J_1) \in \mathcal{D}_1(\rho, \delta, P)$. Let $x = x(O)$, where $O \in \mathcal{Q}$, and let $i \in \{0, 1, \ldots, p(x) - 1\}$. Let the ratios $r(C : L_i(x))$ for all out-gaps and all 2-leaf segments $C \subset L_i(x)$ of $R(2,i,x)$ be as given in Lemma 9.7. For all $\xi \in \mathrm{Co}(2,i,x)$, letting $I = f_{\iota'}\xi \subset L_i(x)$, we define*

$$J_2(\xi) = r(I : L_i(x)) r(L_i(x) : mI) \rho(I : K_i)^{-1/\delta} e^{-P/\delta} .$$

For all $(C, D) \in \mathrm{Gap}(2,i,x)$, we define

$$\gamma(C : D) = r(f_{\iota'}C : L_i(x)) r(L_i(x) : f_{\iota'}D) .$$

Lemma 9.8. *Let $\mathcal{D}_1(\rho, \delta, P) \neq \emptyset$. The cocycle-gap map $b = b_{\rho,\delta,P} : \mathcal{D}_1(\rho, \delta, P) \to \mathcal{D}_2(\rho, \delta, P)$ is well-defined by $b(\gamma_1, J_1) = (\gamma_2, J_2)$ where γ_2 and J_2 are as given in Definition 24. Furthermore, the cocycle-gap map b is a bijection.*

Proof. Let us check that (γ_2, J_2) satisfies properties (i)-(iii) of $\mathcal{D}_2(\rho, \delta, P)$. By construction of the ratios r, in Lemmas 9.3-9.7, (γ_2, J_2) satisfies properties (i) and (ii) in the definition of $\mathcal{D}_2(\rho, \delta, P)$. Let us check property (iii). Let us denote by A and B the $p(x)$-cylinders of $M(1,x)$ and $M(2,x)$, respectively, containing the point x. By invariance of r, we have that $r(A : B) = r(f_\iota^{p(x)} A : f_\iota^{p(x)} B)$, and so $r(A : f_\iota^{p(x)} A) = r(B : f_\iota^{p(x)} B)$. By invariance of the ι-measure ratio function ρ, we have that $\rho(A : B) = \rho(f_\iota^{p(x)} A : f_\iota^{p(x)} B)$, and so $\rho(A : f_\iota^{p(x)} A) = \rho(B : f_\iota^{p(x)} B)$. Since, by hypotheses $\prod_{l=0}^{p(x)-1} J(m^i f_\iota^{p(x)-i} A) = 1$, we get, from (9.5), that $r(A : f_\iota^{p(x)} A) = \rho(A : f_\iota^{p(x)} A) e^{p(x)P/\delta}$. Therefore,

$$\begin{aligned} r(B : f_\iota^{p(x)} B) &= r(A : f_\iota^{p(x)} A) \\ &= \rho(A : f_\iota^{p(x)} A) e^{p(x)P/\delta} \\ &= \rho(B : f_\iota^{p(x)} B) e^{p(x)P/\delta} \end{aligned}$$

and so, using (9.5), we obtain that $\prod_{i=0}^{p(x)-1} J(\pi_{\mathbf{B}_o^{\iota'}} f_\iota^i B) = 1$. □

Definition 25 *Let $\mathbf{B}^\iota$ be a gap train-track. Let $\delta > 0$ and $P \in \mathbb{R}$. Let ρ be an ι-measure ratio function and $\nu = \nu_\rho$ the corresponding Gibbs measure on Θ. We say that a pair (γ, J) is a (ν, δ, P) ι* cocycle-gap pair *, if (γ, J) has the following properties:*

(i) γ is an ι-gap ratio function.
(ii) J is an ι measure-length ratio cocyle.
(iii) If $\mathbf{B}^{\iota'}$ is a no-gap train-track, then (γ, J) satisfies the following cocyle-gap property*: $b(\gamma|\mathrm{Gap}_1, J|\mathrm{Co}_1) = (\gamma|\mathrm{Gap}_2, J|\mathrm{Co}_2)$, where $b = b_{\nu,\delta,P}$ is the cocyle-gap map.*

Let $\mathcal{JG}^\iota(\nu, \delta, P)$ be the set of all (ν, δ, P) ι cocycle-gap pairs.

Theorem 9.9. *Let $\mathbf{B}^\iota$ be a gap train-track. Let $\delta > 0$ and $P \in \mathbb{R}$. Let ρ be an ι-measure ratio function with corresponding Gibbs measure ν.*

(i) If there is a (ρ, δ, P) ι-measure-length ratio cocycle, then the set $\mathcal{JG}^\iota(\nu, \delta, P)$ is an infinite dimensional space.
(ii) If $\mathcal{S}$ is a C^{1+} self-renormalizable structure with natural geometric measure $\mu_{\mathcal{S},\delta} = i_*\nu$ and pressure P, then $(\gamma_\mathcal{S}, J_{\mathcal{S},\delta}) \in \mathcal{JG}^\iota(\nu, \delta, P)$.
(iii) If the set $\mathcal{JG}^\iota(\nu, \delta, P) \neq \emptyset$, then there is a well-defined injective map $(\gamma, J) \to r(\gamma, J)$ which associates to each cocycle-gap pair $(\gamma, J) \in \mathcal{JG}^\iota(\nu, \delta, P)$ an ι-ratio function $r(\gamma, J)$ satisfying (9.5) and (9.6).

Remark 9.10. Let $0 < \delta < 1$ and $P = 0$. Let ρ be an ι-measure ratio function with corresponding Gibbs measure ν. Since $J = 1$ is a (ρ, δ, P) ι-measure-length ratio cocycle, then, by Theorem 9.9, the set $\mathcal{JG}^\iota(\nu, \delta, P)$ is an infinite dimensional space.

Proof of Theorem 9.9. Proof of (i). Choose a map $\gamma_1 : \mathrm{Gap}_1 \to \mathbb{R}^+$. Let J_0 be a (ρ, δ, P) ι-measure-length ratio cocycle, and let $J_1 = J_0|\mathrm{Co}_1$. Since $(\gamma_1, J_1) \in \mathcal{D}_1(\rho, \delta, P)$, by Lemma 9.8, the pair $(\gamma_2, J_2) = b_{\rho,\delta,P}(\gamma_1, J_1) \in \mathcal{D}_2(\rho, \delta, P)$ is well-defined. Let k_0 and k_2 be maps such that $J_0 = k_0/(k_0 \circ f_{\iota'})$ and $J_2 = k_2/(k_2 \circ f_{\iota'})$. For every $x(O) \in \mathcal{Q}$, let $x = x(O)$, and let B be the $p(x)$-cylinder of $M(2, x)$ containing the point x. Recall that the points $\pi_{\mathbf{B}_o^{\iota'}} B, \pi_{\mathbf{B}_o^{\iota'}} f_\iota B, \ldots, \pi_{\mathbf{B}_o^{\iota'}} f_\iota^{p(x)-1} B$ in $\mathbf{B}_o^{\iota'}$ form a periodic orbit under f_ι, with period $p(x)$, and that the primary cylinders contained in the set $\mathrm{Co}(2, i, x)$ are pre-orbits of the points $\pi_{\mathbf{B}_o^{\iota'}} f_\iota^i B$ in $\mathbf{B}_o^{\iota'}$, under f_ι. Therefore, there is a small neighbourhood V of Co_2 in $\mathbf{B}_o^{\iota'}$, there is $\varepsilon > 0$, small enough, and there is an Hölder continuous map $k : \mathbf{B}_o^{\iota'} \to \mathbb{R}^+$ with the following properties:

(i) $k|\mathrm{Co}_2 = k_2$, $k|(\mathbf{B}_o^{\iota'} \setminus V) = k_0$ and $\mathrm{Co}_1 \subset \mathbf{B}_o^{\iota'} \setminus V$.
(ii) Let $a = \min_{\xi \in \mathrm{Co}_2}\{J_0(\xi), J_2(\xi)\}$ and $b = \max_{\xi \in \mathrm{Co}_2}\{J_0(\xi), J_2(\xi)\}$, and let $J = k/(k \circ f_{\iota'})$. For every $\xi \in V$, we have that $a - \varepsilon \leq J(\xi) \leq b + \varepsilon$, and, so, J satisfies the cocycle-gap property.

Choosing an Hölder continuous map $\gamma : \mathcal{G}^{\iota'} \to \mathbb{R}^+$ such that $\gamma|\mathrm{Gap}_1 = \gamma_1$ and $\gamma|\mathrm{Gap}_2 = \gamma_2$ and by property (i) above, the pair (γ, J) satisfies (9.1). Therefore, the pair (γ, J) is contained in $\mathcal{JG}^\iota(\nu, \delta, P)$. Using that (9.1) is an open condition, the above construction allows us to construct an infinite set of ι-measure-length ratio cocycles and an infinite set of gap ratio functions such that the corresponding pairs are contained in $\mathcal{JG}^\iota(\nu, \delta, P)$.

Proof of (ii). Let $\mathcal{S}$ be a C^{1+} self-renormalizable structure with natural geometric measure $\mu_{\mathcal{S},\delta} = i_* \nu$ and pressure P. By Lemma 9.1, $J_{\mathcal{S},\delta}$ is a (ρ, δ, P) ι-measure-length ratio cocycle and, by (9.4), $\gamma_{\mathcal{S}}$ is an ι-gap ratio function. If $\mathbf{B}^{\iota'}$ is a no-gap train-track, using (9.5) and (9.6), the pair $(\gamma_{\mathcal{S}}, J_{\mathcal{S},\delta})$ satisfies the cocycle-gap condition because the ratio function $r_{\mathcal{S}}$ associated to $\mathcal{S}$ is well-defined along the ι-boundaries of the Markov rectangles.
Proof of (iii). The equations (9.5) and (9.6) give us an inductive construction, on the level n of the n-cylinders and n-gaps, of a ratio function r in terms of $(\rho, J, \gamma, \delta, P)$ with the property that the ratio between a leaf n-cylinder segment C and a leaf n-cylinder or n-gap segment D with a common endpoint with C is bounded away from zero and infinity independent of n and of the cylinders and gaps considered. The construction gives that r is invariant under f. The Hölder continuity of γ, J and ρ implies that r satisfies (2.2). If $\mathbf{B}^{\iota'}$ is a no-gap train-track, by the construction of the cocycle-gap condition, the ratio function r is well-defined along the ι-boundaries of the Markov rectangles. Hence, r is an ι-ratio function. □

9.5 Further literature

The HR structures of Pinto and Rand [163] and the measure solenoid functions inspired the development of the notion of cocycle-gap pairs. This chapter is based on Pinto and Rand [166].

10

Hausdorff realizations

We present a construction of all hyperbolic basic sets of diffeomorphisms on surfaces which have an invariant measure that is absolutely continuous with respect to Hausdorff measure. These C^{1+} hyperbolic diffeomorphisms are C^{1+} realizations of Gibbs measures. The cocycle-gap pairs form a moduli space for the C^{1+} conjugacy classes of C^{1+} hyperbolic realizations of Gibbs measures.

10.1 One-dimensional realizations of Gibbs measures

Let $\mathcal{S}$ be a C^{1+} self-renormalizable structure on a train-track $\mathbf{B}^\iota$. In Theorem 8.15 we have shown that the map

$$(\mathcal{S}, \delta) \rightarrow \rho_{\mathcal{S},\delta} \tag{10.1}$$

is well-defined where $\rho_{\mathcal{S},\delta}$ is the ι-measure ratio function associated to a Gibbs measure $\nu_{\mathcal{S},\delta} = \nu$ such that $\mu_{\mathcal{S},\delta} = (i_\iota)_* \nu_\iota$ is a natural geometric measure of $\mathcal{S}$.

Lemma 10.1. *(Rigidity) Let $\mathbf{B}^\iota$ be a no gap train-track (and $\delta = 1$). The map $\mathcal{S} \rightarrow \rho_{\mathcal{S},\delta}$ is a one-to-one correspondence between C^{1+} self-renormalizable structures on $\mathbf{B}^\iota$ and ι-measure ratio functions. Furthermore, $\rho_{\mathcal{S},\delta} = r_{\mathcal{S}}$ where $r_{\mathcal{S}}$ is the ratio function determined by $\mathcal{S}$.*

However, if $\mathbf{B}^\iota$ is a gap train-track, then the set of pre-images of the map $(\mathcal{S}, \delta) \rightarrow \rho_{\mathcal{S},\delta}$ is an infinite dimensional space (see Lemma 10.3 below).

Proof. By Lemma 8.7, the C^{1+} self-renormalizable structure $\mathcal{S}$ realizes a Gibbs measure $\nu = \nu_{\mathcal{S},\delta}$. By Theorem 8.15, we get that $\rho_{\mathcal{S},\delta} = r_{\mathcal{S}}$. Since, by Lemma 4.4, the ratio function $r_{\mathcal{S}}$ determines uniquely the C^{1+} self-renormalizable structure $\mathcal{S}$, the map $\mathcal{S} \rightarrow \rho_{\mathcal{S},\delta}$ is a one-to-one correspondence. □

Definition 26 *Let $\mathbf{B}_\iota$ and $\mathbf{B}_{\iota'}$ be (gap or no-gap) train-tracks. Let ρ be an ι-measure ratio function and $\nu = \nu_\rho$ on Θ the corresponding Gibbs measure (see Remark 8.6). Let us denote by $\mathcal{D}^\iota(\nu, \delta, P)$ the set of all C^{1+} self-renormalizable structures $\mathcal{S}$ with geometric natural measure $\mu_{\mathcal{S},\delta} = (i_\iota) * \nu_\iota$ and pressure P.*

By Lemma 10.1, if $\mathbf{B}^\iota$ is a no-gap train-track, and $\delta = 1$ and $P = 0$, the set $\mathcal{D}^\iota(\nu, \delta, P)$ is a singleton.

Let $\mathbf{B}^\iota$ be a gap train-track and $\mathcal{S}$ a C^{1+} self-renormalizable structure in $\mathcal{D}^\iota(\nu, \delta, P)$. In Lemma 9.2, we associate to the C^{1+} self-renormalizable structure $\mathcal{S}$ a measure-length ratio cocycle $J_\mathcal{S}$, and, in § 9.1, we associate to the C^{1+} self-renormalizable structure $\mathcal{S}$ a gap ratio function $\gamma_\mathcal{S}$. By Theorem 9.9, if $\mathbf{B}^{\iota'}$ is a no-gap train-track, then the cylinder-gap condition of $r_\mathcal{S}$ implies that the pair $(\gamma_\mathcal{S}, J_{\mathcal{S},\delta})$ satisfies the cocycle-gap condition. Therefore, the map

$$\mathcal{S} \to (\gamma_\mathcal{S}, J_{\mathcal{S},\delta}) \tag{10.2}$$

between C^{1+} self-renormalizable structures contained in $\mathcal{D}^\iota(\nu, \delta, P)$ and pairs contained in $\mathcal{JG}^\iota(\nu, \delta, P)$ is well-defined.

Definition 27 *The* (δ_ι, P_ι)-bounded solenoid equivalence class *of a Gibbs measure ν is the set of all solenoid functions σ_ι with the following properties: There is $C = C(\sigma_\iota) > 0$ such that for every pair $(\xi, D) \in \mathrm{msc}_\iota$*

$$|\delta_\iota \log s_\iota(D_\Lambda \cap \xi_\Lambda : \xi_\Lambda) - \log \rho_\xi(D) - nP_\iota| < C \ ,$$

where (i) ρ is the ι-extended measure scaling function of ν, (ii) s_ι is the scaling function determined by σ_ι, (iii) $\xi_\Lambda = i(\pi_{\iota'}^{-1}\xi)$ is an ι'-leaf primary cylinder segment and (iv) $D_\Lambda = i(\pi_\iota^{-1}D)$ and so $D_\Lambda \cap \xi_\Lambda$ is an ι-leaf n-cylinder segment.

Remark 10.2. Let $\sigma_{1,\iota}$ and $\sigma_{2,\iota}$ be two solenoid functions in the same (δ_ι, P_ι)-bounded solenoid equivalence class of a Gibbs measure ν. Using the fact that $\sigma_{1,\iota}$ and $\sigma_{2,\iota}$ are bounded away from zero, we obtain that the corresponding scaling functions also satisfy inequality (3.4) for all pairs $(J, m^i J)$ where J is an ι-leaf $(i+1)$-cylinder. Hence, the solenoid functions $\sigma_{1,\iota}$ and $\sigma_{2,\iota}$ are in the same bounded equivalence class (see Definiton 10).

By Lemma 10.3, below, the set $\mathcal{JG}^\iota(\nu, \delta, P)$ gives a parametrization of all C^{1+} self-renormalizable structures $\mathcal{S}$ which are pre-images of the ι-measure ratio function $\rho_{\nu,\iota}$, under the map $\mathcal{S} \to \rho_{\mathcal{S},\delta}$ given in (10.1), with a natural geometric measure $\mu_\iota = (i_\iota)_* \nu_\iota$ and pressure $P(\mathcal{S}, \delta) = P$. Hence, $\mathcal{JG}^\iota(\nu, \delta, P)$ forms a moduli space for the set of all C^{1+} self-renormalizable structures in $\mathcal{D}^\iota(\nu, \delta, P)$.

Lemma 10.3. *(Flexibility) Let $\mathbf{B}_\iota$ be a gap train-track. Let ρ be an ι-measure ratio function and $\nu = \nu_\rho$ the corresponding Gibbs measure on Θ.*

(i) Let $\delta > 0$ and $P \in \mathbb{R}$ be such that $\mathcal{JG}^\iota(\nu, \delta, P) \neq \emptyset$. The map $\mathcal{S} \to (\gamma_\mathcal{S}, J_{\mathcal{S},\delta})$ determines a one-to-one correspondence between C^{1+} self-renormalizable structures in $\mathcal{D}^\iota(\nu, \delta, P)$ and cocycle-gap pairs in $\mathcal{JG}^\iota(\nu, \delta, P)$.
(ii) A C^{1+} self-renormalizable structure $\mathcal{S}$ is contained in $\mathcal{D}^\iota(\nu, \delta, P)$ if, and only if, the ι-solenoid function $\sigma_\mathcal{S}$ is contained in the (δ, P)-bounded solenoid equivalence class of ν (see Definition 27).

Proof. Proof of (i). Let us prove that $(J, \gamma) \in \mathcal{JG}^\iota(\nu, \delta, P)$ determines a unique C^{1+} self-renormalizable structure $\mathcal{S}$ with a natural geometric measure $\mu_{\mathcal{S},\delta} = (i_\iota)_*\nu_\iota$. By Theorem 9.9, the pair (J, γ) determines a unique ι-ratio function $r = r_\iota(J, \gamma)$. By Lemma 4.4, the ι-ratio function r determines a unique C^{1+} self-renormalizable structure $\mathcal{S}$ with an atlas $\mathcal{B}(r)$. Let us prove that $\mu_\iota = (i_\iota)_*\nu_\iota$ is a natural geometric measure of $\mathcal{S}$ with the given δ and P. Let ρ be the ι-measure ratio function associated to the Gibbs measure ν. By Lemma 7.5, for every leaf n-cylinder or n-gap segment I we obtain that

$$\mu_\iota(I) = \mathcal{O}(\rho(I \cap \xi : \xi)) \tag{10.3}$$

for every $\xi \in \pi_{\iota'}(I)$. On the other hand, by construction of the ratio function r_ι and using (9.5), we get

$$\rho(I \cap \xi : \xi) = e^{-nP} r(I \cap \xi : \xi)^\delta \prod_{j=0}^{n-1} \left(J\left(\tau_{\iota'}^j(\xi)\right)\right)^{-\delta} .$$

Since J is a Hölder cocycle, it follows that $\prod_{j=0}^{n-1} J\left(\tau_{\iota'}^j(\xi)\right) = k(\xi)/k(\tau_{\iota'}^n(\xi))$ is uniformly bounded away from zero and infinity, where k is an Hölder continuous positive function. By (4.1), we get that

$$r(I \cap \xi : \xi) = \mathcal{O}\left(|I|_j\right) \tag{10.4}$$

where $j \in \mathcal{B}(r)$ and I is contained in the domain of j. Hence,

$$\rho(I \cap \xi : \xi) = \mathcal{O}\left(|I|_j^\delta e^{-nP}\right) . \tag{10.5}$$

Putting together equations (10.3) and (10.5), we deduce that $\mu_\iota(I) = \mathcal{O}\left(|I|_j^\delta e^{-nP}\right)$, and so $\mu_\iota = (i_\iota)^*\nu_\iota$ is a natural geometric measure of $\mathcal{S}$ with the given δ and P.
Proof of (ii). Let $\mathcal{S}$ be a C^{1+} self-renormalizable structure in $\mathcal{D}^\iota(\nu, \delta, P)$. Then, putting together (10.4) and (10.5), there is $\kappa > 0$ such that

$$|\delta \log r_\iota(I \cap \xi : \xi) - \log \rho(I \cap \xi : \xi) - np| < \kappa$$

for every leaf n-cylinder I and $\xi \in \pi_{\iota'}(I)$. Thus the solenoid function $r|\mathrm{sol}^\iota$ is in the (δ, P)-bounded solenoid equivalence class of ν.

Conversely, let $\mathcal{S}$ be a C^{1+} self-renormalizable structure in the (δ, P)-bounded solenoid equivalence class of ν and $\mu_\iota = (i_\iota)_* \nu_\iota$, i.e. there is $\kappa > 0$ such that

$$|\delta \log r_\iota(I \cap \xi : \xi) - \log \rho(I \cap \xi : \xi) - np| < \kappa \tag{10.6}$$

for every leaf n-cylinder I and $\xi \in \pi_{\iota'}(I)$. Hence, using (10.3) and (10.4) in (10.6), we get $\mu_\iota(I) = \mathcal{O}\left(|I|_j^\delta e^{-nP}\right)$. Since $\mu_\iota = (i_\iota)_* \nu_\iota$ we get that $\mathcal{S}$ is contained in $\mathcal{D}^\iota(\nu, \delta, P)$. □

Lemma 10.4. *Let $\mathbf{B}^\iota$ be a (gap or a no-gap) train-track. Let $\delta > 0$ and $P \in \mathbb{R}$. Let $\mathcal{S}_1 \in \mathcal{D}^\iota(\nu_1, \delta, P)$ and $\mathcal{S}_2 \in \mathcal{D}^\iota(\nu_2, \delta, P)$ be C^{1+} self-renormalizable structures The following statements are equivalent:*

(i) The C^{1+} self-renormalizable structures $\mathcal{S}_1$ and $\mathcal{S}_2$ are Lipschitz conjugate;
(ii) The Gibbs measures ν_1 and ν_2 are equal;
(iii) The solenoid functions $s_{\mathcal{S}_1}$ and $s_{\mathcal{S}_2}$ are in the same bounded equivalence class (Definition 10).

Proof. Proof that (i) is equivalent to (ii). Using (8.2), if $\nu_1 = \nu_2$, then the C^{1+} self-renormalizable structure $\mathcal{S}_1$ is Lipschitz conjugate to $\mathcal{S}_2$. Conversely, if $\mathcal{S}_1$ is Lipschitz conjugate to $\mathcal{S}_2$, then the C^{1+} self-renormalizable structure $\mathcal{S}_1$ (and $\mathcal{S}_2$) satisfies (8.2) with respect to the measures $\mu_{\iota,1} = (i_\iota)_* \nu_{\iota,1}$ and $\mu_{\iota,2} = (i_\iota)_* \nu_{\iota,2}$. By Lemma 8.7, there is a unique τ-invariant Gibbs measure satisfying (8.2) and so $\nu_1 = \nu_2$.
Proof that (ii) is equivalent to (iii). Using (3.4) and (4.1), we obtain that the C^{1+} self-renormalizable structures $\mathcal{S}$ and $\mathcal{S}'$ on $\mathbf{B}^\iota$ are in the same Lipschitz equivalence class if, and only if, the corresponding solenoid functions $r_{\mathcal{S},\iota}|\mathrm{sol}^\iota$ and $r_{\mathcal{S}',\iota}|\mathrm{sol}^\iota$ are in the same bounded equivalence class. Hence, statement (ii) is equivalent to statement (iii). □

10.2 Two-dimensional realizations of Gibbs measures

We start by giving the definition of a natural geometric measure for a C^{1+} hyperbolic diffeomorphism.

Definition 28 *For $\iota \in \{s, u\}$, if $\mathbf{B}^\iota$ is a gap train-track assume $0 < \delta_\iota < 1$, and if $\mathbf{B}^\iota$ is a no-gap train-track take $\delta_\iota = 1$.*

(i) Let g be a C^{1+} hyperbolic diffeomorphism in $\mathcal{T}(f, \Lambda)$. We say that g has a natural geometric measure *$\mu = \mu_{g,\delta_s,\delta_u}$ with pressures $P_s = P_s(g, \delta_s, \delta_u)$ and $P_u = P_u(g, \delta_s, \delta_u)$ if, there is $\kappa > 1$ such that for all leaf n_s-cylinder I_s, for all leaf n_u-cylinder I_u,*

$$\kappa^{-1} < \frac{\mu(R)}{|I_u|^{\delta_u} |I_s|^{\delta_s} e^{-n_s P_s - n_u P_u}} < \kappa \ , \tag{10.7}$$

where R is the (n_s, n_u)-rectangle $[I_s, I_u]$ and where the lengths $|\cdot|$ are measured in the stable and unstable C^{1+}foliated lamination atlases $\mathcal{A}_s(g,\rho)$ and $\mathcal{A}_u(g,\rho)$ of g with respect to some Riemannian metric ρ.
(ii) We say that a C^{1+} hyperbolic diffeomorphism with a natural geometric measure $\mu = \mu_{g,\delta_s,\delta_u}$ with pressures $P_s = P_s(g, \delta_s, \delta_u)$ and $P_u = P_u(g, \delta_s, \delta_u)$ is a C^{1+} realization of a Gibbs measure $\nu = \nu_{g,\delta_s,\delta_u}$ *if $\mu = i_*\nu$. We denote by $\mathcal{T}(\nu, \delta_s, P_s, \delta_u, P_u)$ the set of all these C^{1+} hyperbolic diffeomorphisms $g \in \mathcal{T}(f, \Lambda)$.*

A C^{1+} hyperbolic diffeomorphism $g \in \mathcal{T}(f, \Lambda)$ determines a unique pair $(\mathcal{S}(g, s), \mathcal{S}(g, u))$ of C^{1+} stable and unstable self-renormalizable structures (see Lemma 4.5). By Lemma 8.7, for $\delta_s > 0$ and $\delta_u > 0$, the pair $(\mathcal{S}(g, s), \mathcal{S}(g, u))$ of self-renormalizable structures determines a unique pair of natural geometric measures $(\mu_{\mathcal{S}(g,s),\delta_s}, \mu_{\mathcal{S}(g,u),\delta_u})$ corresponding to a unique pair of Gibbs measures $(\nu_{\mathcal{S}(g,s),\delta_s}, \nu_{\mathcal{S}(g,u),\delta_u})$. Furthermore, by Theorem 8.15, the self-renormalizable structures $(\mathcal{S}(g, s), \mathcal{S}(g, u))$ determine a pair of measure ratio functions $(\rho_{\mathcal{S}(g,s),\delta_s}, \rho_{\mathcal{S}(g,u),\delta_u})$ of $(\nu_{\mathcal{S}(g,s),\delta_s}, \nu_{\mathcal{S}(g,u),\delta_u})$.

Lemma 10.5. *Let g be a C^{1+} hyperbolic diffeomorphism contained in $\mathcal{T}(f, \Lambda)$. The following statements are equivalent:*

(i) g has a natural geometric measure $\mu_{g,\delta_s,\delta_u}$;
(ii) g is a C^{1+} realization of a Gibbs measure $\nu_{g,\delta_s,\delta_u}$;
(iii) $\nu_{\mathcal{S}(g,s),\delta_s} = \nu_{\mathcal{S}(g,u),\delta_u}$;
(iv) The s-measure ratio function $\rho_{\mathcal{S}(g,s),\delta_s}$ is dual to the u-measure ratio function $\rho_{\mathcal{S}(g,u),\delta_u}$.

Furthermore, if g has a natural geometric measure $\mu_{g,\delta_s,\delta_u}$, then $(\pi_s)_*\mu_{g,\delta_s,\delta_u} = \mu_{\mathcal{S}(g,s),\delta_s}$ and $(\pi_u)_*\mu_{g,\delta_s,\delta_u} = \mu_{\mathcal{S}(g,u),\delta_u}$.

Proof. By Theorem 8.17, (iii) is equivalent to (iv). By definition if g is a C^{1+} realization of a Gibbs measure $\nu_{g,\delta_s,\delta_u}$, then $\mu_{g,\delta_s,\delta_u} = i_*\nu_{g,\delta_s,\delta_u}$ is a natural geometric measure of g, and so (ii) implies (i). Let us prove first that (i) implies (ii) and (iii), and secondly that (iii) implies (i). Then, the last paragraph of this lemma follows from (10.10) below which ends the proof.
(i) implies (ii) and (iii). Let $\mu_{g,\delta_s,\delta_u}$ be the natural geometric measure of g. Since the stable and unstable lamination atlases $\mathcal{A}_s(g,\rho)$ and $\mathcal{A}_u(g,\rho)$ of g are C^{1+} foliated (see § 1.7) and by construction of the C^{1+} train-track atlases $\mathcal{B}_s(g,\rho)$ and $\mathcal{B}_u(g,\rho)$, in § 4.2, we obtain that there is $\kappa_1 \geq 1$ with the property that, (for $\iota = s$ and u) and for every ι-leaf n_ι-cylinder I,

$$\kappa_1^{-1}|I|_\rho \leq |I'|_j \leq \kappa_1 |I|_\rho \tag{10.8}$$

where $I' = \pi_{\mathbf{B}^\iota}(I)$, where $|I'|_j$ is measured in any chart $j \in \mathcal{B}_\iota(g,\rho)$ and where $|I|_\rho$ is the length in the Riemannian metric ρ of the minimal full ι-leaf containing I. Let I'_Λ be the $(1, n_\iota)$-rectangle in Λ such that $\pi_{\mathbf{B}^\iota}(I'_\Lambda) = I'$. Noting that $(\pi_{\mathbf{B}^\iota})_*\mu_{g,\delta_s,\delta_u}(I') = \mu_{g,\delta_s,\delta_u}(I'_\Lambda)$, by (10.7) and (10.8), there is $\kappa_2 \geq 1$ such that

$$\kappa_2^{-1} \leq \frac{(\pi_{\mathbf{B}^\iota})_* \mu_{g,\delta_s,\delta_u}(I')}{|I'|_j^{\delta_\iota} e^{-n_\iota P_\iota}} \leq \kappa_2 , \tag{10.9}$$

for every n_ι-cylinder I' on the train-track. By Lemma 8.7, the natural geometric measure determined by the C^{1+} self-renormalizable structure $\mathcal{S}(g,\iota)$ and by $\delta_\iota > 0$ is uniquely determined by (10.9). Hence,

$$(\pi_{\mathbf{B}^s})_* \mu_{g,\delta_s,\delta_u} = \mu_{\mathcal{S}(g,s),\delta_s} \quad \text{and} \quad (\pi_{\mathbf{B}^u})_* \mu_{g,\delta_s,\delta_u} = \mu_{\mathcal{S}(g,u),\delta_u} . \tag{10.10}$$

Therefore, the Gibbs measures $\nu_{\mathcal{S}(g,s),\delta_s}$ and $\nu_{\mathcal{S}(g,u),\delta_u}$ on Θ are equal which proves (iii), and $\mu_{g,\delta_s,\delta_u} = i_* \nu_{\mathcal{S}(g,s),\delta_s} = i_* \nu_{\mathcal{S}(g,u),\delta_u}$ which proves (ii).
(iii) implies (i). Let us denote $\nu_{\mathcal{S}(g,s),\delta_s} = \nu_{\mathcal{S}(g,u),\delta_u}$ by ν. Let $\mu = i_*\nu$. For $\iota \in \{s,u\}$, by definition of a C^{1+} realization of a Gibbs measure as a self-renormalizable structure $\mathcal{S}(g,\delta_\iota)$, for every ι-leaf n_ι-cylinder I_ι, there is $\kappa_3 \geq 1$ such that

$$\kappa_3^{-1} \leq \frac{\mu_\iota(I'_\iota)}{|I'_\iota|_j^{\delta_\iota} e^{-n_\iota P_\iota}} \leq \kappa_3 ,$$

where $I'_\iota = \pi_{\mathbf{B}^\iota}(I)$ and $|I'_\iota|_j$ is measured in any chart $j \in \mathcal{B}_\iota(g,\rho)$. Hence, by (10.8), for $\iota = s$ and u, we obtain that

$$\mu_\iota(I'_\iota) = \mathcal{O}\left(|I_\iota|_\rho^{\delta_\iota} e^{-n_\iota P_\iota}\right) . \tag{10.11}$$

Let R be the rectangle $[I_s, I_u]$. By Lemma 8.11,

$$\mu(R) = \int_{I'_{\iota'}} \rho_{\iota,\xi}(R : M) \mu_{\iota'}(d\xi) ,$$

where M is the Markov rectangle containing R. By Lemma 8.7 (i) and (ii), we get that $\rho_{\iota,\xi}(R : M) = \mathcal{O}(\mu_\iota(I'_\iota))$ for every $\xi \in \pi_{\mathbf{B}^{\iota'}}(R)$. Hence

$$\mu(R) = \mathcal{O}(\mu_s(I'_s)\mu_u(I'_u)) . \tag{10.12}$$

Putting together (10.11) and (10.12), we get

$$\mu(R) = \mathcal{O}\left(|I_u|_\rho^{\delta_u} |I_s|_\rho^{\delta_s} e^{-n_s P_s - n_u P_u}\right) \tag{10.13}$$

and so μ is a natural geometric measure. □

Lemma 10.6. *The map $g \to (\mathcal{S}(g,s), \mathcal{S}(g,u))$ gives a one-to-one correspondence between C^{1+} conjugacy classes of hyperbolic diffeomorphisms contained in $\mathcal{T}(\nu, \delta_s, P_s, \delta_u, P_u)$ and pairs of C^{1+} self-renormalizable structures contained in $\mathcal{D}^s(\nu, \delta_s, P_s) \times \mathcal{D}^u(\nu, \delta_u, P_u)$.*

Proof. By Lemma 10.5, if $g \in \mathcal{T}(\nu, \delta_s, P_s, \delta_u, P_u)$, then, for $\iota \in \{s,u\}$, $\mathcal{S}(g,\iota) \in \mathcal{D}^\iota(\nu, \delta_\iota, P_\iota)$. Conversely, by Lemma 4.5, a pair $(\mathcal{S}_s, \mathcal{S}_u) \in \mathcal{D}^s(\nu, \delta_s, P_s) \times \mathcal{D}^u(\nu, \delta_u, P_u)$ determines a C^{1+} hyperbolic diffeomorphism g such that $\mathcal{S}(g,s) = \mathcal{S}_s$ and $\mathcal{S}(g,u) = \mathcal{S}_u$ and $\nu_{\mathcal{S}(g,s),\delta_s} = \nu_{\mathcal{S}(g,u),\delta_u} = \nu$. Therefore, by Lemma 10.5, we obtain that g is a C^{1+} realization of the Gibbs measure ν.
□

Lemma 10.7. *(Dual-rigidity) Let $\mathbf{B}^{\iota'}$ be a no-gap train-track (and so $\delta_{\iota'} = 1$ and $P_{\iota'} = 0$). For every $\delta_\iota > 0$ and every C^{1+} ι-self-renormalizable structure $\mathcal{S}_\iota$ there is a unique C^{1+} ι'-self-renormalizable structure $\mathcal{S}_{\iota'}$ such that the C^{1+} hyperbolic diffeomorphism g corresponding to the pair $(\mathcal{S}_s, \mathcal{S}_u) = (\mathcal{S}(g,s), \mathcal{S}(g,u))$ has a natural geometric measure $\mu_{g,\delta_s,\delta_u}$. Furthermore, $\mu_{\mathcal{S}_s,\delta_s} = (\pi_{\mathbf{B}^s})_* \mu_{g,\delta_s,\delta_u}$ and $\mu_{\mathcal{S}_u,\delta_u} = (\pi_{\mathbf{B}^u})_* \mu_{g,\delta_s,\delta_u}$.*

Proof. By Lemma 8.7, a C^{1+} self-renormalizable structure $\mathcal{S}_\iota$ and $\delta_\iota > 0$ determine a unique Gibbs measure $\nu = \nu_{\mathcal{S}_\iota,\delta_\iota}$ and $P_\iota \in \mathbb{R}$ such that $\mathcal{S}_\iota \in \mathcal{D}^\iota(\nu, \delta_\iota, P_\iota)$ is a C^{1+} realization of ν. By Theorem 8.15, the C^{1+} self-renormalizable structure $\mathcal{S}_\iota$ determines an ι-measure ratio function $\rho_{\mathcal{S}_\iota,\delta_\iota}$ for the Gibbs measure ν. By Theorem 8.17, the ι-measure ratio function $\rho_{\mathcal{S}_\iota,\delta_\iota}$ determines a unique ι'-measure ratio function $\rho_{\iota'}$ of ν on Θ. By Lemma 10.1, there is a unique C^{1+} self-renormalizable structure $\mathcal{S}_{\iota'}$, with ι'-measure ratio function $\rho_{\mathcal{S}_{\iota'},1} = \rho_{\iota'}$, which is a C^{1+} realization of the Gibbs measure ν. By Lemma 4.5, the pair $(\mathcal{S}_s, \mathcal{S}_u)$ determines a C^{1+} hyperbolic diffeomorphism g such that $\mathcal{S}(g,s) = \mathcal{S}_s$ and $\mathcal{S}(g,u) = \mathcal{S}_u$. Hence, $\nu_{\mathcal{S}(g,s),\delta_s} = \nu$ and $\nu_{\mathcal{S}(g,u),\delta_u} = \nu$ which implies that $\nu_{\mathcal{S}(g,s),\delta_s} = \nu_{\mathcal{S}(g,u),\delta_u}$. Therefore, by Lemma 10.5, g is a C^{1+} realization of the Gibbs measure ν with natural geometric measure $\mu_{g,\delta_s,\delta_u} = i_*\nu$. Thus, $\mu_{\mathcal{S}_s,\delta_s} = (\pi_{\mathbf{B}^s})_* \mu_{g,\delta_s,\delta_u}$ and $\mu_{\mathcal{S}_u,\delta_u} = (\pi_{\mathbf{B}^u})_* \mu_{g,\delta_s,\delta_u}$. □

Recall the definition of the maps $g \to (\mathcal{S}(g,s), \mathcal{S}(g,u))$ and $\mathcal{S}(g,\iota) \to (\gamma_{\mathcal{S}(g,\iota)}, J_{\mathcal{S}(g,\iota),\delta_\iota})$ for ι equal to s and u.

Theorem 10.8. *(Flexibility) Let $\mathbf{B}_\iota$ be a gap train-track. Let ν be a Gibbs measure determining an ι-measure ratio function. Let $\delta_\iota > 0$ and $P_\iota \in \mathbb{R}$ be such that $\mathcal{JG}^\iota(\nu, \delta_\iota, P_\iota) \neq \emptyset$.*

(i) (Smale horseshoes) Let $\delta_{\iota'} > 0$ and $P_{\iota'} \in \mathbb{R}$ be such that $\mathcal{JG}^{\iota'}(\nu, \delta_{\iota'}, P_{\iota'}) \neq \emptyset$. The map

$$g \to (\gamma_{\mathcal{S}(g,s)}, J_{\mathcal{S}(g,s),\delta_s}, \gamma_{\mathcal{S}(g,u)}, J_{\mathcal{S}(g,u),\delta_u})$$

gives a one-to-one correspondence between C^{1+} conjugacy classes of hyperbolic diffeomorphisms in $\mathcal{T}(\nu, \delta_s, P_s, \delta_u, P_u)$ and pairs of stable and unstable cocycle-gap pairs in $\mathcal{JG}^s(\nu, \delta_s, P_s) \times \mathcal{JG}^u(\nu, \delta_u, P_u)$.
(ii) (Codimension one attractors and repellors) Let $\delta_{\iota'} = 1$ and $P_{\iota'} = 0$. The map $g \to (\gamma_{\mathcal{S}(g,\iota)}, J_{\mathcal{S}(g,\iota),\delta_\iota})$ gives a one-to-one correspondence between C^{1+} conjugacy classes of hyperbolic diffeomorphisms in $\mathcal{T}(\nu, \delta_s, P_s, \delta_u, P_u)$ and pairs of stable and unstable cocycle-gap pairs in $\mathcal{JG}^\iota(\nu, \delta_\iota, P_\iota)$.

Proof. Statement (i) follows from putting together the results of lemmas 10.3 and 10.6. Statement (ii) follows as statement (i) using the fact that, by Lemma 10.1, the C^{1+} self-renormalizable structure $\mathcal{S}(g,\iota)$ uniquely determines $\mathcal{S}(g,\iota')$ in this case. □

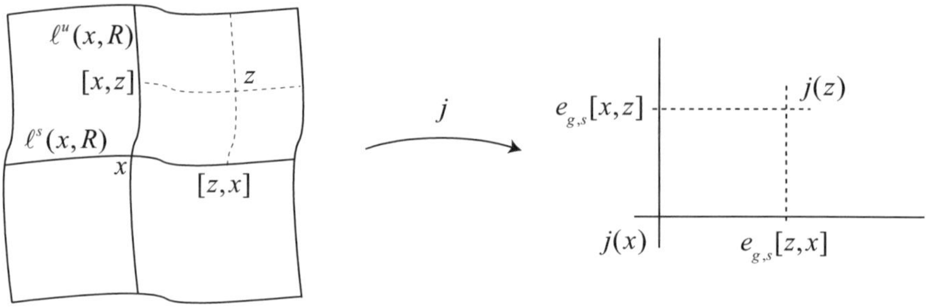

Fig. 10.1. An orthogonal chart.

Lemma 10.9. *Let g_1 and g_2 be C^{1+} hyperbolic diffeomorphisms. The following statements are equivalent:*

(i) The diffeomorphism g_1 is Lipschitz conjugate to g_2.
(ii) For ι equal to s and u, $\mathcal{S}(g_1, \iota)$ is Lipschitz conjugate to $\mathcal{S}(g_2, \iota)$.
(iii) For ι equal to s and u, the solenoid functions $s_{g_1,\iota}$ and $s_{g_2,\iota}$ are in the same bounded equivalence class (Definition 10).

Proof. Proof that (i) is equivalent to (ii). For all $x \in \Lambda$, let A be a small open set of M containing x, and let R be a rectangle (not necessarily a Markov rectangle) such that $A \cap \Lambda \subset R$. We construct an orthogonal chart $j : R \to \mathbb{R}^2$ as follows. Let $e_{g,s} : \ell^s(x, R) \to \mathbb{R}$ be a chart contained in $\mathcal{A}^s(g, \rho)$ and $e_{g,u} : \ell^u(x, R) \to \mathbb{R}$ be a chart contained in $\mathcal{A}^u(g, \rho)$. The orthogonal chart j on R is now given by $j(z) = (e_{g,s}[z, x]), e_{g,u}[x, z])) \in \mathbb{R}^2$ (see Figure 10.1). By Pinto and Rand [163], the orthogonal chart $j : R \to \mathbb{R}^2$ has an extension $\hat{j} : B \to \mathbb{R}^2$ to an open set B of the surface such that $\hat{j}$ is C^{1+} compatible with the charts in the C^{1+} structure $\mathcal{C}(g)$ of the surface M. Hence, using the orthogonal charts, any two C^{1+} hyperbolic diffeomorphisms g_1 and g_2 are Lipschitz conjugate if, and only if the charts in $\mathcal{A}^\iota(g_1, \rho_1)$ are bi-Lipschitz compatible with the charts in $\mathcal{A}^\iota(g_2, \rho_2)$ for ι equal to s and u. By construction of the train-track atlases $\mathcal{B}^\iota(g_1, \rho_1)$ and $\mathcal{B}^\iota(g_2, \rho_2)$ from the lamination atlases $\mathcal{A}^\iota(g_1, \rho_1)$ and $\mathcal{A}^\iota(g_2, \rho_2)$, the charts in $\mathcal{A}^\iota(g_1, \rho_1)$ are bi-Lipschitz compatible with the charts in $\mathcal{A}^\iota(g_2, \rho_2)$ if, and only if, the charts in $\mathcal{B}^\iota(g_1, \rho_1)$ are bi-Lipschitz compatible with the charts in $\mathcal{B}^\iota(g_2, \rho_2)$. Hence, the C^{1+} hyperbolic diffeomorphisms g_1 and g_2 are Lipschitz conjugate if, and only if, for ι equal to s and u, the corresponding C^{1+} self-renormalizable structures $\mathcal{S}(g_1, \iota)$ and $\mathcal{S}(g_2, \iota)$ are Lipschitz conjugate. Therefore, statement (i) is equivalent to statement (ii).
Proof that (ii) is equivalent to (iii). Follows from Lemma 10.4. □

Lemma 10.10. *Let $\delta_s > 0$, $\delta_u > 0$ and $P_s, P_u \in \mathbb{R}$.*

(i) A C^{1+} hyperbolic diffeomorphism g is contained in $\mathcal{T}(\nu, \delta_s, P_s, \delta_u, P_u)$ if, and only if, for ι equal to s and u, the ι-solenoid function $\sigma_{g,\iota}$ is

contained in the (δ_ι, P_ι)-bounded solenoid equivalence class of ν (see Definition 27).

(ii) If $g_1 \in \mathcal{T}(\nu_1, \delta_s, P_s, \delta_u, P_u)$ and $g_2 \in \mathcal{T}(\nu_2, \delta_s, P_s, \delta_u, P_u)$ are C^{1+} hyperbolic diffeomorphisms, then g_1 is Lipschitz conjugate to g_2 if, and only if, $\nu_1 = \nu_2$.

Proof. Proof of (i). By Lemma 4.5, the C^{1+} hyperbolic diffeomorphism g determines a unique pair $(\mathcal{S}(g,s), \mathcal{S}(g,u))$ of C^{1+} self-renormalizable structures such that $\sigma_{g,s} = \sigma_{\mathcal{S}(g,s),s}$ and $\sigma_{g,u} = \sigma_{\mathcal{S}(g,u),u}$. By Lemma 10.6, $g \in \mathcal{T}(\nu, \delta_s, P_s, \delta_u, P_u)$ if, and only if, $(\mathcal{S}(g,s), \mathcal{S}(g,u)) \in \mathcal{D}^s(\nu, \delta_s, P_s) \times \mathcal{D}^u(\nu, \delta_u, P_u)$. By Lemma 10.3 (ii), for ι equal to s and u, $\mathcal{S}(g,\iota) \in \mathcal{D}^\iota(\nu, \delta_\iota, P_\iota)$ if, and only if, $\mathcal{S}(g,\iota)$ is contained in the (δ_ι, P_ι)-bounded solenoid equivalence class of ν which ends the proof.
Proof of (ii). By Lemma 10.6, $g_1 \in \mathcal{T}(\nu_1, \delta_s, P_s, \delta_u, P_u)$ and $g_2 \in \mathcal{T}(\nu_2, \delta_s, P_s, \delta_u, P_u)$ if, and only if, for ι equal to s and u, $\mathcal{S}(g_1,\iota) \in \mathcal{D}^\iota(\nu_1, \delta_\iota, P_\iota)$ and $\mathcal{S}(g_2,\iota) \in \mathcal{D}^\iota(\nu_2, \delta_\iota, P_\iota)$. By Lemma 10.4, $\mathcal{S}(g_1,\iota)$ and $\mathcal{S}(g_2,\iota)$ are Lipschitz conjugate if, and only if, $\nu_1 = \nu_2$. Since, by Lemma 10.9, g_1 and g_2 are Lipschitz conjugate if, and only if, for ι equal to s and u, $\mathcal{S}(g_1,\iota)$ and $\mathcal{S}(g_2,\iota)$ are Lipschitz conjugate, we get that g_1 and g_2 are Lipschitz conjugate if, and only if, $\nu_1 = \nu_2$. □

10.3 Invariant Hausdorff measures

Let $\mathcal{S}_\iota$ be a C^{1+} ι self-renormalizable structure. By Remark 8.9, a natural geometric measure $\mu_{\mathcal{S}_\iota,\delta_\iota}$ with pressure $P(\mathcal{S}_\iota, \delta_\iota) = 0$ is an invariant measure absolutely continuous with respect to the Hausdorff measure of $\mathbf{B}^\iota$ and δ_ι is the Hausdorff dimension of $\mathbf{B}^\iota$ with respect to the charts of $\mathcal{S}_\iota$. Let us denote $\mathcal{D}^\iota(\nu, \delta_\iota, 0)$ and $\mathcal{JG}^\iota(\nu, \delta_\iota, 0)$ respectively by $\mathcal{D}^\iota(\nu, \delta_\iota)$ and $\mathcal{JG}^\iota(\nu, \delta_\iota)$. By Lemma 8.7, for every C^{1+} ι self-renormalizable structure $\mathcal{S}_\iota$ there is a unique Gibbs measure $\nu_{\mathcal{S}_\iota}$ such that $\mathcal{S}_\iota \in \mathcal{D}^\iota(\nu, \delta_\iota)$. Using Lemma 10.6, we obtain that the sets $[\nu] \subset \mathcal{T}_{f,\Lambda}(\delta_s, \delta_u)$ defined in the introduction are equal to the sets $\mathcal{T}(\nu, \delta_s, 0, \delta_u, 0)$ (see Definition 28).

Theorem 10.11. *The map $g \to (\mathcal{S}_s(g), \mathcal{S}_u(g))$ gives a 1-1 correspondence between C^{1+} conjugacy classes in $[\nu] \subset \mathcal{T}_{f,\Lambda}(\delta_s, \delta_u)$ and pairs in $\mathcal{D}^s(\nu, \delta_s) \times \mathcal{D}^u(\nu, \delta_u)$.*

Hence, if $g \in \mathcal{T}_{f,\Lambda}(\delta_s, \delta_u)$, then $\delta(\mathcal{S}_s(g)) = \delta_s$ and $\delta(\mathcal{S}_u(g)) = \delta_u$. Let $\mathcal{S}_\iota$ be a C^{1+} ι self-renormalizable structure. If $\delta(\mathcal{S}_\iota) = 1$ we call $\mathbf{B}^\iota$ a *no-gap train-track.* If $0 < \delta(\mathcal{S}_\iota) < 1$ we call $\mathbf{B}^\iota$ a *gap train-track.* Let ι' denote the element of $\{s,u\}$ which is not $\iota \in \{s,u\}$.

Proof of Theorem 10.11. Theorem 10.11 follows from Lemma 10.6. □

Theorem 10.12. *There is a natural map $g \to (\mathcal{S}_s(g), \mathcal{S}_u(g))$ which gives a one-to-one correspondence between C^{1+} conjugacy classes in $\mathcal{T}(f,\Lambda)$ and pairs of stable and unstable C^{1+} self-renormalizable structures.*

Hence, for a pair $(\mathcal{S}_s, \mathcal{S}_u)$ of C^{1+} self-renormalizable structures to be realizable by a C^{1+} hyperbolic diffeomorphism in $\mathcal{T}(f,\Lambda)$, the unstable C^{1+} self-renormalizable structure does not impose any restriction in the stable C^{1+} self-renormalizable structure, and *vice-versa.* The same is no longer true if we ask $g \in \mathcal{T}(f,\Lambda)$ to be a C^{1+}-Hausdorff realization of a Gibbs measure as we describe in the next section.

Proof of Theorem 10.12. Theorem 10.12 follows from Lemma 4.5. □

Theorem 10.13. *(i) Any two elements of $[\nu] \subset \mathcal{T}_{f,\Lambda}(\delta_s, \delta_u)$ have the same set of stable and unstable eigenvalues and these sets are a complete invariant of $[\nu]$ in the sense that if $g_1, g_2 \in \mathcal{T}_{f,\Lambda}(\delta_s, \delta_u)$ have the same eigenvalues if, and only if, they are in the same subset $[\nu]$.*

(ii) The map $\nu \to [\nu] \subset \mathcal{T}_{f,\Lambda}(\delta_s, \delta_u)$ gives a $1-1$ correspondence between C^{1+}-Hausdorff realizable Gibbs measures ν and Lipschitz conjugacy classes in $\mathcal{T}_{f,\Lambda}(\delta_s, \delta_u)$.

Proof. Proof of statement (i). By Lemma 10.10 (ii), the sets $[\nu] \subset \mathcal{T}_{f,\Lambda}(\delta_s, \delta_u)$ are Lipschitz conjugacy classes in $\mathcal{T}_{f,\Lambda}(\delta_s, \delta_u)$, and the map $\nu \to \mathcal{T}(\nu, \delta_s, \delta_u)$ is injective. If $g \in \mathcal{T}_{f,\Lambda}(\delta_s, \delta_u)$, then g has a natural geometric measure $\mu_{g,\delta_s,\delta_u}$ with pressures $P_s(g, \delta_s, \delta_u)$ and $P_u(g, \delta_s, \delta_u)$ equal to zero. By Lemma 10.5, there is a Gibbs measure $\nu = \nu_{g,\delta_s,\delta_u}$ on Θ such that $i_*\nu = \mu_{g,\delta_s,\delta_u}$ and so $g \in [\nu] \subset \mathcal{T}_{f,\Lambda}(\delta_s, \delta_u)$. Hence, the map $\nu \to \mathcal{T}(\nu, \delta_s, \delta_u)$ is surjective into the Lipschitz conjugacy classes in $\mathcal{T}_{f,\Lambda}(\delta_s, \delta_u)$.
Proof of statement (ii). By Theorem 11.3 (ii), the set of stable and unstable eigenvalues of all periodic orbits of a C^{1+} hyperbolic diffeomorphisms $g \in \mathcal{T}_{f,\Lambda}(\delta_s, \delta_u)$ is a complete invariant of each Lipschitz conjugacy class, and by statement (i) of this lemma the sets $\mathcal{T}(\nu, \delta_s, \delta_u)$ are the Lipschitz conjugacy classes in $\mathcal{T}_{f,\Lambda}(\delta_s, \delta_u)$. □

Theorem 10.14. *Let $\mathbf{B}^s$ and $\mathbf{B}^u$ be the stable and unstable train-tracks determined by a C^{1+} hyperbolic diffeomorphism (f,Λ). The set $\mathcal{D}^\iota(\nu, \delta_\iota)$ is non-empty if, and ony if, the ι-measure solenoid function $\sigma_\nu : \mathrm{msol}^\iota \to \mathbb{R}^+$ of the Gibbs measure ν has the following properties:*

(i) If $\mathbf{B}^\iota$ and $\mathbf{B}^{\iota'}$ are no-gap train-tracks, then σ_ν has a non-vanishing Hölder continuous extension to the closure of msol^ι *satisfying the boundary condition.*
(ii) If $\mathbf{B}^\iota$ is a no-gap train-track and $\mathbf{B}^{\iota'}$ is a gap train-track, then σ_ν has a non-vanishing Hölder continuous extension to the closure of msol^ι.

(iii) If $\mathbf{B}^\iota$ is a gap train-track and $\mathbf{B}^{\iota'}$ is a no-gap train-track, then σ_ν has a non-vanishing Hölder continuous extension to the closure of msol^ι *satisfying the cylinder-cylinder condition.*

(iv) If $\mathbf{B}^\iota$ and $\mathbf{B}^{\iota'}$ are gap train-tracks, then σ_ν does not have to satisfy any extra-condition.

Furthermore, $\mathcal{D}^\iota(\nu, \delta_\iota) \neq \emptyset$ if, and only if, $\mathcal{D}^{\iota'}(\nu, \delta_{\iota'}) \neq \emptyset$

Proof. We will separate the proof in three parts. In part (i), we prove that if $\mathcal{S}_\iota \in \mathcal{D}^\iota(\nu, \delta_\iota)$, then $\sigma_{\nu,\iota}$ satisfies the properties indicated in Theorem 10.14. In part (ii), we prove the converse of part (i). In part (iii), we prove that $\mathcal{D}^\iota(\nu, \delta_\iota) \neq \emptyset$ if, and ony if, $\mathcal{D}^{\iota'}(\nu, \delta_{\iota'}) \neq \emptyset$.

Part (i). Let $\mathcal{S}_\iota \in \mathcal{D}^\iota(\nu, \delta_\iota)$. By Theorem 8.15, $\mathcal{S}_\iota$ and δ_ι determine a unique ι-measure ratio function $\rho_{\nu,\iota}$ of the Gibbs measure ν. Hence, the function $\rho_{\nu,\iota}|\mathrm{msol}^{\iota'}$ is the ι-measure solenoid function $\sigma_{\nu,\iota}$ of ν and, by Lemma 8.5, $\sigma_{\nu,\iota}$ satisfies the properties indicated in Theorem 10.14.

Part (ii). Conversely, if ν has an ι-solenoid function $\sigma_{\nu,\iota}$ satisfying the properties indicated in Theorem 10.14, by lemmas 8.2 and 8.5, $\sigma_{\nu,\iota}$ determines a unique ι-measure ratio function $\rho_{\nu,\iota}$ of ν. If $\mathbf{B}^\iota$ is a no-gap train-track, by Lemma 10.1, there is a C^{1+} self-renormalizable structure $\mathcal{S}_\iota \in \mathcal{D}^\iota(\nu, \delta_\iota)$ with $\delta_\iota = 1$. If $\mathbf{B}^\iota$ is a gap train-track, then, by Remark 9.10, the set $\mathcal{JG}^\iota(\nu, \delta_\iota)$ is non-empty (in fact it is an infinite dimensional space). Hence, by Lemma 10.3, the set $\mathcal{D}(\nu, \delta_\iota)$ is also non-empty which ends the proof.

Part (iii). To prove that $\mathcal{D}^\iota(\nu, \delta_\iota) \neq \emptyset$ if, and ony if, $\mathcal{D}^{\iota'}(\nu, \delta_{\iota'}) \neq \emptyset$, it is enough to prove one of the implications. Let us prove that if $\mathcal{D}^\iota(\nu, \delta_\iota) \neq \emptyset$, then $\mathcal{D}^{\iota'}(\nu, \delta_{\iota'}) \neq \emptyset$, Let $\mathcal{S}_\iota \in \mathcal{D}^\iota(\nu, \delta_\iota)$. By Theorem 8.15, $\mathcal{S}_\iota$ and δ_ι determine a unique ι-measure ratio function $\rho_{\nu,\iota}$ of the Gibbs measure ν. By Theorem 8.17, the ι-measure ratio function $\rho_{\nu,\iota}$ determines a unique dual ι'-measure ratio function $\rho_{\nu,\iota'}$ of ν. Hence, the function $\rho_{\nu,\iota'}|\mathrm{msol}^{\iota'}$ is the ι-measure solenoid function $\sigma_{\nu,\iota'}$ of ν and, by Lemma 8.5, $\sigma_{\nu,\iota'}$ satisfies the properties indicated in Theorem 10.14. Now the proof follows as in part (ii), with ι changed by ι', which shows that $\sigma_{\nu,\iota'}$ determines a non-empty set $\mathcal{D}^{\iota'}(\nu, \delta_{\iota'})$. □

Theorem 10.15. *(Anosov diffeomorphisms) Suppose that f is a C^{1+} Anosov diffeomorphism of the torus Λ. Fix a Gibbs measure ν on Θ. Then, the following statements are equivalent:*

(i) The set ν, $[\nu] \subset \mathcal{T}_{f,\Lambda}(1,1)$ is non-empty and is precisely the set of $g \in \mathcal{T}_{f,\Lambda}(1,1)$ such that (g, Λ_g, ν) is a C^{1+} Hausdorff realization. In this case $\mu = (i_g)_*\nu$ is absolutely continuous with respect to Lesbegue measure.

(ii) The stable measure solenoid function $\sigma_{\nu,s} : \mathrm{msol}^s \to \mathbb{R}^+$ has a non-vanishing Hölder continuous extension to the closure of msol^s *satisfying the boundary condition.*

(iii) The unstable measure solenoid function $\sigma_{\nu,u} : \mathrm{msol}^u \to \mathbb{R}^+$ has a non-vanishing Hölder continuous extension to the closure of msol^s *satisfying the boundary condition.*

The treatment of codimension one attractors has a number of extra-dificulties due to the fact that the invariant set Λ is locally a Cartesian product of a Cantor set with an interval but the stable and unstable measure solenoid functions are built in a similar way to the construction for Anosov diffeomorphisms. In the case of codimension one attractors, the continuous extension of the stable measure solenoid functions have to satisfy the *cylinder-cylinder condition* for the corresponding Gibbs measures to be C^{1+}-Hausdorff realizable (see § 8.1) . The cylinder-cylinder condition, like the boundary condition, consists of a finite set of simple algebraic equalities and is needed because the Markov rectangles have common boundaries along the stable laminations. Hence, the cylinder-cylinder condition just applies to the stable measure function.

Theorem 10.16. *(Codimension one attractors) Suppose that f is a C^{1+} surface diffeomorphism and Λ is a codimension one hyperbolic attractor. Fix a Gibbs measure ν on Θ. Then, the following statements are equivalent:*

(i) For all $0 < \delta_s < 1$, $[\nu] \subset \mathcal{T}_{f,\Lambda}(\delta_s, 1)$ is non-empty and is precisely the set of $g \in \mathcal{T}_{f,\Lambda}(\delta_s, 1)$ such that (g, Λ_g, ν) is a C^{1+} Hausdorff realization. In this case $\mu = (i_g)_*\nu$ is absolutely continuous with respect to the Hausdorff measure on Λ_g.
(ii) The stable measure solenoid function $\sigma_{\nu,s} : \mathrm{msol}^s \to \mathbb{R}^+$ has a non-vanishing Hölder continuous extension to the closure of msol^s *satisfying the cylinder-cylinder condition.*
(iii) The unstable measure solenoid function $\sigma_{\nu,u} : \mathrm{msol}^u \to \mathbb{R}^+$ has a non-vanishing Hölder continuous extension to the closure of msol^u.

In the case of Smale horseshoes, there are no extra conditions that the measure solenoid functions have to satisfy for the corresponding Gibbs measures to be C^{1+}-Hausdorff realizable.

Proof of Theorem 10.15 and Theorem 10.16. Proof that statement (i) implies statements (ii) and (iii). If $g \in [\nu] \subset \mathcal{T}_{f,\Lambda}(\delta_s, \delta_u)$, by Lemma 10.6, the sets $\mathcal{D}^s(\nu, \delta_s)$ and $\mathcal{D}^u(\nu, \delta_u)$ are both non-empty. Hence, by Theorem 10.14, the stable measure solenoid function of the Gibbs measure ν satisfies (ii) and the unstable measure solenoid function of the Gibbs measure ν satisfies (iii).
Proof that statement (ii) implies statement (i), and that statement (iii) implies statement (i). By Theorem 10.14, the properties of the the ι-solenoid function $\sigma_{\nu,\iota}$ indicated in this theorem imply that $\mathcal{D}^\iota(\nu, \delta_\iota) \neq \emptyset$. Again, by Theorem 10.14 and $\mathcal{D}^{\iota'}(\nu, \delta_{\iota'}) \neq \emptyset$. Hence, by Lemma 10.6, the set $[\nu] \subset \mathcal{T}_{f,\Lambda}(\delta_s, \delta_u)$ is non-empty. Therefore, every $g \in \mathcal{T}(\nu, \delta_s, \delta_u)$ is a C^{1+}-Hausdorff realization of ν which ends the proof. □

Theorem 10.17. *(Smale horseshoes) Suppose that (f, Λ) is a Smale horseshoe and ν is a Gibbs measure on Θ. Then, for all $0 < \delta_s, \delta_u < 1$, $[\nu] \subset \mathcal{T}_{f,\Lambda}(\delta_s, \delta_u)$ is non-empty and is precisely the set of $g \in \mathcal{T}_{f,\Lambda}(\delta_s, \delta_u)$*

such that (g, Λ_g, ν) is a C^{1+} Hausdorff realization. In this case $\mu = (i_g)_*\nu$ is absolutely continuous with respect to the Hausdorff measure on Λ_g.

Proof. Let ν be a Gibbs measure. By Theorem 10.14, the set $\mathcal{D}^s(\nu, \delta_s)$ and $\mathcal{D}^u(\nu, \delta_u)$ are both non-empty. Hence, by Lemma 10.6, the set $\mathcal{T}(\nu, \delta_s, \delta_u)$ is also non-empty. Therefore, every $g \in \mathcal{T}(\nu, \delta_s, \delta_u)$ is a C^{1+}-Hausdorff realization of ν which ends the proof. □

10.3.1 Moduli space $\mathcal{SOL}^\iota$

Recall the definiton of the set $\mathcal{SOL}^\iota$ given in § 8.2. By Theorem 10.18, below, the set of all ι-measure solenoid functions σ_ν with the properties indicated in Theorem 10.14 determine an infinite dimensional metric space $\mathcal{SOL}^\iota$ which gives a nice parametrization of all Lipschitz conjugacy classes $\mathcal{D}^\iota(\nu, \delta)$ of C^{1+} self-renormalizable structures $\mathcal{S}_\iota$ with a given Hausdorff dimension δ.

Theorem 10.18. *If $\mathbf{B}^\iota$ is a gap train-track assume $0 < \delta_\iota < 1$ and if $\mathbf{B}^\iota$ is a no-gap train-track assume $\delta_\iota = 1$.*

(i) The map $\mathcal{S} \to \rho_{\mathcal{S},\delta_\iota}$ induces a one-to-one correspondence between the sets $\mathcal{D}^\iota(\nu, \delta_\iota)$ and the elements of $\mathcal{SOL}^\iota$.
(ii) The map $g \to \rho_{\mathcal{S}(g,\iota),\delta_\iota}$ induces a one-to-one correspondence between the sets $[\nu]$ contained in $\mathcal{T}_{f,\Lambda}(\delta_s, \delta_u)$ and the elements of $\mathcal{SOL}^\iota$.

Proof. Proof of (i). If $\mathcal{S} \in \mathcal{D}^\iota(\nu, \delta_\iota)$, then the Hausdorff dimension of $\mathcal{S}$ is δ_ι, and $\mathcal{S}$ determines an ι-measure ratio function $\rho_{\mathcal{S},\delta_\iota} = \rho_{\nu,\iota}$ which does not depend upon $\mathcal{S} \in \mathcal{D}^\iota(\nu, \delta_\iota)$. By Lemma 8.5, $\rho_{\nu,\iota}|\text{Msol}^\iota$ is an element of $\mathcal{SOL}^\iota$. Hence, the map $\mathcal{S} \to \rho_{\mathcal{S},\delta_\iota}$ associates to each set $\mathcal{D}^\iota(\nu, \delta)$ a unique element of $\mathcal{SOL}^\iota$. Conversely, let $\hat{\sigma} \in \text{Msol}^\iota$. By Lemma 8.5, $\hat{\sigma}$ determines a unique ι-measure ratio function $\rho_{\hat{\sigma}}$ such that $\rho_{\hat{\sigma}}|\text{Msol}^\iota = \hat{\sigma}$. By Corollary 6.8, the ι-measure ratio function $\rho_{\hat{\sigma}}$ determines a Gibbs measure $\nu_{\hat{\sigma}}$. If $\mathbf{B}^\iota$ is a no-gap train-track, then, by Lemma 10.1, $\rho_{\hat{\sigma}}$ determines a non-empty set $\mathcal{D}^\iota(\nu_{\hat{\sigma}}, \delta_\iota)$. If $\mathbf{B}^\iota$ is a gap train-track, then, by Remark 9.10, the set $\mathcal{JG}^\iota(\nu, \delta_\iota)$ is non-empty and so, by Lemma 10.3, the set $\mathcal{D}^\iota(\nu_{\hat{\sigma}}, \delta_\iota)$ is also non-empty. Therefore, each element $\hat{\sigma} \in \text{Msol}^\iota$ determines a unique non-empty set $\mathcal{D}^\iota(\nu_{\hat{\sigma}}, \delta_\iota)$ of C^{1+} self-renormalizable structures $\mathcal{S}$ with $\rho_{\mathcal{S},\delta_\iota}|\text{Msol}^\iota = \hat{\sigma}$.

Proof of (ii). By Lemma 10.5, if $g \in [\nu]$, then $\mathcal{S}(g, \iota) \in \mathcal{D}^\iota(\nu, \delta_\iota)$ and so, by statement (i) of this lemma, $\rho_{\mathcal{S}(g,\iota),\delta_\iota}|\text{Msol}^\iota$ is an element of $\mathcal{SOL}^\iota$ which does not depend upon $g \in [\nu]$. Conversely, let $\hat{\sigma} \in \text{Msol}^\iota$. By statement (i) of this lemma, $\hat{\sigma}$ determines an ι-measure ratio function $\rho_{\hat{\sigma},\iota}$, and a non-empty set $\mathcal{D}^\iota(\nu_{\hat{\sigma}}, \delta_\iota)$. By Lemma 10.5, $\rho_{\hat{\sigma},\iota}$ determines a unique dual ι'-ratio function $\rho_{\hat{\sigma},\iota'}$ associated to the Gibbs measure $\nu_{\hat{\sigma}}$. Again, by statement (i) of this lemma, $\rho_{\hat{\sigma},\iota'}|\text{Msol}^{\iota'}$ determines a non-empty set $\mathcal{D}^{\iota'}(\nu_{\hat{\sigma}}, \delta_{\iota'})$. By Lemma 10.6, the set $\mathcal{D}^s(\nu_{\hat{\sigma}}, \delta_s) \times \mathcal{D}^u(\nu_{\hat{\sigma}}, \delta_u)$ determines a unique non-empty set $[\nu_{\hat{\sigma}}] \subset \mathcal{T}_{f,\Lambda}(\delta_s, \delta_u)$ of hyperbolic diffeomorphisms $g \in [\nu_{\hat{\sigma}}]$ such that $\rho_{\mathcal{S}(g,\iota),\delta_\iota}|\text{Msol}^\iota = \hat{\sigma}$. □

10.3.2 Moduli space of cocycle-gap pairs

By Lemma 10.4, each set $\mathcal{D}^\iota(\nu,\delta)$ is a Lipschitz conjugacy class. Hence, by Theorem 10.19 proved below, if $\mathbf{B}^\iota$ is a no-gap train-track, then the Lipschitz conjugacy class consists of a single C^{1+} self-renormalizable structure. Furthermore, by Lemma 11.2, the set of eigenvalues of all periodic orbits of $\mathcal{S}_\iota$ is a complete invariant of each set $\mathcal{D}^\iota(\nu,\delta)$.

Theorem 10.19. *Let us suppose that* $\mathcal{D}^\iota(\nu,\delta)\neq\emptyset$.

(i) (Flexibility) If $\mathbf{B}^\iota$ *is a gap train-track, then* $\mathcal{D}^\iota(\nu,\delta)$ *is an infinite dimensional space parametrized by cocycle-gap pairs contained in* $\mathcal{JG}^\iota(\nu,\delta)$.

(ii) (Rigidity) If $\mathbf{B}^\iota$ *is a no-gap train-track, then* $\mathcal{D}^\iota(\nu,1)$ *consists of a single* C^{1+} *self-renormalizable structure.*

Proof. Statement (i) follows from Lemma 10.1. Now, let us prove statement (ii). By Remark 9.10, the set $\mathcal{JG}^\iota(\nu,\delta)$ is an infinite dimensional space, and by Lemma 10.3, the set $\mathcal{D}^\iota(\nu,\delta)$ is parameterized by the cocycle-gap pairs in $\mathcal{JG}^\iota(\nu,\delta)$ which ends the proof. □

Theorem 10.20. *(Rigidity) If* $\delta_\iota=1$, *the mapping* $g\to\mathcal{S}_{\iota'}(g)$ *gives a 1-1 correspondence between* C^{1+} *conjugacy classes in* $[\nu]\subset\mathcal{T}_{f,\Lambda}(\delta_s,\delta_u)$ *and* C^{1+} *self-renormalizable structures in* $\mathcal{D}^{\iota'}(\nu,\delta_{\iota'})$.

Proof. By Lemma 10.6, if $g\in\mathcal{T}(\nu,\delta_s,\delta_u)$, then $\mathcal{S}_{\iota'}(g)\in\mathcal{D}^{\iota'}(\nu,\delta_{\iota'})$. Conversely, let $\mathcal{S}_{\iota'}$ be a C^{1+} self-renormalizable structure contained in $\mathcal{D}^{\iota'}(\nu,\delta_{\iota'})$. By Lemma 10.6, a pair $(\mathcal{S}_\iota,\mathcal{S}_{\iota'})$ determines a C^{1+} hyperbolic diffeomorphism $g\in\mathcal{T}(\nu,\delta_s,\delta_u)$. if, and only if, $\mathcal{S}_{\iota'}\in\mathcal{D}^{\iota'}(\nu,\delta_{\iota'})$. By Theorem 10.14, the set $\mathcal{D}(\nu,\delta_\iota)$ is non-empty. Noting that $\delta_\iota=1$, it follows from Theorem 10.19 (ii) that the set $\mathcal{D}^\iota(\nu,\delta_\iota)$ contains only one C^{1+} self-renormalizable structure $\mathcal{S}_\iota$ which finishes the proof. □

10.3.3 δ_ι-bounded solenoid equivalence class of Gibbs measures

When we speak of a δ_ι-bounded solenoid equivalence class of ν we mean a $(\delta_\iota,0)$-bounded solenoid equivalence class of a Gibbs measure ν (see Definition 27). In § 9.4, we use the cocycle-gap pairs to construct explicitly the solenoid functions in the δ_ι-bounded solenoid equivalence classes of the Gibbs measures ν. By Theorem 10.21 (ii) proved below, given an ι-solenoid function σ_ι there is a unique Gibbs measure ν such that σ_ι belongs to the δ_ι-bounded solenoid equivalence class of ν.

Theorem 10.21. *(i) There is a natural map* $g\to(\sigma_s(g),\sigma_u(g))$ *which gives a one-to-one correspondence between* C^{1+} *conjugacy classes of* C^{1+} *hyperbolic diffeomorphisms* $g\in\mathcal{T}(\nu,\delta_s,\delta_u)$ *and pairs* $(\sigma_s(g),\sigma_u(g))$ *of stable and unstable solenoid functions such that, for* ι *equal to* s *and* u, $\sigma_\iota(g)$ *is contained in the* δ_ι*-bounded solenoid equivalence class of* ν.

(ii) There is a natural map $\mathcal{S}_\iota \to \sigma_{\mathcal{S}_\iota}$ which gives a one-to-one correspondence between C^{1+} self-renormalizable structures $\mathcal{S}_\iota$ contained in $\mathcal{D}^\iota(\nu, \delta_\iota)$ and ι-solenoid functions $\sigma_{\mathcal{S}_\iota}$ contained in the δ_ι-bounded equivalence class of ν.

(iii) Let us suppose that $\mathcal{D}^\iota(\nu, \delta_\iota) \neq \emptyset$.

(a) (Rigidity) If $\delta_\iota = 1$, then the δ_ι-bounded solenoid equivalence class of ν is a singleton consisting in the continuous extension of the ι measure solenoid function $\sigma_{\nu,\iota}$ to sol^ι.

(b) (Flexibility) If $0 < \delta_\iota < 1$, then the δ_ι-bounded solenoid equivalence class of ν is an infinite dimensional space of solenoid functions.

Proof. Statement (i) follows from Lemma 10.10 (i). Statement (ii) follows from Lemma 10.1 if $\mathbf{B}^\iota$ is a no-gap train-track, and from Lemma 10.3 (ii) if $\mathbf{B}^\iota$ is a gap train-track. Statement (iii) follows from statement (ii) and Theorem 10.19. □

Theorem 10.22. *Given an ι-solenoid function σ_ι and $0 < \delta_{\iota'} \leq 1$, there is a unique Gibbs measure ν and a unique $\delta_{\iota'}$-bounded equivalence class of ν consisting of ι'-solenoid functions $\sigma_{\iota'}$ such that the C^{1+} conjugacy class of hyperbolic diffeomorphisms $g \in \mathcal{T}_{f,\Lambda}(\delta_s, \delta_u)$ determined by the pair (σ_s, σ_u) have an invariant measure $\mu = (i_g)_*\nu$ absolutely continuous with respect to the Hausdorff measure.*

Proof. By Theorem 10.21 (ii), the ι-solenoid function σ_ι determines a unique C^{1+} self-renormalizable structure $\mathcal{S}_\iota \in \mathcal{D}^\iota(\nu, \delta_\iota)$. By Theorem 10.14, the set $\mathcal{D}^{\iota'}(\nu, \delta_{\iota'})$ is nonempty. Let $\mathcal{S}_{\iota'} \in \mathcal{D}^{\iota'}(\nu, 1)$. By Theorem 10.21 (ii), the C^{1+} self-renormalizable structure $\mathcal{S}_{\iota'}$ determines a unique ι'-solenoid function $\sigma_{\iota'}$ such that, by Theorem 10.21 (i), the pair $(\sigma_\iota, \sigma_{\iota'})$ determines a unique C^{1+} conjugacy class $\mathcal{T}(\nu, \delta_s, \delta_u)$ of hyperbolic diffeomorphisms $g \in \mathcal{T}(\nu, \delta_s, \delta_u)$ with an invariant measure $\mu = i_*\nu$ absolutely continuous with respect to the Hausdorff measure. □

Putting together Theorem 10.21 and Theorem 10.22, we obtain the following implications:

(i) (Flexibility for Smale horseshoes) For $\iota = s$ and u, given a ι-solenoid function σ_ι there is an infinite dimensional space of solenoid functions $\sigma_{\iota'}$ such that the C^{1+} hyperbolic Smale horseshoes determined by the pairs (σ_s, σ_u) have an invariant measure μ absolutely continuous with respect to the Hausdorff measure.

(ii) (Rigidity for Anosov diffeomorphisms) For $\iota = s$ and u, given an ι-solenoid function σ_ι there is a unique ι'-solenoid function such that the C^{1+} Anosov diffeomorphisms determined by the pair (σ_s, σ_u) has an invariant measure μ absolutely continuous with respect to Lebesgue.

(iii) (Flexibility for codimension one attractors) Given an unstable solenoid function σ_u there is an infinite dimensional space of stable solenoid functions σ_s such that the C^{1+} hyperbolic codimension one attractors determined by the pairs (σ_s, σ_u) have an invariant measure μ absolutely continuous with respect to the Hausdorff measure.

(iv) (Rigidity for codimension one attractors) Given an s-solenoid function σ_s there is a unique unstable solenoid function σ_u such that the C^{1+} hyperbolic codimension one attractors determined by the pair (σ_s, σ_u) have an invariant measure μ absolutely continuous with respect to the Hausdorff measure using non-zero stable and unstable pressures.

10.4 Further literature

Cawley [21] characterised all C^{1+}-Hausdorff realizable Gibbs measures as Anosov diffeomorphisms using cohomology classes on the torus. This chapter is based on Pinto and Rand [166].

11

Extended Livšic-Sinai eigenvalue formula

We present an extension of the eigenvalue formula of A. N. Livšic and Ja. G. Sinai for Anosov diffeomorphisms that preserve an absolutely continuous measure to hyperbolic basic sets on surfaces which possess an invariant measure absolutely continuous with respect to Hausdorff measure. We also give a characterization of the Lipschitz conjugacy classes of such hyperbolic systems in a number of ways, for example following De la Llave, Marco and Moriyon, in terms of eigenvalues of periodic points and Gibbs measures.

11.1 Extending the eigenvalues's result of De la Llave, Marco and Moriyon

De la Llave, Marco and Moriyon [70, 71, 75, 76] have shown that the set of stable and unstable eigenvalues of all periodic points is a complete invariant of the C^{1+} conjugacy classes of Anosov diffeomorphisms.

Let $\mathcal{P}$ be the set of all periodic points in Λ under f. Let $p(x)$ be the (smallest) period of the periodic point $x \in \mathcal{P}$. For every $x \in \mathcal{P}$ and $\iota \in \{s, u\}$, let $j : J \to \mathbb{R}$ be a chart in $\mathcal{A}(g, \rho_g)$ such that $x \in J$. The *eigenvalue* $\lambda^{\iota}_{g,\iota}(x)$ *of* x is the derivative of the map $j^{-1} f^p j$ at $j(x)$.

For $\iota \in \{s, u\}$, by construction of the train-tracks, $\mathcal{P}^{\iota} = \pi_{\mathbf{B}^{\iota}}(\mathcal{P})$ is the set of all periodic points in $\mathbf{B}^{\iota}$ under the Markov map f_{ι}. Furthermore, $\pi_{\mathbf{B}^{\iota}}|\mathcal{P}$ is an injection and the periodic points $x \in \Lambda$ and $\pi_{\mathbf{B}^{\iota}}(x) \in \mathbf{B}^{\iota}$ have the same period $p(x) = p(\pi_{\mathbf{B}^{\iota}}(x))$. Let us denote $\pi_{\mathbf{B}^{\iota}}(x)$ by x_{ι}. Let $\mathcal{S}_{\iota}$ be a C^{1+} self-renormalizable structure. Let $j : J \to \mathbb{R}$ be a train-track chart of $\mathcal{S}_{\iota}$ such that $x_{\iota} \in J$. The *eigenvalue* $\lambda_{\mathcal{S}_{\iota}}(x_{\iota})$ *of* x_{ι} is the derivative of the map $j \circ \tau_{\iota}^{p(x_{\iota})} \circ j^{-1}$ at $j(x_{\iota})$, where τ_{ι} is the Markov map on the train-track $\mathbf{B}^{\iota}$.

For every $x \in \mathcal{P}$, every $\iota \in \{s, u\}$ and every $n \geq 0$, let $I^{\iota}_n(x)$ be an ι-leaf $(np(x) + 1)$-cylinder segment such that $x \in I^{\iota}_n(x)$ and $f_{\iota}^{p(x)} I^{\iota}_{n+1}(x) = I^{\iota}_n(x)$.

Lemma 11.1. *For $\iota \in \{s, u\}$, let $\mathcal{S}_{\iota} \in \mathcal{D}^{\iota}(\nu, \delta_{\iota}, P_{\iota})$ be a C^{1+} ι self-renormalizable structure. For every $x \in \mathcal{P}$,*

$$\lambda_{\mathcal{S}_\iota}(x_\iota) = r_{\mathcal{S}_\iota}(I_0^\iota(x) : I_1^\iota(x)) \tag{11.1}$$
$$= \rho_{\nu,\iota}(I_0^\iota : I_1^\iota)^{-1/\delta_\iota} e^{-p(x)P_\iota/\delta_\iota} \tag{11.2}$$
$$= \rho_{\nu,\iota'}(I_0^{\iota'} : I_1^{\iota'})^{-1/\delta_\iota} e^{-p(x)P_\iota/\delta_\iota} , \tag{11.3}$$

where $r_{\mathcal{S}_\iota}$ is the ι-ratio function of $\mathcal{S}_\iota$, $\rho_{\nu,\iota}$ is the ι-measure ratio function of the Gibbs measure ν, and $\rho_{\nu,\iota'}$ is the ι'-measure ratio function of the Gibbs measure ν.

Proof. For every $x \in \mathcal{P}$, let us denote by p the period $p(x)$ of x, and let us denote by I_n^ι the interval $I_n^\iota(x)$. We note that the p-mother $m^p I_{n+1}^\iota$ of I_{n+1}^ι is I_n^ι, and so $f_\iota^p I_{n+1}^\iota = m^p I_{n+1}^\iota$. By (4.1),

$$r_{\mathcal{S}_\iota}(I_0^\iota : I_1^\iota) = \lim_{n\to\infty} \frac{|I_n^\iota|}{|I_{n+1}^\iota|} .$$

Hence,

$$\begin{aligned}\lambda_{\mathcal{S}_\iota}(x_\iota) &= \lim_{n\to\infty} \frac{|f_\iota^p I_{n+1}^\iota|}{|I_{n+1}^\iota|} \\ &= \lim_{n\to\infty} \frac{|I_n^\iota|}{|I_{n+1}^\iota|} \\ &= r_{\mathcal{S}_\iota}(I_0^\iota : I_1^\iota),\end{aligned}$$

which proves (11.1). By Theorem 8.15, the ι-measure ratio function $\rho_{\mathcal{S}_\iota,\delta_\iota}$ is the ι-measure ratio function $\rho_{\nu,\iota}$ of the Gibbs measure ν. Hence, by (9.5), we get

$$\begin{aligned} r_{\mathcal{S}_\iota}(I_1^\iota : I_0^\iota) &= \prod_{l=0}^{p-1} r_{\mathcal{S}_\iota}(m^l I_1^\iota : m^{l+1} I_1^\iota) \\ &= \prod_{l=0}^{p-1} \left(J_{\mathcal{S},\delta_\iota}(\xi_l) \rho_{\nu,\iota}(m^l I_1^\iota : m^{l+1} I_1^\iota)^{1/\delta_\iota} e^{P_\iota/\delta_\iota} \right), \end{aligned} \tag{11.4}$$

where $\xi_l = f_\iota^{p-l} m^l I_1^\iota \in \mathbf{B}_o^\iota$. We note that $f_{\iota'}\xi_l = \xi_{l+1}$ and $f_{\iota'}\xi_{p-1} = \xi_0$ in $\mathbf{B}_o^{\iota'}$. Since $J_{\mathcal{S}_\iota,\delta_\iota} = \kappa/(\kappa \circ f_{\iota'})$ for some function κ, we get

$$\prod_{l=0}^{p-1} J_{\mathcal{S}_\iota,\delta_\iota}(\xi_l) = \prod_{l=0}^{p-1} \frac{\kappa(\xi_l)}{\kappa(\xi_{l+1})} = 1 . \tag{11.5}$$

Furthermore,

$$\prod_{l=0}^{p-1} \rho_{\nu,\iota}(m^l I_1^\iota : m^{l+1} I_1^\iota) = \rho_{\nu,\iota}(I_1^\iota : I_0^\iota) . \tag{11.6}$$

Using (11.5) and (11.6) in (11.4) we obtain that

$$r_{\mathcal{S}_\iota}(I_1^\iota : I_0^\iota) = \rho_{\nu,\iota}(I_1^\iota : I_0^\iota)^{1/\delta_\iota} e^{pP_\iota/\delta_\iota} \ .$$

Therefore, by (11.1), we have

$$\begin{aligned} \lambda_{\mathcal{S}_\iota}(x_\iota) &= r_{\mathcal{S}_\iota}(I_0^\iota : I_1^\iota) \\ &= \rho_{\nu,\iota}(I_0^\iota : I_1^\iota)^{-1/\delta_\iota} e^{-pP_\iota/\delta_\iota} , \end{aligned}$$

which proves (11.2). By Lemma 8.14, there is $0 < \varepsilon < 1$ such that, for every $n \geq 0$,

$$\rho_{\nu,s}(I_{n+1}^s : I_n^s) \in (1 \pm \varepsilon^n) \frac{\mu([I_{n+1}^s, I_1^u])}{\mu([I_n^s, I_1^u])} \tag{11.7}$$

and

$$\rho_{\nu,u}(I_{n+1}^u : I_n^u) \in (1 \pm \varepsilon^n) \frac{\mu([I_1^s, I_{n+1}^u])}{\mu([I_1^s, I_n^u])} \ . \tag{11.8}$$

Since $f^{np}\left([I_1^s, I_{n+1}^u]\right) = \left([I_{n+1}^s, I_1^u]\right)$ and by invariance of μ, we obtain that

$$\frac{\mu([I_{n+1}^s, I_1^u])}{\mu([I_n^s, I_1^u])} = \frac{\mu([I_1^s, I_{n+1}^u])}{\mu([I_1^s, I_n^u])} . \tag{11.9}$$

Putting together (11.7), (11.8) and (11.9), we obtain that

$$\rho_{\nu,s}(I_{n+1}^s : I_n^s) \in (1 \pm \varepsilon^n) \rho_{\nu,u}(I_{n+1}^u : I_n^u) \ .$$

Hence, by invariance of $\rho_{\mathcal{S}_s,s}$ and $\rho_{\mathcal{S}_s,u}$ under f, we obtain

$$\begin{aligned} \rho_{\nu,s}(I_1^s : I_0^s) &= \lim_{n\to\infty} \rho_{\nu,s}(I_{n+1}^s : I_n^s) \\ &= \lim_{n\to\infty} \rho_{\nu,u}(I_{n+1}^u : I_n^u) \\ &= \rho_{\nu,u}(I_1^u : I_0^u), \end{aligned}$$

which proves (11.3). □

Lemma 11.2. *Let $\mathbf{B}^\iota$ be a (gap or a no-gap) train-track.*

(i) The C^{1+} self-renormalizable structures $\mathcal{S}_1 \in \mathcal{D}^\iota(\nu_1, \delta, P)$ and $\mathcal{S}_2 \in \mathcal{D}^\iota(\nu_2, \delta, P)$ have the same eigenvalues for all periodic orbits if, and only if, ν_1 is equal to ν_2.
(ii) The set of eigenvalues of all periodic orbits of a C^{1+} self-renormalizable structure is a complete invariant of each Lipschitz conjugacy class.

Statement (ii) of the above lemma for Markov maps is also in Sullivan [231].

Proof. Proof of (i). By Lemma 10.4, the C^{1+} self-renormalizable structures $\mathcal{S}_1 \in \mathcal{D}^\iota(\nu_1, \delta, P)$ and $\mathcal{S}_2 \in \mathcal{D}^\iota(\nu_2, \delta, P)$ are Lipschitz conjugate if, and only if, the Gibbs measures ν_1 and ν_2 are equal. By Lemma 11.1, if the Gibbs measures ν_1 and ν_2 are equal, then $\mathcal{S}_1$ and $\mathcal{S}_2$ have the same eigenvalues for all periodic orbits. Hence, to finish the proof of statement (i), we are going to prove that if the C^{1+} self-renormalizable structures $\mathcal{S}_1$ and $\mathcal{S}_2$ have the same eigenvalues for all periodic orbits, then the C^{1+} self-renormalizable structures $\mathcal{S}_1$ and $\mathcal{S}_2$ are Lipschitz conjugate.

Without loss of generality, let us assume that $\mathcal{S}_1$ and $\mathcal{S}_2$ are unstable C^{1+} self-renormalizable structures. For $j \in \{1,2\}$, the (restricted) u-scaling function $z_{u,j} : \Theta^u \to \mathbb{R}^+$ of $\mathcal{S}$ is well-defined by (see § 4.6)

$$z_{u,j}(w_0 w_1 \ldots) = \lim_{n\to\infty} \frac{|\pi_{\mathbf{B}^s} \circ f^{n+1} \circ \pi_{\mathbf{B}^u}^{-1} \circ i_u(w_0 w_1 \ldots)|_{k_n}}{|\pi_{\mathbf{B}^s} \circ f^{n} \circ \pi_{\mathbf{B}^u}^{-1} \circ i_u(w_1 w_2 \ldots)|_{k_n}} ,$$

where k_n is a train-track chart in a C^{1+} self-renormalizable atlas $\mathcal{B}_j$ determined by $\mathcal{S}_j$ such that the domain of the chart k_n contains $\pi_{\mathbf{B}^s} \circ f^n \circ \pi_{\mathbf{B}^u}^{-1} \circ i_u(w_1 w_2 \ldots)$. For every stable-leaf $(i+1)$-cylinder J, let $w(J) \in \Theta^u$ be such that $i_u(w(J)) = \pi_{\mathbf{B}^u}(J)$. Hence, for every $l \in \{0, \ldots, i-1\}$, we have that

$$\pi_{\mathbf{B}^u}^{-1} \circ i_u(f_u^l w(J)) = f^{-i+l}(m^l J) ,$$

where $f^{-i+l}(m^l J)$ are stable-leaf primary cylinders. By construction of the (restricted) u-scaling function $z_{u,j}$ and of the u-scaling function $s_{u,j}$ of $\mathcal{S}_j$, we have that

$$s_{u,j}(J : m^i J) = \prod_{l=0}^{i-1} z_{u,j}(f_u^l(w(J))) . \tag{11.10}$$

Let $\mathcal{P}_{\Theta^u}$ be the set of all periodic point under the shift. For every $w = w_0 w_1 \ldots \in \mathcal{P}_{\Theta^u}$ let $p(w)$ be the smallest period of w. By construction of the train-tracks, for every w, there is a unique periodic point $x(w) \in \Lambda$ with period $p(w)$ with respect to the map f such that $i_u(w) = \pi_{\mathbf{B}^u} x(w)$. Furthermore, there is a unique periodic point $\pi_s x(w) \in \mathbf{B}^s$ with period $p(w)$ for the Markov map. By (11.10), for every $w \in \mathcal{P}_{\Theta^u}$, we have that

$$\prod_{i=0}^{p(w)-1} z_{u,j}(f_u^i(w)) = \lambda_{\mathcal{S}_j}(\pi_{\mathbf{B}^s} x(w)) . \tag{11.11}$$

Since the C^{1+} self-renormalizable structures $\mathcal{S}_1$ and $\mathcal{S}_2$ have the same eigenvalues for all periodic orbits, by (11.11), we have that

$$\prod_{i=0}^{p(w)-1} \frac{z_{u,1}(f_u^i(w))}{z_{u,2}(f_u^i(w))} = 1 , \tag{11.12}$$

for every $w \in \mathcal{P}_{\Theta^u}$. From Livšic's theorem (e.g. see Katok and Hasselblatt [65]), we get that

$$\frac{z_{u,1}(w)}{z_{u,2}(w)} = \frac{\kappa(w)}{\kappa \circ f_u(w)}, \tag{11.13}$$

where $\kappa : \Theta^u \to \mathbb{R}^+$ is a positive Hölder continuous function. By (11.10) and (11.13), for every stable-leaf $(i+1)$-cylinder J we obtain that

$$\frac{s_{u,1}(J:m^iJ)}{s_{u,2}(J:m^iJ)} = \prod_{l=0}^{i-1} \frac{z_{u,1}(f_u^l(w(J)))}{z_{u,2}(f_u^l(w(J)))}$$
$$= \frac{\kappa(w)}{\kappa \circ f_u^i(w)} \,. \tag{11.14}$$

Since κ is bounded away from zero and infinity, there is $C > 1$ such that, for all $w \in \Theta^u$ and $i \geq 1$, we have that

$$C^{-1} < \frac{\kappa(w)}{\kappa \circ \tau_u^i(w)} < C \,. \tag{11.15}$$

Putting together (11.14) and (11.15), we obtain that

$$\frac{s_{u,1}(J:m^iJ)}{s_{u,2}(J:m^iJ)} \,.$$

Therefore, the ι-solenoid functions $\sigma_{u,1} : \mathrm{sol}^\iota \to \mathbb{R}^+$ and $\sigma_{u,2} : \mathrm{sol}^\iota \to \mathbb{R}^+$ corresponding to the C^{1+} self-renormalizable structures $\mathcal{S}_1$ and $\mathcal{S}_2$ are in the same bounded equivalence class (see Definition 10). Hence, by Lemma 10.4, the C^{1+} self-renormalizable structures $\mathcal{S}_1$ and $\mathcal{S}_2$ are Lipschitz conjugate.
Proof of (ii). Statement (ii) follows from putting together Lemma 10.4 and Statement (i) of this lemma with $P = 0$. □

Lemma 11.3. *(i) If $g \in \mathcal{T}(f, \Lambda)$, then $\lambda_{g,s}(x) = \lambda_{\mathcal{S}_s}(x_s)^{-1}$ and $\lambda_{g,u}(x) = \lambda_{\mathcal{S}_s}(x_u)$ where, for $\iota \in \{s, u\}$, $\lambda_{g,\iota}(x)$ is the eigenvalue of the C^{1+} hyperbolic diffeomorphism g and $\lambda_{\mathcal{S}_\iota}$ is the eigenvalue of the C^{1+} self-renormalizable structure $\mathcal{S}_\iota = \mathcal{S}(g, \iota)$.*
(ii) The set of stable and unstable eigenvalues of all periodic orbits of a C^{1+} hyperbolic diffeomorphism $g \in \mathcal{T}(f, \Lambda)$ is a complete invariant of each Lipschitz conjugacy class.

Proof. By construction of the train-track atlas $\mathcal{B}^\iota(g, \rho_g)$ from the lamination atlas $\mathcal{A}^\iota(g, \rho_g)$ in § 4.2, if $\lambda_{g,\iota}(x)$ is the eigenvalue of $x \in \mathcal{P}$, then the eigenvalue of $x_\iota \in \mathcal{P}_\iota$ is either $\lambda_{g,\iota}(x)$ if $\iota = u$, or $\lambda_{g,\iota}^{-1}$ if $\iota = s$. □

Theorem 11.4. *(i) If $g \in \mathcal{T}(f, \Lambda)$, then $\lambda_{g,s}(x) = \lambda_{\mathcal{S}_s}(x_s)^{-1}$ and $\lambda_{g,u}(x) = \lambda_{\mathcal{S}_s}(x_u)$ where, for $\iota \in \{s, u\}$, $\lambda_{g,\iota}(x)$ is the eigenvalue of the C^{1+} hyperbolic diffeomorphism g and $\lambda_{\mathcal{S}_\iota}$ is the eigenvalue of the C^{1+} self-renormalizable structure $\mathcal{S}_\iota = \mathcal{S}(g, \iota)$.*
(ii) The set of stable and unstable eigenvalues of all periodic orbits of a C^{1+} hyperbolic diffeomorphism $g \in \mathcal{T}(f, \Lambda)$ is a complete invariant of each Lipschitz conjugacy class.

Proof. By Lemma 11.2, the set of eigenvalues of all periodic orbits of a C^{1+} self-renormalizable structure is a complete invariant of each Lipschitz conjugacy class of C^{1+} self-renormalizable structures. Hence, using Lemma 10.9, we get that the set of stable and unstable eigenvalues of all periodic orbits of a C^{1+} hyperbolic diffeomorphism g is a complete invariant of each Lipschitz conjugacy class. □

11.2 Extending the eigenvalue formula of A. N. Livšic and Ja. G. Sinai

We show an extension of the eigenvalue formula of A. N. Livšic and Ja. G. Sinai for Anosov diffeomorphisms to C^{1+} hyperbolic diffeomorphisms.

Theorem 11.5. *A C^{1+} hyperbolic diffeomorphism $g \in \mathcal{T}(f,\Lambda)$ has a natural geometric measure $\mu_{g,\delta_s,\delta_u}$ with pressures $P_s = P_s(g,\delta_s,\delta_u)$ and $P_u = P_u(g,\delta_s,\delta_u)$ if, and only if, for all $x \in \Lambda$*

$$\lambda_{g,s}(x_s)^{-\delta_s} e^{p(x)P_s} = \lambda_{g,u}(x_u)^{\delta_u} e^{p(x)P_u} . \tag{11.16}$$

From Theorem 11.5, we get the following corollary.

Corollary 11.6. *A C^{1+} hyperbolic diffeomorphism $g \in \mathcal{T}(f,\Lambda)$ has a g-invariant probability measure which is absolutely continuous to the Hausdorff measure on Λ_g if and only if for every periodic point x of $g|\Lambda_g$,*

$$\lambda_{g,s}(x)^{\delta_{g,s}} \lambda_{g,u}(x)^{\delta_{g,u}} = 1 .$$

Proof of Theorem 11.5. By Lemma 8.7, the C^{1+} self-renormalizable structures $S(g,s)$ and $S(g,u)$ are C^{1+} realizations of Gibbs measures $\nu_1 = \nu_{S(g,s),\delta_s}$ and $\nu_2 = \nu_{S(g,u),\delta_s}$. By Lemmas 11.1 and 11.3, for all $x \in \mathcal{P}$, we have

$$\begin{aligned} \lambda_{g,u}(x_u) &= \lambda_{S(g,u)}(x_u) \\ &= \rho_{\nu_2,u}(I_0^u : I_1^u)^{-1/\delta_u} e^{-p(x)P_u/\delta_u} \end{aligned} \tag{11.17}$$

and

$$\begin{aligned} \lambda_{g,s}(x_s) &= \lambda_{S(g,s)}(x_s)^{-1} \\ &= \rho_{\nu_1,u}(I_0^u : I_1^u)^{1/\delta_s} e^{p(x)P_s/\delta_s} . \end{aligned} \tag{11.18}$$

Let us prove that if the C^{1+} hyperbolic diffeomorphism g has a natural geometric measure, then (11.16) holds. Hence, by Lemma 10.5, the Gibbs measures ν_1 and ν_2 are equal. By (11.17), we have

$$\begin{aligned} \rho_{\nu_1,u}(I_0^u : I_1^u) &= \rho_{\nu_2,u}(I_0^u : I_1^u) \\ &= \lambda_{g,u}(x_\iota)^{-\delta_u} e^{-p(x)P_u} . \end{aligned} \tag{11.19}$$

By (11.18), we obtain that

$$\rho_{\nu_1,u}(I_0^u : I_1^u) = \lambda_{g,s}(x_\iota)^{\delta_s} e^{-p(x)P_s} \ . \tag{11.20}$$

Putting together (11.19) and (11.20), we obtain that

$$\lambda_{g,s}(x_s)^{-\delta_s} e^{p(x)P_s} = \lambda_{g,u}(x_u)^{\delta_u} e^{p(x)P_u} \ ,$$

and so (11.16) holds. Conversely, let us prove that if (11.16) holds, then the C^{1+} hyperbolic diffeomorphism g has a natural geometric measure. Putting together (11.16) and (11.18), we obtain that

$$\lambda_{g,u}(x_u) = \rho_{\nu_1,u}(I_0^u : I_1^u)^{-1/\delta_u} e^{-nP_u/\delta_u} \ .$$

Hence, the Gibbs measure ν_1 determines the same set of eivenvalues for all periodic orbits of self-renormalizable structures in $\mathbf{B}^u$ as the Gibbs measure ν_2. Therefore, by Lemma 11.2, $\nu_1 = \nu_2$ and consequently, by Lemma 10.5, the C^{1+} hyperbolic diffeomorphism g has a natural geometric measure. □

11.3 Further literature

Livšic and Sinai [69] proved that an Anosov diffeomorphism f admits an f-invariant measure that is absolutely continuous with respect to the Lebesgue measure on M if, and only if, $\lambda_{f,s}(x)\lambda_{f,u}(x) = 1$ for every periodic point $x \in \mathcal{P}$. De la Llave [70], De la Llave, Marco and Moriyon [71], and Marco and Moriyon [75, 76] have shown that the set of stable and unstable eigenvalues of all periodic points is a complete invariant of the C^{1+} conjugacy classes of Anosov diffeomorphisms. This chapter is based on Pinto and Rand [166].

12

Arc exchange systems and renormalization

We describe the construction of stable arc exchange systems from the stable laminations of hyperbolic diffeomorphisms. A one-to-one correspondence is established between (i) Lipshitz conjugacy classes of C^{1+H} stable arc exchange systems that are C^{1+H} fixed points of renormalization and (ii) Lipshitz conjugacy classes of C^{1+H} diffeomorphisms f with hyperbolic basic sets Λ that admit an invariant measure absolutely continuous with respect to the Hausdorff measure on Λ. Let $HD^s(\Lambda)$ and $HD^u(\Lambda)$ be, respectively, the Hausdorff dimension of the stable and unstable leaves intersected with the hyperbolic basic set Λ. If $HD^u(\Lambda) = 1$, then the Lipschitz conjugacy is in fact a C^{1+H} conjugacy in (i) and (ii). We prove that if the stable arc exchange system is a $C^{1+HD^s+\alpha}$ fixed point of renormalization with bounded geometry, then the stable arc exchange system is smoothly conjugate to an affine stable arc exchange system.

12.1 Arc exchange systems

Recall that a *train-track* $\mathbb{T} = \bigsqcup_{j=1}^{n} I_j / \sim$ is the disjoint union of non-trivial sets I_j, topologically nontrivial closed intervals, with a given endpoints equivalence relation. Let $\bigsqcup_{j=1}^{n} I_j$ be a finite disjoint union of non-trivial compact intervals. An *endpoints equivalence relation* consists in fixing pairwise disjoint equivalence classes $E_1, \ldots, E_i$ such that $\cup_{j=1}^{i} E_j$ is equal to the set of all endpoints of the intervals $I_1, \ldots, I_n$, and any two endpoints x and y are equivalent if, and only if, they belong to a same set E_j. We allow the case where some, or all, equivalence classes are singletons. If all the equivalence classes are singletons, then the endpoints equivalence relation is trivial.

The closed (resp., open) intervals contained in $\bigsqcup_{j=1}^{n} I_j$ are called *closed* (resp., *open*) *arcs* of the train-track $\mathbb{T}$. If $\mathbb{T}$ has junctions, then one fix a *set of junction arcs* $K_1, \ldots K_m \subset \mathbb{T}$ that are images of intervals $J_1, \ldots, J_m \subset \mathbb{R}$ by homeomorphisms $k_i : J_i \to K_i$ with the property that $k_i(\mathrm{int} J_i)$ intersects only one junction. From now on, a train-track $\mathbb{T}$ has always associated to a fixed set of junction arcs allowed. If I is closed (respectively, open), we say that $k(I)$ is a closed (respectively, open) arc in $\mathbb{T}$. A *chart* in $\mathbb{T}$ is the inverse of a parametrization. A *topological atlas* $\mathcal{B}$ on the train-track $\mathbb{T}$ is a given set of charts $\{(j, J)\}$ on the train-track covering locally every arc. A $C^{1+\alpha}$, with $\alpha > 0$, atlas $\mathcal{B}$ on the train-track $\mathbb{T}$ is a topological atlas such that the overlap maps are $C^{1+\alpha}$ and have uniformly $C^{1+\alpha}$ bounded norm. A C^{1+H} atlas $\mathcal{B}$ is a $C^{1+\alpha}$ atlas, for some $\alpha > 0$.

Definition 29 *The quadruple $(\Phi, \mathcal{J}_\Phi, \mathbb{T}_\Phi, \mathcal{B}_\Phi)$ is a C^{1+H}* arc exchange system *if the following properties are satisfied:*

(i) $\mathbb{T}_\Phi$ is a train-track with a set $\{L_{\Phi,1}, \ldots, L_{\Phi,m}\}$ of junction arcs, and $\mathcal{B}_\Phi$ is a $C^{1+\alpha}$ train-track atlas, for some $\alpha > 0$.
(ii) Φ is a set of homeomorphisms $\phi_i : I_{\Phi,i} \to J_{\Phi,i}$ such that $\phi_i|\mathrm{int}(I_{\Phi,i})$ is a $C^{1+\alpha}$ diffeomorphism, and $I_{\Phi,i}$ and $J_{\Phi,i}$ are nontrivial closed arcs.
(iii) $\mathcal{J}_\Phi$ is a set of $C^{1+\alpha}$ diffeomorphisms $e_j = e_{\Phi,j} : L_{\Phi,j} \to K_{\Phi,j}$, for $j = 1, \ldots, m$, with the following properties: (a) $L_{\Phi,j}$ is a junction arc, (b) there are closed arcs $I^L_{\Phi,j}$ and $I^R_{\Phi,j}$ such that $I^L_{\Phi,j} \cup I^R_{\Phi,j} = L_{\Phi,j}$ and $I^L_{\Phi,j} \cap I^R_{\Phi,j}$ is a junction, and (c) there are maps $\phi^L_{j,i_1}, \ldots, \phi^L_{j,i_{n(j,R)}}$ and $\phi^R_{j,i_1}, \ldots, \phi^R_{j,i_{n(j,R)}}$ in Φ such that $e_j|I^L_{\Phi,j} = \phi^L_{j,i_{n(j,L)}} \circ \ldots \circ \phi^L_{j,i_1}$ and $e_j|I^R_{\Phi,j} = \phi^R_{j,i_{n(j,R)}} \circ \ldots \circ \phi^R_{j,i_1}$.

For simplicity, (a) we assume that if $\phi_i : I_{\Phi,i} \to J_{\Phi,i}$ is in Φ, then there is $\phi_j : I_{\Phi,j} \to J_{\Phi,j}$ in Φ such that $I_{\Phi,j} = J_{\Phi,i}$, $J_{\Phi,j} = I_{\Phi,i}$ and $\phi_j = \phi_i^{-1}$, and (b) for every $x \in \mathbb{T}_\Phi$, there exist at most two distinct intervals $I_{\Phi,i}$ and $I_{\Phi,j}$ containing x. For simplicity of notation, we will denote by Φ the C^{1+H} exchange system $(\Phi, \mathcal{J}_\Phi, \mathbb{T}_\Phi, \mathcal{B}_\Phi)$. We will call $\mathcal{J}_\Phi$ the *junction exchange set* of the C^{1+H} arc exchange system Φ.

We say that a finite sequence $\{\phi_{i_n} \in \Phi\}_{n=1}^m$ or an infinite sequence $\{\phi_{i_n} \in \Phi\}_{n \geq 1}$ is *admissible* with respect to x, if $\phi_{i_n} \circ \ldots \circ \phi_{i_1}(x) \in I_{\Phi,i_{n+1}}$ and $\phi_{i_n} \neq \phi^{-1}_{i_{n-1}}$, for all $n > 1$. We define the invariant set Ω_Φ of Φ as being the set of all points $x \in \mathbb{T}_\Phi$ for which there are two distinct infinite admissible sequences $\{\phi^F_{i_n} \in \Phi\}_{n \geq 1}$ and $\{\phi^B_{i_n} \in \Phi\}_{n \geq 1}$ with respect to x. The forward orbit $\mathcal{O}^F(x)$ of a point $x \in \Omega_\Phi$ is the set $\{\phi^F_{i_n}(x) : n \geq 1\}$, and the backward orbit $\mathcal{O}^B(x)$ of x is the set $\{\phi^B_{i_n}(x) : n \geq 1\}$. We will assume that the invariant set Ω_Φ is *minimal*, i.e, for every $x \in \Omega_\Phi$, the closure $\overline{\mathcal{O}^F(x)}$ is equal to the invariant set Ω_Φ and that the closure $\overline{\mathcal{O}^B(x)}$ is also equal to the invariant set Ω_Φ. Furthermore, we will assume that the endpoints of the intervals $I_{\Phi,1}, \ldots, I_{\Phi,n}$ belong to the invariant set Ω_Φ and $\Omega_\Phi \subset \cup_{i=1}^n I_{\Phi,i}$. We denote the Hausdorff dimension of Ω_Φ by $HD(\Omega_\Phi)$. If $0 < HD(\Omega_\Phi) < 1$, we call Φ a C^{1+H} *arc exchange system.* If $HD(\Omega_\Phi) = 1$, we call Φ a C^{1+H} *interval exchange system.*

We say that an arc exchange system Φ is determined by a map $\phi : I_\phi \to J_\phi$ if all the maps $\phi_i : I_{\Phi,i} \to J_{\Phi,i}$ contained in Φ are the restriction of the map ϕ or its inverse ϕ^{-1} to $I_{\Phi,i}$. In this case, we call ϕ an *arc exchange map.* We note that not all arc exchange systems are determined by arc exchange maps.

Let $\Phi = \{\phi_i : I_{\Phi,i} \to J_{\Phi,i}; i = 1, \ldots, n\}$ and $\Psi = \{\psi_i : I_{\Psi,i} \to J_{\Psi,i}; i = 1, \ldots, n\}$ be $C^{1+\alpha}$ arc exchange systems with junction sets $\mathcal{J}_\Phi = \{e_{\Phi,j} : L_{\Phi,j} \to K_{\Phi,j}; j = 1, \ldots, m\}$ and $\mathcal{J}_\Psi = \{e_{\Psi,j} : L_{\Psi,j} \to K_{\Psi,j}; j = 1, \ldots, m\}$, respectively. We say that Φ and Ψ are C^0 *conjugate*, if there is a homeomorphism $h : \Omega_\Phi \to \Omega_\Psi$ with the following properties:

(i) h has a homeomorphic extension $\xi : \mathbb{T}_\Phi \to \mathbb{T}_\Psi$ such that $I_{\Psi,i} = \xi\,(I_{\Phi,i})$, $J_{\Psi,i} = \xi\,(J_{\Phi,i})$, $L_{\Psi,i} = \xi\,(L_{\Phi,i})$ and $K_{\Psi,i} = \xi\,(K_{\Phi,i})$.
(ii) For every $1 \leq i \leq n$, $h \circ \phi_i(x) = \psi_i \circ h(x)$, where $x \in \Omega_\Phi \cap I_{\Phi,i}$.
(iii) For every $1 \leq j \leq m$, $h \circ e_{\Phi,j}(x) = e_{\Psi,j} \circ h(x)$, where $x \in \Omega_\Phi \cap L_{\Phi,i}$.

By minimality of Ω_Φ, h is uniquely determined and the arcs $\xi\,(I_{\Phi,i})$, $\xi\,(J_{\Phi,i})$, $\xi\,(L_{\Phi,i})$ and $\xi\,(K_{\Phi,i})$ do not depend upon the extension ξ of h. We say that Φ and Ψ are *Lipschitz conjugate*, if there is a Lipschitz homeomorphic extension $\xi : \mathbb{T}_\Phi \to \mathbb{T}_\Psi$ of h satisfying property (i) above. We say that Φ and Ψ are $C^{1+\alpha}$ *conjugate*, for some $\alpha > 0$, if there is a $C^{1+\alpha}$ homeomorphic extension $\xi : \mathbb{T}_\Phi \to \mathbb{T}_\Psi$ of h satisfying property (i) above. We say that Φ and Ψ are C^{1+H} *conjugate*, if Φ and Ψ are $C^{1+\alpha}$ conjugate, for some $\alpha > 0$. We denote by $[\Phi]_{C^{1+\alpha}}$ the set of all $C^{1+\alpha}$ arc exchange systems that are $C^{1+\alpha}$ conjugate to Φ, and we denote by $[\Phi]_{C^{1+H}}$ the set $\bigcup_{\alpha>0}[\Phi]_{C^{1+\alpha}}$.

12.1.1 Induced arc exchange systems

Let $g \in \mathcal{F}$. Suppose that M and N are Markov rectangles of g, and $x \in M$ and $y \in N$. We say that x and y are *stable holonomically related* if (i) there is an unstable leaf segment $\ell^u(x, y)$ such that $\partial\ell^u(x, y) = \{x, y\}$, and (ii) $\ell^u(x, y) \subset \ell^u(x, M) \cup \ell^u(y, N)$. Let $P = P_\mathcal{M}$ be the set of all pairs (M, N) such that there are points $x \in M$ and $y \in N$ stable holonomically related.

For every Markov rectangle $M \in \mathcal{M}$, choose a spanning leaf segment ℓ_M in M. Let $\mathcal{I} = \{\ell_M : M \in \mathcal{M}\}$. For every pair $(M, N) \in \mathcal{P}$, there are maximal leaf segments $\ell^D_{(M,N)} \subset \ell_M$, $\ell^C_{(M,N)} \subset \ell_N$ such that the holonomy $h_{(M,N)} : \ell^D_{(M,N)} \to \ell^C_{(M,N)}$ is well-defined (see §1.2 and §1.5). We call such holonomies $h_{(M,N)} : \ell^D_{(M,N)} \to \ell^C_{(M,N)}$ the (stable) primitive holonomies associated to the Markov partition $\mathcal{M}$. The complete set $\mathcal{H}^s$ of stable holonomies consists of all primitive holonomies $h_{(M,N)}$ and their inverses $h^{-1}_{(M,N)}$, for every $(M, N) \in \mathcal{P}^s$. The complete set $\mathcal{H}^u$ is defined similarly to $\mathcal{H}^s$ (see §5.1).

Let $f : T \to T$ be the Anosov automorphism defined by $f(x, y) = (x + y, y)$, where $T = \mathbf{R}^2 \setminus (\mathbf{Zv} \times \mathbf{Zw})$. We exhibit the complete set of holonomies $\mathcal{H}_{f,\mathcal{M}} = \{h_{(A,A)}, h_{(A,B)}, h_{(B,A)}, h^{-1}_{(A,A)}, h^{-1}_{(A,B)}, h^{-1}_{(B,A)}\}$ associated to the Markov partition $\mathcal{M} = \{A, B\}$ of f. We consider a derived-Anosov diffeomorphism $g : T \to T$ semi-conjugated, by a map $\pi : T \to T$, to the Anosov automorphism f. The derived-Anosov diffeomorphism g admits a Markov partition $\mathcal{M}_g = \{A_1, A_2, B_1\}$ with the property that $A = \pi(A_1) \cup \pi(A_2)$ and $B = \pi(B_1)$. The complete sets of holonomies $\mathcal{H}_{g,\mathcal{M}_g}$ and $\mathcal{H}_{f,\mathcal{M}}$ are related by the following equalities: $h_{(A,B)} \circ \pi|\pi(\ell^D_{(A_1,B_1)}) = \pi \circ h_{(A_1,B_1)}$,

$h_{(A,A)} \circ \pi|\pi(\ell^D_{(A_2,A_1)}) = \pi \circ h_{(A_2,A_1)}$, $h_{(B,A)} \circ \pi|\pi(\ell^D_{(B_1,A_1)}) = \pi \circ h_{(B_1,A_1)}$ and $h_{(B,A)} \circ \pi|\pi(\ell^D_{(B_1,A_2)}) = \pi \circ h_{(B_1,A_2)}$ (see Figure 12.1).

Lemma 12.1. *The triple* $(f, \Lambda, \mathcal{M})$ *induces a train-track* $\mathbb{T}^\iota_f$ *with a set of junction arcs. Furthermore, the atlas* $\mathcal{A}^\iota(f,\rho)$ *induces a* $C^{1+\alpha}$ *atlas* $\mathcal{B}^\iota(f,\rho)$ *on* $\mathbb{T}^\iota_f$.

Proof. For every ι-leaf segment $\ell^\iota_M \in \mathcal{I}^\iota$, let $\hat{\ell}^\iota_M$ be the smallest full ι-leaf segment containing ℓ^ι_M (see definition in §1.1). If $HD(\Lambda^\iota) = 1$, then $\ell^\iota_M = \hat{\ell}^\iota_M$. By the Stable Manifold Theorem, there are C^{1+H} diffeomorphisms $j_{\iota,M} : \hat{\ell}^\iota_M \to \hat{J}^\iota_M$. We choose the C^{1+H} diffeomorphisms $j_{\iota,M} : \hat{\ell}^\iota_M \to \hat{J}^\iota_M$ with the extra property that their images are pairwise disjoint, i.e. $\hat{J}^\iota_M \cap \hat{J}^\iota_N = \emptyset$ for all $M, N \in \mathcal{M}$ such that $M \neq N$. Let

$$\hat{L}^\iota_{\mathcal{M}} = \bigsqcup_{i=1}^{n} \hat{\ell}^\iota_{M_i} \quad \text{and} \quad L^\iota_{\mathcal{M}} = \hat{L}^\iota_{\mathcal{M}} \bigcap \Lambda^\iota_f. \tag{12.1}$$

Let $j_\iota : \hat{L}^\iota_{\mathcal{M}} \to \hat{J}^\iota_{\mathcal{M}}$ be the map defined by $j_\iota|\hat{\ell}^\iota_M = j_{\iota,M}$, for every $M \in \mathcal{M}$. Let $\ell^{\iota'}_M(x)$ be the spanning ι'-leaf segment of the Markov rectangle $M \in \mathcal{M}$ passing through x. Let

$$\pi_\iota : \bigcup_{i=1}^{n} M_i \to L^\iota_{\mathcal{M}} \tag{12.2}$$

be the projection defined by $\pi_\iota(x_i) = y_i$, where $y_i \in \ell^{\iota'}_{M_i}(x_i) \cap L^\iota_{\mathcal{M}}$, for every $x_i \in M_i$. If $HD(\Lambda^\iota) < 1$, then the endpoints equivalence relation is trivial. If $HD(\Lambda^\iota) = 1$, then the endpoints equivalence relation is non trivial, as we now describe. The endpoints $\hat{x}_i \in \hat{\ell}^\iota_{M_i}$ and $\hat{x}_j \in \hat{\ell}^\iota_{M_j}$ are in the same endpoints equivalence class, if $\ell^{\iota'}_{M_i}(x_i) \cap \ell^{\iota'}_{M_j}(x_j)$ is non-empty. The endpoints equivalence class in $\hat{L}^\iota_{\mathcal{M}}$ is the minimal equivalence class satisfying the above property. Let the ι-train-track $\mathbb{T}^\iota_f = \hat{L}^\iota_{\mathcal{M}} / \sim$ be the set $\hat{L}^\iota_{\mathcal{M}}$ with the endpoints equivalence class as defined above.

If $HD(\Lambda^\iota) < 1$, the charts $k_{\iota,M}$, for every $M \in \mathcal{M}$, form a $C^{1+\alpha}$ atlas $\mathcal{B}^\iota(f,\rho)$ for the train-track $\mathbb{T}^\iota_f$.

If $HD(\Lambda^\iota) = 1$, for every pair $(M,N) \in \mathcal{P}^{\iota'}$, we define $\hat{L}^\iota_{(M,N)} = \hat{\ell}^\iota_M \cup \hat{\ell}^\iota_N \subset \mathbb{T}^\iota_f$ as a *junction arc*. We fix an ι-leaf segment $L^\iota_{(M,N)}$ that is the union of two spanning ι-leaf segments L^ι_M and L^ι_N. For every ι-leaf segment $L^\iota_{(M,N)}$, let $\tilde{L}^\iota_{(M,N)}$ be the smallest full ι-leaf segment containing $L^\iota_{(M,N)}$, and a chart $\tilde{\jmath}_{(M,N)} : \tilde{L}^\iota_{(M,N)} \to J^\iota_{(M,N)}$ in the atlas $\mathcal{A}^\iota(f,\rho)$. By Pinto and Rand [164], the holonomies $h_M : \ell^\iota_M \to \ell^\iota_{(M,N)} \cap M$ and $h_N : \ell^\iota_N \to \ell^\iota_{(M,N)} \cap N$ have $C^{1+\alpha}$ extensions $\tilde{h}_M : \hat{\ell}^\iota_M \to \tilde{L}^\iota_{(M,N)}$ and $\tilde{h}_N : \hat{\ell}^\iota_N \to \tilde{L}^\iota_{(M,N)}$ onto their images. We define the *junction stable chart* $j_{(M,N)} : \hat{L}^\iota_{(M,N)} \to J^\iota_{(M,N)}$ in $\mathcal{B}^\iota(f,\rho)$ by

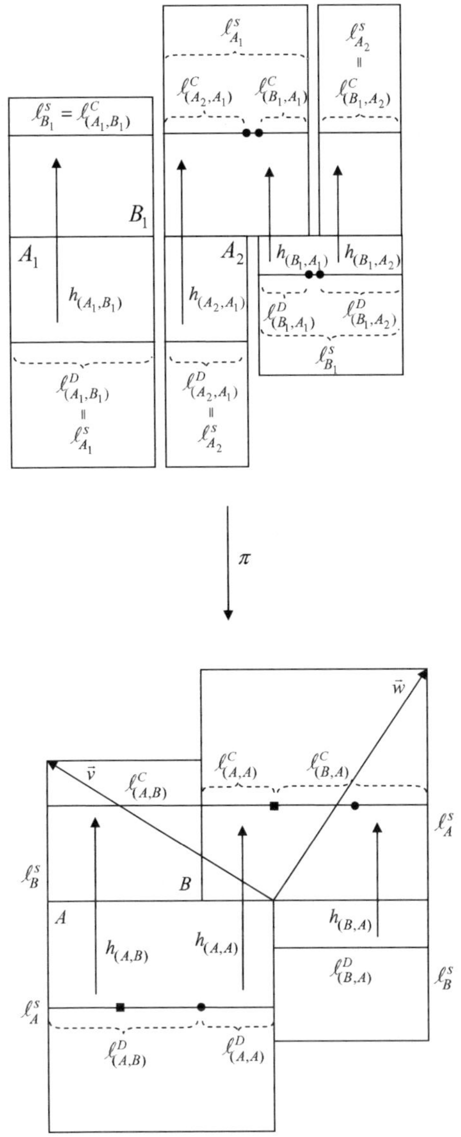

Fig. 12.1. The complete set of holonomies $\mathcal{H}_{g,\mathcal{M}_g} = \{h_{(A_1,B_1)}, h_{(A_2,A_1)}, h_{(B_1,A_1)}, h_{(B_1,A_2)}, h^{-1}_{(A_1,B_1)}, h^{-1}_{(A_2,A_1)}, h^{-1}_{(B_1,A_1)}, h^{-1}_{(B_1,A_2)}\}$ for the derived-Anosov diffeomorphism $g : T \to T$ semi-conjugated, by a map $\pi : T \to T$, to the Anosov automorphism $f : T \to T$ defined by $f(x,y) = (x+y,y)$. The complete set of holonomies for the Anosov automorphism $f : T \to T$ associated to the Markov partition $\mathcal{M} = \{A, B\}$ is given by $\mathcal{H}_{f,\mathcal{M}} = \{h_{(A,A)}, h_{(A,B)}, h_{(B,A)}, h^{-1}_{(A,A)}, h^{-1}_{(A,B)}, h^{-1}_{(B,A)}\}$. The complete set of holonomies $\mathcal{H}_{g,\mathcal{M}_g}$ is related to $\mathcal{H}_{f,\mathcal{M}}$ as follows: $h_{(A,B)} \circ \pi|\pi(\ell^D_{(A_1,B_1)}) = \pi \circ h_{(A_1,B_1)}$, $h_{(A,A)} \circ \pi|\pi(\ell^D_{(A_2,A_1)}) = \pi \circ h_{(A_2,A_1)}$, $h_{(B,A)} \circ \pi|\pi(\ell^D_{(B_1,A_1)}) = \pi \circ h_{(B_1,A_1)}$ and $h_{(B,A)} \circ \pi|\pi(\ell^D_{(B_1,A_2)}) = \pi \circ h_{(B_1,A_2)}$.

$j_{(M,N)}|\hat{\ell}^{\iota}_M = \tilde{j}_{(M,N)} \circ \tilde{h}_M$ and $j_{(M,N)}|\hat{\ell}^{\iota}_N = \tilde{j}_{(M,N)} \circ \tilde{h}_N$. By construction, the charts k_M, for every $M \in \mathcal{M}$, and $k_{(M,N)}$, for every $(M,N) \in \mathcal{P}^{\iota'}$, form a $C^{1+\alpha}$ atlas $\mathcal{B}^{\iota}(f,\rho)$ for $\mathbb{T}^{\iota}_f$. □

Let $A_{(M,N)}, B_{(M,N)} \in \mathcal{M}$ be the Markov rectangles such that there is a ι'-leaf segment $L^{\iota'}_{(M,N)}$ that (i) passes through x, (ii) has endpoints $a = a_{(M,N)} \in \mathrm{int}A_{(M,N)}$ and $b = b_{(M,N)} \in \mathrm{int}B_{(M,N)}$, and (iii) $L^{\iota'}_{(M,N)} \setminus (\ell^{\iota'}(a, A_{(M,N)}) \cup \ell^{\iota'}(b, B_{(M,N)}))$ is contained in the ι'-boundaries of Markov rectangles, where $\ell^{\iota'}(a, A_{(M,N)})$ is the spanning leaf of $A_{(M,N)}$ passing through a, and $\ell^{\iota'}(b, B_{(M,N)})$ is the spanning leaf of $B_{(M,N)}$ passing through b. Let $\ell_{(A,M,N)}$ be an ι-spanning leaf of $A_{(M,N)}$ passing through a, and let $\ell_{(B,M,N)}$ be an ι-spanning leaf of $B_{(M,N)}$ passing through b. For $i \in \{A,B\}$, fix $K_{(i,M,N)} \subset \ell_{(i,M,N)}$ and $L_{(i,M,N)} \subset L_{(M,N)}$ such that the basic holonomy $h_{(i,M,N)} : K_{(i,M,N)} \cap \Lambda \to L_{(i,M,N)}$ is well-defined. Let $\hat{K}_{(i,M,N)}$, $\hat{L}_{(i,M,N)}$, $\hat{\ell}^D_{(M,N)}$ and $\hat{\ell}^C_{(M,N)}$ be the smallest full ι-leaf segments that contain $K_{(i,M,N)}$, $L_{(i,M,N)}$, $\ell^D_{(M,N)}$ and $\ell^C_{(M,N)}$, respectively. The set of all basic holonomies $h_{(i,M,N)} : \hat{K}_{(i,M,N)} \to L_{(i,M,N)}$, with $i \in \{A,B\}$ and $(M,N) \in \mathcal{P}^{\iota'}$, form the *$\iota$-primitive junction set* (see Figure 12.2).

Lemma 12.2. *The triple $(f, \Lambda, \mathcal{M})$ induces a C^{1+H} ι-arc exchange system*

$$(\Phi^{\iota}_{f,\mathcal{M}}, \mathcal{J}^{\iota}_{\Phi}, \mathbb{T}^{\iota}_{\Phi}, \mathcal{B}^{\iota}_{\Phi}(f,\rho)),$$

with the following properties:

(i) The set $\Phi^{\iota} = \Phi^{\iota}_{f,\mathcal{M}}$ consists of all $C^{1+\alpha}$ diffeomorphisms $\phi^{\iota}_{(M,N)} : \hat{\ell}^D_{(M,N)} \to \hat{\ell}^C_{(M,N)}$, with $i \in \{A,B\}$ and $(M,N) \in \mathcal{P}^{\iota}$ such that $\phi^{\iota}_{(M,N)}|\ell^D_{(M,N)} = h_{(M,N)}$.

(ii) The junction set $\mathcal{J}_{\Phi}$ consists of all $C^{1+\alpha}$ diffeomorphisms $e_{(i,M,N)} : \hat{K}_{(i,M,N)} \to \hat{L}_{(i,M,N)}$, with $i \in \{A,B\}$ and $(M,N) \in \mathcal{P}^{\iota'}$, such that $e_{(i,M,N)}|K_{(i,M,N)} = h_{(i,M,N)}$, for every $i \in \{A,B\}$ and $(M,N) \in \mathcal{P}^{\iota'}$.

Proof. Since the holonomies are $C^{1+\alpha}$ diffeomorphisms with respect to $\mathcal{A}^{\iota}(f,\rho)$, (a) there are $C^{1+\alpha}$ diffeomorphic extensions $\phi^{\iota}_{(M,N)} : \hat{\ell}^D_{(M,N)} \to \hat{\ell}^C_{(M,N)}$ of the holonomies $h^{\iota}_{(M,N)} : \ell^D_{(M,N)} \to \ell^C_{(M,N)}$ with respect to the atlas $\mathcal{B}^{\iota}(f,\rho)$, for $(M,N) \in \mathcal{P}^{\iota}$, and (b) there are $C^{1+\alpha}$ diffeomorphic extensions $e^{\iota}_{(i,M,N)} : \hat{K}_{(i,M,N)} \to \hat{L}_{(i,M,N)}$ of the holonomies $h^{\iota}_{(i,M,N)} : K_{(i,M,N)} \to L_{(i,M,N)}$ with respect to the atlas $\mathcal{B}^{\iota}(f,\rho)$, for $(M,N) \in \mathcal{P}^{\iota'}$ and $i \in \{A,B\}$. □

12.2 Renormalization of arc exchange systems

Let $\Phi = \{\phi_i : \tilde{I}_{\Phi,i} \to \tilde{J}_{\Phi,i} : i = 1,\ldots,n\}$ and $\Psi = \{\psi_i : \tilde{I}_{\Psi,i} \to \tilde{I}_{\Psi,i} : i = 1,\ldots,m\}$ be C^{1+H} arc exchange systems. We say that Ψ is a *renormalization*

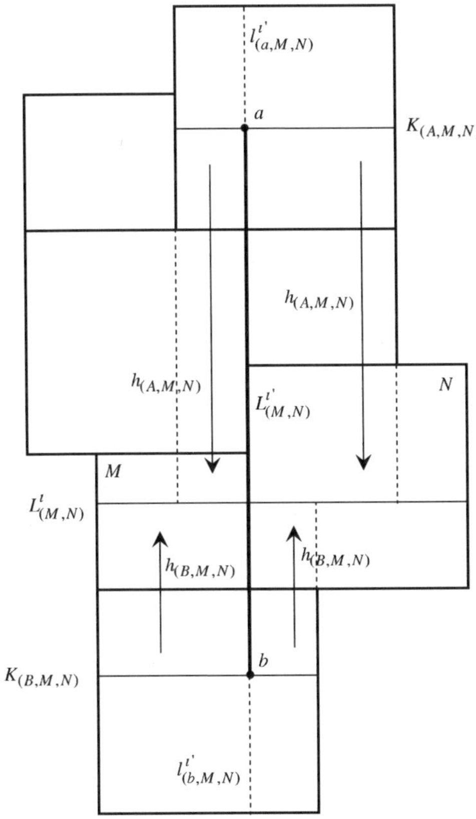

Fig. 12.2. The construction of the elements of the junction set.

of Φ if there is a *renormalization sequence set* $\mathcal{S} = \mathcal{S}(\Phi, \Psi) = \{\underline{s}^1, \ldots, \underline{s}^m\}$ with the following properties:

(i) For every $i \in \{1, \ldots, n\}$, we have that

$$\psi_i = \phi_{s^i_{k(\underline{s}^i)}} \circ \ldots \circ \phi_{s^i_1} | I_{\Psi,i},$$

where $\underline{s}^i = s^i_{k(\underline{s}^i)} \ldots s^i_1 \in \mathcal{S}$. In particular, $\Omega_\Psi \subset \Omega_\Phi$ and $I_{\psi_i} \subset I_{\Phi_{s,1^i}}$.

(ii) For every $x \in \Omega_\Phi \backslash \Omega_\Psi$, there are exactly two distinct sequences $\underline{s}^i, \underline{s}^j \in \mathcal{S}$ with the property that there are points $y_i \in I_{\Psi,i}$, $y_j \in I_{\Psi,j}$ such that

$$x = \phi_{s^i_{k(x,i)}} \circ \ldots \circ \phi_{s^i_1}(y_i) \quad \text{and} \quad x = \phi_{s^j_{k(x,j)}} \circ \ldots \circ \phi_{s^j_1}(y_j),$$

for some $0 < k(x,i) < k(\underline{s}^i)$ and $0 < k(x,j) < k(\underline{s}^j)$.

For every $\underline{\Phi} \in [\Phi]_{C^0}$, let $\xi_{\underline{\Phi}} : \mathbb{T}_\Phi \to \mathbb{T}_{\underline{\Phi}}$ be an extension of the topological conjugacy h between the C^{1+H} arc exchange systems Φ and $\underline{\Phi}$. Since h is unique, by minimality of Ω_Φ, for every $\underline{s}^i \in \mathcal{S}$, $\xi(I_{\psi_i})$ and $\xi(J_{\psi_i})$ are the smallest closed arcs containing $h(I_{\psi_i})$ and $h(J_{\psi_i})$, respectively, and, so, are uniquely determined. Define the C^{1+H} arc exchange system $\underline{\Psi}$ by

$$\underline{\Psi} = \left\{ \underline{\psi}_i = \underline{\phi}_{s^i_{k(\underline{s}^i)}} \circ \ldots \circ \underline{\phi}_{s^i_1} : \xi(\tilde{I}_{\psi_i}) \to \xi(\tilde{J}_{\psi_i}), \text{ for every } \underline{s}^i \in \mathcal{S}(\Phi, \Psi) \right\}.$$

For every $e_{\Phi,j} : L_{\phi_j} \to K_{\phi_j}$, let I^L_Ψ, I^R_Ψ, $\psi^L_{j,1}, \ldots, \psi^L_{j,n(j,L)}$ and $\psi^R_{j,1}, \ldots, \psi^R_{j,n(j,L)}$ be as in property (ii) of definition of $C^{1+\alpha}$ arc exchange system, in §12.1. We define the junction set $\mathcal{J}_{\underline{\Psi}} = \{e_{\underline{\Psi},1}, \ldots, e_{\underline{\Psi},m}\}$ of $\underline{\Psi}$ as follows: $e_{\underline{\Psi},j} : L_{\underline{\Psi},j} \to K_{\underline{\Psi},j}$ is given by $e_{\underline{\Psi},j}|\xi(\phi(I^L_{\underline{\Psi},j})) = \underline{\Psi}^L_{j,i_{n(j,L)}} \circ \ldots \circ \underline{\Psi}^L_{j,i_1}$ and $e_{\underline{\Psi},j}|\xi(\phi(I^R_{\underline{\Psi},j})) = \underline{\Psi}^R_{j,i_{n(j,R)}} \circ \ldots \circ \underline{\Psi}^R_{j,i_1}$. By construction, $\underline{\Psi}$ is topologically conjugate to Ψ and does not depend on the extension ξ of h considered in the sets $\xi(I_{\psi_1}), \ldots, \xi(I_{\psi_n})$. Furthermore, $\underline{\Psi}$ is a C^{1+H} arc exchange system that is a renormalization of $\underline{\Phi}$ with respect to the renormalization sequence set $\mathcal{S}(\underline{\Phi}, \underline{\Psi}) = \mathcal{S}(\Phi, \Psi)$. Hence, the *renormalization operator* R is well-defined by $R\underline{\Phi} = \underline{\Psi}$.

Definition 30 *Let* $R : [\Phi]_{C^0} \to [\Psi]_{C^0}$ *be a renormalization operator. We say that a* $C^{1+\alpha}$ *arc exchange system* $\Gamma \in [\Phi]_{C^0}$ *is a* $C^{1+\alpha}$ fixed point of the renormalization operator R, *if* $R\Gamma$ *is* $C^{1+\alpha}$ *conjugated to* Γ, *i.e* $[R\Gamma]_{C^{1+\alpha}} = [\Gamma]_{C^{1+\alpha}}$. *We say that a* C^{1+H} *arc exchange system* $\Gamma \in [\Phi]_{C^0}$ *is a* C^{1+H} fixed point of the renormalization operator R, *if* Γ *is* $C^{1+\alpha}$ *fixed point of the renormalization operator* R, *for some* $\alpha > 0$.

12.2.1 Renormalization of induced arc exchange systems

We present an explicit construction of a renormalization operator $R = R_{f,\mathcal{M}}$ acting on the topological conjugacy class of the C^{1+H} arc exchange system $\Phi_{f,\mathcal{M}}$ induced by $(f, \Lambda, \mathcal{M})$. Let the Markov partition $\mathcal{N} = f_*\mathcal{M}$ be the push-forword of the Markov partition $\mathcal{M}$, i.e, for every $M \in \mathcal{M}$, $N = f(M) \in \mathcal{N}$.

Lemma 12.3. *Let* $\Phi_{f,\mathcal{M}}$ *and* $\Phi_{f,\mathcal{N}}$ *be the* C^{1+H} *arc exchange systems induced (as in Lemma 12.2), respectively, by* $(f, \Lambda, \mathcal{M})$ *and* $(f, \Lambda, \mathcal{N})$.

(a) There is a well-defined renormalization operator

$$R = R_{f,\mathcal{M}} : [\Phi_{f,\mathcal{M}}]_{C^0} \to [\Phi_{f,\mathcal{N}}]_{C^0}.$$

(b) Let $\Psi = R\Phi$. *For every* $e_{\Phi,j} : L_{\phi_j} \to K_{\phi_j}$ *and* $e_{\Psi,j} : L_{\psi_j} \to K_{\psi_j}$, *let* $I^L_{\phi_j}$, $I^R_{\phi_j}$, $I^L_{\psi_j}$ *and* $I^R_{\psi_j}$ *be as in property (iii) of the Definition 29. If* $e_{\Phi,j}|I^L_{\phi_j} = \phi^L_{j,i_n} \circ \ldots \circ \phi^L_{j,i_1}$ *and* $e_{\Phi,j}|I^R_{\phi_j} = \phi^R_{j,i_n} \circ \ldots \circ \phi^R_{j,i_1}$, *then* $e_{\Psi,j}|I^L_{\psi_j} = \psi^L_{j,i_n} \circ \ldots \circ \psi^L_{j,i_1}$ *and* $e_{\Psi,j}|I^R_{\psi_j} = \psi^R_{j,i_n} \circ \ldots \circ \psi^R_{j,i_1}$.

Proof. For simplicity of notation, let us denote $k_{\mathcal{M}}$ by k (see (12.1)). We choose a map

$$\sigma : \{1, \ldots, n\} \to \{1, \ldots, n\} \tag{12.3}$$

with the property that $N_i \cap M_{\sigma(i)} \neq \emptyset$, where $N_i \in \mathcal{N}$ and $M_{\sigma(i)} \in \mathcal{M}$. For each $N_i \in \mathcal{N}$, let ℓ_{N_i} be the stable spanning leaf segment $\ell_{M_{\sigma(i)}} \cap \pi(N_i)$, and let $\hat{\ell}_{N_i}$ be the corresponding full stable spanning leaf (i.e $\hat{\ell}_{N_i} \cap \Lambda = \ell_{N_i}$), where $\pi : \bigcup_{i=1}^{n} M_i \to L_{\mathcal{M}}$ is the natural projection as defined in (12.1). Set

$$\mathcal{L}_{\mathcal{N}} = \bigcup_{i=1}^{n} \ell_{N_i} \quad \text{and} \quad \hat{\mathcal{L}}_{\mathcal{N}} = \bigcup_{i=1}^{n} \hat{\ell}_{N_i}.$$

The set $\hat{\mathcal{L}}_{\mathcal{N}}$ determines the train-track $\mathbb{T}_{\mathcal{N}}$ with atlas $\mathcal{B}(f, \rho)$ as constructed in Lemma 12.1. Let $\mathcal{H}_{\mathcal{N}} = \{h_{(N_i,N_j)} : \ell^D_{(N_i,N_j)} \to \ell^C_{(N_i,N_j)} | (N_i, N_j) \in \mathcal{P}_{\mathcal{N}}\}$ be the (stable) primitive holonomic system associated to the Markov partition $\mathcal{N}$. By construction, for every $(N_i, N_j) \in \mathcal{P}_{\mathcal{N}}$ there is a sequence $h_{\alpha_1}, \ldots, h_{\alpha_n}$ of holonomies in $\mathcal{H}_{\mathcal{M}}$ such that

$$h_{(N_i,N_j)} = h_{\alpha_n} \circ \ldots \circ h_{\alpha_1} | \hat{\ell}^D_{N_i}.$$

Let

$$\psi_{(N_i,N_j)} : \hat{\ell}^D_{(N_i,N_j)} \to \hat{\ell}^D_{(N_i,N_j)}$$

be given by $\psi_{(N_i,N_j)} = \phi_{\alpha_n} \circ \ldots \circ \phi_{\alpha_1}$, where $\phi_{\alpha_i} \in \Phi_{f,\mathcal{M}}$ and $\phi_{\alpha_i} | \ell^D_{(N_i,N_j)} = h_{\alpha_i} | \ell^D_{(N_i,N_j)}$. Set

$$\Psi = \left\{ \psi_{(N_i,N_j)} : \hat{\ell}^D_{(N_i,N_j)} \to \hat{\ell}^C_{(N_i,N_j)} | (N_i, N_j) \in \mathcal{P}_{\mathcal{N}} \right\}.$$

Let $\Phi_{f,\mathcal{N}}$ be as constructed in Lemma 12.2. Hence, $\Psi = \Phi_{f,\mathcal{N}}$, and, so, Ψ is a C^{1+H} arc exchange system. Since the set $\mathcal{S}(\Phi_{f,\mathcal{M}}, \Phi_{f,\mathcal{N}})$ of all sequences $\alpha_1 \ldots \alpha_n$ such that $\psi_{(N_i,N_j)} = \phi_{\alpha_n} \circ \ldots \circ \phi_{\alpha_1}$, for some $(N_i, N_j) \in P_{\mathcal{N}}$, form a renormalizable sequence set, the C^{1+H} arc exchange system $\Phi_{f,\mathcal{N}}$ is a renormalization of $\Phi_{f,\mathcal{M}}$. Therefore, by §12.2, there is a well-defined renormalization operator $R = R_{f,\mathcal{M}} : [\Phi_{f,\mathcal{M}}]_{C^0} \to [\Phi_{f,\mathcal{N}}]_{C^0}$. Since $\mathcal{N} = f_*\mathcal{M}$ and $R\Phi_{f,\mathcal{M}} = \Phi_{f,\mathcal{N}}$, property (b) holds. □

Lemma 12.4. *The C^{1+H} arc exchange system $\Phi_{f,\mathcal{M}}$ is a C^{1+H} fixed point of renormalization, i.e $[R\Phi_{f,\mathcal{M}}]_{C^0} = [\Phi_{f,\mathcal{M}}]_{C^0}$, where $R = R_{f,\mathcal{M}} : [\Phi_{f,\mathcal{M}}]_{C^0} \to [\Phi_{f,\mathcal{N}}]_{C^0}$ is the renormalization operator.*

Proof. We construct a $C^{1+\alpha}$ conjugacy $\Theta : \mathbb{T}_{\mathcal{N}} \to \mathbb{T}_{\mathcal{M}}$ between $\Phi_{f,\mathcal{M}}$ and $\Phi_{f,\mathcal{N}}$. For every $N \in \mathcal{N}$ and $M = f^{-1}(N)$, there is a holonomy θ_N between the spanning leaf segments $f^{-1}(\ell_N)$ and ℓ_M. By Theorem 1.6 (see also Pinto and Rand [164]), the holonomy θ_N has a $C^{1+\alpha}$ diffeomorphic extension $\hat{\theta}_N : f^{-1}(\hat{\ell}_N) \to \hat{\ell}_M$. Let $\Theta : \mathbb{T}_{\mathcal{N}} \to \mathbb{T}_{\mathcal{M}}$ be the $C^{1+\alpha}$ diffeomorphism given by

$$\Theta | \hat{\ell}_N = \hat{\theta}_N \circ f^{-1}, \tag{12.4}$$

for every $N \in \mathcal{N}$. We observe that each pair

$$(N_i, N_j) \in \mathcal{P}_{\mathcal{N}}$$

determines a unique pair $(M_i, M_j) = (f^{-1}(N_i), f^{-1}(N_j)) \in \mathcal{P}_{\mathcal{M}}$, and vice-versa. By Lemma 12.3(b), it is enough to prove that Θ conjugates $\phi_{(N_i,N_j)}|\ell^D_{(N_i,N_j)}$ with $\phi_{(M_i,M_j)}|\ell^D_{(M_i,M_j)}$, for every $(N_i, N_j) \in \mathcal{P}_{\mathcal{N}}$, to show that $\Phi_{f,\mathcal{M}}$ is a C^{1+H} fixed point of renormalization.

By construction of the maps θ_{N_i} and θ_{N_j}, we have that

$$h_{(M_i,M_j)}|\ell^D_{(M_i,M_j)} = \theta_{N_i} \circ f^{-1} \circ h_{(N_i,N_j)} \circ f \circ \theta^{-1}_{N_j},$$

and so

$$\begin{aligned}\Theta \circ \psi_{(N_i,N_j)} \circ \Theta^{-1}|\ell^D_{(M_i,M_j)} &= \theta_{N_j} \circ f^{-1} \circ h_{(N_i,N_j)} \circ f \circ \theta^{-1}_{N_j} \\ &= h_{(M_i,M_j)} \\ &= \psi_{(M_i,M_j)},\end{aligned}$$

which ends the proof. □

12.3 Markov maps versus renormalization

The map $F : \hat{\mathbb{T}} \subset \mathbb{T} \to \mathbb{T}$ determines a $C^{1+\alpha}$ *Markov map*, with respect to the atlas $\mathcal{B}$ and with invariant set $\Omega \subset \hat{\mathbb{T}}$, if the following properties are satisfied:

(i) $\hat{\mathbb{T}} = \mathbb{T}$ or $\hat{\mathbb{T}}$ is a union of closed intervals.
(ii) $F : \mathbb{T} \to \mathbb{T}$ is a $C^{1+\alpha}$ diffeomorphism, for every (small) arc, with respect to the $C^{1+\alpha}$ atlas $\mathcal{B}$ on the train-track $\mathbb{T}$.
(iii) There exist $c > 0$ and $\lambda > 1$ such that, for every $x \in \Omega$,
$$|d(j_n \circ F^n \circ i^{-1})(x)| > c\lambda^n, \tag{12.5}$$
with respect to charts $i, j_n \in \mathcal{B}$.
(iv) The map F admits a Markov partition $\{K_1, \ldots, K_m\}$, i.e. there exists a finite set of arcs $\{\hat{K}_1, \ldots, \hat{K}_m\}$ such that (a) $K_i = \hat{K}_i \cap \Omega$, (b) $\cup_{i=1}^m \partial \hat{K}_i \subset \Omega$ and (c) $F\left(\partial \hat{K}_i\right) \subset \cup_{i=1}^m \partial \hat{K}_i$, for every $j = 1, \ldots, m$.

Let $\underline{F} : L_{\mathcal{M}} \to L_{\mathcal{M}}$ be the map induced by the action of f^{-1} on stable leaf segments, i.e. $\underline{F}(x) = \pi \circ f^{-1}(x)$ for every $x \in L$ (see (12.2)). Since f is a local diffeomorphism, the map $\underline{F}$ is a local homeomorphism. Let $\tilde{F} : k_{\mathcal{M}}(L_{\mathcal{M}}) \to k_{\mathcal{M}}(L_{\mathcal{M}})$ be the map defined by $\tilde{F} = k_{\mathcal{M}} \circ \underline{F} \circ k^{-1}_{\mathcal{M}}$. Since the holonomies have $C^{1+\alpha}$ extensions (see Theorem 1.6 and also Pinto and Rand [164]), and the map f is $C^{1+\alpha}$, for some $\alpha > 0$, the map $\tilde{F}$ has a $C^{1+\alpha}$ extension $F_{f,\mathcal{M}} : \mathbb{T}_f \to \mathbb{T}_f$, with respect to the atlas $\mathcal{B}^\iota(f, \rho)$, (not uniquely determined) that is a $C^{1+\alpha}$ Markov map with Markov partition $\{k_{\mathcal{M}} \circ \pi(M_1), \ldots, k_{\mathcal{M}} \circ \pi(M_l)\}$, where $\mathcal{M} = \{M_1, \ldots, M_l\}$ is the Markov partition of f (see also Pinto and Rand [163]). Hence, the map $F_{f,\mathcal{M}} : \mathbb{T}_f \to \mathbb{T}_f$ constructed above is a $C^{1+\alpha}$ Markov map.

Definition 31 *Let $h : \Omega_\Phi \to \Omega_\Psi$ be the topological conjugacy between a C^{1+H} arc exchange system $\Psi = \{\psi_i : I_{\psi_i} \to J_{\psi_i}; i = 1, \ldots, m\}$ and $\Phi_{f,\mathcal{M}} = \{\phi_i : I_{\phi_i} \to J_{\phi_i}; i = 1, \ldots, n\}$. We say that Ψ induces a C^{1+H} Markov map*

$$F_\Psi : \mathbb{T}_\Psi \to \mathbb{T}_\Psi,$$

if F_Ψ is a $C^{1+\alpha}$ Markov map, for some $\alpha > 0$, and $F_\Psi \circ h(x) = h \circ F_{f,\mathcal{M}}(x)$, for every $x \in \Omega_\Psi$.

Let us suppose that the C^{1+H} arc exchange system Ψ is a C^0 fixed point of renormalization $[R\Psi]_{C^0} = [\Psi]_{C^0}$. In this case, Ψ is an infinitely renormalizable C^{1+H} arc exchange system, i.e there is an infinite sequence

$$\left(R^m\Psi = \left\{\psi_i^{(m)} : \tilde{I}_{\psi_i}^{(m)} \to \tilde{J}_{\psi_i}^{(m)}; i = 1, \ldots, n(m)\right\}\right)_{m \geq 1}$$

of arc exchange systems inductively determined, for every $m \geq 1$, by $R^m\Psi = R(R^{m-1}\Psi)$.

Set

$$L_m^{(1)} = \left\{\psi_{s_k^i}^{(m)} \circ \ldots \circ \psi_{s_1^i}^{(m)} \left(I_{\psi_i}^{(m+1)}\right) : I_{\psi_i}^{(m+1)} \subset I_{\psi_{s_1^i}}^{(m)}, 0 \leq k \leq k(\underline{s}_i), \underline{s}_i \in \mathcal{S}\right\}.$$

Set, inductively on $j \geq 1$, the sets

$$L_m^{(j)} = \left\{\psi_{s_k^i}^{(m)} \circ \ldots \circ \psi_{s_1^i}^{(m)}(I) : I \in L_{m+1}^{(j-1)}, I \subset I_{\psi_{s_1^i}}^{(m)}, 0 \leq k \leq k(\underline{s}_i), \underline{s}_i \in \mathcal{S}\right\}.$$

By construction, $L_m^{(j+1)} \subset L_m^{(j)}$ and $\Omega_{R^m\Psi} = \cap_{j \geq 1} L_m^{(j)}$. We call $L_m^{(j)}$ the *j-th level of the partition of $R^m\Psi$*. Let the j-gap set $G_m^{(j)}$ of $R^m\Psi$ be the set of all maximal closed intervals I such that $I \subset J$ for some $J \in L_m^{(j-1)}$ and $\mathrm{int} I \cap K = \emptyset$, for every $K \in L_m^{(j)}$. We say that the C^{1+H} arc exchange system Ψ *has bounded geometry*, if there are constants $0 < c_1, c_2 < 1$ such that, for all $j \geq 1$ and all intervals $I \in L_0^{(j)} \cup G_0^{(j)}$ contained in a same interval $K \in L_0^{(j-1)}$, we have $c_1 < |\zeta(I)|/|\zeta(K)| < c_2$, where the length is measured with respect to any chart ζ in the $C^{1+\alpha}$ atlas $\mathcal{B}_\Psi$.

Lemma 12.5. *Let $\Phi_{f,\mathcal{M}}$ be a C^{1+H} arc exchange system induced by $(f, \Lambda, \mathcal{M})$. A C^{1+H} arc exchange system $\Psi \in [\Phi_{f,\mathcal{M}}]_{C^0}$, with bounded geometry, determines a C^{1+H} Markov map F_Ψ topologically conjugate to $F_{f,\mathcal{M}}$ if, and only if, Ψ is a C^{1+H} fixed point of the renormalization operator $R_{f,\mathcal{M}}$.*

Remark 12.6. Lemma 12.5 also holds for $C^{1,\alpha}$ regularities.

Proof of Lemma 12.5. For simplicity of notation, let us denote $k_{\mathcal{M}}$ by k (see (12.1)). Let $\Theta : K_{\mathcal{N}} \to K_{\mathcal{M}}$ be the $C^{1+\alpha}$ diffeomorphism as constructed in (12.4). For every $N \in \mathcal{N}$, let $M = f^{-1}(N) \in \mathcal{M}$. Recall that $\ell_N \subset \ell_{M_{\sigma(i)}} \subset L_{\mathcal{M}}$ (see (12.3)). By construction of $F = F_{f,\mathcal{M}}$ and Θ, the spanning leaf segment $\ell_N \subset L_{\mathcal{N}}$ has the property that $F \circ k(\ell_N) = k(\ell_M)$ and $F|k(\ell_N) = \Theta$. Therefore,

$$F|K_{\mathcal{N}} = \Theta. \tag{12.6}$$

Every leaf segment $\ell \subset L_{\mathcal{M}}$ with the property that $F \circ k(\ell) = k(\ell_M)$ is a spanning leaf segment of N. Therefore, there is a sequence $e_{\alpha_1}, \ldots, e_{\alpha_p}$ of arc exchange maps in $\Phi = \Phi_{f,\mathcal{M}}$ such that

$$e_{\alpha_p} \circ \ldots \circ e_{\alpha_1}(k(\ell)) = k(\ell_N).$$

Furthermore,

$$F|k(\ell) = \Theta \circ e_{\alpha_p} \circ \ldots \circ e_{\alpha_1}. \tag{12.7}$$

Let $\xi : \cup_{i=1}^n I_{\phi_i} \to \cup_{i=1}^n I_{\psi_i}$ be a homeomorphic extension of the conjugacy between Φ and Ψ. For every $e \in \Phi$, there is a unique $\underline{e} \in \Psi$ such that $\underline{e} = \xi \circ e \circ \xi^{-1}$. Since F_Ψ is topologically conjugate to F, by (12.6), we have that

$$F_\Psi|\xi(K_{\mathcal{N}}) = \Theta_\Psi, \tag{12.8}$$

where $\Theta_\Psi : \xi(K_{\mathcal{N}}) \to \xi(K_{\mathcal{M}})$ is a homeomorphic extension of the conjugacy between Ψ and its renormalization $R\Psi$. Letting $\tilde{\ell}_N$, $\tilde{\ell}$ and $e_{\alpha_1}, \ldots, e_{\alpha_p}$ be as above, by (12.7), we obtain that

$$F_\Psi|\xi \circ k(\tilde{\ell}) = \Theta_\Psi \circ \underline{e}_{\alpha_p} \circ \ldots \circ \underline{e}_{\alpha_1}. \tag{12.9}$$

By (12.8), if F_Ψ is $C^{1+\alpha}$, then Θ_Ψ is $C^{1+\alpha}$ (also along arcs containing junctions). By (12.9), if Θ_Ψ is $C^{1+\alpha}$, then F_Ψ is locally a $C^{1+\alpha}$ diffeomorphism.

Let $L_0^{(j)}$ be the j-th level of the partition of Ψ. By construction, every interval $I \in L_0^{(j)}$ has the property that $F_\Psi^{j-1}(I)$ is an element of the Markov partition of F_Ψ (this property characterizes $L_0^{(j)}$). In particular, the map F_Ψ sends each interval $I \in L_0^{(j)}$ onto an interval $F_\Psi(I) \in L_0^{(j-1)}$ for every $j > 0$. Hence, if Ψ has bounded geometry we obtain that the length of the sets in $L_0^{(j)}$ converge exponentially fast to 0 when j tends to infinity. Therefore, using the Mean Value Theorem, we obtain that if Ψ has bounded geometry, then F_Ψ satisfies property (ii) and, conversely, if F_Ψ satisfies property (ii) we obtain that Ψ has bounded geometry. So, we conclude that if Ψ is a $C^{1+\alpha}$ arc exchange system, with bounded geometry, then F_Ψ is a $C^{1+\alpha}$ Markov map, and vice-versa. □

12.4 C^{1+H} flexibility

Let $(f, \Lambda, \mathcal{M})$ be a C^{1+H} hyperbolic diffeomorphism. Let $\mathcal{C}^\iota_{f,\mathcal{M}}$ be the topological conjugacy class of $\Phi^\iota_{f,\mathcal{M}}$. Let $\mathcal{F}$ be the set of all C^{1+H} hyperbolic diffeomorphisms topologically conjugate to f (see §2.1).

Theorem 12.7. *There is a unique map*

$$\mathcal{T}^\iota_{f,\mathcal{M}} : \mathcal{F} = \{[g]_{C^{1+H}} : g \in \mathcal{F}\} \to \mathcal{C}^\iota = \left\{[\Phi^\iota]_{C^{1+H}} : \Phi^\iota \in \mathcal{C}^\iota_{f,\mathcal{M}}\right\}$$

defined by $\mathcal{T}^\iota_{f,\mathcal{M}}([g]_{C^{1+H}}) = [\Phi^\iota_{g,\mathcal{M}_g}]_{C^{1+H}}$*, where* $\mathcal{M}_g$ *is the pushforword of the Markov partition* $\mathcal{M}$ *of* f *by the topological conjugacy between* f *and* g*. The map* $\mathcal{T}_\iota = \mathcal{T}^\iota_{f,\mathcal{M}} : \mathcal{F} \to \mathcal{C}$ *has the following properties:*

(a) If $[\Phi^\iota]_{C^{1+H}} = \mathcal{T}_\iota[g]_{C^{1+H}}$*, then* $HD(\Omega^\iota_\Phi) = HD(\Lambda^\iota_g)$*;*
(b) $\mathcal{T}_\iota(\mathcal{F}) = \mathcal{C}^\iota_R$*, where* $\mathcal{C}^\iota_R \subset \mathcal{C}$ *is the set of all* C^{1+H} *conjugacy classes* $[\Phi^\iota]_{C^{1+H}} \in \mathcal{C}$ *that are* C^{1+H} *fixed points of renormalization,* $[R^\iota\Phi^\iota]_{C^{1+H}} = [\Phi^\iota]_{C^{1+H}}$*;*
(c) For every pair $([\Phi^s]_{C^{1+H}}, [\Phi^u]_{C^{1+H}}) \in \mathcal{C}^s_R \times \mathcal{C}^u_R$*, there is a unique* C^{1+H} *conjugacy class of* C^{1+H} *hyperbolic diffeomorphisms*

$$g \in \mathcal{T}_s^{-1}([\Phi^s]_{C^{1+H}}) \cap \mathcal{T}_u^{-1}([\Phi^u]_{C^{1+H}});$$

(d) For every $[\Phi^\iota]_{C^{1+H}} \in \mathcal{C}^\iota_R$ *there is a unique Lipschitz conjugacy class of* C^{1+H} *hyperbolic diffeomorphisms* $g \in \mathcal{T}_\iota^{-1}([\Phi^\iota]_{C^{1+H}})$ *that admits an invariant measure absolutely continuous with respect to the Hausdorff measure on* Λ_g*;*
(e) The set $\mathcal{C}^\iota_R$ *is characterized by a moduli space consisting of solenoid functions;*
(f) The set $\mathcal{C}^\iota_L$ *consisting of all Lipschitz conjugacy classes in* $\mathcal{C}^\iota_R$ *is also characterized by a moduli space consisting of measure solenoid functions.*

The above solenoid functions and measure solenoid functions are introduced in Pinto and Rand [163, 167], where they are used to construct moduli spaces for the set of all C^{1+H} and Lipschitz conjugacy classes of C^{1+H} hyperbolic diffeomorphisms (see Chapter 3). If $HD(\Lambda^{\iota'}) = 1$, then, in Theorem 12.7, the Lipschitz conjugacy classes coincide with the C^{1+H} conjugacy classes, and, so, $\mathcal{C}^\iota_L = \mathcal{C}^\iota_R$.

Remark 12.8. We note that in Theorem 12.7, if the ι-lamination of the hyperbolic basic set Λ is orientable, then the ι-arc exchange systems in $\mathcal{C}^\iota_{f,\mathcal{M}}$ are determined by ι-arc exchange maps.

Proof of Theorem 12.7. By Theorem 1.6 (see also Pinto and Rand [164]), the basic holonomies are $C^{1+\alpha}$ diffeomorphisms with respect to the $C^{1+\alpha}$ atlases $\mathcal{A}^\iota(g_1, \rho_1)$ and $\mathcal{A}^\iota(g_2, \rho_2)$, for some $\alpha > 0$. Hence, there is a $C^{1+\alpha}$

diffeomorphism $\underline{u} : \mathbb{T}_{g_1} \to \mathbb{T}_{g_2}$, with respect to the atlases $\mathcal{B}^\iota(g_1, \rho_1)$ on $\mathbb{T}_{g_1}$ and $\mathcal{A}^\iota(g_2, \rho_2)$ on $\mathbb{T}_{g_2}$, such that $\underline{u} \circ \pi_{g_1} = \pi_{g_2} \circ \underline{u}$, where $\pi_{g_1} : \Lambda_{g_1} \to \mathbb{T}_{g_1}$ and $\pi_{g_2} : \Lambda_{g_2} \to \mathbb{T}_{g_2}$ are the natural projections. Hence, the $C^{1+\alpha}$ induced arc exchange system Φ_{g_1} is $C^{1+\alpha}$ conjugate to the $C^{1+\alpha}$ induced arc exchange system Φ_{g_1}.

Proof of statement (a). Since the holonomies are $C^{1+\alpha}$ (see Theorem 1.6 and also Pinto and Rand [164]), the Hausdorff dimension of the stable leaf segments ℓ is the same independently of the stable leaf segment considered, and so equal to $HD(\Lambda_g^s)$. In particular, all leaf segments $\ell_{M_g} \in \mathcal{I}_g$ have the same Hausdorff dimension which is equal to the Hausdorff dimension of L_g. Since the arc invariant set $\mathbb{T}_{\Phi_{g,\mathcal{M}_g}}$ is equal to $k(L_g)$, the Hausdorff dimension $HD\left(\mathbb{T}_{\Phi_{g,\mathcal{M}_g}}\right)$ is equal to $HD(\Lambda_g^s)$.

Proof of statement (b). By Lemma 12.4, if $g \in \mathcal{F}$, then the C^{1+H} arc exchange system $\Phi_{g,\mathcal{M}_g}$ is a fixed point of the renormalization operator $R_{g,\mathcal{M}_g}$ that, by construction, is the same as $R_{f,\mathcal{M}}$. Hence, $\mathcal{T}(\mathcal{F}) \subset \mathcal{C}_R$.

The proof that $\mathcal{T}(\mathcal{F}) \supset \mathcal{C}_R$ follows from the proof of the statement (c) below.

Proof of statement (c). Let Φ be a C^{1+H} arc exchange system such that $[R\Phi]_{C^{1+H}} = [\Phi]_{C^{1+H}}$. Since $[R\Phi]_{C^{1+H}} = [\Phi]_{C^{1+H}}$, by Lemma 12.5, the C^{1+H} arc exchange system Φ induces a Markov map F_Φ. Therefore, (Φ, F_Φ) is equivalent to a $C^{1+\alpha}$ self-renormalizable structure as defined in Chapter 4.

By Theorem 1.6 (see also Pinto and Rand [164, 167]), there is a one-to-one correspondence between C^{1+H} conjugacy classes of (Φ, F_Φ) and C^{1+H} conjugacy classes of C^{1+H} diffeomorphisms $g(\Phi, F_\Phi)$ with hyperbolic invariant set Λ_g, and with an invariant measure absolutely continuous with respect to the Hausdorff measure.

Proof of statement (d). Let Φ be a C^{1+H} arc exchange system such that $[R\Phi]_{C^{1+H}} = [\Phi]_{C^{1+H}}$. Since $[R\Phi]_{C^{1+H}} = [\Phi]_{C^{1+H}}$, by Lemma 12.5, the C^{1+H} arc exchange system Φ induces a Markov map F_Φ. Let $\mathcal{C}_F$ be the set of all C^{1+H} conjugacy classes of pairs (Φ, F_Φ). Hence, there is a one-to-one map $m_1 : \mathcal{C}_R \to \mathcal{C}_F$ given by $m_1(\Phi) = (\Phi, F_\Phi)$. By Lemma 9.2 (see also Pinto and Rand [166, 167]), there is a well-defined Teichmüller space TS consisting of solenoid functions, and a one-to-one map $m_2 : TS \to \mathcal{C}_F$ given by $m_2(s) = (\Phi, F_\Phi)$. Therefore, $m_1^{-1} \circ m_2 : TS \to \mathcal{C}_R$ is a one-to-one map. □

12.5 $C^{1,HD}$ rigidity

Let us present the following notion of $C^{1,HD}$ regularity of a function (see §5.1).

Definition 32 *Let $\phi : I \to J$ be a homeomorphism between open sets $I \subset \mathbb{R}$ and $J \subset \mathbb{R}$. If $0 < \alpha < 1$, then ϕ is said to be $C^{1,\alpha}$ if ϕ is differentiable and for all points $x, y \in I$*

$$|\phi'(y) - \phi'(x)| \leq \chi_\phi(|y - x|), \tag{12.10}$$

where the positive function $\chi_\phi(t)$ *satisfies* $\lim_{t\to 0} \chi_\phi(t)/t^\alpha = 0$. ϕ *is said to be* $C^{1,\alpha}$, *if, for all points* $x, y \in I$,

$$\left|\log \phi'(x) + \log \phi'(y) - 2\log \phi'\left(\frac{x+y}{2}\right)\right| \leq \chi(|y - x|),$$

where the positive function $\chi(t)$ *satisfies* $\lim_{t\to 0} \chi(t)/t = 0$.

In particular, for every $\beta > \alpha > 0$, a $C^{1+\beta}$ diffeomorphism is $C^{1,\alpha}$, and, for every $\gamma > 0$, a $C^{2+\gamma}$ diffeomorphism is $C^{1,1}$. We note that the regularity $C^{1,1}$ (also denoted by $C^{1+zigmund}$) of a diffeomorphism θ used in this chapter is stronger than the regularity $C^{1+Zigmund}$ (see de Melo and van Strien [99]). The importance of these $C^{1,\alpha}$ smoothness classes for a diffeomorphism $\theta : I \to J$ follows from the fact that if $0 < \alpha < 1$, then the map θ will distort ratios of lengths of short intervals in an interval $K \subset I$ by an amount that is $o(|I|^\alpha)$, and if $\alpha = 1$ the map θ will distort the cross-ratios of quadruples of points in an interval $K \subset I$ by an amount that is $o(|I|)$ (see Chapter 5).

An *arc exchange system* $(\Phi, \mathcal{J}_\Phi, \mathbb{T}_\Phi, \mathcal{B}_\Phi)$ is *affine*, if $\mathcal{B}_\Phi$ is an affine atlas and the maps in Φ and in $\mathcal{J}_\Phi$ are affine with respect to the charts in $\mathcal{B}_\Phi$.

Theorem 12.9. *Let* $\mathcal{C}^\iota_{f,\mathcal{M}}$ *be the topological conjugacy class of* C^{1+H} ι*-arc exchange systems determined by a* C^{1+H} *hyperbolic diffeomorphism* $(f, \Lambda, \mathcal{M})$ *(as in Theorem 12.7). Every* $C^{1,HD(\Lambda^\iota)}$ *arc exchange system* $\Phi \in \mathcal{C}_{f,\mathcal{M}}$, *with bounded geometry, that is a* $C^{1,HD(\Omega_\Phi)}$ *fixed point of renormalization operator, i.e* $[R_{f,\mathcal{M}}\Phi]_{C^{1,HD(\Omega_\Phi)}} = [\Phi]_{C^{1,HD(\Omega_\Phi)}}$, *is* $C^{1,HD(\Lambda^\iota)}$ *conjugate to an affine* ι*-arc exchange system that is an affine fixed point of renormalization. Furthermore, the* $C^{1,HD(\Lambda^\iota)}$ *arc exchange system* $\Phi \in \mathcal{C}_{f,\mathcal{M}}$ *determines stable transversely affine ratio functions* r_Φ.

Corollary 12.10. *Let* $\mathcal{C}_{f,\mathcal{M}}$ *be the topological conjugacy class of* C^{1+H} *Cantor exchange systems determined by a* C^{1+H} *diffeomorphism* f *with codimension 1 hyperbolic attractor* Λ *and with a Markov partition* $\mathcal{M}$ *satisfying the disjointness property (as in Theorem 12.7). There is no* $C^{1,HD(\Omega_\Phi)}$ *Cantor exchange system* $\Phi \in \mathcal{C}_{f,\mathcal{M}}$, *with bounded geometry, that is a* $C^{1,HD(\Omega_\Phi)}$ *fixed point of renormalization operator, i.e* $[R_{f,\mathcal{M}}\Phi]_{C^{1,HD(\Omega_\Phi)}} = [\Phi]_{C^{1,HD(\Omega_\Phi)}}$.

Proof. By Theorem 12.9, we obtain that r_Φ is a stable transversely affine ratio function. However, putting together Theorem 5.9 and Lemma 5.11, there are no stable transversely affine ratio functions with respect to the stable lamination of Λ_f, and so we get a contradiction. □

Proof of Theorem 12.9. Let us suppose that the arc exchange system Ψ is a $C^{1,\alpha}$ fixed point of the renormalization operator $R_{f,\mathcal{M}}$ with $\alpha = HD(\mathbb{T}_\Psi)$ and with bounded geometry. Hence, by Lemma 12.5, Ψ induces a $C^{1,\alpha}$ Markov map F_Ψ. Let ξ be the homeomorphic extension of the conjugacy between Φ

and Ψ, and set $\eta = \xi \circ k \circ \pi$. We will consider the following two distinct cases: (a) $HD(\Lambda^\iota) < 1$ and (b) $HD(\Lambda^\iota) = 1$.

Case $HD(\Lambda^\iota) < 1$. Let T_n be the set of all pairs (I, J) such that (i) I is a stable leaf n-cylinder, (ii) J is a stable leaf n-gap cylinder, and (iii) I and J have a unique common endpoint. Using the Mean Value Theorem and that F_Ψ is a $C^{1,\alpha}$ Markov map, the function $r : \cup_{n \geq 1} T_n \to \mathbb{R}^+$ given by

$$r(I, J) = \lim_{m \to +\infty} \frac{|\eta \circ f^m(J)|}{|\eta \circ f^m(I)|}$$

is well-defined, where $|L|$ means the length of the smallest interval containing $L \subset \mathbb{R}$. By bounded geomatry of Ψ, we obtain that r is bounded away from zero. Furthermore, using that F_Ψ is a $C^{1,\alpha}$ Markov map, for every pair $(I, J) \in T_n$, we get

$$\frac{|\eta(J)|}{|\eta(I)|} \left(1 - C_n \left(|\eta(I \cup J)|^\alpha\right)\right) \leq r(I, J) \leq \frac{|\eta(J)|}{|\eta(I)|} \left(1 + C_n \left(|\eta(I \cup J)|^\alpha\right)\right), \tag{12.11}$$

where $C_n \in \mathbb{R}_0^+$ converges to zero when n tends to infinity.

Let $h = h_{(M,N)} : \tilde{\ell}^D_{(M,N)} \to \tilde{\ell}^C_{(M,N)}$ be a ι-primitive holonomy. Since the arc exchange system is $C^{1,\alpha}$, for every $(I, J) \in T_n$ such that $I \cup J \subset \tilde{\ell}^D_{(M,N)}$, we get

$$1 - C_n |\eta(I \cup J)|^\alpha \leq \frac{|\eta(I)|}{|\eta(J)|} \frac{|\eta \circ h(J)|}{|\eta \circ h(I)|} \leq 1 + C_n |\eta(I \cup J)|^\alpha, \tag{12.12}$$

where $C_n \in \mathbb{R}_0^+$ converges to zero when n tends to infinity.

From (12.11), we obtain that

$$\frac{|\eta(I)|}{|\eta(J)|} \frac{|\eta \circ h(J)|}{|\eta \circ h(I)|} \left(1 - C_n |\eta(I \cup J)|^\alpha\right) \leq \frac{r(h(I), h(J))}{r(I, J)} \leq$$

$$\leq \frac{|\eta(I)|}{|\eta(J)|} \frac{|\eta \circ h_\alpha(J)|}{|\eta \circ h_\alpha(I)|} \left(1 + C_n |\eta(I \cup J)|^\alpha\right).$$

Thus, using (12.12) we get

$$1 - C'_n |\eta(I \cup J)|^\alpha \leq \frac{r(h(I), h(J))}{r(I, J)} \leq 1 + C'_n |\eta(I \cup J)|^\alpha,$$

where $C'_n \in \mathbb{R}_0^+$ converges to zero when n tends to infinity.

Since $\alpha = HD(\mathbb{T}_\Psi)$, by Theorem 5.4 (see also Pinto and Rand [165]), we obtain that r is a stable transversely affine ratio function.

Case $HD(\Lambda^\iota) = 1$. Let J_0, J_1 and J_2 be distinct leaf segments such that J_0 and J_1 have a common endpoint, and J_1 and J_2 have also a common endpoint. Let the *cross-ratio* $cr(J_0, J_1, J_2)$ be given by

$$cr(J_0, J_1, J_2) = \frac{1 + r(J_1, J_0)}{r(J_2, J_0 \cup J_1 \cup J_2)}.$$

A similar argument to the one above gives that

$$1 - C_n|\eta(J_0 \cup J_1 \cup J_2)| \leq \frac{cr(h(J_0), h(J_1), h(J_2))}{cr(J_0, J_1, J_2)} \leq 1 + C_n|\eta(J_0 \cup J_1 \cup J_2)|,$$

where $C_n \in \mathbb{R}_0^+$ converges to zero when n tends to infinity. Hence, by Theorem 5.4 (see also Pinto and Rand [165]), we obtain that r is a stable transversely affine ratio function. Therefore, the ratio function r determines an affine atlas $\mathcal{A}(r)$ on the ι-leaf segments such that the holonomies and f are affine. Thus, the atlas $\mathcal{B}(r)$, on the train-track $\mathbb{T}_f$, induced by $\mathcal{A}(r)$ is an affine atlas such that the arc exchange system is affine and the Markov map is also affine. Therefore, the arc exchange system is an affine fixed point of renormalization. □

12.6 Further literature

The works of Masur [80], Penner [149], Thurston [234] and Veech [235] show a strong link between affine interval exchange maps and Anosov and pseudo-Anosov maps. E. Ghys and D. Sullivan (see Cawley [21]) observed that Anosov diffeomorphisms on the torus determine circle diffeomorphisms that have an associated renormalization operator. Denjoy [25] has shown the existence of upper bounds for the smoothness of Denjoy maps. Harrison [45] has conjectured that there are no $C^{1+\gamma}$ Denjoy maps with $\gamma > HD$. This conjecture has been proved, partially, by Norton in [105] and by Kra and Schmeling in [67]. This chapter is based on Pinto, Rand and Ferreira [171] and [172].

13

Golden tilings (in collaboration with J.P. Almeida and A. Portela)

We prove a one-to-one correspondence between: (i) Pinto's golden tilings; (ii) smooth conjugacy classes of golden diffeomorphisms of the circle that are fixed points of renormalization; (iii) smooth conjugacy classes of Anosov difeomorphisms, with an invariant measure absolutely continuous with respect to the Lebesgue measure, that are topologically conjugated to the Anosov automorphism $G_A(x, y) = (x + y, x)$; and (iv) solenoid functions. The solenoid functions give a parametrization of the infinite dimensional space consisting of the mathematical objects described in the above equivalences.

13.1 Golden difeomorphisms

We will denote by $\mathbb{S}$ a *clockwise oriented circle* homeomorphic to the circle $\mathbb{S}^1 = \mathbb{R}/(1+\gamma)\mathbb{Z}$, with γ equal to the inverse of the golden number $\left(1+\sqrt{5}\right)/2$. An *arc* in $\mathbb{S}$ is the image of a non trivial interval I in $\mathbb{R}$ by an homeomorphism $\alpha : I \to \mathbb{S}$. If I is closed (resp. open) we say that $\alpha(I)$ is a *closed* (resp. *open*) *arc* in $\mathbb{S}$. We denote by (a, b) (resp. $[a, b]$) the positively oriented open (resp. closed) arc on $\mathbb{S}$ starting at the point $a \in \mathbb{S}$ and ending at the point $b \in \mathbb{S}$. A C^{1+} atlas $\mathcal{A}$ of $\mathbb{S}$ is a set of charts such that (i) every small arc of $\mathbb{S}$ is contained in the domain of some chart in A, and (ii) the overlap maps are $C^{1+\alpha}$ compatible, for some $\alpha > 0$.

A C^{1+} *golden diffeomorphism* is a triple $(g, \mathbb{S}, \mathcal{A})$ where g is a C^{1+} diffeomorphism, with respect to the $C^{1+\alpha}$ atlas $\mathcal{A}$, for some $\alpha > 0$, and g is quasi-symmetric conjugated to the rigid rotation $r_\gamma : \mathbb{S}^1 \to \mathbb{S}^1$, with rotation number equal to γ. In order to simplify the notation, we will denote the C^{1+} golden diffeomorphism $(g, \mathbb{S}, \mathcal{A})$ only by g.

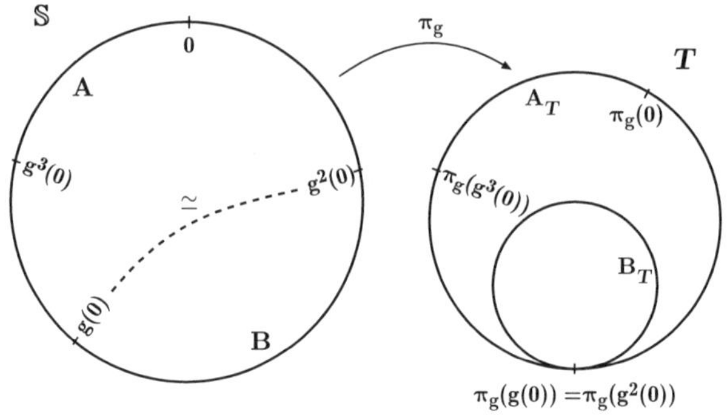

Fig. 13.1. The equivalence relation in $\mathbb{S}$ that gives rise to the train-track T.

13.1.1 Golden train-track

Let us mark a point in $\mathbb{S}$, that we will denote by $0 \in \mathbb{S}$, from now on. Let $\mathrm{A} = \left[g(0), g^2(0)\right]$ be the oriented closed arc in $\mathbb{S}$, with endpoints $g(0)$ and $g^2(0)$ and containing the point 0. Let $\mathrm{B} = \left[g^2(0), g(0)\right]$ be the oriented closed arc in $\mathbb{S}$, with endpoints $g(0)$ and $g^2(0)$ and not containing the point 0. We introduce an *equivalence relation* $\sim$ in $\mathbb{S}$ by identifying the points $g(0)$ and $g^2(0)$. We call the oriented topological space $T(\mathbb{S}, g) = \mathbb{S}/\sim$ by *train-track* (see Figure 13.1). We consider $T = T(\mathbb{S}, g)$ equipped with the quotient topology. Let $\pi_g : \mathbb{S} \to T$ be the natural projection. We call the point $\pi_g(g(0)) = \pi_g(g^2(0)) \in T$ the junction ξ of the train-track T. Let $\mathrm{A}_T = \mathrm{A}_T(\mathbb{S}, g) \subset T$ be the projection by π_g of the closed arc A, and let $\mathrm{B}_T = \mathrm{B}_T(\mathbb{S}, g) \subset T$ be the projection by π_g of the closed arc B. A *parametrization* in T is the image of a non trivial interval I in $\mathbb{R}$ by a homeomorphism $\alpha : I \to T$ satisfying the following restrictions:

(i) if $\xi \in \alpha(I)$, there exists $\delta_0 > 0$ such that for all $0 < \delta < \delta_0$, the points $\alpha(x - \delta)$ and $\alpha(x + \delta)$ do not belong simultaneously to B_T, where $x = \alpha^{-1}(\xi)$.

If I is closed (resp. open) we say that $\alpha(I)$ is a *closed* (resp. *open*) *arc* in T. A *chart* in T is the inverse of a parametrization. A *topological atlas* $\mathcal{B}$ on the train-track T is a set of charts $\{(j, J)\}$ on the train-track with the property that every small arc is contained in the domain of a chart in $\mathcal{B}$, i.e. for any open arc K on the train-track and any $x \in K$ there exists a chart $\{(j, J)\} \in \mathcal{B}$ such that $J \cap K$ is a non trivial open arc on the train-track and $x \in J \cap K$. A C^{1+} *atlas* $\mathcal{B}$ *in* T is a topological atlas $\mathcal{B}$ such that the overlap maps are $C^{1+\alpha}$ and have uniformly $C^{1+\alpha}$ bounded norm, for some $\alpha > 0$.

13.1.2 Golden arc exchange systems

The construction of the arc exchange systems, that we now present, is inspired in Rand's commuting pairs (see Rand [189]) and in Pinto-Rand's complete set of holonomies (see Pinto and Rand [165] and §13.2.2).

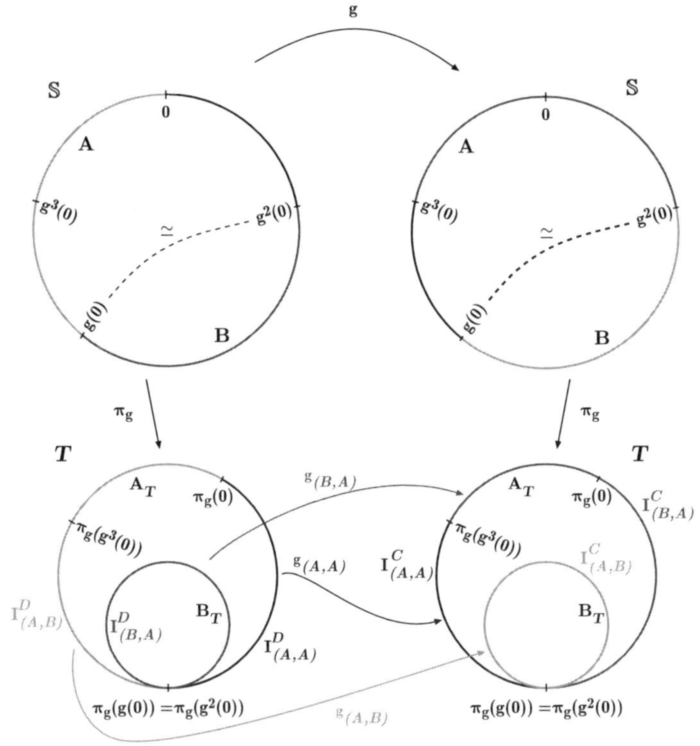

Fig. 13.2. The arc exchange maps for the train-track $T = T(\mathbb{S}, g)$.

The C^{1+} golden diffeomorphism $g : \mathbb{S} \to \mathbb{S}$ determines three maximal diffeomorphisms $g_{(A,A)}$, $g_{(A,B)}$ and $g_{(B,B)}$, on the train-track, with the property that the domain and the counterdomain of each diffeomorphism are either contained in A or in B, as we now describe: let $I^D_{(A,A)}$ be the arc $\pi_g([0, g^2(0)])$, let $I^D_{(A,B)}$ be the arc $\pi_g([g(0), 0])$, and let $I^D_{(B,B)}$ be the arc $\pi_g([g^2(0), g(0)])$. Let $I^C_{(A,A)}$ be the arc $\pi_g([g(0), g^3(0)])$, let $I^C_{(A,B)}$ be the arc $\pi_g([g^2(0), g(0)])$, and let $I^C_{(B,B)}$ be the arc $\pi_g([g^3(0), g^2(0)])$. Let $g_{(A,A)} : I^D_{(A,A)} \to I^C_{(A,A)}$ be the homeomorphism determined by $g_{(A,A)} \circ \pi_g = \pi_g \circ g$, let $g_{(A,B)} : I^D_{(A,B)} \to I^C_{(A,B)}$ be the homeomorphism determined by $g_{(A,B)} \circ \pi_g = \pi_g \circ g$, and let $g_{(B,B)}$: $I^D_{(B,B)} \to I^C_{(B,B)}$ be the homeomorphism determined by $g_{(B,B)} \circ \pi_g = \pi_g \circ g$. We call these maps and their inverses by *arc exchange maps*. The *arc exchange system*

$$E(g) = E(\mathbb{S}, g) = \left\{ g_{(A,A)}, g^{-1}_{(A,A)}, g_{(A,B)}, g^{-1}_{(A,B)}, g_{(B,B)}, g^{-1}_{(B,B)} \right\}$$

is the union of all arc exchange maps defined with respect to the train-track $T(\mathbb{S}, g)$ (see Figure 13.2).

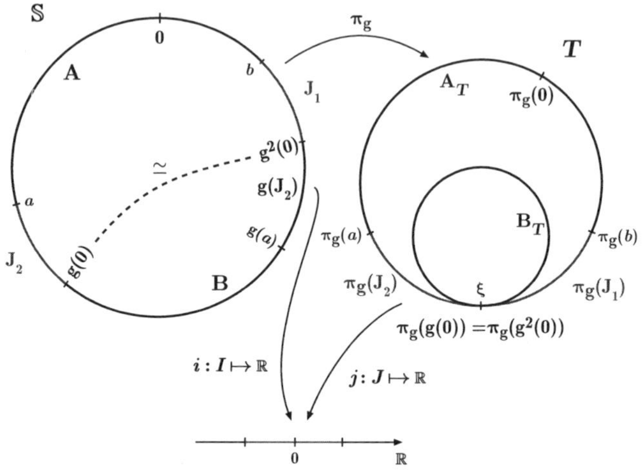

Fig. 13.3. Construction of the chart $j : J \to \mathbb{R}$ in case (ii).

Let $\mathcal{A}$ be an atlas in $\mathbb{S}$ for which g is C^{1+}. We are going to construct an atlas $\mathcal{B}$ in the golden train-track that is the extended pushforward $\mathcal{B}_{\mathcal{A}} = (\pi_g)_* \mathcal{A}$ of the atlas $\mathcal{A}$ in $\mathbb{S}$. If $x \in T \backslash \{\xi\}$, then there exists a sufficiently small open arc J, containing x, such that $\pi_g^{-1}(J)$ is contained in the domain of some chart (I, i) in $\mathcal{A}$. In this case, we define $(J, i \circ \pi_g^{-1})$ as a chart in $\mathcal{B}$. If $x = \xi$ and J is a small arc containing ξ, then either (i) $\pi_g^{-1}(J)$ is an arc in $\mathbb{S}$ or (ii) $\pi_g^{-1}(J)$ is a disconnected set that consists of a union of two connected components. In case (i), $\pi_g^{-1}(J)$ is connected and we define $\left(g, i \circ \pi_g^{-1}\right)$ as a chart in $\mathcal{B}$. In case (ii), $\pi_g^{-1}(J)$ is a disconnected set that is the union of two connected arcs J_1 and J_2 of the form $\left(b, g^2(0)\right]$ and $[g(0), a)$, respectively (see Figure 13.3). Let $(I, i) \in \mathcal{A}$ be a chart such that $I \supset (b, g(a))$. We define $j : J \to \mathbb{R}$ as follows:

$$j(x) = \begin{cases} i \circ \pi_g^{-1}(x), & \text{if } x \in \pi_g((b, g^2(0)]) \\ i \circ g \circ \pi_g^{-1}(x), & \text{if } x \in \pi_g([g(0), a)) \end{cases}$$

We call the atlas determined by these charts, the *extended pushforward atlas of* $\mathcal{A}$ and, by abuse of notation, we will denote it by $\mathcal{B}_{\mathcal{A}} = (\pi_g)_* \mathcal{A}$.

Definition 13.1. *An arc exchange system E is C^{1+} in the train-track T, with respect to a C^{1+} atlas $\mathcal{B}$, if the following properties are satisfied:*

(i) There is a quasi-symmetric homeomorphism $h : T\left(\mathbb{S}^1, r_\gamma\right) \to T$ that conjugates the exchange maps $e \in E$ with the exchange maps $e \in E\left(\mathbb{S}^1, r_\gamma\right)$, with respect to the atlas $\mathcal{B}_{iso}$.
(ii) If $e \in E$, then e is a $C^{1+\alpha}$ diffeomorphism, with respect to the charts in $\mathcal{B}$, for some $\alpha > 0$.
(iii) If $e_1 : I_1 \to J_1$ and $e_2 : I_2 \to J_2$ in E are such that (a) $I = I_1 \cup I_2$ and $J = J_1 \cup J_2$ are arcs, (b) $I_1 \cap I_2$ is a single point $\{p\}$ and (c) $e_1(p) = e_2(p)$, then the map $e : I \to J$ defined by $e|_{I_1} = e_1$ and $e|_{I_2} = e_2$ is a $C^{1+\alpha}$ diffeomorphism with respect the charts in $\mathcal{B}$, for some $\alpha > 0$. (It follows that $J = J_1 \cap J_2$ is the single point $e_1(p) = e_2(p)$.)

Let us consider the rigid rotation $r_\gamma : \mathbb{S}^1 \to \mathbb{S}^1$ with the atlas $\mathcal{A}_{iso}$ given by the local isometries with respect to the natural metric in $\mathbb{S}^1$ induced by the Euclidean metric in $\mathbb{R}$. The arc exchange system $E\left(\mathbb{S}^1, r_\gamma\right)$ is *rigid* with respect to the extended pushforward atlas $\mathcal{B}_{iso} = \left(\pi_{r_\gamma}\right)_* \mathcal{A}_{iso}$, i.e. the maps $e \in E(\mathbb{S}^1, r_\gamma)$ are translations in $\mathcal{B}_{iso}$.

Lemma 13.2. *(i) If g is a C^{1+} golden diffeomorphism with respect to a C^{1+} atlas $\mathcal{A}$, then the arc exchange system $E(g)$ is C^{1+} with respect to the extended pushforward $\mathcal{B}_{\mathcal{A}} = (\pi_g)_* \mathcal{A}$ of the C^{1+} atlas $\mathcal{A}$.*
(ii) If E is a C^{1+} arc exchange system with respect to a C^{1+} atlas $\mathcal{B}$, then the golden homeomorphism $g(E)$ is C^{1+} with respect to the pullback $\mathcal{A}_{\mathcal{B}} = (\pi_g)^* \mathcal{B}$ of the C^{1+} atlas $\mathcal{B}$.

Proof. Lemma 13.2 follows from the above construction of the arc exchange system E, and the definition of the extended pushforward atlas $\mathcal{B}_{\mathcal{A}} = (\pi_g)_* \mathcal{A}$. □

13.1.3 Golden renormalization

Feigenbaum [33, 34] and Coullet and Tresser [23] introduced renormalization for unimodal maps. The operator for general rotations was first defined in Rand et al. [196]. Sullivan pointed out that R_g has a smooth atlas, corresponding to the fact that the renormalization operator acts on the space of commuting pairs as introduced in Rand [188, 191]. Here, we follow a new, but equivalent, construction.

The *renormalization of* $(g, \mathbb{S}, \mathcal{A})$ is the triple $(R_g, \mathrm{A}_T, \mathcal{B}|_{\mathrm{A}_T})$ (see Figure 13.4), where (i) the circle $\mathrm{A}_T = [g(0), g^2(0)]/\sim$ is taken with the orientation of $[g(0), g^2(0)]$, from right to left, i.e. with the original orientation in the train-track reversed, (ii) $\mathcal{B}|_{\mathrm{A}_T}$ is the restriction of the atlas $\mathcal{B}$ to A_T, and (iii) $R_g : \mathrm{A}_T \to \mathrm{A}_T$ is the map given by

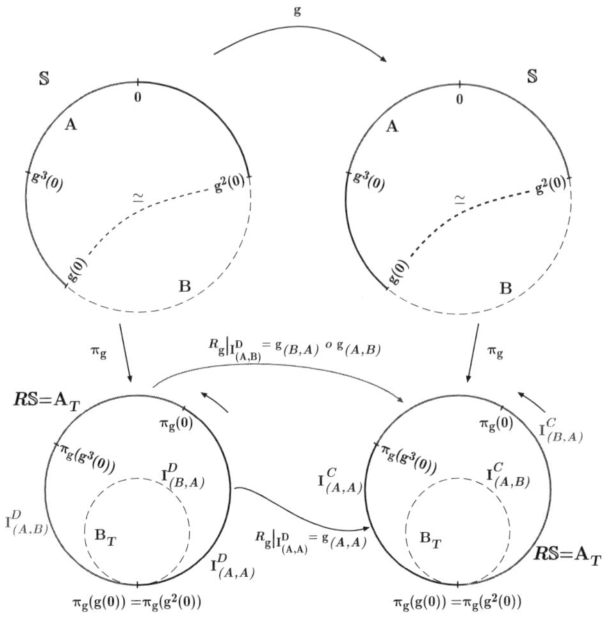

Fig. 13.4. The renormalization $(R_g, \mathrm{A}_T, \mathcal{B}|_{\mathrm{A}_T})$.

$$R_g(x) = \begin{cases} g_{(A,A)}(x), & \text{if } x \in I^D_{(A,A)} \\ g_{(B,A)} \circ g_{(A,B)}(x), & \text{if } x \in I^D_{(A,B)} \end{cases} \tag{13.1}$$

For simplicity, we will refer to the renormalization $(R_g, R\mathbb{S}, R\mathcal{A}) = (R_g, \mathrm{A}_T, \mathcal{B}|_{\mathrm{A}_T})$ by renormalization of g, and we will denote it, for simplicity of notation, by R_g.

Let $\mathcal{F}$ be the set of all C^{1+} golden diffeomorphisms $(g, \mathbb{S}, \mathcal{A})$.

Lemma 13.3. *The renormalization R_g of a C^{1+} golden diffeomorphism g is a C^{1+} golden diffeomorphism, i.e. there is a well defined map $R : \mathcal{F} \to \mathcal{F}$ given by $R(g) = R_g$. In particular, the renormalization R_{r_γ} of the rigid rotation is the rigid rotation r_γ.*

Proof. Let us consider the rigid rotation $r_\gamma : \mathbb{S}^1 \to \mathbb{S}^1$ with the atlas $\mathcal{A}_{iso}$ given by the local isometries, with respect to the natural metric in $\mathbb{S}^1$ induced by the Euclidean metric in $\mathbb{R}$. Then, there is an affine map $h : \mathbb{S}^1 \to \mathrm{A}_T$, with respect to the atlas $\mathcal{A}_{iso}$ in $\mathbb{S}^1$ and the atlas $\mathcal{B}_{iso}|_{\mathrm{A}}$ in A_T, uniquely determined by $h(0) = \pi_{r_\gamma}(0) \in \mathrm{A}_T$. The map h is an affine conjugacy between $(r_\gamma, \mathbb{S}^1, \mathcal{A}_{iso})$ and $(R_{r_\gamma}, \mathrm{A}_T, \mathcal{B}_{iso}|_{\mathrm{A}_T})$. If $g : \mathbb{S} \to \mathbb{S}$ is a C^{1+} golden diffeomorphism, then there is a unique quasi-symmetric homeomorphism $\psi : \mathbb{S} \to \mathbb{S}^1$ conjugating g with the golden rigid rotation such that $\psi(0) = [0] \in \mathbb{S}^1$. Hence, $\pi_g \circ \psi|_{\mathrm{A}}$ is a topological conjugacy between R_g and R_{r_γ}. Since R_{r_γ} is the golden rigid rotation, we get that R_g is also quasi-symmetric conjugated to the golden rigid rotation. Since R_g is C^{1+} with respect to the atlas $\mathcal{B}|_{\mathrm{A}}$, we get that R_g is a C^{1+} golden diffeomorphism. □

The marked point $0 \in \mathbb{S}$ determines a marked point $\pi_g(0)$ in the circle $A_T = R\mathbb{S}$. Since R_g is homeomorphic to a golden rigid rotation, there exists $h : \mathbb{S} \to R\mathbb{S}$, with $h(0) = \pi_g(0)$, such that h conjugates g and R_g.

Definition 13.4. *If $h : \mathbb{S} \to R\mathbb{S}$ is C^{1+}, we call g a C^{1+}* fixed point of renormalization. *We will denote by $\mathcal{R} \subset \mathcal{F}$ the set of all C^{1+} fixed points of renormalization.*

We note that the rigid rotation r_γ, with respect to the atlas $\mathcal{A}_{iso}$, is an affine fixed point of renormalization. Hence, $r_\gamma \in \mathcal{R}$.

13.1.4 Golden Markov maps

Let $(g, \mathbb{S}, \mathcal{A})$ be a C^{1+} golden diffeomorphism and $(R_g, R\mathbb{S}, R\mathcal{A}) = (R_g, \mathrm{A}_T, \mathcal{B}|_{\mathrm{A}_T})$ its renormalization. Let $T = T(\mathbb{S}, g)$ and $RT = T(R\mathbb{S}, R_g)$ be the golden train-tracks determined by the C^{1+} golden diffeomorphisms g and R_g, respectively. Let $\mathcal{B}$ and $R\mathcal{B}$ be the atlas in the train-tracks T and RT, that are the extended pushforwards of the atlases $\mathcal{A}$ and $R\mathcal{A}$, respectively. Let $E(g)$ be the arc exchange system determined by the golden diffeomorphism g.

Let the map $\mathrm{d}\tilde{M}_{\mathrm{A}} : \mathrm{A}_T \subset T \to RT$ be defined by $\tilde{M}_{\mathrm{A}}(x) = \pi_{R_g}(x)$. The image $\tilde{M}_{\mathrm{A}}\,(\mathrm{A}_T)$ is the set RT. Let the map $\mathrm{d}\tilde{M}_{\mathrm{B}} : \mathrm{B}_T \subset T \to RT$ be defined by $\tilde{M}_{\mathrm{B}}(x) = \pi_{R_g} \circ g_{(B,B)}(x)$. The image of the transformation $\tilde{M}_{\mathrm{B}(\mathbb{S},g)}$ is the set A_{RT}. The map $\tilde{M}_g : T \to RT$ is defined as follows:

$$\tilde{M}_g(x) = \begin{cases} \tilde{M}_{\mathrm{A}}(x), \text{ if } x \in \mathrm{A}_T \\ \tilde{M}_{\mathrm{B}}(x), \text{ if } x \in \mathrm{B}_T \end{cases} .$$

The map $\tilde{M}_g$ is a local homeomorphism, and $\tilde{M}_g$ is C^{1+} with respect to the atlas $\mathcal{B}$ in T and the atlas $R\mathcal{B}$ in RT.

Let h be the homeomorphism that conjugates g and R_g sending the marked point 0 of g in the marked point 0 of R_g. This homeomorphism induces a homeomorphism $\tilde{h} : T \to RT$ such that $\tilde{h} \circ \pi_g(x) = \pi_{R_g} \circ h(x)$, for all $x \in \mathbb{S}$. Let the *Markov map $M_g : T \to T$ associated to $g \in \mathcal{F}$* be defined by $M_g = \tilde{h}^{-1} \circ \tilde{M}_g$. In particular, M_{r_γ} is an affine map with respect to the atlas $\mathcal{B}_{iso}$ (see Figure 13.5).

Lemma 13.5. *The diffeomorphism g is a fixed point of renormalization if, and only if, the Markov map M_g associated to $(g, \mathbb{S}, \mathcal{A})$ is a C^{1+} local diffeomorphism with respect to the atlas $\mathcal{B} = (\pi_g)_* \mathcal{A}$.*

Proof. If g is a C^{1+} fixed point of renormalization, then the conjugacy $h : \mathbb{S} \to R\mathbb{S}$ is a C^{1+} diffeomorphism. Hence, $\tilde{h} : T \to RT$ is a C^{1+} diffeomorphism.

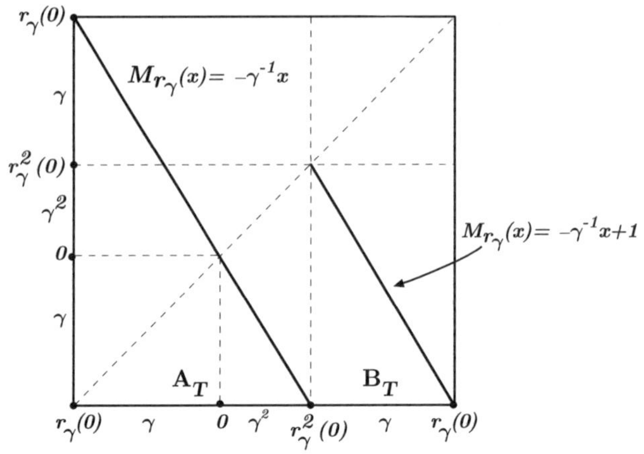

Fig. 13.5. The golden Markov map M_{r_γ} with respect to the atlas $\mathcal{B}_{iso}$.

Since $\tilde{M} : T \to RT$ is a C^{1+} local diffeomorphism, we obtain that $M_g = \tilde{h}^{-1} \circ \tilde{M}$ is a C^{1+} local diffeomorphism. Conversely, let $\tilde{M}$ be a local diffeomorphism. For every small enough train-track arc $J \subset T$, let M_J^{-1} be the inverse of $M|_J$. Hence, the map $\tilde{h}|_J = \tilde{M} \circ M_J^{-1}$ is a C^{1+} diffeomorphism onto its image. Therefore, $\tilde{h}$ is a C^{1+} diffeomorphism with respect to the atlas $\mathcal{B}$ in T and $R\mathcal{B}$ in RT, which implies that the map $h : \mathbb{S} \to R\mathbb{S}$, defined by $h \circ \pi_g(x) = \pi_{R_g} \circ h(x)$ is also a C^{1+} diffeomorphism with respect to the atlases $\mathcal{A}$ in $\mathbb{S}$ and $R\mathcal{A}$ in $R\mathbb{S}$. $\square$

13.2 Anosov diffeomorphisms

The *(golden) Anosov automorphism* $G_A : \mathbb{T} \to \mathbb{T}$ is given by $G_A(x, y) = (x + y, x)$, where $\mathbb{T}$ is equal to $\mathbb{R}^2/(v\mathbb{Z}\times w\mathbb{Z})$ with $v = (\gamma, 1)$ and $w = (-1, \gamma)$. Let $\pi : \mathbb{R}^2 \to \mathbb{T}$ be the natural projection. Let $\mathbf{A}$ and $\mathbf{B}$ be the rectangles $[0, 1] \times [0, 1]$ and $[-\gamma, 0] \times [0, \gamma]$ respectively (see Figure 13.6). A Markov partition $\mathcal{M}_A$ of G_A is given by $\pi(\mathbf{A})$ and $\pi(\mathbf{B})$. The unstable manifolds of G_A are the projection by π of the vertical lines of the plane, and the stable manifolds of G_A are the projection by π of the horizontal lines of the plane.

A C^{1+} *(golden) Anosov diffeomorphism* $G : \mathbb{T} \to \mathbb{T}$ is a $C^{1+\alpha}$, with $\alpha > 0$, diffeomorphism such that (i) G is topologically conjugated to G_A; (ii) the tangent bundle has a $C^{1+\alpha}$ hyperbolic splitting into a stable direction and an unstable direction. We denote by $\mathcal{C}_G$ the C^{1+} structure on $\mathbb{T}$ in which G is a C^{1+} diffeomorphism. A Markov partition $\mathcal{M}_G$ of G is given by $h(\pi(\mathbf{A}))$ and $h(\pi(\mathbf{A}))$, where h is the topological conjugacy between G_A and G. Let d_ρ be the distance on the torus $\mathbb{T}$, determined by a Riemannian metric ρ.

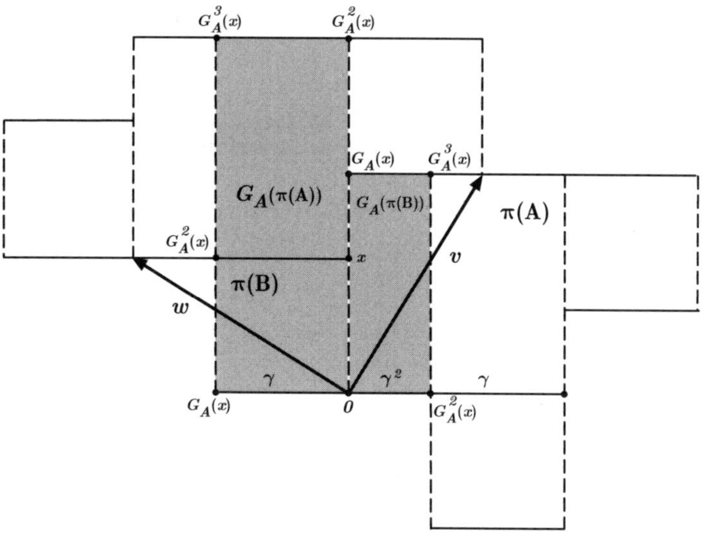

Fig. 13.6. The golden automorphism G_A.

13.2.1 Golden diffeomorphisms

Let G be a C^{1+} Anosov diffeomorphism. For each Markov rectangle R, let t_R^s be the set of all unstable spanning leaf segments of R. Thus, by the local product structure, one can identify t_R^s with any stable spanning leaf segment $\ell^s(x, R)$ of R. We form the space $\mathbb{S}_G$ by taking the disjoint union $\bigsqcup_{\pi(\mathbf{A}),\pi(\mathbf{B})} t_R^s$ (where $\pi(\mathbf{A})$ and $\pi(\mathbf{B})$ are the Markov rectangles of the Markov partition $\mathcal{M}_G$) and identifying two points $I \in t_R^s$ and $J \in t_{R'}^s$ if (i) $R \neq R^\iota$, (ii) the unstable leaf segments I and J are unstable boundaries of Markov rectangles, and (iii) $int(I \cap J) \neq \emptyset$. The space $\mathbb{S}_G$ is topologically a *clockwise oriented circle.* Let $\pi_{\mathbb{S}_G} : \bigsqcup_{R\in\mathcal{M}_G} R \to \mathbb{S}_G$ be the natural projection sending $x \in R$ to the point $\ell^u(x, R)$ in $\mathbb{S}_G$.

Let $I_\mathbb{S}$ be an arc of $\mathbb{S}_G$ and I a leaf segment such that $\pi_{\mathbb{S}_g}(I) = I_\mathbb{S}$. The chart $i : I \to \mathbb{R}$ in $\mathcal{L} = \mathcal{L}^s(G, \rho)$ determines a *circle chart* $i_\mathbb{S} : I_\mathbb{S} \to \mathbb{R}$ for $I_\mathbb{S}$ given by $i_\mathbb{S} \circ \pi_{\mathbb{S}_G} = i$. We denote by $\mathcal{A}(G, \rho)$ the set of all circle charts $i_\mathbb{S}$ determined by charts i in $\mathcal{L} = \mathcal{L}^s(G, \rho)$. Given any circle charts $i_\mathbb{S} : I_\mathbb{S} \to \mathbb{R}$ and $j_\mathbb{S} : J_\mathbb{S} \to \mathbb{R}$, the overlap map $j_\mathbb{S} \circ i_\mathbb{S}^{-1} : i_\mathbb{S}(I_\mathbb{S} \cap J_\mathbb{S}) \to j_\mathbb{S}(I_\mathbb{S} \cap J_\mathbb{S})$ is equal to $j_\mathbb{S} \circ i_\mathbb{S}^{-1} = j \circ \theta \circ i^{-1}$, where $i = i_\mathbb{S} \circ \pi_{\mathbb{S}_G} : I \to \mathbb{R}$ and $j = j_\mathbb{S} \circ \pi_{\mathbb{S}_G} : J \to \mathbb{R}$ are charts in $\mathcal{L}$, and

$$\theta : i^{-1}(i_\mathbb{S}(I_\mathbb{S} \cap J_\mathbb{S})) \to j^{-1}(j_\mathbb{S}(I_\mathbb{S} \cap J_\mathbb{S}))$$

is a basic stable holonomy. By Lemma 4.1, there exists $\alpha > 0$ such that, for all circle charts $i_\mathbb{S}$ and $j_\mathbb{S}$ in $\mathcal{A}(G, \rho)$, the overlap maps $j_\mathbb{S} \circ i_\mathbb{S}^{-1} = j \circ \theta \circ i^{-1}$ are $C^{1+\alpha}$ diffeomorphisms with a uniform bound in the $C^{1+\alpha}$ norm, for some $\alpha > 0$. Hence, $\mathcal{A}(G, \rho)$ is a C^{1+} atlas.

Suppose that I and J are stable leaf segments and $\theta : I \to J$ a holonomy such that, for every $x \in I$, the unstable leaf segments with endpoints x and $\theta(x)$ cross once, and only once, an stable boundary of a Markov rectangle. We define the *arc rotation map* $\tilde{\theta}_G : \pi_{\mathbb{S}}(I) \to \pi_{\mathbb{S}}(J)$, associated to θ, by $\tilde{\theta}_G(\pi_{\mathbb{S}}(x)) = \pi_{\mathbb{S}}(\theta(x))$. By Theorem 1.6 (see also Pinto and Rand [164]), there exists $\alpha > 0$ such that the holonomy $\theta : I \to J$ is a $C^{1+\alpha}$ diffeomorphism, with respect to the C^{1+} lamination atlas $\mathcal{L}(G, \rho)$. Hence, the arc rotation maps $\tilde{\theta}_G$ are C^{1+} diffeomorphisms, with respect to the C^{1+} atlas $\mathcal{A}(G, \rho)$.

Lemma 13.6. *There is a well-defined C^{1+} golden diffeomorphism g_G, with respect to the C^{1+} atlas $\mathcal{A}(G, \rho)$, such that $g|\pi_{\mathbb{S}_G} = \tilde{\theta}$, for every arc rotation map $\tilde{\theta}$. In particular, if G_A is the Anosov automorphism, then g is the golden rigid rotation r_γ, with respect to the isometric atlas $\mathcal{A}_{iso} = \mathcal{A}(G_A, E)$, where E corresponds to the Euclidean metric in the plane.*

Proof. Let us consider the Anosov automorphism G_A and lamination atlas $\mathcal{L}_{iso} = \mathcal{L}^s(G_A, E)$. Let $\mathcal{A}_{iso} = \mathcal{A}(G_A, E)$ be the atlas in $\mathbb{S}_A$ determined by $\mathcal{L}_{iso}$. The overlap maps of the charts in $\mathcal{A}_{iso}$ are translations, and the arc rotation maps $\tilde{\theta}_A : \pi_{\mathbb{S}_A}(I) \to \pi_{\mathbb{S}_A}(J)$, as defined above, are also translations, with respect to the charts in $\mathcal{A}_{iso}$. Furthermore, the rigid golden rotation $r_\gamma : \mathbb{S}_A \to \mathbb{S}_A$, with respect to the atlas $\mathcal{A}_{iso}$, has the property that $r_\gamma|_{\pi_{\mathbb{S}_A}(I)} = \tilde{\theta}_A$. Hence, for every Anosov diffeomorphism G, let $h : \mathbb{T} \to \mathbb{T}$ be the topological conjugacy between G_A and G. Let $g : \mathbb{S}_G \to \mathbb{S}_G$ be the map determined by $g \circ \pi_G \circ h(x) = r_\gamma \circ \pi_{G_A}(x)$, with rotation number γ. Since the arc rotation maps $\tilde{\theta}_G = \pi_{\mathbb{S}_G}(I) \to \pi_{\mathbb{S}_G}(J)$ are C^{1+}, with respect to the atlas $\mathcal{A}(G, \rho)$ and $g|_{\pi_{\mathbb{S}_G}(I)} = \tilde{\theta}_G$, we obtain that g is a C^{1+} diffeomorphism. □

13.2.2 Arc exchange system

Roughly speaking, train-tracks are the optimal leaf-quotient spaces on which the unstable and stable Markov maps induced by the action of G on leaf segments are local homeomorphisms.

Let G be a C^{1+} Anosov diffeomorphism. For each Markov rectangle R, let t^s_R be the set of unstable spanning leaf segments of R. Thus, by the local product structure one can identify t^s_R with any stable spanning leaf segment $\ell^s(x, R)$ of R. We form the space $\mathbf{T}_G$ by taking the disjoint union $\bigsqcup_{\pi(\mathbf{A}),\pi(\mathbf{B})} t^s_R$ (where $\pi(\mathbf{A})$ and $\pi(\mathbf{B})$ are the Markov rectangles of the Markov partition $\mathcal{M}_G$) and identifying two points $I \in t^s_R$ and $J \in t^s_{R'}$ if (i) the unstable leaf segments I and J are unstable boundaries of Markov rectangles and (ii) $int(I \cap J) = \emptyset$. This space is called the *stable train-track* and it is denoted by $\mathbf{T}_G$.

Let $\pi_{\mathbf{T}_G} : \bigsqcup_{R \in \mathcal{M}_G} R \to \mathbf{T}_G$ be the natural projection sending $x \in R$ to the point $\ell^u(x, R)$ in $\mathbf{T}_G$. A *topologically regular point* I in $\mathbf{T}_G$ is a point with a unique preimage under $\pi_{\mathbf{T}_G}$ (that is the preimage of I is not a union of distinct unstable boundaries of Markov rectangles). If a point has more than one preimage by $\pi_{\mathbf{T}_G}$, then we call it a *junction*. Hence, there is only one junction.

By construction, the Anosov train-track $\mathbf{T}_G$ is topologically equivalent to the golden train-track T_{g_G} determined by the C^{1+} golden diffeomorphism $g_G \in \mathcal{F}$.

We say that I_T is a *stable train-track segment* of $\mathbf{T}_G$, if there is an stable leaf segment I, not intersecting stable boundaries of Markov rectangles, such that $\pi_{\mathbf{T}_G}|_I$ is an injection and $\pi_{\mathbf{T}_G}(I) = I_T$.

A chart $i : I \to \mathbb{R}$ in $\mathcal{L}^s(G, \rho)$ determines a *train-track chart* $i_{\mathbf{B}} : I_T \to \mathbb{R}$ for I_T given by $i_T \circ \pi_{\mathbf{T}_G} = i$. We denote by $\mathcal{B} = \mathcal{B}(G, \rho)$ the set of all train-track charts i_T determined by charts i in $\mathcal{L} = \mathcal{L}(G, \rho)$. Given any train-track charts $i_T : I_T \to \mathbb{R}$ and $j_T : J_T \to \mathbb{R}$ in $\mathcal{B}$, the overlap map $j_T \circ i_T^{-1} : i_T(I_T \cap J_T) \to j_T(I_T \cap J_T)$ is equal to $j_T \circ i_T^{-1} = j \circ \theta \circ i^{-1}$, where $i = i_T \circ \pi_{\mathbf{T}_G} : I \to \mathbb{R}$ and $j = j_T \circ \pi_{\mathbf{T}_G} : J \to \mathbb{R}$ are charts in $\mathcal{L}$, and

$$\theta : i^{-1}(i_T(I_T \cap J_T)) \to j^{-1}(j_T(I_T \cap J_T))$$

is a basic stable holonomy. By Lemma 4.1, there exists $\alpha > 0$ such that, for all train-track charts i_T and j_T in $\mathcal{B}(G, \rho)$, the overlap maps $j_T \circ i_T^{-1} = j \circ \theta \circ i^{-1}$ have $C^{1+\alpha}$ diffeomorphic extensions with a uniform bound in the $C^{1+\alpha}$ norm. Hence, $\mathcal{B}(G, \rho)$ is a $C^{1+\alpha}$ atlas in $\mathbf{T}_G$.

Suppose that M and N are Markov rectangles, and $x \in int(M)$ and $y \in int(N)$. We say that x and y are *stable holonomically related*, if (i) there is an stable leaf segment $\ell^u(x, y)$ such that $\partial \ell^u(x, y) = \{x, y\}$, and (ii) $\ell^u(x, y) \subset \ell^u(x, M) \cup \ell^u(y, N)$. Let $P = P_{\mathcal{M}}$ be the set of all pairs (M, N) such that there are points $x \in int(M)$ and $y \in int(N)$ unstable holonomically related.

For every Markov rectangle $M \in \mathcal{M}_G$, choose a stable spanning leaf segment $\ell(x, M)$ in M for some $x \in M$. Let $\mathcal{I} = \{\ell_M : M \in \mathcal{M}\}$. For every pair $(M, N) \in P$, there are maximal leaf segments $\ell^D_{(M,N)} \subset \ell_M$, $\ell^C_{(M,N)} \subset \ell_N$ such that the stable holonomy $h_{(M,N)} : \ell^D_{(M,N)} \to \ell^C_{(M,N)}$ is well-defined. We call such holonomies $h_{(M,N)} : \ell^D_{(M,N)} \to \ell^C_{(M,N)}$ the *stable primitive holonomies* associated to the Markov partition $\mathcal{M}_G$. The *complete set of stable holonomies* $\mathcal{H}_G$ consists of all stable primitive holonomies and their inverses.

Definition 13.7. *A complete set of stable holonomies $\mathcal{H}_G$ is $C^{1+zygmund}$ if, and only if, all holonomies in $\mathcal{H}_G$ are $C^{1+zygmund}$, with respect to the atlas $\mathcal{L}^s(G, \rho)$.*

Let $h_G : \mathbb{T} \to \mathbb{T}$ be the topological conjugacy between the Anosov automorphism G_A and G. The rectangles $h_G \circ \pi(\mathbf{A})$ and $h_G \circ \pi(\mathbf{B})$ form a Markov partition for G. In Figure 13.7, we exhibit the complete set of stable holonomies

$$\mathcal{H}_G = \left\{ h_{(\mathbf{A},\mathbf{A})}, h^{-1}_{(\mathbf{A},\mathbf{A})}, h_{(\mathbf{A},\mathbf{B})}, h^{-1}_{(\mathbf{A},\mathbf{B})}, h_{(\mathbf{B},\mathbf{A})}, h^{-1}_{(\mathbf{B},\mathbf{A})} \right\}$$

associated to the Markov partition $\mathcal{M}_G = \{\pi(\mathbf{A}), \pi(\mathbf{B})\}$ of G.

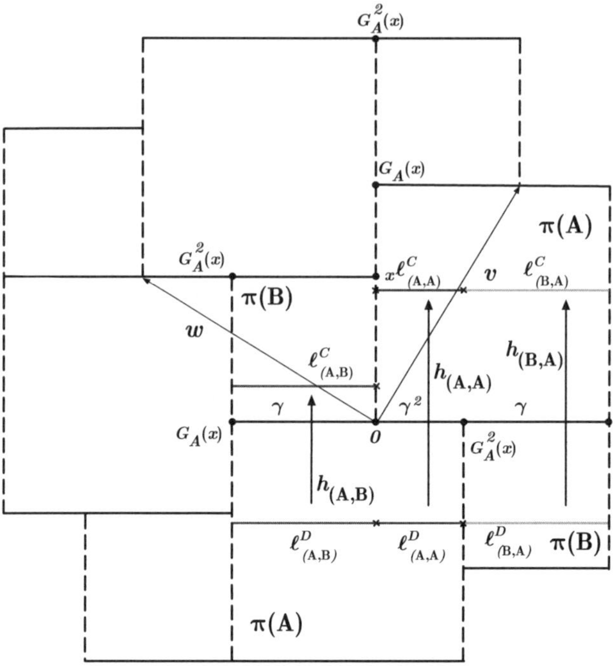

Fig. 13.7. A complete set of stable holonomies $\mathcal{H}_G$ associated to the Markov partition $\mathcal{M}_G$.

For every $h_{(M,N)} : \ell^D_{(M,N)} \to \ell^C_{(M,N)}$ in $\mathcal{H}_G$, let $I^D_{(M,N)} = \pi_{\mathbf{T}_G}(\ell^D_{(M,N)})$ and $I^C_{(M,N)} = \pi_{\mathbf{T}_G}(\ell^C_{(M,N)})$. Let $e_{(M,N)} : I^D_{(M,N)} \to I^C_{(M,N)}$ be the *arc exchange map* determined by $\pi_{\mathbf{T}_G} \circ h_{(M,N)} = e_{(M,N)} \circ \pi_{\mathbf{T}_G}$. We denote by E_G the set of all arc exchange maps and their inverses,

$$E_G = \left\{ e_{(\mathbf{A},\mathbf{A})}, e^{-1}_{(\mathbf{A},\mathbf{A})}, e_{(\mathbf{A},\mathbf{B})}, e^{-1}_{(\mathbf{A},\mathbf{B})}, e_{(\mathbf{B},\mathbf{A})}, e^{-1}_{(\mathbf{B},\mathbf{A})} \right\} .$$

Lemma 13.8. *For every $G \in \mathcal{G}$, the arc exchange system $E(g_G)$, with respect to the atlas $\mathcal{B} = (\pi_{g_G})_* \mathcal{A}(G,\rho)$, is C^{1+} conjugate to E_G, with respect to the atlas $\mathcal{B}(G,\rho)$.*

Proof. The construction, in §13.1.2, of the extended pushforward atlas $\mathcal{B} = (\pi_{g_G})_* \mathcal{A}$ of $\mathcal{A}(G,\rho)$ coincides, up to smooth equivalence of charts, with the construction, in this section, of the atlas $\mathcal{B}(G,\rho)$. □

13.2.3 Markov maps

The *(stable) Markov map* $M_G : \mathbf{T}_G \to \mathbf{T}_G$ is the mapping induced by the action of G on unstable spanning leaf segments, that it is defined as follows: if

$I \in \mathbf{T}_G$, $M_G(I) = \pi_{\mathbf{T}_G}(G(I))$ is the unstable spanning leaf segment containing $G(I)$. This map M_G is a local homeomorphism because G sends short stable leaf segments homeomorphically onto short stable leaf segments.

For $n \geq 1$, an *n-cylinder* is the projection into $\mathbf{T}_G$ of an stable leaf n-cylinder segment. Thus, each Markov rectangle in $\mathbb{T}$ projects in a unique primary stable leaf segment in $\mathbf{T}_G$.

Given a topological chart (e, U) on the train-track $\mathbf{T}_G$ and a train-track segment $C \subset U$, we denote by $|C|_e$ the length of $e(C)$. We say that M_G has *bounded geometry* in a C^{1+} atlas $\mathcal{B}$, if there is $\kappa_1 > 0$ such that, for every n-cylinder C_1 and n-cylinder C_2 with a common endpoint with C_1, we have $\kappa_1^{-1} < |C_1|_e/|C_2|_e < \kappa_1$, where the lengths are measured in any chart (e, U) of the atlas such that $C_1 \cup C_2 \subset U$. We note that M_G has bounded geometry, with respect to a C^{1+} atlas $\mathcal{B}$, if, and only if, there are $\kappa_2 > 0$ and $0 < \nu < 1$ such that $|C|_e \leq \kappa_2 \nu^n$, for every n-cylinder and every $e \in \mathcal{B}$.

By Lemma 4.2, we obtain that M_G is a C^{1+} local diffeomorphism and has *bounded geometry* in $\mathcal{B}(G, \rho)$.

Lemma 13.9. *For every $G \in \mathcal{G}$, the C^{1+} golden diffeomorphism g_G is a C^{1+} fixed point of renormalization, with respect to the atlas $\mathcal{B}(\mathcal{G}, \rho)$.*

Proof. For every $G \in \mathcal{G}$, let g_G be the C^{1+} golden diffeomorphism, with respect to the atlas $\mathcal{B}(G, \rho)$. Since M_G is a C^{1+} Markov map with bounded geometry, with respect to the atlas $\mathcal{B}(\mathcal{G}, \rho)$, by Lemma 13.5, we obtain that g_G is a C^{1+} fixed point of renormalization. □

13.2.4 Exchange pseudo-groups

The elements $\tilde{\theta}$ of the *stable exchange pseudo-group on* $\mathbf{T}_G$ are the mappings defined as follows: suppose that I and J are stable leaf segments and $\theta : I \to J$ a holonomy. Then, it follows from the definition of the stable train-track $\mathbf{T}_G$ that the map $\tilde{\theta} : \pi_{\mathbf{B}}(I) \to \pi_{\mathbf{B}}(J)$ given by $\tilde{\theta}(\pi_{\mathbf{B}}(x)) = \pi_{\mathbf{B}}(\theta(x))$ is well-defined. The collection of all such local mappings forms the *basic stable exchange pseudo-group in* $\mathbf{T}_G$.

Lemma 13.10. *The elements of the exchange pseudo-group in $\mathbf{T}_G$ are C^{1+}, with respect to an atlas $\mathcal{B}$, if, and only if, the arc exchange system is C^{1+}, with respect to the atlas $\mathcal{B}$.*

Proof. The elements of the exchange pseudo-group in $\mathbf{T}_G$ can be written as compositions of elements of the arc exchange system, using property (iii) in Definition 13.1. Hence, if the exchange pseudo-group is C^{1+}, then the elements of the arc exchange system are C^{1+}, with respect to an atlas $\mathcal{B}$, and vice-versa.
□

13.2.5 Self-renormalizable structures

The C^{1+} structure $\mathcal{S}$ on $\mathbf{T}_G$ is an *stable self-renormalizable structure*, if there is a C^{1+} atlas $\mathcal{B}$ in this structure, with the following properties:

(i) the Markov map M_G is a $C^{1+\alpha}$ local diffeomorphism, for some $\alpha > 0$, and has bounded geometry with respect to $\mathcal{B}$.
(ii) The elements of the basic exchange pseudo-group are $C^{1+\alpha}$ local diffeomorphisms, for some $\alpha > 0$, with respect to $\mathcal{B}$.

Lemma 13.11. *There is a one-to-one correspondence between C^{1+} golden diffeomorphisms, that are C^{1+} fixed points of renormalization, and C^{1+} self renormalizable structures $\mathcal{S}$.*

Proof. Let $\mathcal{S}$ be a C^{1+} self-renormalizable structure and $\mathcal{B}$ a C^{1+} atlas of $\mathcal{S}$. By Lemma 13.10, the C^{1+} self-renormalizable structure $\mathcal{S}$ determines a C^{1+} arc exchange system $E_{\mathcal{S}}$, with respect to $\mathcal{B}$. By Lemma 13.2, $g(E_{\mathcal{S}})$ is a C^{1+} golden diffeomorphism, with respect to the pullback atlas $\mathcal{A} = (\pi_g)^* \mathcal{B}$ of the C^{1+} atlas $\mathcal{B}$. The C^{1+} self-renormalizable structure $\mathcal{S}$ determines, also, a C^{1+} Markov map $M_{\mathcal{S}}$ with respect to $\mathcal{B}$. Hence, by Lemma 13.5, $g(E_{\mathcal{S}})$ is a C^{1+} fixed point of renormalization. Conversely, let us suppose that g is a C^{1+} fixed point of renormalization, with respect to a C^{1+} atlas $\mathcal{A}$ of $\mathbb{S}$. Since g is a C^{1+} fixed point of renormalization, by Lemma 13.5, g determines a C^{1+} Markov map M_g, with respect to the extended pushforward atlas $\mathcal{B} = (\pi_g)_* \mathcal{A}$ of the C^{1+} atlas $\mathcal{A}$. By Lemma 13.2, g determines a C^{1+} arc exchange system $E(g)$, with respect to the atlas $\mathcal{B}$. By Lemma 13.10, the C^{1+} arc exchange system $E(g)$ determines a C^{1+} exchange pseudo-group, and so the C^{1+} atlas $\mathcal{B}$ determines a C^{1+} self-renormalizable structure $\mathcal{S}$ that contains $\mathcal{B}$. □

Lemma 13.12. *The map $G \to g(G)$ is a one-to-one correspondence between C^{1+} conjugacy classes of C^{1+} Anosov diffeomorphisms $G \in \mathcal{G}$ and C^{1+} conjugacy classes of C^{1+} golden diffeomporphisms $g_G \in \mathcal{R}$ that are C^{1+} fixed points of renormalization.*

Proof. By Theorem 10.19, the map $G \to \mathcal{S}(G)$ determines a one-to-one correspondence between C^{1+} Anosov diffeomorphisms, with an invariant measure absolutely continuous with respect to the Lebesgue measure, and C^{1+} self-renormalizable structures on $\mathbf{T}_G$. By Lemma 13.11, there is a one-to-one correspondence between C^{1+} golden diffeomorphisms $g(\mathcal{S}(G))$, that are C^{1+} fixed points of renormalization and C^{1+} self-renormalizable structures $\mathcal{S}(G)$. □

13.3 HR structures

Pinto-Rand's *HR structure* associates an affine structure to each stable and unstable leaf segment in such a way that these vary Hölder continuously with the leaf and are invariant under G_A.

Let G be a C^{1+} Anosov diffeomorphism, and let $\mathcal{L}^u(G,\rho)$ be an unstable lamination atlas associated to a Riemannian metric ρ. If I is a stable leaf segment, then by $|I| = |I|_\rho$, we mean the length of the stable leaf containing I, measured using the Riemannian metric ρ. Let $h : \mathbb{T} \to \mathbb{T}$ be the topological conjugacy between the automorphism G_A and the Anosov diffeomorphism G. Using the mean value theorem and the fact that G is $C^{1+\alpha}$, for some $\alpha > 0$, for all short unstable leaf segments K of G_A and all leaf segments I and J contained in K, the unstable realized ratio function r_G given by

$$r_G(I:J) = \lim_{n\to\infty} \frac{|G^{-n}(h(I))|}{|G^{-n}(h(J))|}$$

is well-defined. By Theorem 10.16, we get the following equivalence:

Theorem 13.13. *The map $G \to r_G$ determines a one-to-one correspondence between C^{1+} conjugacy classes of Anosov diffeomorphisms, with an invariant measure that is absolutely continuous with respect to the Lebesgue measure, and unstable ratio functions.*

Let sol denote the set of all ordered pairs (I, J) of unstable spanning leaf segments of Markov rectangles, such that the intersection of I and J consists of a single endpoint.

By Lemma 3.3, the map $r \to r|\mathrm{sol}$ gives a one-to-one correspondence between unstable ratio functions and unstable solenoid functions.

Let $\mathcal{SOL}$ be the set consisting of all unstable solenoid functions. The set $\mathcal{SOL}$ has a natural metric. Combining Theorem 13.13 with Lemma 3.3, we obtain the following corollary.

Corollary 13.14. *The map $G \to r_G|\mathrm{sol}$ determines a one-to-one correspondence between C^{1+} conjugacy classes of Anosov diffeomorphisms, with an invariant measure that is absolutely continuous with respect to the Lebesgue measure, and unstable solenoid functions in $\mathcal{SOL}$.*

13.4 Fibonacci decomposition

The Fibonacci numbers F_1, F_2, $F_3, \ldots$, are inductively given by the well-known relation $F_{n+2} = F_{n+1} + F_n$, $n \geq 1$, where F_1 and F_2 are both equal to 1. We say that a finite sequence $F_{n_0}, \ldots, F_{n_p}$ is a *Fibonacci decomposition* of a natural number $i \in \mathbb{N}$, if the following properties are satisfied:

(i) $i = F_{n_p} + \cdots + F_{n_0}$;
(ii) F_{n_k} is the biggest Fibonacci number smaller than $i - \left(F_{n_p} + \cdots + F_{n_{k+1}}\right)$ for every $0 \leq k \leq p$;
(iii) If $F_{n_0} = F_1$ then n_1 is even, and if $F_{n_0} = F_2$ then n_1 is odd.

Like this, every natural number $i \in \mathbb{N}$ has a unique Fibonacci decomposition.

We define the *Fibonacci shift* $\sigma_F : \mathbb{N} \to \mathbb{N}$ as follows: For every $i \in \mathbb{N}$ let $F_{n_0}, F_{n_1}, \ldots, F_{n_p}$ be the Fibonacci decomposition associated to i, i.e. $i = F_{n_p} + \ldots + F_{n_0}$. We define $\sigma_F(i) = F_{n_p+1} + \cdots + F_{n_0+1}$. Hence, letting $F_{n_0}, F_{n_1}, \ldots, F_{n_p}$ be the Fibonacci decomposition associated to $i \in \mathbb{N}$, if $F_{n_0} \neq F_1$ then $\sigma_F^{-1}(i) = F_{n_p-1} + \cdots + F_{n_0-1}$, and if $F_{n_0} = F_1$ then $\sigma_F^{-1}(i) = \emptyset$. For simplicity of notation, we will denote $\sigma_F(i)$ by $\sigma(i)$.

13.4.1 Matching condition

The matching condition is linked to the invariance under the Anosov dynamics of the affine structures along the unstable leaves, as we will make it clear in §13.3 (see the geometric interpretation of the matching condition in Figure 13.10). Let $\mathbb{L} = \{i \in \mathbb{N} : i \geq 2\}$. We say that a sequence $(a_i)_{i \in \mathbb{L}}$ satisfies the *matching condition*, if, for every $i = F_{n_p} + \cdots + F_{n_0}$, the following conditions hold:

(i) If $F_{n_0} = F_1$ or, $F_{n_0} = F_3$ and n_1 odd, then

$$a_{\sigma(i)} = a_i \left(a_{\sigma(i)+1} + 1\right)^{-1} .$$

(ii) If $F_{n_0} = F_2$ or, $n_0 > 3$ and even, then

$$a_{\sigma(i)} = a_i \left(a_{\sigma(i)-1}^{-1} + 1\right) .$$

(iii) If $F_{n_0} = F_3$ and n_1 even or $n_0 > 3$ and odd, then

$$a_{\sigma(i)} = \frac{a_i \left(1 + a_{\sigma(i)-1}\right)}{a_{\sigma(i)-1} \left(1 + a_{\sigma(i)+1}\right)} .$$

Therefore, every sequence $(b_i)_{i \in \mathbb{L} \setminus \sigma(\mathbb{L})}$ determines, uniquely, a sequence $(a_i)_{i \in \mathbb{L}}$ as follows: for every $i \in \mathbb{L} \setminus \sigma(\mathbb{L})$, we define $a_i = b_i$ and, for every $i \in \sigma(\mathbb{L})$, we define $a_{\sigma(i)}$ using the matching condition and the elements a_j of the sequence with $j \in \{j : 2 \leq j < \sigma(i) \vee j \in \mathbb{L}\}$ already determined.

13.4.2 Boundary condition

Similarly to the matching condition, the boundary condition is linked to the affine structures along the boundaries of a Markov partition for the Anosov dynamics, as we will make it clear in §13.3 (see the geometric interpretation of the boundary condition in Figure 13.9). A sequence $(a_i)_{i \in \mathbb{L}}$ satisfies the *boundary condition*, if the following limits are well-defined and satisfy the inequalities:

(i) $\lim_{i \to +\infty} a_{F_i+2}^{-1} \left(1 + a_{F_i+1}^{-1}\right) \neq 0$;
(ii) $\lim_{i \to +\infty} a_{F_i} \left(1 + a_{F_i+1}\right) \neq 0$.

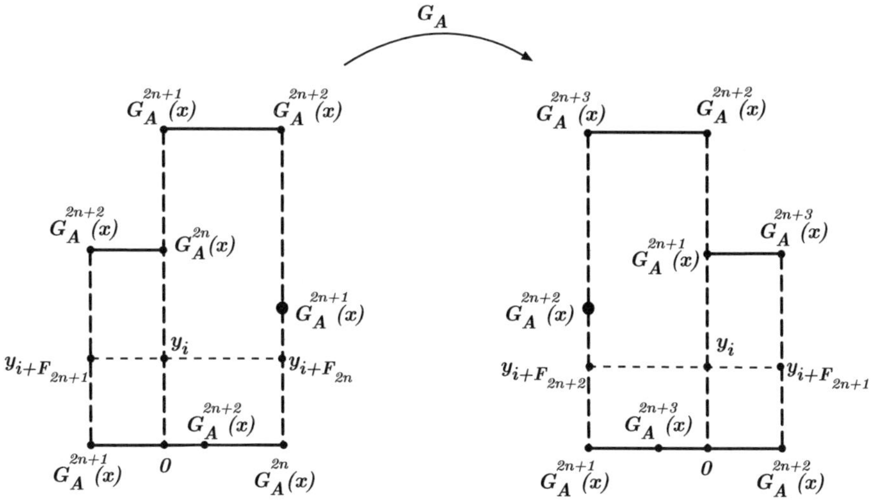

Fig. 13.8. The exponentially fast Fibonacci repetitive condition.

13.4.3 The exponentially fast Fibonacci repetitive property

The exponentially fast Fibonacci repetitive property is linked to the Hölder continuity along transversals of the affine structures of the unstable leaves of the Anosov diffeomorphism (see the geometric interpretation of the exponentially fast Fibonacci repetitive property in Figure 13.8).

A sequence $(a_i)_{i\in\mathbb{L}}$ is said to be *exponentially fast Fibonacci repetitive*, if there exist constants $C \geq 0$ and $0 < \mu < 1$ such that

$$|a_{i+F_n} - a_i| \leq C\mu^n,$$

for every $n \geq 3$ and $2 \leq i < F_{n+1}$.

13.4.4 Golden tilings

A *tiling* $\mathcal{T} = \{I_i \subset \mathbb{R} : i \in \mathbb{L}\}$ of the positive real line is a collection of tiling intervals I_i, with the following properties:

(i) the tiling intervals are closed intervals;
(ii) the union $\cup_{i\in\mathbb{L}} I_i$ is equal to the positive real line;
(iii) any two distinct intervals have disjoint interiors;
(iv) for every $i \in \mathbb{L}$ the intersection of the tiling intervals I_i and I_{i+1} is only a point, which is an endpoint, simultaneously, of both intervals.

The tilings $\mathcal{T}_1 = \{I_i \subset \mathbb{R} : i \in \mathbb{L}\}$ and $\mathcal{T}_2 = \{J_i \subset \mathbb{R} : i \in \mathbb{L}\}$ of the positive real line are in the same affine class, if there exists an affine map $h : \mathbb{R} \to \mathbb{R}$ such that $h(I_i) = J_i$, for every $i \in \mathbb{L}$. Thus, every positive sequence $(a_i)_{i\in\mathbb{L}}$

determines a unique affine class of tilings $\mathcal{T} = \{I_i \subset \mathbb{R} : i \in \mathbb{L}\}$ such that $a_i = |I_{i+1}| \,/\, |I_i|$, and vice-versa.

Definition 13.15. *A* golden sequence $(a_i)_{i\in\mathbb{L}}$ *is an exponentially fast Fibonacci repetitive sequence that satisfies the matching and the boundary conditions. A tiling* $\mathcal{T} = \{I_i \subset \mathbb{R} : i \in \mathbb{L}\}$ *of the positive real line is* golden, *if the corresponding sequence* $(a_i = |I_{i+1}|/|I_i|)_{i\in\mathbb{L}}$ *is a golden sequence.*

13.4.5 Golden tilings versus solenoid functions

Let W_0 be the positive vertical axis. Hence, $W = \pi(W_0)$ is the unstable leaf with only one endpoint $z = \pi(0,0)$ that is the fixed point of G_A, and W passes through all the unstable boundaries of the Markov rectangles **A** and **B**.

Recall that $\mathbb{L} = \{i \in \mathbb{N} : i \geq 2\}$. Let $K_1 \in W$ be the union of all the unstable boundaries of the Markov rectangles. Let $K_2, K_3, \ldots \in W$ be the unstable leaves with the following properties: (i) K_i is an unstable spanning leaf of a Markov rectangle, for every $i \geq 1$; (ii) $K_i \cap K_{i+1} = \{y_i\}$ is a common boundary point of both K_i and K_{i+1}, for every $i \geq 1$. By construction, the set

$$\mathcal{L} = \{(K_i, K_{i+1}),\, i \geq 2\}$$

is contained in sol and it is dense in sol.

For every golden tiling $\mathcal{T} = \{I_i \subset \mathbb{R} : i \in \mathbb{L}\}$ with associated golden sequence $(a_i)_{i\in\mathbb{L}}$, let $\sigma_{\mathcal{T}} : \mathcal{L} \to \mathbb{R}^+$ be defined by $\sigma_{\mathcal{T}}((I_i, I_{i+1})) = a_i$.

Theorem 13.16. *The map* $\mathcal{T} \to \sigma_{\mathcal{T}}$ *gives a one-to-one correspondence between golden tilings and solenoid functions. In particular, if* $\mathcal{T}_R$ *is the rigid golden tiling, then* $\sigma_{\mathcal{T}_R}$ *is the solenoid function corresponding to the* C^{1+} *conjugacy class of the Anosov automorphism* G_A, *i.e.* $\sigma_{G_A} = \sigma_{\mathcal{T}_R}$.

Proof. Let $m : \mathbb{N}_0 \to \mathbb{T}$ be the *marking* defined by $m(0) = G^{-1}(y_1)$ and $m(i) = y_i$, for every $i \geq 1$. Let $J_{\mathbf{A}}$ and $J_{\mathbf{B}}$ be the boundaries of the rectangles **A** and **B** in $\mathbb{R}^2$, contained in the horizontal axis. There is a natural inclusion $inc : \pi(J_{\mathbf{A}} \cup J_{\mathbf{B}}) \to \mathbb{S}_A$ that associates to each point $x \in \pi(J_{\mathbf{A}})$ the point $\ell^s(x, \mathbf{A}) \in \mathbb{S}_A$, and to each point $x \in \pi(J_{\mathbf{B}})$ the point $\ell^s(x, \mathbf{B}) \in \mathbb{S}_A$. we observe that (i) $m(\mathbb{N}_0) \subset \pi(J_{\mathbf{A}} \cup J_{\mathbf{B}})$, (ii) $inc \circ m(0) = inc \circ m(1)$, and (iii) $inc \circ m(i) = g_A^i(0)$, where $0 = \pi_A(\ell(z, \mathbf{A})) = \pi_A(\ell(z, \mathbf{B}))$ and g_A is the golden rigid rotation determined by the Anosov automorphism G_A, with respect to the atlas $\mathcal{A}(G_A, E)$. The closest returns of g_A to 0 are given by the sequence $g_A^{F_2}(0), g_A^{F_3}(0), \ldots$, where $F_2, F_3, F_4, \ldots$ is the Fibonacci sequence. Hence, if $K_i, K_{i+1} \in \pi_A(\mathbf{A})$, then i satisfies the condition (i) of the rigid golden tiling; if $K_i \in \pi_A(\mathbf{A})$ and $K_{i+1} \in \pi_A(\mathbf{B})$, then i satisfies the condition (ii) of the rigid golden tiling; if $K_i \in \pi_A(\mathbf{B})$ and $K_{i+1} \in \pi_A(\mathbf{A})$ then i satisfies the condition (iii) of the rigid golden tiling. Hence, the golden sequence $(a_i)_{i\in\mathbb{N}}$ associated to the rigid golden tiling $\mathcal{T}_R$ has the property $r_i = K_{i+1}/K_i$. Hence, $\sigma_{G_A} = \sigma_{\mathcal{T}_R}$.

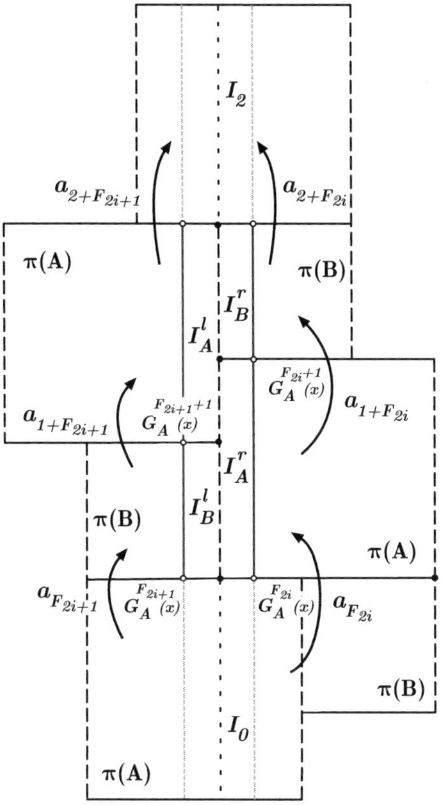

Fig. 13.9. The boundary condition for the sequence A.

Now, let $\mathcal{T} = \{I_i \subset \mathbb{R} : i \in \mathbb{L}\}$ be a golden tiling with associated golden sequence $A = (a_i)_{i\in\mathbb{L}}$. Since the tiling $\mathcal{T}$ satisfies the exponentially fast Fibbonaci repetitive property, we get that $\sigma_{\mathcal{T}}$ has a Hölder continuous extension $\hat{\sigma}_{\mathcal{T}}$ to sol. Since the golden sequence A satisfies the matching condition (see Figure 13.10), we get that $\sigma_{\mathcal{T}}$ satisfies the matching condition and, by continuity, its extension $\hat{\sigma}_{\mathcal{T}}$ also satisfies the matching condition. Let I_M^l and I_M^r be the left and right boundaries of the Markov rectangle $M \in \{\mathbf{A}, \mathbf{B}\}$ (see Figure 13.9). The leaf I^1 is equal to $I_A^l \cup I_B^l$ and to $I_A^r \cup I_B^r$. Let I_0 be the primary leaf segment with a single common endpoint with the primary leaf segment I_1. By the above construction, we have

$$\begin{aligned}\frac{|I_A^r| + |I_B^r|}{|I_0|} &= \sigma(I_0 : I_A^r)\left(1 + \sigma(I_A^r : I_B^r)\right) \\ &= \lim_{i\to\infty} a_{F_{2i}}\left(1 + a_{F_{2i}+1}\right)\end{aligned}$$

and

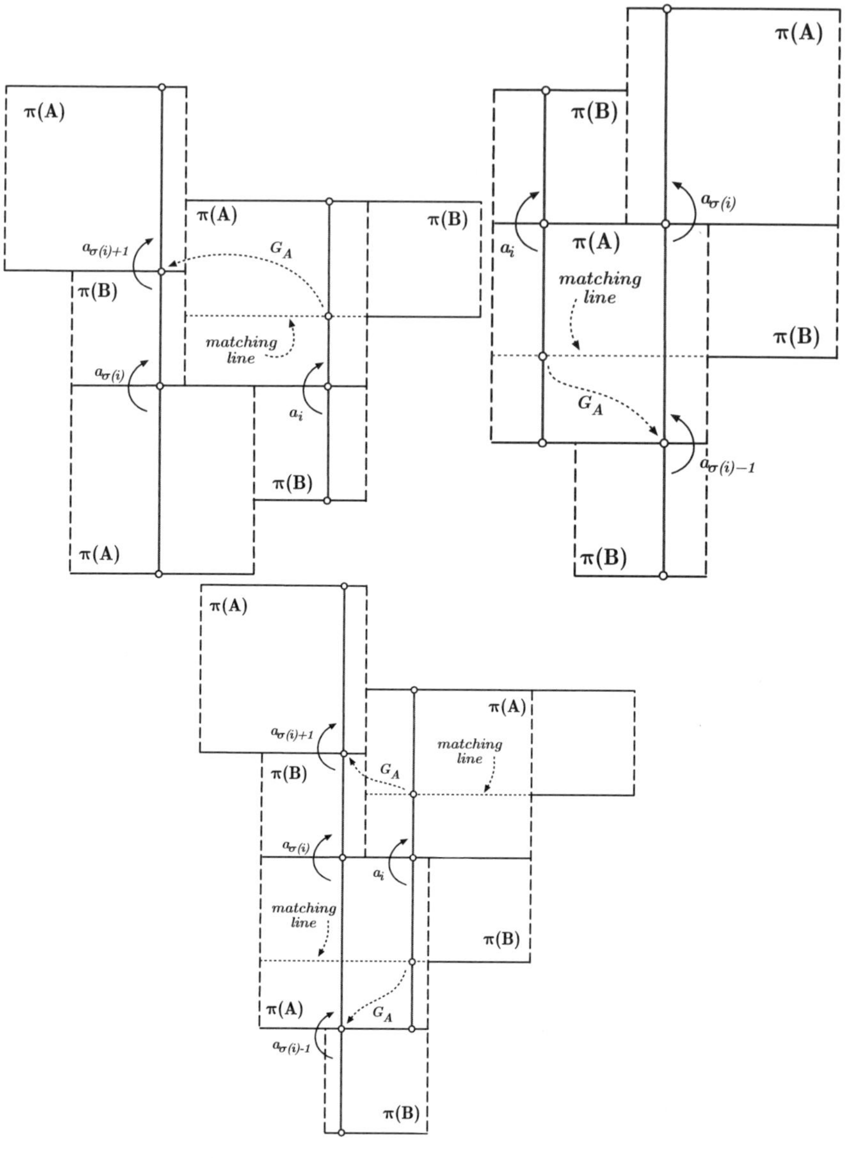

Fig. 13.10. The matching condition for the sequence A for the three possible cases: condition (i) corresponds to $I_{i-1} \in \mathbf{B}$ and $I_i \in \mathbf{A}$; condition (ii) corresponds to $I_{i-1} \in \mathbf{A}$ and $I_i \in \mathbf{B}$; condition (iii) corresponds to $I_{i-1} \in \mathbf{A}$ and $I_i \in \mathbf{A}$;

$$\begin{aligned} \frac{|I_B^l| + |I_A^l|}{|I_0|} &= \sigma\left(I_0 : I_B^l\right)\left(1 + \sigma\left(I_B^l : I_A^l\right)\right) \\ &= \lim_{i \to \infty} a_{F_{2i+1}}\left(1 + a_{F_{2i+1}+1}\right) \ , \end{aligned}$$

where $(|I_A^r|+|I_B^r|)\,/\,|I_0|$ and $\left(|I_B^l|+|I_A^l|\right)/\,|I_0|$ mean the ratios of these leaf segments given by the solenoid function. Since the tiling $\mathcal{T}$ satisfies the boundary condition (see Figure 13.9), we get that

$$\frac{|I_A^r|+|I_B^r|}{|I_0|} = \frac{|I_B^l|+|I_A^l|}{|I_0|} . \tag{13.2}$$

By the above construction, we have

$$\begin{aligned}\frac{|I_B^r|+|I_A^r|}{|I_2|} &= \sigma\left(I_2 : I_B^r\right)\left(1+\sigma\left(I_B^r : I_A^r\right)\right) \\ &= \lim_{i\to\infty} a_{F_{2i}+2}^{-1}\left(1+a_{F_{2i}+1}^{-1}\right)\end{aligned}$$

and

$$\begin{aligned}\frac{|I_A^l|+|I_B^l|}{|I_2|} &= \sigma\left(I_2 : I_A^l\right)\left(1+\sigma\left(I_A^l : I_B^l\right)\right) \\ &= \lim_{i\to\infty} a_{F_{(2i+1)}+2}^{-1}\left(1+a_{F_{(2i+1)}+1}^{-1}\right) .\end{aligned}$$

Since the tiling $\mathcal{T}$ satisfies the boundary condition (see Figure 13.9) we get that

$$\frac{|I_B^r|+|I_A^r|}{|I_2|} = \frac{|I_A^l|+|I_B^l|}{|I_0|} . \tag{13.3}$$

By the equalities (13.2) and (13.3), we obtain that sol is well-defined in the unstable spanning leaf segments of the unstable boundaries of the Markov rectangle and satisfy the boundary condition. Hence, a golden tiling $\mathcal{T}$ determines a Hölder solenoid function $\hat{\sigma}_{\mathcal{T}}$: sol $\to \mathbb{R}^+$, and vice-versa. □

13.4.6 Golden tilings versus Anosov diffeomorphisms

Let $\mathcal{G}$ be the set of all smooth Anosov difeomorphisms, with an invariant measure absolutely continuous with respect to the Lebesgue measure, that are topologically conjugated to the Anosov automorphism $G(x,y)=(x+y,x)$. Pinto et al. [154] proved that there is a one-to-one correspondence between

(i) golden tilings;
(ii) smooth conjugacy classes of golden diffeomorphism of the circle that are fixed point of renormalization;
(iii) smooth conjugacy classes of Anosov difeomorphisms in $\mathcal{G}$;
(iv) Pinto-Rand's solenoid functions.

Pinto et al. [154] proved the existence of an infinite dimensional space of golden tilings. However, we are only able to construct explicitly the following golden tiling $\mathcal{T}_R=\{I_m\subset\mathbb{R} : m\in\mathbb{L}\}$: for every $i=F_{n_p}+\ldots+F_{n_0}$,

(i) if $F_{n_0} = F_1$ or, $F_{n_0} = F_3$ and n_1 odd, then $a_i = \gamma^{-1}$;
(ii) if $F_{n_0} = F_2$ or, $n_0 > 3$ and even, then $a_i = \gamma$;
(iii) if $F_{n_0} = F_3$ and n_1 even or $n_0 > 3$ and odd, then $a_i = 1$.

We call $\mathcal{T}_R$ the *golden rigid tiling.* Pinto et al. [154] proved that an Anosov diffeomorphism $G \in \mathcal{G}$ with a $C^{1+zygmund}$ complete system of unstable holonomies corresponds to the rigid golden tiling.

13.5 Further literature

A. Pinto and D. Sullivan [175] proved a related result for C^{1+} conjugacy classes of expanding circle maps (see also Apendix C). Pinto et al. [153] extend the results of this chapter to Anosov diffeomorphisms. This chapter is based on Pinto and Rand [161] and Pinto, Almeida and Portela [154].

14

Pseudo-Anosov diffeomorphisms in pseudo-surfaces

There are diffeomorphisms on a compact surface S with uniformly hyperbolic 1 dimensional stable and unstable foliations if and only if S is a torus: the Anosov diffeomorphisms. What is happening on the other surfaces? This question leads to the study of pseudo-Anosov maps. Both Anosov and pseudo-Anosov maps appear as periodic points of the geodesic Teichmüller flow T_t on the unitary tangent bundle of the moduli space over S. We observe that the points of pseudo-Anosov maps are regular (the foliations are like the ones for the Anosov automorphisms) except for a finite set of points, called singularities, which are characterized by their number of prongs k. The stable and unstable foliations near the singularities are determined by the real and the imaginary parts of the quadratic differential $\sqrt{z^{k-2}(dz)^2}$. By a coordinate change $u(z) = z^{k/2}$ the quadratic differential $z^{k-2}(dz)^2$ gives rise to the quadratic differential $(du)^2$ and, in this new coordinates, the pseudo-Anosov maps are uniform contractions and expansions of the stable and unstable foliations. This fact inspired the construction of Pinto-Rand's pseudo-smooth structures, near the singularities, such that the pseudo-Anosov maps are smooth for this pseudo-smooth structures, and have the property that the stable and unstable foliations are uniformly contracted and expanded by the pseudo-Anosov dynamics. We define a pseudo-linear algebra, the first step in constructing the notion of the derivative of a map at a singularity. In this way, we obtain a pseudo-smooth structure at the singularity, leading to Pinto-Rand's pseudo-smooth manifolds, pseudo-smooth submanifolds, pseudo-smooth splittings and pseudo-smooth diffeomorphisms. The Stable Manifold Theorem, for pseudo-smooth manifolds, is presented giving the associated pseudo-Anosov diffeomorphisms.

14.1 Affine pseudo-Anosov maps

Let A_c be a conformal structure on a compact surface S. Two conformal structures A_c and B_c are equivalent if, and only if, there is a conformal map

h such that $A_c = h^*(B_c)$. The moduli space $M_S = \{[A_c]\}$ has a natural metric given by the minimal quasi-conformal distortion of the maps from the elements of a class $[A_c]$ to the elements of the other class $[B_c]$.

The geodesic (Teichmüller) flow T_t on the unitary tangent bundle of the moduli space has a dense set of periodic orbits. If the surface S is a torus, then the periodic points correspond to Anosov automorphisms. If the surface S is not a torus, then the periodic points correspond to pseudo-Anosov maps.

All the points of an Anosov automorphism are regular. The points of a pseudo-Anosov maps are regular, except for a finite set of points called singularities. A regular point is locally characterized by a quadratic differential $(dz)^2$. The stable and unstable foliations are determined by the real and the imaginary parts of $\sqrt{(dz)^2} = \pm dz$.

The singularities of pseudo-Anosov maps are characterized by their number of prongs k. A k-prong singularity is locally characterized by a quadratic differential $z^{k-2}(dz)^2$. The stable and unstable foliations are determined by the real and the imaginary parts of $\sqrt{z^{k-2}(dz)^2}$. If the pseudo-Anosov map has a singularity with an odd number of prongs, then the stable and unstable foliations are non-orientable.

By a coordinate change $u(z) = z^{k/2}$, the quadratic differential $z^{k-2}(dz)^2$ gives rise to the quadratic differential $(du)^2$. In this new coordinates, the pseudo-Anosov maps are locally affine contractions and expansions of the stable and unstable foliations by λ^{-1} and λ, respectively.

How can we regard the image of $u(z) = z^{k/2}$? The answer to this question leads us to the construction of Pinto-Rand's paper models, where the pseudo-Anosov maps constructed above are affine.

14.2 Paper models Σ_k

Let $\mathbb{H} = \{(x, y) \in \mathbb{R}^2 : y \geq 0\}$ denote the upper half plane with the Euclidean metric d_E. Consider the space $\sqcup_{j \in \mathbb{Z}_k} \mathbb{H}_{j\pi}$ which is the disjoint union of k copies of $\mathbb{H}$, with $\mathbb{Z}_k = \mathbb{Z}/k\mathbb{Z}$. Let the *paper models* Σ_k be the space obtained from $\sqcup_{j \in \mathbb{Z}_k} \mathbb{H}_{j\pi}$ by identifying $(x, 0) \in \mathbb{H}_{(j+1)\pi}$ with $(-x, 0) \in \mathbb{H}_{j\pi}$, for all $x \geq 0$. Let $s \in \Sigma_k$ be the point determined by $(0, 0) \in \mathbb{H}_{j\pi}$ for every $j \in \mathbb{Z}_k$. The Euclidean metric d_E on the upper half planes $\mathbb{H}_{j\pi}$ naturally define a flat metric on $\Sigma_k \setminus \{s\}$ which extends to a *continuous* metric d_k on Σ_k (see Figure 14.1).

The map $i : \mathbb{R} \to \Sigma_k$ is an *isometry* if, and only if, there is an isometry $i_{\mathbb{H}} : \mathbb{H} \to \Sigma_k$ such that $i_{\mathbb{H}}(x, 0) = i(x)$, for all $x \in \mathbb{R}$ (see Figure 14.2).

We say that:

- $l \subset \Sigma_k$ is a *straight line* in Σ_k if, and only if, there is an isometry $i : \mathbb{R} \to \Sigma_k$ such that $l = i(\mathbb{R})$;
- $l_{a \to b} \subset \Sigma_k$ is a *semi-straight line* in Σ_k, with origin at a and passing through b, if, and only if, there is an isometry $i : \mathbb{R} \to \Sigma_k$ such that $l_{a \to b} = i([a', +\infty))$ with $i(a') = a$ and $i(b') = b$, for some points $a' < b'$;

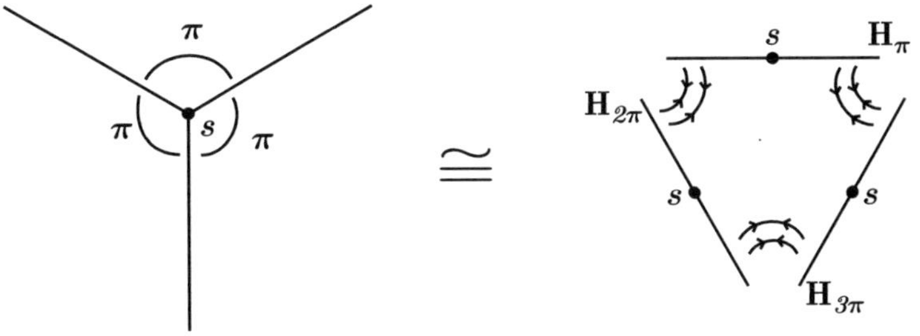

Fig. 14.1. $k = 3$.

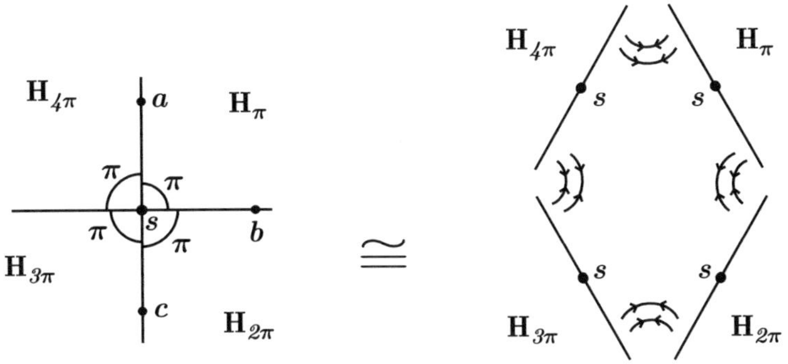

Fig. 14.2. There is a straight line passing through a and b. There is no straight line passing through a and c.

- $l_{a,b} \subset \Sigma_k$ is a *segment straight line* in Σ_k, with endpoints a and b, if, and only if, there is an isometry $i : \mathbb{R} \to \Sigma_k$ such that $l_{a,b} = i([a', b'])$ with $i(a') = a$ and $i(b') = b$, for some points $a' < b'$. The interior $\text{int}l_{a,b}$ of $l_{a,b}$ is equal to $l_{a,b} \setminus \{a, b\}$.

Let $l_{s\to a}$ and $l_{s\to b}$ be two semi-straight lines in Σ_k. To fix ideas, let us suppose that $l_{s\to a} \subset \mathbb{H}_{j\pi}$ and $l_{s\to b} \subset \mathbb{H}_{(j+n)\pi}$, with $j, j+n \in \mathbb{Z}_k$. Let $l_{s\to c}$ be the semi-straight line formed by the points of $\mathbb{H}_{j\pi}$ and $\mathbb{H}_{(j+1)\pi}$ that were identified at the construction of Σ_k. Analogously, let $l_{s\to d}$ be the semi-straight line formed by the points of $\mathbb{H}_{(j+n-1)\pi}$ and $\mathbb{H}_{(j+n)\pi}$ that were identified at the construction of Σ_k. Let $\alpha \in [0, \pi]$ be the angle $\sphericalangle(l_{s\to a}, l_{s\to c})$ between the semi-straight lines $l_{s\to a}$ and $l_{s\to c}$, and let $\beta \in [0, \pi]$ be the angle $\sphericalangle(l_{s\to d}, l_{s\to b})$ between the semi-straight lines $l_{s\to d}$ and $l_{s\to b}$. We say that the *angle* $\sphericalangle(l_{s\to a}, l_{s\to b})$ *between the semi-straight lines* $l_{s\to a}$ *and* $l_{s\to b}$ is given by

$$\sphericalangle(l_{s\to a}, l_{s\to b}) = \alpha + (n-1)\pi + \beta.$$

Given $\alpha \in \mathbb{R}/k\pi\mathbb{R}$ and two points $x, y \in \Sigma_k$, we say that they are in an *α-angular region*, if $\sphericalangle(l_{s\to x}, l_{s\to y}) \leq \alpha$ (see Figure 14.3).

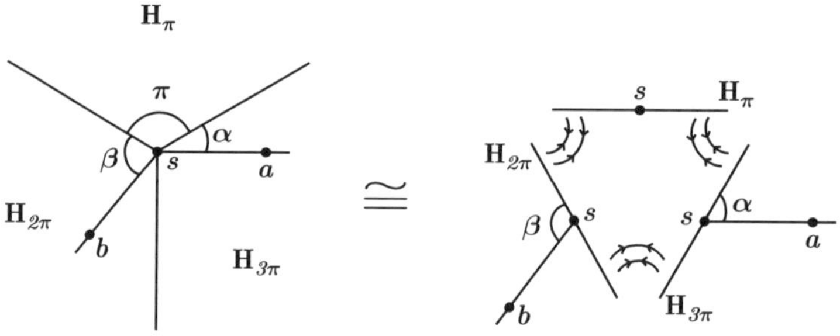

Fig. 14.3. The angle $\sphericalangle(l_{s\to a}, l_{s\to b}) = \alpha + \pi + \beta$.

14.3 Pseudo-linear algebra

Given two points $x, y \in \Sigma_k$, we say that $\mathbf{y} = y - x$ is a *vector* if, and only if, there is a segment straight line $l_{x,y} \subset \Sigma_k$ with endpoints x and y; we call x the *origin* and y the *endpoint* of the vector $y - x$. The *norm* $\|y - x\|$ of the vector $y - x$ is given by $d_k(x, y)$.

Given a vector $\mathbf{y} = y - x$ and a constant $\lambda \in \mathbb{R}$, the vector $w - x = \lambda(y - x)$ is well-defined if, and only if, there is an isometry $i_{\mathbb{H}} : \mathbb{H} \to \Sigma_k$ with the following property: there are points $x_{\mathbb{H}}, y_{\mathbb{H}}, w_{\mathbb{H}} \in \mathbb{H}$ such that

(i) $x = i_{\mathbb{H}}(x_{\mathbb{H}})$, $y = i_{\mathbb{H}}(y_{\mathbb{H}})$ and $w = i_{\mathbb{H}}(w_{\mathbb{H}})$;
(ii) $w_{\mathbb{H}} - x_{\mathbb{H}} = \lambda(y_{\mathbb{H}} - x_{\mathbb{H}})$;
(iii) if $s \in \text{int} l_{x,w}$, then $s \in \text{int} l_{x,y}$;
(iv) if $s = x$, then $\lambda \geq 0$.

We note that the vector $\lambda(y - x)$ is well-defined, for all $0 \leq \lambda \leq 1$. The above conditions (iii) and (iv) imply that the vector $w - x$ does not depend upon the isometry considered, and so $w - x$ is uniquely determined.

Given two vectors $\mathbf{y} = y - x$ and $\mathbf{z} = z - x$ with the same origin, the vector $\mathbf{w} = w - x$, with $\mathbf{w} = \mathbf{y} + \mathbf{z}$, is equal to the *sum* of the vectors $y - x$ with $z - x$ if, and only if, there is an isometry $i_{\mathbb{H}} : \mathbb{H} \to \Sigma_k$ with the following property: there exists a constant $\lambda > 0$ and there are points $x_{\mathbb{H}}, y_{\mathbb{H}}, z_{\mathbb{H}}, w_{\mathbb{H}} \in \mathbb{H}$ such that (see Figure 14.4)

(i) the vectors $y' - x = \lambda(y - x)$, $z' - x = \lambda(z - x)$ and $w' - x = \lambda(w - x)$ are well-defined;
(ii) $x = i_{\mathbb{H}}(x_{\mathbb{H}})$, $y' = i_{\mathbb{H}}(y_{\mathbb{H}})$, $z' = i_{\mathbb{H}}(z_{\mathbb{H}})$ and $w' = i_{\mathbb{H}}(w_{\mathbb{H}})$;
(iii) $w_{\mathbb{H}} = y_{\mathbb{H}} + z_{\mathbb{H}} - x_{\mathbb{H}}$;
(iv) if $s \in \text{int} l_{x,w}$, then $s \in \text{int} l_{x,y} \cup \text{int} l_{x,z}$.

The above condition (iv) implies that the vector $\mathbf{w} = w - x$ does not depend upon the isometry considered. If s is a singularity, with order k, then there are k distinct vectors $x_1 - s, \ldots, x_k - s$, all with norm equal to one, such that $x_i - s + x_{i+1} - s = s - s$, for all $i \in \mathbb{Z}_k$ (see Figure 14.5).

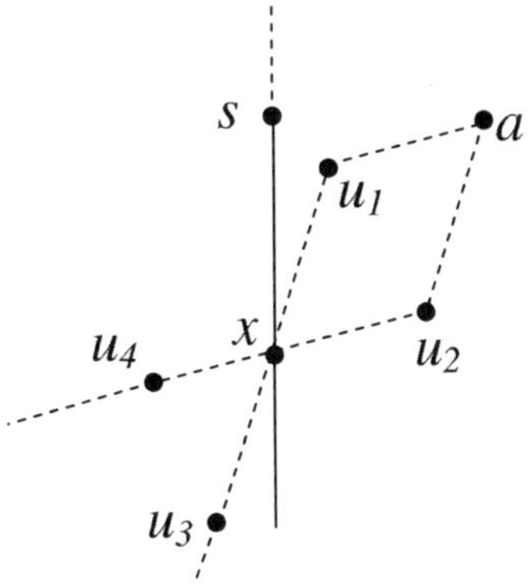

Fig. 14.4. $u_1 + u_2 = a$ and $\langle(u_1, u_3), (u_2, u_4)\rangle$ is a basis of $\mathbb{V}_x$.

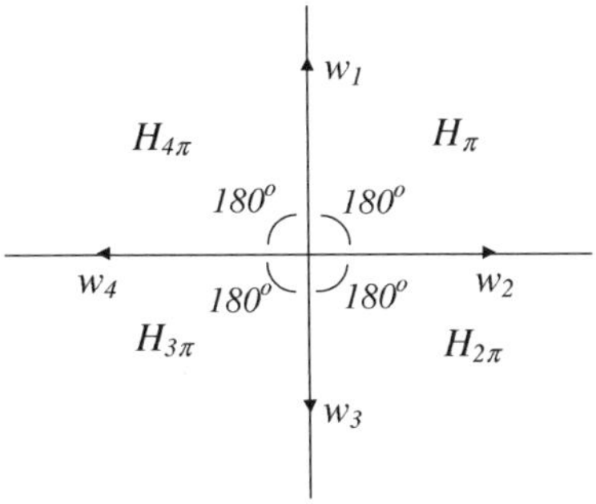

Fig. 14.5. + is not associative: $(\mathbf{w_1}+\mathbf{w_2})+\mathbf{w_3} = \mathbf{w_3}$; $\mathbf{w_1}+(\mathbf{w_2}+\mathbf{w_3}) = \mathbf{w_1}$. There is not a unique "inverse": $\mathbf{w_1} + \mathbf{w_2} = \mathbf{0}$; $\mathbf{w_1} + \mathbf{w_4} = \mathbf{0}$, where $\mathbf{0} = s - s$. $\mathbf{w_2} + \mathbf{w_4}$ is not well-defined.

The *pseudo-linear space* $\mathbb{V}_x$ at x is the set of all vectors with origin at x, together with the operations of addition of vectors and of multiplication of a vector by a constant, as constructed above. Let l_x be either (i) the empty set or (ii) a semi-straight line contained in a semi-straight line with origin at x. The *branched linear space* $\mathbb{V}_{l_x}$ is given by $\mathbb{V}_x \setminus \text{int}l_x$ (see Figure 14.6).

A *pseudo-linear subspace* $\mathbb{S}_x$ of a pseudo-linear space $\mathbb{V}_x$ (see Figure 14.7) is a subset of $\mathbb{V}_x$ with the following properties:

(i) For all $\mathbf{u}, \mathbf{v} \in \mathbb{S}_x$ such that $\mathbf{u}+\mathbf{v}$ is well-defined, we have that $\mathbf{u}+\mathbf{v} \in \mathbb{S}_x$;
(ii) For all $\lambda \in \mathbb{R}$ and $\mathbf{u} \in \mathbb{S}_x$ such that $\lambda\mathbf{u}$ is well-defined, we have that $\lambda\mathbf{u} \in \mathbb{S}_x$.

A *full pseudo-linear space* $\mathbb{S}_x$ is a pseudo-linear subspace $\mathbb{S}_x$ with the following property: If $\mathbf{u} \in \mathbb{S}_x$ and $\mathbf{v} \in \mathbb{V}_x$ are such that $\mathbf{u}+\mathbf{v} = \mathbf{0}$, then $\mathbf{v} \in \mathbb{S}_x$. Hence, a full pseudo-linear subspace $\mathbb{S}_s$, $\mathbb{S}_s \neq \mathbb{V}_s$, at the singularity s, with order k, is the image of an isometry $i : \Sigma_k^1 \to \mathbb{V}_s$.

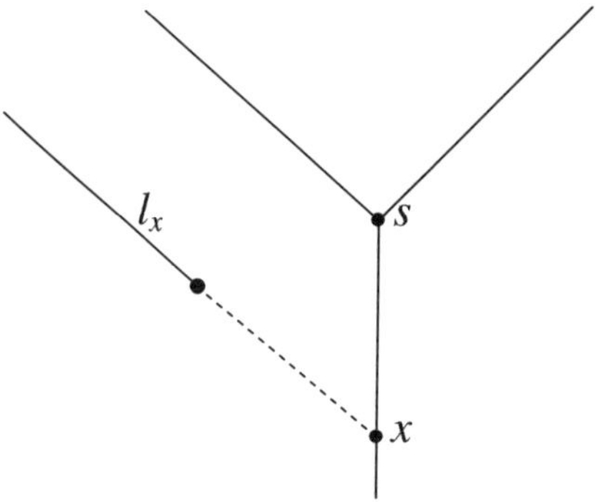

Fig. 14.6. The branched linear space $\mathbb{V}_{l_x}$.

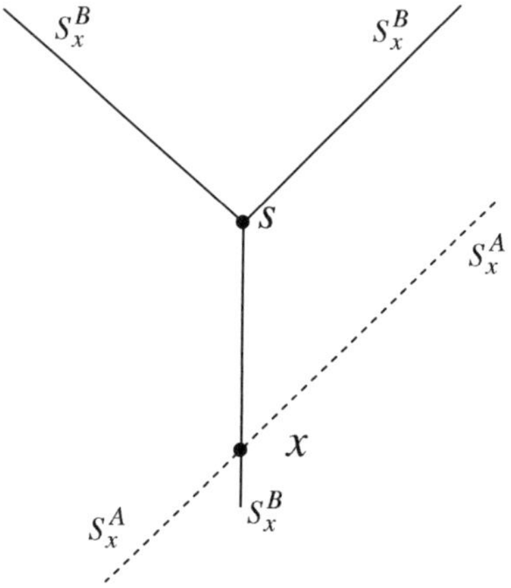

Fig. 14.7. Pseudo-linear subspaces $\mathbb{S}_x^A$ and $\mathbb{S}_x^B$ at x.

A pseudo-affine subspace $\mathbb{S}$ at a point $x \in \Sigma_k \setminus \{s\}$, with $\mathbb{S}_x \neq \mathbb{V}_x$, is the image of an isometry $i : A \to \mathbb{V}_x$ with A equal either $\mathbb{R}$ or Σ_k^1.

A map $L : \mathbb{V}_{l_x} \to \mathbb{V}_y$ is *linear* (see Figures 14.8 and 14.9), if the set $\mathbb{V}_{l_x}$ is a branched linear space in Σ_k, $\mathbb{V}_y$ is a pseudo-linear space in $\Sigma_{k'}$ and L satisfies the following properties:

(i) For every $\mathbf{v}, \mathbf{w} \in \mathbb{V}_{x,l}$ such that the vectors $\mathbf{v} + \mathbf{w}$ and $L(\mathbf{v}) + L(\mathbf{w})$ are well-defined, we have $L(\mathbf{v} + \mathbf{w}) = L(\mathbf{v}) + L(\mathbf{w})$;
(ii) For every $\lambda \in \mathbb{R}$ and $\mathbf{v} \in \mathbb{V}_{x,l}$ such that the vectors $\lambda\mathbf{v}$ and $L(\lambda\mathbf{v})$ are well-defined, we have $L(\lambda\mathbf{v}) = \lambda L(\mathbf{v})$;
(iii) $L(a - x) = s - y$, where a is the origin of l_x, $a - x \in \mathbb{V}_x$ is the vector with origin at x and $s - y \in \mathbb{V}_y$ is the vector with origin at y.

Given two linear maps $L_1 : \mathbb{V}_{l_x} \to \mathbb{V}_y$ and $L_2 : \mathbb{V}_{l_y} \to \mathbb{V}_z$, there is a unique linear map $L_3 : \mathbb{V}_{l'_x} \to \mathbb{V}_z$ such that $L_3 | \mathbb{V}_{l_x} \cap \mathbb{V}_{l'_x} = L_2 \circ L_1$, where l'_x might be

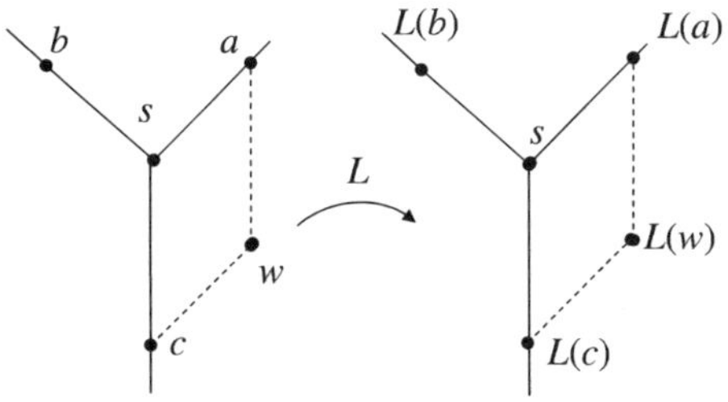

Fig. 14.8. Linear map at the singularity s.

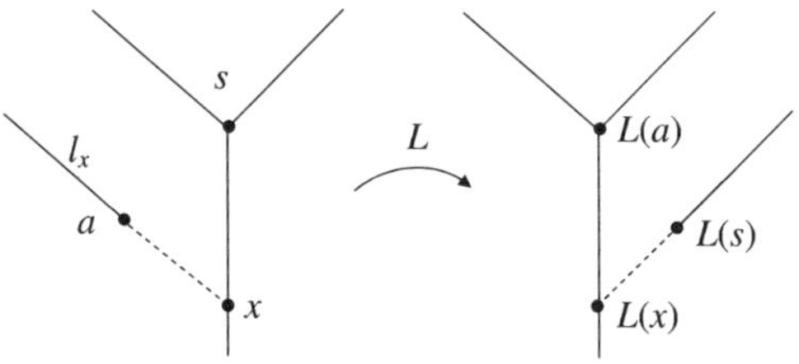

Fig. 14.9. Linear map at the point x.

distinct of l_x (see Figure 14.10). Hence, the *composition* $L_2 \circ L_1$ of two linear maps is well-defined by $L_3 = L_2 \circ L_1$, and so it is a linear map.

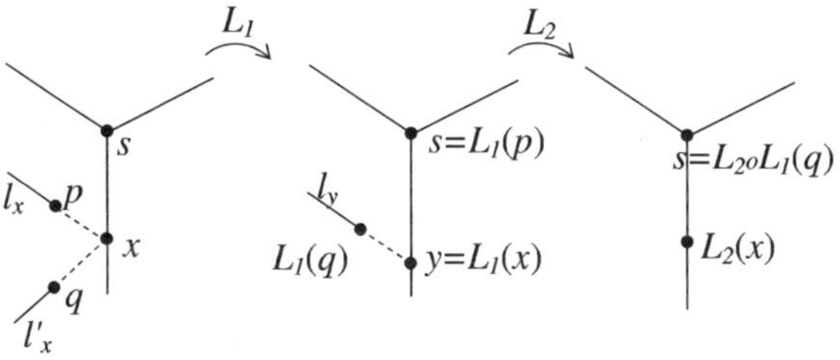

Fig. 14.10. The composition $L_3 = L_2 \circ L_1$ is well-defined.

A map $L_1 : \mathbb{V}_{l_x} \to \mathbb{V}_y$ is an *isomorphism* if, and only if, there is a linear map $L_2 : \mathbb{V}_{l_y} \to \mathbb{V}_x$ such that $L_2 \circ L_1 | \mathbb{V}_{l_x} \cap L_1^{-1}(\mathbb{V}_{l_y})$ and $L_1 \circ L_2 | \mathbb{V}_{l_y} \cap L_2^{-1}(\mathbb{V}_{l_x})$ are the identity maps. We note that if the linear map L_2 exists, then it is unique. Hence, the *inverse map* L_1^{-1} of L_1 is well-defined by $L_1^{-1} = L_2$. The kernel of a linear map $L : \mathbb{V}_{l_x} \to \mathbb{V}_y$ is equal to the intersection $\mathbb{V}_{l_x} \cap \mathbb{S}_x$ of a pseudo-linear subspace $\mathbb{S}_x$ with $\mathbb{V}_{l_x}$.

We say that a vector $y - x$ has a *parallel transport* from x to z (see Figure 14.11), if there are a vector $w - z$, a constant λ, with $|\lambda| \leq 1$, and an isometry $i_{\mathbb{H}} : \mathbb{H} \to \Sigma_k$ with the following property: there are points $x_{\mathbb{H}}, y_{\mathbb{H}}, z_{\mathbb{H}}, w_{\mathbb{H}} \in \mathbb{H}$ such that

(i) $w' - z = \lambda(w - z)$ and $y' - x = \lambda(y - x)$ are well-defined;

(ii) $x = i_{\mathbb{H}}(x_{\mathbb{H}})$, $z = i_{\mathbb{H}}(z_{\mathbb{H}})$, $y' = i_{\mathbb{H}}(y_{\mathbb{H}})$ and $w' = i_{\mathbb{H}}(w_{\mathbb{H}})$;
(iii) $w_{\mathbb{H}} - z_{\mathbb{H}} = y_{\mathbb{H}} - x_{\mathbb{H}}$;
(iv) if $s \in l_{z,w} \setminus \{w\}$, then $s \in \text{int} l_{x,y}$.

The parallel transport is uniquely determined, if $s \notin l_{z,w} \setminus \{w\}$ or if $s \in l_{z,w} \setminus \{w\} \cap \text{int} l_{x,y}$. Let $\mathbb{V}_{x \to z}$ be the set of all vectors that have a parallel transport from x to z. The *parallel transport map* $\mathbb{P}_{x \to z} : \mathbb{V}_{x \to z} \to \mathbb{V}_z$ is well-defined by $\mathbb{P}_{x \to z}(u) = v$, where the vector v is the parallel transport of the vector u from x to z, when $\mathbb{V}_{x \to z}$ is non-empty.

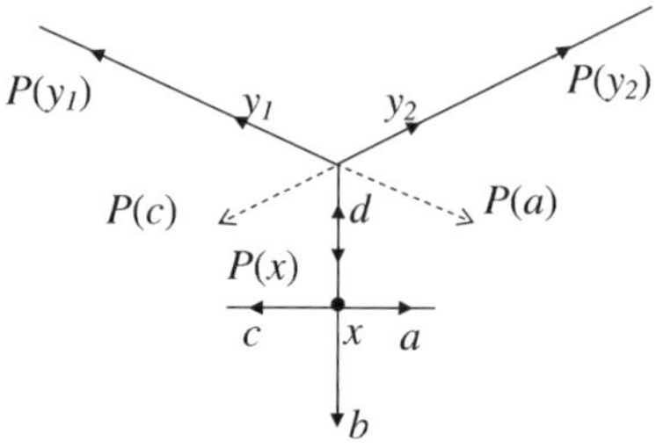

Fig. 14.11. Parallel transport from x to s.

We note that the parallel transport map $\mathbb{P}_{x \to z}$ is a linear map, except in the case where $x = s$ and $z \neq s$, because $\mathbb{P}_{s \to z}$ is just defined in an open 2π-angular region. However, $\mathbb{P}_{z \to s} : \mathbb{V}_{l_z} \to \mathbb{V}_s$ is a linear map and $\mathbb{P}_{s \to z} \circ \mathbb{P}_{z \to z} | \mathbb{V}_{l_z}$ is the identity.

We say that a map $G : \mathbb{V}_{x_1}^m \to \mathbb{V}_y$ is an *m-multilinear map*, if, for every $(\mathbf{a}_1, \ldots, \mathbf{a}_{i-1}, \mathbf{0}, \mathbf{a}_{i+1}, \ldots, \mathbf{a}_m)$, there is $\mathbb{V}_{l_i}$, where l_i depends upon $(\mathbf{a}_1, \ldots, \mathbf{a}_{i-1}, \mathbf{0}, \mathbf{a}_{i+1}, \ldots, \mathbf{a}_m)$, such that the map $g : \mathbb{V}_{l_i} \to \mathbb{V}_y$ defined by $g(\mathbf{a}_i) = G(\mathbf{a}_1, \ldots, \mathbf{a}_i, \ldots, \mathbf{a}_m)$ is a linear map.

Lemma 14.1. *Let $L : \mathbb{V}_{x_1}^m \to \mathbb{V}_{y_1}$ be an m-multilinear map. Let x_2 and y_2 be such that the parallel transport maps $\mathbb{P}_{x_1 \to x_2}$ and $\mathbb{P}_{y_1 \to y_2}$ are well-defined. Suppose that if x_1 is a singularity with order k, then x_2 is a singularity with order $2nk$, for some $n \geq 1$. Then, there is an m-multilinear map $L_P : \mathbb{V}_{x_2}^m \to \mathbb{V}_{y_2}$ such that*

$$L_P \left(P_{x_1 \to x_2}(\mathbf{v}_1), \ldots, P_{x_1 \to x_2}(\mathbf{v}_m) \right) = P_{y_1 \to y_2} \left(L(\mathbf{v}_1, \ldots, \mathbf{v}_m) \right),$$

whenever both sides are well-defined.

We call the above linear map L_P the *parallel transport of L from (x_1, y_1) to (x_2, y_2)*. We note that the parallel transport L_P of L is an isomorphism.

Proof. The map $P_{y_1 \to y_2} \circ L_1 \circ P_{x_1 \to x_2}^{-1}$ has a unique extension to a linear map. □

Let $L_1 : \mathbb{V}^m_{x_1} \to \mathbb{V}_{y_1}$ and $L_2 : \mathbb{V}^m_{x_1} \to \mathbb{V}_{y_2}$ be two m-multilinear maps. Let $0 \leq h \leq 1$ be such that $L_1(\mathbf{v})$ and $L_2(\mathbf{v})$ are well-defined, for all $\mathbf{v}$ with $\|\mathbf{v}\| = h$, and such that there is $\mathbf{w}(\mathbf{v})$ with the property that $\mathbf{w}(\mathbf{v}) + L_1(\mathbf{v}) = L_2(\mathbf{v})$. We define the *distance $d(L_1, L_2)$ between the m-multilinear maps L_1 and L_2* as follows:

$$d(L_1, L_2) = \begin{cases} +\infty, & \text{if } h = 0 \\ \max_{\mathbf{v}} \frac{\|\mathbf{w}(\mathbf{v})\|}{h}, & \text{otherwise} \end{cases}$$

Let $L_1 : \mathbb{V}^m_{x_1} \to \mathbb{V}_{y_1}$ and $L_2 : \mathbb{V}^m_{x_2} \to \mathbb{V}_{y_2}$ be two m-multilinear maps. Let $\mathbb{L}$ be the set of all parallel transport L_P of L_2 from (x_2, y_2) to (x_1, y_1). We define the *distance $d(L_1, L_2)$ between the m-multilinear maps L_1 and L_2* as follows:

$$d(L_1, L_2) = \begin{cases} +\infty, & \text{if } \mathbb{L} = \emptyset \\ \min_{L_P \in \mathbb{L}} d(L_1, L_P), & \text{otherwise} \end{cases}$$

We note that $d(L_1, L_2) = d(L_2, L_1)$.

14.4 Pseudo-differentiable maps

Let $f : A \subset \Sigma_k \to \Sigma_{k'}$ be a map defined on an open neighbourhood A of x in Σ_k. We say that the map f is *pseudo-differentiable at x*, if there is a linear map $D_x f : \mathbb{V}_{l_x} \to \mathbb{V}_{f(x)}$ with the following property: For all $\mathbf{v} \in \mathbb{V}_{l_x}$, there exists a constant $h_0 > 0$ such that there is a unique vector $\mathbf{w}(h, \mathbf{v})$ satisfying

$$\mathbf{w}(h, \mathbf{v}) + f(x) = f(x + h\mathbf{v}),$$

for all $0 < h < h_0$, and

$$D_x f(\mathbf{v}) = \lim_{h \to 0} \frac{\mathbf{w}(h, \mathbf{v})}{h}.$$

By induction, let us suppose that the $(m-1)^{\text{th}}$-derivative $D^{m-1}_x f : \mathbb{V}^m_x \to \mathbb{V}_{f(x)}$ of f is well-defined in an open set A containing x. We say that f is m *pseudo-differentiable at x*, if there is an m-multilinear map

$$D^m_x f : \mathbb{V}^m_x \to \mathbb{V}_{f(x)}$$

with the following property: For all $\mathbf{v} \in \mathbb{V}^m_x$, there exists a constant $h_0(\mathbf{v}) > 0$ such that there is a unique vector $\mathbf{w}(h, \mathbf{v})$ satisfying

$$\mathbf{w}(h, \mathbf{v}_1, \ldots, \mathbf{v}_m) + D^{m-1}_x f(\mathbf{v}_1, \ldots, \mathbf{v}_m) = D^{m-1}_{x+h\mathbf{v}_1} f(\mathbf{v}_2, \ldots, \mathbf{v}_m),$$

for all $0 < h < h_0(\mathbf{v})$, and

$$D^m_x f(\mathbf{v}_1, \ldots, \mathbf{v}_m) = \lim_{h \to 0} \frac{1}{h} \mathbf{w}(h, \mathbf{v}_1, \ldots, \mathbf{v}_m).$$

A map $f : A \to \Sigma_{k'}$ is C^m, with $m \in \mathbb{N}$, in the open set $A \subset \Sigma_k$, if f is m-differentiable for all $x \in A$, and the m-derivative $D_x f$ varies continuously with x. We say that f is a $C^{m+\alpha}$, with $m \in \mathbb{N}$ and $0 < \alpha \leq 1$, if f is C^m and there exists $c > 0$ such that

$$\|D_x f - D_y f\| \leq c\|x - y\|^\alpha,$$

for all $x, y \in A$ with the property that there is a parallel transport L_p from x to y.

We say that $B_\varepsilon = B_\varepsilon(x, s) \subset A$ is an *avoid singularity cone*, if $d(x, y) = \varepsilon d(x, s)$ and $\alpha = d(x, s)/\varepsilon$ (see Figure 14.12).

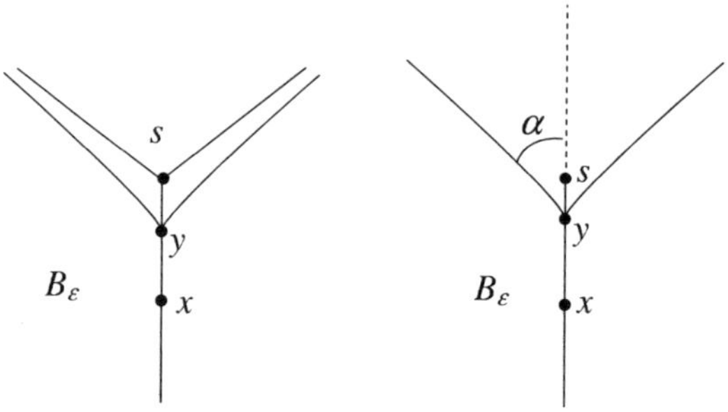

Fig. 14.12. Avoid singularity cone.

Theorem 14.2. *(Taylor's Theorem) Let $f : A \subset \Sigma_k \to \Sigma_k$ be a C^m pseudo-map defined on an open set A. Let $B_\varepsilon \subset A$ be an avoid singularity cone and $0 < \varepsilon < 1$ small. Then, for all $x, y \in B_\varepsilon$ with $\|y - x\| \leq \varepsilon$, the vectors $\mathbf{z}_m(x, y)$ and $\mathbf{w}_m(x, y)$ are well-defined by*

$$\begin{aligned} \mathbf{z}_m(x, y) &= \left(\ldots \left(D_x f(y - x) + D_x^2 f(y - x, y - x)\right) + \ldots\right) + \\ &\quad + \frac{1}{m!} D_x^m f(y - x, \ldots, y - x) \\ f(y) - f(x) &= \mathbf{z}_m(x, y) + \mathbf{w}_m(x, y). \end{aligned}$$

Furthermore,

$$\|\mathbf{w}_m(x, y)\| \leq \chi(\|y - x\|)\|y - x\|^m,$$

where $\chi : \mathbb{R}_0^+ \to \mathbb{R}_0^+$ is a continuous map with $\chi(0) = 0$.

Let $l_1, \ldots, l_{2k}$ be semi-straight lines with origin at s such that $0 < \sphericalangle(l_i, l_{i+1}) < \pi$ and $\sphericalangle(l_i, l_{i+2}) = \pi$ for every $i \in \mathbb{Z}_{2k}$. Then, $S_s^1 = \cup_{i \in \mathbb{Z}_{2k}} l_{2i}$ and $S_s^2 = \cup_{i \in \mathbb{Z}_{2k}} l_{2i+1}$ are pseudo-linear subspaces at the singularity s. We call the *direct sum $S_s^1 \bigoplus S_s^2$ of S_s^1 and S_s^2* to the set of all pairs $(\mathbf{u}, \mathbf{v})$ of vectors with the property that if $\mathbf{u}_i \in l_i$, then $\mathbf{u}_{i+1} \in l_{i+1}$, for all $i \in \mathbb{Z}_{2k}$. By construction, there are one-to-one maps

$$\begin{aligned} \Theta_1 &: \mathbb{V}_s \to S_s^1 \oplus S_s^2 \\ \Theta_2 &: \Sigma_k \to S_s^1 \oplus S_s^2 \end{aligned}$$

given by $\Theta_1^{-1}(\mathbf{u},\mathbf{v}) = \mathbf{u}+\mathbf{v}$ and $\Theta_2^{-1}(\mathbf{u},\mathbf{v}) = (\mathbf{u}+\mathbf{v})+s$. We say that $\langle(\mathbf{u}_1,\ldots,\mathbf{u}_{2k-1}),(\mathbf{u}_2,\ldots,\mathbf{u}_{2k})\rangle$ is a *basis* of $\mathbb{V}_s$, if $\mathbf{u}_i \in l_i$ and $\mathbf{u}_i+\mathbf{u}_{i+2}=\mathbf{0}$, for every $i\in\mathbb{Z}_{2k}$ (see Figures 14.13 and 14.14).

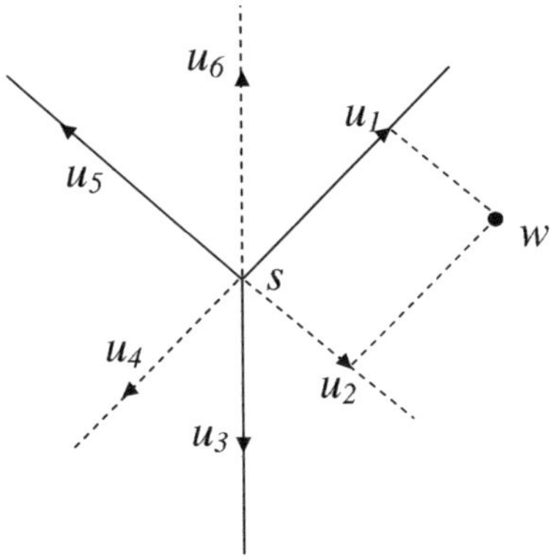

Fig. 14.13. $\mathbf{u_1}+\mathbf{u_2}=\mathbf{w}$ and $\langle(\mathbf{u_1},\mathbf{u_3},\mathbf{u_5}),(\mathbf{u_2},\mathbf{u_4},\mathbf{u_6})\rangle$ is a basis of $\mathbb{V}_s$.

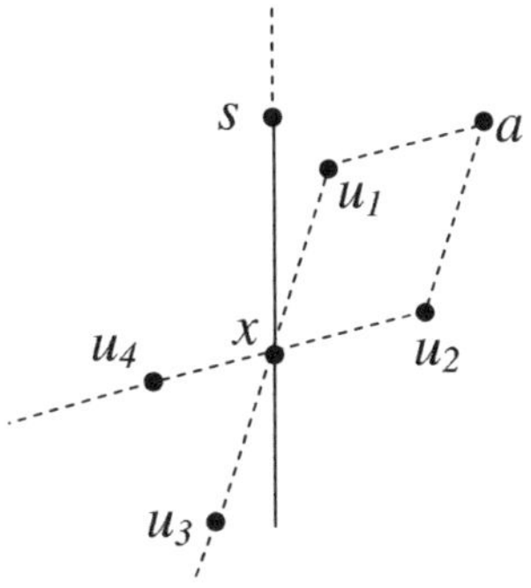

Fig. 14.14. $\mathbf{u_1}+\mathbf{u_2}=\mathbf{a}$ and $\langle(\mathbf{u_1},\mathbf{u_3}),(\mathbf{u_2},\mathbf{u_4})\rangle$ is a basis of $\mathbb{V}_x$.

For every $i\in\mathbb{Z}_{2k}$, let $\mathbf{u}_i\in l_i$ be such that $\|\mathbf{u_i}\|=1$. Let $D_{K_i}=\mathbb{R}^2\setminus((-\infty,0)\times\{0\})$. We define the map $K_i : D_{K_i}\to\mathbb{V}_s$ at the singularity by

$$K_i(a,b)=\begin{cases} a\mathbf{u}_i+b\mathbf{u}_{i+1}, & \text{if } a,b\geq 0\\ a\mathbf{u}_i+b\mathbf{u}_{i-1}, & \text{if } a\geq 0, b\leq 0\\ a\mathbf{u}_{i+2}+b\mathbf{u}_{i+1}, & \text{if } a\leq 0, b>0\\ a\mathbf{u}_{i-2}+b\mathbf{u}_{i-1}, & \text{if } a\leq 0, b<0\end{cases}$$

The set of maps $K_1,\ldots,K_{2k}$ is called a *coordinate system for* $\mathbb{V}_s$ ($\cong \Sigma_k$) given by $S_1\bigoplus S_2$.

Lemma 14.3. *Let* $K_1,\ldots,K_{2k}$ *be a coordinate system for* Σ_k.

(i) Let $L : V_{l_s} \to V_{l'_s}$ *be a linear map at the singularity. Then, there is a unique linear map* $L' : \mathbb{R}^2 \to \mathbb{R}^2$ *such that* $L'(a,b) = K_j^{-1} \circ L \circ K_i = (a,b)$*, where* $j = j(i,a,b)$ *has the property that* $L \circ K_i(a,b) \in D_{K_j}$*.*
(ii) A map $f : A \to \Sigma_{k'}$ *is* C^r *on* $A \subset \Sigma_k$ *if, and only if,* $K_j^{-1} \circ f \circ K_i$ *is* C^r*, where* $j = j(i,a,b)$ *has the property that* $f \circ K_i(a,b) \in D_{K_j}$*.*

14.4.1 C^r pseudo-manifolds

Let M be a topological space. A *chart* $c : U \to \Sigma_k$ is a homeomorphism onto its image defined on an open set U of M (recall that $\Sigma_2 = \mathbb{R}^2$). If $k \neq 2$, then we call $c : U \to \Sigma_k$ a *singular chart*. A *topological atlas* $\mathcal{A}$ of M is a collection of charts

$$c_x : U_x \to \Sigma_{k_x}$$

such that the union $\cup_{x \in M} U_x$ of the open sets cover M. A C^r *pseudo-atlas* $\mathcal{A}$ of M is a topological atlas $\mathcal{A}$ of M with the following properties: (i) $\mathcal{A}$ has just a finite set of singular charts; (ii) the overlap maps

$$c_x \circ c_y^{-1} : c_y(U_x \cap U_y) \to c_x(U_x \cap U_y)$$

are C^r diffeomorphisms. A topological space M with a C^r pseudo-atlas $\mathcal{A}$ is called a C^r *pseudo-manifold*, that we will denote by the pair $(M, \mathcal{A})$. A topological space N contained in a C^r manifold $(M, \mathcal{A})$ is a *pseudo-submanifold* of M, if there is a collection $\mathcal{B}$ of charts

$$e_x : V_x \to \Sigma_{k_x}$$

with the following properties (see Figure 14.15):

(i) The set N is contained in the union $\cup_{x \in N} V_x$;
(ii) For all $x \in N$, $e_x(N \cap V_x)$ is the intersection of a pseudo-linear subspace $S_{e_x(x)}$ at $e_x(x)$ with an open set of M;
(iii) The dimension of $S_{e_x(x)}$ is 1;
(iv) The overlap maps

$$e_x \circ c_x^{-1} : c_x(U_x \cap V_x) \to e_x(U_x \cap V_x)$$

between the charts $c_x \in \mathcal{A}$ and $e_x \in \mathcal{B}$ are C^r diffeomorphisms.

Hence, the first derivative at every point is locally a bijection over a corresponding pseudo-linear subspace with dimension 1. We call the above charts e_x the *submanifold charts* of N.

Definition 14.4. *Let* $(M, \mathcal{A})$ *and* $(M', \mathcal{A}')$ *be* C^r *manifolds. The map* $f : M \to M'$ *is pseudo* C^r *if, and only if, the maps* $c_x \circ f \circ e_y^{-1}$ *are* C^r *with respect to charts* $c_x \in \mathcal{A}$ *and* $e_y \in \mathcal{A}'$*. The map* $f : M \to M'$ *is* C^r pseudo-diffeomorphism *if, and only if,* $f : M \to M'$ *is a homeomomorphism and the maps* $c_x \circ f \circ c_y^{-1}$ *are* C^r *with respect to charts* $c_x \in \mathcal{A}$ *and* $c_y \in \mathcal{A}'$*.*

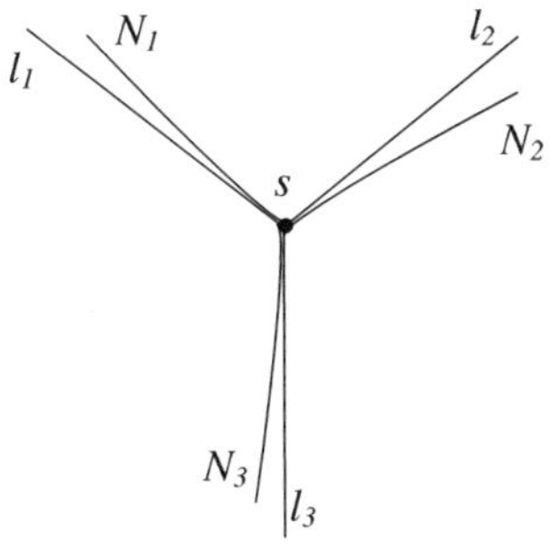

Fig. 14.15. The full subspace $\mathbb{S}_s = \cup_{i=1}^3 l_i$ at the singularity, and the pseudo-submanifold $N = \cup_{i=1}^3 N_i$.

14.4.2 Pseudo-tangent spaces

The *pseudo-tangent fiber bundle* $T\Sigma_k$ *of* Σ_k is the set $\cup_{x \in \Sigma_k} \mathbb{V}_x$, with the natural induced topology by Σ_k. We also call the pseudo-linear space $\mathbb{V}_x$ at x the *pseudo-tangent space* $T_x\Sigma_k$ *at* x ($T_x\Sigma_k \cong \mathbb{V}_x$).

The *pseudo-tangent space* T_xM *at* $x \in M$ of a C^r pseudo-manifold $(M, \mathcal{A})$ is a pseudo-linear space isomorphic to $T_{c_x(x)}\Sigma_{k_x}$, where $c_x : U_x \to \Sigma_{k_x}$ is a chart in $\mathcal{A}$ with $x \in U_x$. The *tangent fiber bundle* TM *of a* C^r *manifold* $(M, \mathcal{A})$ is the topological set $\cup_{x \in M} T_xM$, with the induced topology by the topological sets

$$\cup_{x \in U_x} T_{c_x(x)} \Sigma_{k_x}.$$

The *tangent space* T_xN *at* $x \in N$ *of a* C^r *submanifold* N of M is a pseudo-linear subspace $T_xN \subset T_xM$ isomorphic to the pseudo-linear subspace $S_{e_x(x)}$ at $e_x(x)$. The *tangent fiber subbundle* $TN \subset TM$ of a C^r submanifold N of M is the topological set $\cup_{x \in N} T_xN$.

14.4.3 Pseudo-inner product on Σ_k

Let $I_x \subset \mathbb{V}_x \times \mathbb{V}_x$ be the set of all pairs $(\mathbf{u}, \mathbf{v}) \in I_x$ such that $|\sphericalangle(\mathbf{u}, \mathbf{v})| \leq \pi$. A *pseudo-inner product*

$$i : I_x \to \mathbb{R}$$

at a point $x \in \Sigma_k$ is a bi-linear map with the following properties:

- $i(\mathbf{u}, \mathbf{v}) = i(\mathbf{v}, \mathbf{u})$, for all $(\mathbf{u}, \mathbf{v}) \in I_x$;
- $i(\mathbf{u}, \mathbf{u}) \geq 0$, for all $(\mathbf{u}, \mathbf{u}) \in I_x$;
- $i(\mathbf{u}, \mathbf{u}) = 0$ if, and only if, $\mathbf{u} = 0$ $(= x - x)$.

A C^r *pseudo-Riemannian metric* in an open set $U \subset \Sigma_k$ is a map

$$<, >: \cup_{x \in U} I_x \to \mathbb{R}$$

with the following properties:

- $<,>_x = <,> |I_x$ is an inner product;
- For every isometry $i_{\mathbb{H}} : \mathbb{H} \to \Sigma_k$, the pullback by $i_{\mathbb{H}}$

$$< y - x, z - x >_{x,\mathbb{H}} = < i_{\mathbb{H}}(y) - i_{\mathbb{H}}(x), i_{\mathbb{H}}(z) - i_{\mathbb{H}}(x) >_{i_{\mathbb{H}}(x)}$$

  of the inner products $<,>_{i_{\mathbb{H}}(x)}$ in U induces a C^r Riemannian metric in $i_{\mathbb{H}}^{-1}(U)$.

Let $(M, \mathcal{A})$ be a C^r manifold. Let $J_x \subset T_xM \times T_xM$ be the pull-back by the derivative of the chart $c_i : U_i \to \Sigma_{k_i}$ in $\mathcal{A}$ of $I_{c_i(x)}$. A *C^r pseudo-Riemannian metric in a C^r manifold* $(M, \mathcal{A})$ is a map

$$<,>: \cup_{x \in M} J_x \to \mathbb{R}$$

such that, for every chart $c_i : U_i \to \Sigma_{k_x}$ in $\mathcal{A}$, the push-forward of $<,>$ is a C^r Riemannian metric $<,>_{c_i(U_i)}$ in $c_i(U_i)$.

We say that $(x, \mathbf{u}_x)$ and $(x, \mathbf{v}_x)$ are *direction equivalent* $(x, \mathbf{u}_x) \sim (x, \mathbf{v}_x)$ if, and only if, $\mathbf{u}_x$ and $\mathbf{v}_x$ belong to a same dimension 1 full subspace $\mathbb{S}_x$. $T\Sigma_k / \sim$ is the [direction set. A C^r *direction field* is a continuous map

$$\phi : \Sigma_k \to T\Sigma_k / \sim$$

such that for every isometry $i_{\mathbb{H}} : \mathbb{H} \to \Sigma_k$, the map $\hat{\phi} : \mathbb{H} \to T\mathbb{H}/ \sim$ given by $\hat{\phi} = di_{\mathbb{H}}^{-1} \circ \phi \circ i_{\mathbb{H}}$ is C^r.

A C^r *splitting* is a pair (ϕ_s, ϕ_u) of C^r direction fields such that, for every $x \in \Sigma_k$, we have

$$\mathbb{V}_x = \mathbb{S}_{\phi_s(x)} \oplus \mathbb{S}_{\phi_u(x)},$$

where $\mathbb{S}_{\phi_\iota(x)}$ is a dimension 1 full subspace containing $\phi_\iota(x)$.

Definition 14.5. *Let $(M, \mathcal{A})$ be a C^r pseudo-manifold with a pseudo-Riemannian metric. A C^r pseudo-diffeomorphism $f : M \to M$ is a C^r* pseudo-Anosov diffeomorphism, *if M has a 1 dimensional smooth splitting $E^s \oplus E^u$ of the tangent bundle, with the following properties: (i) the splitting is invariant under Tf, and (ii) Tf expands uniformly E^u and contracts uniformly E^s.*

The set of all C^r pseudo-Anosov diffeomorphisms on M is an open set.

Theorem 14.6. *(Stable Manifold Theorem) If $f : M \to M$ is a C^r pseudo-Anosov diffeomorphism, then the stable and unstable sets at the points of Λ are C^r pseudo-submanifolds with dimension 1.*

Proof. First, we prove that the stable and unstable sets through the singularities are C^r pseudo-submanifolds. Then, we prove that the stable and unstable sets through the other points are also C^r pseudo-submanifolds. The singularities are periodic points, because f is a pseudo-diffeomorphism and so the image of a singularity is a singularity with the same order. Let us construct the unstable manifold at the singularity s (for simplicity $f(s) = s$). Let

$E_{1,cut}, \ldots, E_{k,cut}$ at a singularity s be the *cut sets* represented in Figure 14.16. By the Whitney's extension theorem, there is a C^r diffeomorphism F_1 on the plane such that $F_1|_{E_{1,cut}} = f$. By the Hirsch and Pugh [48] Stable Manifold Theorem, the unstable set passing through $(0,0)$ of F_1 is a C^r submanifold $W^u = W_1^u \cup W_2^u$. Doing the same with respect to $E_{i,cut}$, we get that the unstable set

$$W^u(s) = \bigcup_{i=1}^{k} W_i^u$$

at the singularity is a C^r submanifold tangent to the unstable subspace (see Figure 14.17).

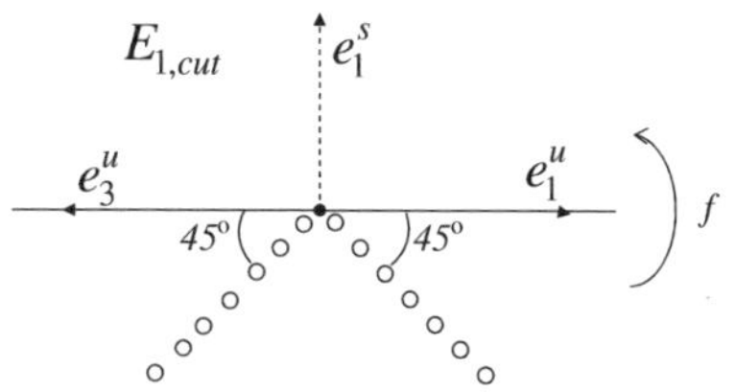

Fig. 14.16. A $E_{1,cut}$ cut set at a singularity.

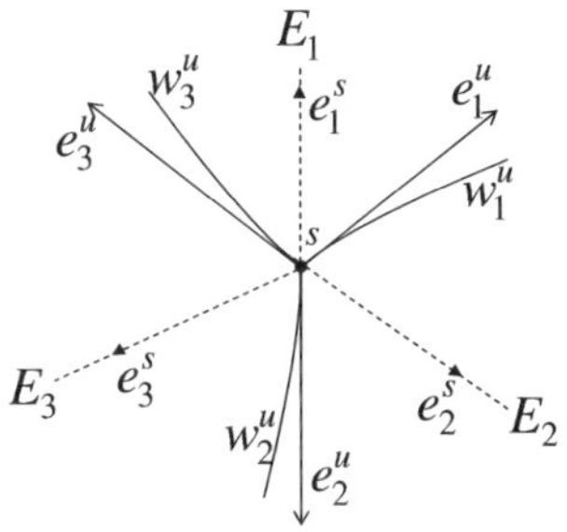

Fig. 14.17. The unstable set at a singularity $s \in \Sigma_3$.

Away from the singularities, let $(x_n)_{n \in \mathbb{Z}}$ be an orbit of f. If $x_n \in E_{i_n,cut}$, then we take the C^r diffeomorphism F_{i_n} such that $F_{i_n}|E_{i_n,cut} = f$ in a neighbourhood of x_n. Applying the Hirsch and Pugh [48] Stable Manifold Theorem to this orbit, we get that the unstable set at every point of the orbit is a C^r submanifold tangent to the unstable subspace. □

14.5 C^r foliations

A C^{1+} pseudo-foliation satisfies the properties of a C^r foliation with the extra *turntable condition* that we now describe. If s is a singularity, with order $k = k(s)$, then a singular leaf W^ι on M, containing s, is such that $W^\iota \setminus \{s\}$ is the union of k disjoint leaves ℓ^ι_j, $j \in \mathbb{Z}_k$, whose closures intersect in s. The components $\ell^\iota_1, \ldots, \ell^\iota_k$ of $W^\iota(s, \varepsilon) \setminus s$ are called *separatrices* of s. We call W^ι a *singular spinal set* and call the sets ℓ^ι_j emph the separatrices of s.

A C^{1+} foliation satisfies the *turntable condition*: if for all singular spinal sets W^ι with separatrices ℓ^ι_j, $j \in \mathbb{Z}_k$, there are leaf charts (i_j, ℓ^ι_j), such that the maps defined by $i_{j,l}|\ell^\iota_j = -i_j$ and $i_{j,l}|\ell^\iota_l = i_l$ are smooth. A C^{1+} foliation induced by a C^{1+} pseudo-Anosov diffeomorphism satisfies the turntable condition (see Pinto and Rand [160]).

The HR structures and the solenoid functions also apply to C^r pseudo-Anosov diffeomorphisms with the extra turntable condition that we now describe.

For any triple (v_1, v_2, v_3) of points v_1, v_2 and v_3 contained in same ι-leaf, we define the *solenoid limit* $s^z_\iota(v_1, v_2, v_3)$ as follows. For all $i \geq 0$, let

$$(z^i_1, z^i_2, z^i_3), (z^i_2, z^i_3, z^i_4), \ldots, (z^i_{n_i-2}, z^i_{n_i-1}, z^i_{n_i}) \in \mathrm{sol}^\iota$$

be a sequence of triples such that for some $1 < j_i < n_i$

$$v_1 = \lim_{i \to \infty} f^i_\iota(z^i_1) \quad , \quad v_2 = \lim_{i \to \infty} f^i_\iota(z^i_{j_i}) \ \text{ and } \ v_3 = \lim_{i \to \infty} f^i_\iota(z^i_{n_i}).$$

The solenoid *limit* $s^z_\iota(v_1, v_2, v_3)$ is equal to

$$s^z_\iota(v_1, v_2, v_3) = \frac{\sum_{j=j_i-1}^{n_i-2} (s_\iota(z_1, z_2, z_3) \ldots s_\iota(z_j, z_{j+1}, z_{j+2}))}{\sum_{j=1}^{j_i-2} (s_\iota(z_1, z_2, z_3) \ldots s_\iota(z_j, z_{j+1}, z_{j+2}))}.$$

For all singularities s, with order $k = k(s)$, and for all $i \in \mathbb{Z}_k$, let $a_i = (v_i, s, v_{i+1})$ be a triple contained in a leaf ℓ^ι_i which intersects an ι' boundary of a Markov rectangle just in the points v_i and v_{i+1} or in the points v_i, s and v_{i+1}. The limit solenoids $s^{z_i}_\iota(a_i)$ satisfy the following *turntable condition*:

$$\prod_{i=1}^{k} s^{z_i}_\iota(a_i) = 1.$$

If $k(s) = 1$ and $v_1 = v_2$, then $s^z_\iota(v_1, s, v_2) = 1$.

The solenoid functions determined by C^r pseudo-Anosov diffeomorphisms satisfy the turntable condition (see Pinto and Rand [160]).

The train-tracks and the self-renormalizable structures also apply to C^r pseudo-Anosov diffeomorphisms with the extra turntable condition that we now describe.

A C^{1+} atlas $\mathcal{B}$ satisfies the *turntable condition* at a singularity s, with order $k = k(s)$: if for all singular spinal sets on the train-track with separatrices ℓ_j^ι, $j \in \mathbb{Z}_k$, there are leaf charts (i_j, ℓ_j^ι), such that the maps defined by $i_{j,l}|\ell_j^\iota = -i_j$ and $i_{j,l}|\ell_l^\iota = i_l$ are smooth.

A C^{1+} foliation induced by a C^{1+} pseudo-Anosov determines a C^{1+} train-track atlas satisfying the turntable condition that comes from the turntable condition of a C^{1+} foliation. For example, let s be a singularity with order 3, as in Figure 14.1. The Markov partition determines a singular spinal set S^ι with separatrices l_j^ι, $j \in \mathbb{Z}_k$, such that there are train-track charts (i_j, ℓ_j^ι), whose maps defined by $i_{j,l}|\ell_j^\iota = -i_j$ and $i_{j,l}|\ell_l^\iota = i_l$ are smooth.

14.6 Further literature

The theory developed in this book has a natural extension to C^r pseudo-Anosov diffeomorphisms using the turntable conditions (see Pinto and Rand [160]). Pinto and Pujals [155] relate the pseudo-Anosov diffeomorphisms with the Pujals and Sambarino [181, 184] non-uniformly hyperbolic diffeomorphisms. The sympletic forms are defined similarly to the Riemannian metric. Let $(M, \mathcal{A})$ be a C^r pseudo-manifold with a pseudo-volume form ω. Pinto and Viana [176] proved that there is a residual set $\mathcal{R}$ contained in the set of all C^1 pseudo-diffeomorphisms, preserving the volume form, such that if $f \in \mathcal{R}$, then either f is a C^1 pseudo-diffeomorphism or has almost everywhere both Lyapunov exponents zero. In that way we recover the duality given by Mañé-Bochi Theorem in the torus to the other surfaces. This chapter is based on Pinto [152] and Pinto and Rand [160].

A

Appendix A: Classifying C^{1+} structures on the real line

We demonstrate the relations proposed by Sullivan between distinct degrees of smoothness of a homeomorphism of a real line and distinct bounds of the ratio and cross-ratio distortions of intervals of a fixed grid. We emphasize that to prove these relations, we do not have to check the distinct bounds of the ratio and cross-ratio distortions for all intervals, but just for the intervals belonging to a fixed grid.

A.1 The grid

Given $B \geq 1$, $M > 1$ and $\Omega : \mathbb{N} \to \mathbb{N}$, a (B, M) *grid*

$$\mathcal{G}_\Omega = \{I_\beta^n \subset I : n \geq 1 \text{ and } \beta = 1, \ldots, \Omega(n)\}$$

of a closed interval I is a collection of *grid intervals* I_β^n *at level* n with the following properties: (i) The grid intervals are closed intervals; (ii) For every $n \geq 1$, the union $\cup_{\beta=1}^{\Omega(n)} I_\beta^n$ of all grid intervals I_β^n, at level n, is equal to the interval I; (iii) For every $n \geq 1$, any two distinct grid intervals at level n have disjoint interiors; (iv) For every $1 \leq \beta < \Omega(n)$, the intersection of the grid intervals I_β^n and $I_{\beta+1}^n$ is only an endpoint common to both intervals; (v) For every $n \geq 1$, the set of all endpoints of the intervals I_β^n at level n is contained in the set of all end points of the intervals I_β^{n+1} at level $n+1$; (vi) For every $n \geq 1$ and for every $1 \leq \beta < \Omega(n)$, we have $B^{-1} \leq |I_{\beta+1}^n|/|I_\beta^n| \leq B$; (vii) For every $n \geq 1$ and for every $1 \leq \alpha \leq \Omega(n)$, the grid interval I_α^n contains at least two grid intervals at level $n+1$, and contains at most M grid intervals also at level $n+1$.

Let $h : I \to J$ be a homeomorphism between two compact intervals I and J on the real line, and let $\mathcal{G}_\Omega$ be a grid of I. We say that two closed intervals I_β and $I_{\beta'}$ are *adjacent* if their intersection $I_\beta \cap I_{\beta'}$ is only an endpoint common to both intervals. The *logarithmic ratio distortion* $lrd(I_\beta, I_{\beta'})$ between two adjacent intervals I_β and $I_{\beta'}$ is given by

$$lrd(I_\beta, I_{\beta'}) = \log\left(\frac{|I_\beta|}{|I_{\beta'}|}\frac{|h(I_{\beta'})|}{|h(I_\beta)|}\right) .$$

Let I_β, $I_{\beta'}$ and $I_{\beta''}$ be contained in the real line, such that I_β is adjacent to $I_{\beta'}$, and $I_{\beta'}$ is adjacent to $I_{\beta''}$. The *cross-ratio* $cr(I_\beta, I_{\beta'}, I_{\beta''})$ is determined by

$$cr(I_\beta, I_{\beta'}, I_{\beta''}) = \log\left(1 + \frac{|I_{\beta'}|}{|I_\beta|}\frac{|I_\beta| + |I_{\beta'}| + |I_{\beta''}|}{|I_{\beta''}|}\right) .$$

The *cross-ratio distortion* $crd(I_\beta, I_{\beta'}, I_{\beta''})$ is given by

$$crd(I_\beta, I_{\beta'}, I_{\beta''}) = cr(h(I_\beta), h(I_{\beta'}), h(I_{\beta''})) - cr(I_\beta, I_{\beta'}, I_{\beta''}) .$$

A.2 Cross-ratio distortion of grids

Let I_β and $I_{\beta'}$ be two intervals contained in the real line. We define the *ratio* $r(I_\beta, I_{\beta'})$ between the intervals I_β and $I_{\beta'}$ by

$$r(I_\beta, I_{\beta'}) = \frac{|I_{\beta'}|}{|I_\beta|} .$$

Let I_β, $I_{\beta'}$ and $I_{\beta''}$ be contained in the real line, such that I_β is adjacent to $I_{\beta'}$, and $I_{\beta'}$ is adjacent to $I_{\beta''}$. Recall that the *cross-ratio* $cr(I_\beta, I_{\beta'}, I_{\beta''})$ is given by

$$cr(I_\beta, I_{\beta'}, I_{\beta''}) = \log\left(1 + \frac{|I_{\beta'}|}{|I_\beta|}\frac{|I_\beta| + |I_{\beta'}| + |I_{\beta''}|}{|I_{\beta''}|}\right) .$$

We note that

$$cr(I_\beta, I_{\beta'}, I_{\beta''}) = \log\left((1 + r(I_\beta, I_{\beta'}))(1 + r(I_{\beta''}, I_{\beta'}))\right) .$$

Let $h : I \subset \mathbb{R} \to J \subset \mathbb{R}$ be a homeomorphism, and let $\mathcal{G}_\Omega$ be a grid of the compact interval I. We will use the following definitions and notations throughout this section:

(i) We will denote by J^n_β the interval $h(I^n_\beta)$ where I^n_β is a grid interval. We will denote by $r(n,\beta)$ the ratio $r(I^n_\beta, I^n_{\beta+1})$ between the grid intervals I^n_β and $I^n_{\beta+1}$, and we will denote by $r_h(n,\beta)$ the ratio $r(J^n_\beta, J^n_{\beta+1})$.
(ii) Let I_β be an interval contained in I (not necessarily a grid interval). The *average derivative* $dh(I_\beta)$ is given by

$$dh(I_\beta) = \frac{|h(I_\beta)|}{|I_\beta|} .$$

We will denote by $dh(n,\beta)$ the average derivative $dh(I^n_\beta)$ of the grid interval I^n_β.

(iii) The *logarithmic average derivative* $ldh(I_\beta)$ is given by

$$ldh(I_\beta) = \log(dh(I_\beta)) .$$

We will denote by $ldh(n, \beta)$ the logarithmic average derivative $ldh(I^n_\beta)$ of the grid interval I^n_β.

(iv) Let I_β and $I_{\beta'}$ be intervals contained in I (not necessarily grid intervals). We recall that the logarithmic ratio distortion $lrd(I_\beta, I_{\beta'})$ is given by

$$lrd(I_\beta, I_{\beta'}) = \log\left(\frac{|I_\beta|}{|I_{\beta'}|}\frac{|h(I_{\beta'})|}{|h(I_\beta)|}\right) .$$

Hence, we have

$$lrd(I_\beta, I_{\beta'}) = \log\frac{r(J_\beta, J_{\beta'})}{r(I_\beta, I_{\beta'})} = \log\frac{dh(I_{\beta'})}{dh(I_\beta)} .$$

We will denote by $lrd(n, \beta)$ the logarithmic ratio distortion $lrd(I^n_\beta, I^n_{\beta+1})$ of the grid intervals I^n_β and $I^n_{\beta+1}$.

(v) Let the intervals I_β, $I_{\beta'}$ and $I_{\beta''}$ in I (not necessarily grid intervals) be such that I_β is adjacent to $I_{\beta'}$ and $I_{\beta'}$ is adjacent to $I_{\beta''}$. We recall that the cross-ratio distortion $crd(I_\beta, I_{\beta'}, I_{\beta''})$ is given by

$$crd(I_\beta, I_{\beta'}, I_{\beta''}) = cr(h(I_\beta), h(I_{\beta'}), h(I_{\beta''})) - cr(I_\beta, I_{\beta'}, I_{\beta''}) .$$

We note that

$$crd(I_\beta, I_{\beta'}, I_{\beta''}) = \log\left(\frac{1 + r(h(I_\beta), h(I_{\beta'}))}{1 + r(I_\beta, I_{\beta'})}\frac{1 + r(h(I_{\beta''}), h(I_{\beta'}))}{1 + r(I_{\beta''}, I_{\beta'})}\right) . \tag{A.1}$$

For all grid intervals I^n_β, $I^n_{\beta+1}$ and $I^n_{\beta+2}$, we will denote by $cr(n, \beta)$ and $cr_h(n, \beta)$ the cross-ratios $cr(I^n_\beta, I^n_{\beta+1}, I^n_{\beta+2})$ and $cr(J^n_\beta, J^n_{\beta+1}, J^n_{\beta+2})$ respectively. We will denote by $crd(n, \beta)$ the cross-ratio distortion given by $cr_h(n, \beta) - cr(n, \beta)$.

Remark A.1. (a) We will call properties (vi) and (vii) of a (B,M) grid $\mathcal{G}_\Omega$ of an interval I, the *bounded geometry property* of the grid.

(b) By the bounded geometry property of a (B,M) grid $\mathcal{G}_\Omega$, there are constants $0 < B_1 < B_2 < 1$, just depending upon B and M, such that

$$B_1 < \frac{|I^{n+1}_\beta|}{|I^n_\alpha|} < B_2 ,$$

for all $n \geq 1$ and for all grid intervals I^n_α and I^{n+1}_β such that $I^{n+1}_\beta \subset I^n_\alpha$.

(c) We call a $(1, 2)$ grid $\mathcal{G}_\Omega$ of I a *symmetric grid of* I, i.e. (i) all the intervals at the same level n have the same length, and (ii) each grid interval at level n is equal to the union of two grid intervals at level $n + 1$.

A.3 Quasisymmetric homeomorphisms

The definition of a quasisymmetric homeomorphism that we present in this appendix is more adapted to our problem and, apparently, is stronger than the usual one, where the constant d of the quasisymmetric condition in Definition 33, below, is taken to be equal to 1. However, in Lemma A.3, we will prove that they are equivalent.

Definition 33 *Let $d \geq 1$ and $k \geq 1$. The homeomorphism $h : I \to J$ satisfies the (d, k)* quasisymmetric condition, *if*

$$|lrd(I_\beta, I_{\beta'})| \leq \log(k) , \tag{A.2}$$

for all intervals $I_\beta, I_{\beta'} \subset I$ with $d^{-1} \leq |I_{\beta'}|/|I_\beta| \leq d$. The homeomorphism h is quasisymmetric, *if for every $d \geq 1$ there exists $k_d \geq 1$ such that h satisfies the (d, k_d) quasisymmetric condition.*

Lemma A.2. *Let $h : I \to J$ be a homeomorphism and let $\mathcal{G}_\Omega$ be a grid of a compact interval I. The following statements are equivalent:*

(i) The homeomorphism $h : I \to J$ is quasisymmetric.
(ii) There is $k(\mathcal{G}_\Omega) > 1$ such that

$$|r_h(n, \beta)| \leq k(\mathcal{G}_\Omega) , \tag{A.3}$$

for every $n \geq 1$ and every $1 \leq \beta \leq \Omega(n)$.

Let $\mathcal{G}_\Omega$ be a grid of I. From Lemma A.2, we obtain that a homeomorphism $h : I \to J$ is quasisymmetric if, and only if, the set of all intervals J^n_β form a (B, M) grid for some $B \geq 1$ and $M > 1$.
Proof of Lemma A.2. Let us prove that statement (i) implies statement (ii). For every level $n \geq 1$ and every $1 \leq \beta < \Omega(n)$, let $x - \delta_1, x, x + \delta_2 \in I$ be such that $I^n_\beta = [x - \delta_1, x]$ and $I^n_{\beta+1} = [x, x + \delta_2]$. Hence,

$$\frac{r_h(n, \beta)}{r(n, \beta)} = \frac{h(x + \delta_2) - h(x)}{h(x) - h(x - \delta_1)} .$$

Since $h : I \to J$ is (k, B) quasisymmetric, for some $k = k(B)$, we have

$$k^{-1} < \frac{h(x + \delta_2) - h(x)}{h(x) - h(x - \delta_1)} \frac{\delta_1}{\delta_2} < k ,$$

and so, we get

$$k^{-1} < r_h(n, \beta)/r(n, \beta) < k . \tag{A.4}$$

Since, by the bounded geometry property of a grid $\mathcal{G}_\Omega$, there is $B \geq 1$ such that $B^{-1} \leq r(n, \beta) \leq B$, we get $k^{-1}B^{-1} \leq r_h(n, \beta) \leq kB$.

Let us prove that statement (ii) implies statement (i). Let $B \geq 1$ and $M > 1$ be as in the bounded geometry property of a grid. Let $d \geq 1$. Let $x - \delta_1, x, x + \delta_2 \in$

I, be such that $\delta_1 > 0$, $\delta_2 > 0$ and $d^{-1} \leq \delta_2/\delta_1 \leq d$. Let L_1, L_2, R_1 and R_2 be the intervals as constructed in Lemma A.7 with the constant $\alpha = 2$ in Lemma A.7. Hence, we have that

$$|L_1| = |I_l^{n_0+n_1}| \left(1 + \sum_{i=l+1}^{m-2} \prod_{j=l}^{i} r(n_0 + n_1, j) \right) ,$$

$$|L_2| = |I_l^{n_0+n_1}| \left(1 + \sum_{i=l}^{m-1} \prod_{j=l}^{i} r(n_0 + n_1, j) \right) ,$$

$$|R_1| = |I_l^{n_0+n_1}| \left(\sum_{i=m}^{r-2} \prod_{j=l}^{i} r(n_0 + n_1, j) \right) ,$$

$$|R_2| = |I_l^{n_0+n_1}| \left(\sum_{i=m-1}^{r-1} \prod_{j=l}^{i} r(n_0 + n_1, j) \right) .$$

Hence, by monotonicity of the homeomorphism h, we obtain that

$$\frac{|h(R_1)|}{|h(L_2)|} \frac{|L_1|}{|R_2|} \leq \frac{h(x+\delta_2) - h(x)}{h(x) - h(x-\delta_1)} \frac{\delta_1}{\delta_2} \leq \frac{|h(R_2)|}{|h(L_1)|} \frac{|L_2|}{|R_1|} . \tag{A.5}$$

Since, by the bounded geometry property of a grid, $B^{-1} < r(n_0 + n_1, j) < B$ and, by Lemma A.7, $l < m < r$ and $r - l \leq n_2(B, M, d)$, we get that there is $C_1 = C_1(B, n_2) > 1$ such that

$$C_1^{-1} \leq \frac{|L_1|}{|R_2|} = \frac{1 + \sum_{i=l+1}^{m-2} \prod_{j=l+1}^{i} r(n_0 + n_1, j)}{\sum_{i=m-1}^{r-1} \prod_{j=l+1}^{i} r(n_0 + n_1, j)} \leq C_1 ,$$

$$C_1^{-1} \leq \frac{|L_2|}{|R_1|} = \frac{1 + \sum_{i=l}^{m-1} \prod_{j=l}^{i} r(n_0 + n_1, j)}{\sum_{i=m}^{r-2} \prod_{j=l}^{i} r(n_0 + n_1, j)} \leq C_1 . \tag{A.6}$$

By inequality (A.3) of statement (ii), there is $k = k(\mathcal{G}_\Omega) > 1$ such that $k^{-1} < r_h(n_0 + n_1, j) < k$ for every $1 \leq j < \Omega(n_0 + n_1)$. Hence, there is $C_2 = C_2(k, n_2) > 1$ such that

$$C_2^{-1} \leq \frac{|h(R_1)|}{|h(L_2)|} = \frac{\sum_{i=m}^{r-2} \prod_{j=l}^{i} r_h(n_0 + n_1, j)}{1 + \sum_{i=l}^{m-1} \prod_{j=l}^{i} r_h(n_0 + n_1, j)} \leq C_2 ,$$

$$C_2^{-1} \leq \frac{|h(R_2)|}{|h(L_1)|} = \frac{\sum_{i=m}^{r-2} \prod_{j=l}^{i} r_h(n_0 + n_1, j)}{1 + \sum_{i=l}^{m-1} \prod_{j=l}^{i} r_h(n_0 + n_1, j)} \leq C_2 . \tag{A.7}$$

Putting together equalities (A.5), (A.6) and (A.7), we obtain that

$$C_1^{-1} C_2^{-1} \leq \frac{|h(R_1)|}{|h(L_2)|} \frac{|L_1|}{|R_2|} \leq \frac{h(x+\delta_2) - h(x)}{h(x) - h(x-\delta_1)} \frac{\delta_1}{\delta_2} \leq \frac{|h(R_2)|}{|h(L_1)|} \frac{|L_2|}{|R_1|} \leq C_1 C_2 .$$

□

Lemma A.3. *If, for some $d_0 \geq 1$ and $k_0 \geq 1$, a homeomorphism $h : I \to J$ satisfies the (d_0, k_0) quasisymmetric condition, then h is quasisymmetric.*

Proof. If a homeomorphism $h : I \to J$ satisfies the (d_0, k_0) quasisymmetric condition for some $d_0 \geq 1$ and $k_0 \geq 1$, then h satisfies statement (ii) of Lemma A.2 with respect to a symmetric grid (see definition of a symmetric grid in Remark A.1). Hence, by statement (i) of Lemma A.2, the homeomorphism h is quasisymmetric. □

Lemma A.4. *Let $h : I \to J$ be a homeomorphism and $\mathcal{G}_\Omega$ a grid of the compact interval I.*

(i) If $h : I \to J$ is quasisymmetric, then there is $C_0 \geq 0$ such that

$$cr_h(n, \beta) \leq C_0 \ ,$$

for every $n \geq 1$ and every $1 \leq \beta < \Omega(n) - 1$.
(ii) If there is $C_0 > 1$ such that, for every $n \geq 1$ and every $1 \leq \beta < \Omega(n) - 1$,

$$cr_h(n, \beta) \leq C_0 \ ,$$

then, for every closed interval K contained in the interior of I, the homeomorphism h restricted to K is quasisymmetric.

Proof. Let us prove statement (i). By Lemma A.2, there is $C_1 \geq 1$ such that $C_1^{-1} \leq r_h(n, \beta) \leq C_1$ for every level n and every $1 \leq \beta < \Omega(n)$. Therefore, there is $C_2 > 0$ such that, for every level n and every $1 \leq \beta < \Omega(n) - 1$,

$$|cr_h(n, \beta)| = \left|\log\left((1 + r_h(n, \beta))(1 + r_h(n, \beta + 1))^{-1}\right)\right| \leq C_2 \ . \tag{A.8}$$

Let us prove statement (ii). By the bounded geometry property of a grid, there is $n_0 \geq 1$ large enough such that the grid intervals $I_1^{n_0}$ and $I_{\Omega(n)-1}^{n_0}$ do not intersect the interval L. The grid $\mathcal{G}_\Omega$ of I induces, by restriction, a grid of the interval $L' = \cup_{\beta=2}^{\Omega(n)-2} I_\beta^{n_0}$ which contains L. Hence, by Lemma A.2, it is enough to prove that there is $C_1 \geq 1$ such that $C_1^{-1} \leq r_h(n, \beta) \leq C_1$ for every grid interval $I_\beta^n \subset L'$. Now, we will consider separately the following two possible cases: either (i) $r_h(n, \beta) \leq 1$ or (ii) $r_h(n, \beta) > 1$.
Case (i). Let $r_h(n, \beta) = |J_{\beta+1}^n|/|J_\beta^n| \leq 1$. By hypotheses of statement (ii), there is $C_2 > 1$ such that

$$cr_h(n, \beta - 1) = \log\left(1 + \frac{|J_\beta^n|}{|J_{\beta-1}^n|}\frac{|J_{\beta-1}^n| + |J_\beta^n| + |J_{\beta+1}^n|}{|J_{\beta+1}^n|}\right) \leq C_2 \ .$$

Hence, there is $C_3 > 1$ such that

$$1 \leq \frac{|J_\beta^n|}{|J_{\beta+1}^n|} \leq \frac{|J_\beta^n|}{|J_{\beta+1}^n|}\frac{|J_{\beta-1}^n| + |J_\beta^n| + |J_{\beta+1}^n|}{|J_{\beta-1}^n|} \leq C_3 \ ,$$

and so $C_3^{-1} \leq r_h(n,\beta) \leq 1$.
Case (ii). Let $r_h(n,\beta) = |J^n_{\beta+1}|/|J^n_\beta| > 1$. By hypotheses, there is $C_2 > 1$ such that

$$cr_h(n,\beta) = \log\left(1 + \frac{|J^n_{\beta+1}|}{|J^n_\beta|}\frac{|J^n_\beta| + |J^n_{\beta+1}| + |J^n_{\beta+2}|}{|J^n_{\beta+2}|}\right) \leq C_2 .$$

Hence, there is $C_3 > 1$ such that

$$1 \leq \frac{|J^n_{\beta+1}|}{|J^n_\beta|} \leq \frac{|J^n_{\beta+1}|}{|J^n_\beta|}\frac{|J^n_\beta| + |I^n_{\beta+1}| + |I^n_{\beta+2}|}{|J^n_{\beta+2}|} \leq C_3 ,$$

and so $1 < r_h(n,\beta) \leq C_3$. □

A.4 Horizontal and vertical translations of ratio distortions

Lemmas A.5 and A.6 are the key to understand the relations between ratio and cross-ratio distortions. We will use them in the following subsections. In what follows, we will use the following notations:

$$\begin{aligned}
L_1(n,\beta,p) &= \max_{0\leq i\leq p}\{lrd(n,\beta+i)^2\} \\
L_2(n,\beta,p) &= \max_{0\leq i_1\leq i_2<p}\{|lrd(n,\beta+i_1)lrd(n,\beta+i_2)|\} \\
C(n,\beta,p) &= \max_{0\leq i<p}\{|crd(n,\beta+i)|\} \\
M_1(n,\beta,p) &= \max\{L_1(n,\beta,p), C(n,\beta,p)\} \\
M_2(n,\beta,p) &= \max\{L_2(n,\beta,p), C(n,\beta,p)\} .
\end{aligned}$$

Lemma A.5. *Let $h : I \subset \mathbb{R} \to J \subset \mathbb{R}$ be a quasisymmetric homeomorphism and let $\mathcal{G}_\Omega$ be a grid of the closed interval I. Then, the logarithmic ratio distortion and the cross-ratio distortion satisfy the following estimates:*

(i) (cross-ratios distortion versus ratio distortion)

$$\begin{aligned}
crd(n,\beta) &= \frac{lrd(n,\beta)}{1+r(n,\beta)^{-1}} - \frac{lrd(n,\beta+1)}{1+r(n,\beta+1)} \pm \mathcal{O}(L_1(n,\beta,1)) \\
&= \frac{|I^n_{\beta+1}|lrd(n,\beta)}{|I^n_\beta| + |I^n_{\beta+1}|} - \frac{|I^n_{\beta+1}|lrd(n,\beta+1)}{|I^n_{\beta+1}| + |I^n_{\beta+2}|} \pm \mathcal{O}(L_1(n,\beta,1)). \quad \text{(A.9)}
\end{aligned}$$

(ii) (lrd-horizontal translations) There is a constant $C(i) > 0$, not depending upon the level n and not depending upon $1 \leq \beta \leq \Omega(n)$, such that

$$
\begin{aligned}
lrd(n,\beta+i) &= \left(\prod_{k=0}^{i-1} r(n,\beta+k)\right)\frac{1+r(n,\beta+i)}{1+r(n,\beta)}lrd(n,\beta) \pm C(i)M_1(n,\beta,i)\\
&= \frac{|I^n_{\beta+i}|+|I^n_{\beta+i+1}|}{|I^n_\beta|+|I^n_{\beta+1}|}lrd(n,\beta) \pm C(i)M_1(n,\beta,i)\ . \qquad (A.10)
\end{aligned}
$$

(iii) (lrd-vertical translations) Let I^{n-1}_α and $I^{n-1}_{\alpha+1}$ be two adjacent grid intervals. Take $\beta=\beta(n,\alpha)$ and $p=p(n,\alpha)$ such that $I^n_\beta,\dots,I^n_{\beta+p}$ are all the grid intervals contained in the union $I^{n-1}_\alpha \cup I^{n-1}_{\alpha+1}$. Then, for every $0\le i<p$, we have

$$
lrd(n-1,\alpha) = \frac{|I^{n-1}_\alpha|+|I^{n-1}_{\alpha+1}|}{|I^n_{\beta+i}|+|I^n_{\beta+i+1}|}lrd(n,\beta+i)\pm\mathcal{O}(M_2(n,\beta,p))\ . \quad (A.11)
$$

Proof. By Taylor series expansion, we get

$$
\frac{r_h(n,\beta)}{r(n,\beta)} = 1+lrd(n,\beta)\pm\mathcal{O}(lrd(n,\beta)^2) \qquad (A.12)
$$

$$
\frac{r(n,\beta)}{r_h(n,\beta)} = 1-lrd(n,\beta)\pm\mathcal{O}(lrd(n,\beta)^2)\ . \qquad (A.13)
$$

Let us prove equality (A.9). By definition of cross-ratio distortion, we have

$$
crd(n,\beta) = \log\frac{1+r_h(n,\beta)}{1+r(n,\beta)} + \log\frac{1+r_h(n,\beta+1)^{-1}}{1+r(n,\beta+1)^{-1}}\ .
$$

Using equality (A.12), we get

$$
\begin{aligned}
\log\frac{1+r_h(n,\beta)}{1+r(n,\beta)} &= \log\left(1+\frac{r_h(n,\beta)r(n,\beta)^{-1}-1}{1+r(n,\beta)^{-1}}\right)\\
&= \frac{lrd(n,\beta))}{1+r(n,\beta)^{-1}} \pm\mathcal{O}(lrd(n,\beta)^2)\ . \qquad (A.14)
\end{aligned}
$$

Similarly, using equality (A.13), we obtain

$$
\begin{aligned}
\log\frac{1+r_h(n,\beta+1)^{-1}}{1+r(n,\beta+1)^{-1}} &= \log\left(1+\frac{r(n,\beta+1)r_h(n,\beta+1)^{-1}-1}{1+r(n,\beta+1)}\right)\\
&= \frac{-lrd(n,\beta+1)}{1+r(n,\beta+1)} \pm\mathcal{O}(lrd(n,\beta+1)^2)\ . \qquad (A.15)
\end{aligned}
$$

Putting together equations (A.14) and (A.15), we get

$$
\begin{aligned}
crd(n,\beta) &= \log\frac{1+r_h(n,\beta)}{1+r(n,\beta)} + \log\frac{1+r_h(n,\beta+1)^{-1}}{1+r(n,\beta+1)^{-1}}\\
&= \frac{lrd(n,\beta))}{1+r(n,\beta)^{-1}} - \frac{lrd(n,\beta+1)}{1+r(n,\beta+1)} \pm\mathcal{O}(lrd(n,\beta)^2, lrd(n,\beta+1)^2)\ .
\end{aligned}
$$

Let us prove equality (A.10). Using equality (A.9), we get

$$\begin{aligned} lrd(n,\beta+i+1) &= lrd(n,\beta+i)\frac{1+r(n,\beta+i+1)}{1+r(n,\beta+i)^{-1}} \pm \mathcal{O}(M_1(n,\beta+i,1)) \\ &= lrd(n,\beta+i)r(n,\beta+i)\frac{1+r(n,\beta+i+1)}{1+r(n,\beta+i)} \\ &\quad \pm \mathcal{O}(M_1(n,\beta+i,1)) \ . \end{aligned}$$

Hence, we obtain

$$\begin{aligned} lrd(n,\beta+i) &= lrd(n,\beta)\prod_{k=0}^{i-1}\left(r(n,\beta+k)\frac{1+r(n,\beta+k+1)}{1+r(n,\beta+k)}\right) \\ &\quad \pm C(i)M_1(n,\beta,i) \\ &= lrd(n,\beta)\frac{1+r(n,\beta+i)}{1+r(n,\beta)}\prod_{k=0}^{i-1} r(n,\beta+k) \pm C(i)M_1(n,\beta,i) \ , \end{aligned}$$

where the constant $C(i) > 0$ does not depend upon n and upon $1 \leq \beta \leq \Omega(n)$. Since

$$\frac{1+r(n,\beta+i)}{1+r(n,\beta)}\prod_{k=0}^{i-1} r(n,\beta+k) = \frac{|I^n_{\beta+i}| + |I^n_{\beta+i+1}|}{|I^n_\beta| + |I^n_{\beta+1}|} \ ,$$

we get

$$\begin{aligned} lrd(n,\beta+i) &= lrd(n,\beta)\frac{1+r(n,\beta+i)}{1+r(n,\beta)}\prod_{k=0}^{i-1} r(n,\beta+k) \pm C(i)M_1(n,\beta,i) \\ &= lrd(n,\beta)\frac{|I^n_{\beta+i}| + |I^n_{\beta+i+1}|}{|I^n_\beta| + |I^n_{\beta+1}|} \pm C(i)M_1(n,\beta,i) \ . \end{aligned}$$

Let us prove equality (A.11). Let $0 < m = m(n,\alpha) < p$ be such that $I^n_\beta, \ldots, I^n_{\beta+m}$ are all the grid intervals contained in I^{n-1}_α and $I^n_{\beta+m+1}, \ldots, I^n_{\beta+p}$ are all the grid intervals contained in I^{n-1}_α. For simplicity of exposition, we introduce the following definitions:

(i) We define $a_0 = 0$, $a_{h,0} = 0$ and, for every $0 < j < p$, we define

$$a_j = \frac{|I^n_{\beta+j}|}{|I^n_\beta|} = \prod_{i=0}^{j-1} r(n,\beta+i) \quad \text{and} \quad a_{h,j} = \frac{|J^n_{\beta+j}|}{|J^n_\beta|} = \prod_{i=0}^{j-1} r_h(n,\beta+i) \ .$$

(ii) We define

$$R = \frac{|I^{n-1}_\alpha|}{|I^n_\beta|} \ , \quad R' = \frac{|I^{n-1}_{\alpha+1}|}{|I^n_\beta|} \ , \quad R_h = \frac{|J^{n-1}_\alpha|}{|J^n_\beta|} \ , \quad R'_h = \frac{|J^{n-1}_{\alpha+1}|}{|J^n_\beta|} \ .$$

Thus,

$$R = \sum_{j=0}^{m-1} a_j \ , \quad R' = \sum_{j=m}^{p-1} a_j \ , \quad R_h = \sum_{j=0}^{m-1} a_{h,j} \ , \quad R'_h = \sum_{j=m}^{p-1} a_{h,j} \ .$$

(iii) We define

$$E = \sum_{j=1}^{m-1} a_j \left(\sum_{i=0}^{j-1} lrd(n, \beta + i) \right) \text{ and } E' = \sum_{j=m}^{p-1} a_j \left(\sum_{i=0}^{j-1} lrd(n, \beta + i) \right) .$$

We will separate the proof of equality (A.11) in three parts. In the first part, we will prove that

$$lrd(n-1, \alpha) = \frac{E'}{R'} - \frac{E}{R} \pm \mathcal{O}(L_2(n, \beta, p)) \ . \tag{A.16}$$

In the second part, we will prove that

$$\frac{E'}{R'} - \frac{E}{R} = lrd(n, \beta) \frac{|I^n_\alpha| + |I^n_{\alpha+1}|}{|I^n_\beta| + |I^n_{\beta+1}|} \pm \mathcal{O}(M_1(n, \beta, p)) \ . \tag{A.17}$$

In the third part, we will use the previous parts to prove equality (A.11) in the case where $i = 0$. Then, we will use equality (A.10) to extend, for every $0 \le i < p$, the proof of equality (A.11).

First part. By equality (A.12), we have that

$$r_h(n, \beta + i) = r(n, \beta + i)(1 + lrd(n, \beta + i)) \pm \mathcal{O}(lrd((n, \beta + i)^2) \ .$$

Hence, for every $1 \le j < p$, we get

$$\begin{aligned}
a_{h,j} &= \prod_{i=0}^{j-1} r_h(n, \beta + i) \\
&= \prod_{i=0}^{j-1} \left(r(n, \beta + i)(1 + lrd(n, \beta + i)) \pm \mathcal{O}(lrd((n, \beta + i)^2)) \right) \\
&= \prod_{i=0}^{j-1} r(n, \beta + i) \left(1 + \sum_{i=0}^{j-1} lrd(n, \beta + i) \pm \mathcal{O}(L_2(n, \beta + 1, j)) \right) \\
&= a_j + a_j \sum_{i=0}^{j-1} lrd(n, \beta + i) \pm \mathcal{O}\left(a_j L_2(n, \beta + 1, j) \right) \ .
\end{aligned}$$

Thus,

$$\begin{aligned}
R_h &= \sum_{j=0}^{m-1} a_{h,j} \\
&= \sum_{j=0}^{m-1} a_j + \sum_{j=1}^{m-1} a_j \sum_{i=0}^{j-1} lrd(n, \beta + i) \pm \mathcal{O}\left(\sum_{j=0}^{m-1} a_j L_2(n, \beta, j) \right) \\
&= R + E \pm \mathcal{O}(R L_2(n, \beta, m))) \ .
\end{aligned} \tag{A.18}$$

Similarly, we have

$$\begin{aligned}
R'_h &= \sum_{j=m}^{p-1} a_{h,j} \\
&= \sum_{j=m}^{p-1} a_j + \sum_{j=m}^{p-1} a_j \sum_{i=0}^{j-1} lrd(n,\beta+i) \pm \mathcal{O}\left(\sum_{j=m}^{n-1} a_j L_2(n,\beta,j)\right) \\
&= R' + E' \pm \mathcal{O}(R' L_2(n,\beta,p)) \ . \qquad \text{(A.19)}
\end{aligned}$$

By equalities (A.18) and (A.19), we obtain that

$$\begin{aligned}
lrd(n-1,\alpha) &= \log \frac{R'_h}{R'} \frac{R}{R_h} \\
&= \log \frac{R' + E' \pm \mathcal{O}(R' L_2(n,\beta,p))}{R'} - \log \frac{R + E \pm \mathcal{O}(R L_2(n,\beta,m))}{R} \\
&= \frac{E'}{R'} - \frac{E}{R} \pm \mathcal{O}(L_2(n,\beta,p)) \ .
\end{aligned}$$

Second part. By equality (A.10), for every $1 \le j < p$, we obtain

$$\begin{aligned}
\sum_{i=0}^{j-1} lrd(n,\beta+i) &= \sum_{i=0}^{j-1} \left(\frac{a_i(1+r(n,\beta+i))}{1+r(n,\beta)} lrd(n,\beta) \pm \mathcal{O}(M_1(n,\beta,i)) \right) \\
&= \frac{lrd(n,\beta)}{1+r(n,\beta)} \sum_{i=0}^{j-1} (a_i + a_{i+1}) \pm \mathcal{O}(M_1(n,\beta,j)) \ .
\end{aligned}$$

Hence, we obtain that

$$\begin{aligned}
E &= \sum_{j=1}^{m-1} a_j \sum_{i=0}^{j-1} lrd(n,\beta+i) \\
&= \sum_{j=1}^{m-1} a_j \left(\frac{lrd(n,\beta)}{1+r(n,\beta)} \sum_{i=0}^{j-1} (a_i + a_{i+1}) \pm \mathcal{O}(M_1(n,\beta,j)) \right) \\
&= \frac{lrd(n,\beta)}{1+r(n,\beta)} \sum_{j=1}^{m-1} a_j \sum_{i=0}^{j-1} (a_i + a_{i+1}) \pm \mathcal{O}\left(\sum_{j=1}^{m-1} a_j M_1(n,\beta,j) \right) \\
&= \frac{lrd(n,\beta)}{1+r(n,\beta)} R(a_1 + \ldots + a_{m-1}) \pm \mathcal{O}(R M_1(n,\beta,m)) \ . \qquad \text{(A.20)}
\end{aligned}$$

Similarly, we have

$$E' = \sum_{j=m}^{p-1} a_j \sum_{i=0}^{j-1} lrd(n,\beta_i)$$

$$= \frac{lrd(n,\beta)}{1+r(n,\beta)} \sum_{j=m}^{p-1} a_j \sum_{i=0}^{j-1} (a_i + a_{i+1}) \pm \mathcal{O}\left(\sum_{j=m}^{p-1} a_j M_1(n,\beta,j) \right)$$

$$= \frac{lrd(n,\beta)}{1+r(n,\beta)} R'(1 + 2a_1 + \ldots + 2a_{m-1} + a_m + \ldots + a_{p-1})$$
$$\pm \mathcal{O}(R' M_1(n,\beta,p)) \ . \qquad \text{(A.21)}$$

Putting together equalities (A.20) and (A.21), we obtain that

$$\frac{E'}{R'} - \frac{E}{R} = \frac{lrd(n,\beta)}{1+r(n,\beta)}(1 + a_1 + \ldots + a_{p-1}) \pm \mathcal{O}(M_1(n,\beta,p))$$
$$= lrd(n,\beta) \frac{|I^n_\alpha| + |I^n_{\alpha+1}|}{|I^n_\beta| + |I^n_{\beta+1}|} \pm \mathcal{O}(M_1(n,\beta,p)) \ .$$

Third part. In the case where $i = 0$, equality (A.11) follows, from putting together equalities (A.16) and (A.17), since

$$lrd(n-1,\alpha) = \frac{E'}{R'} - \frac{E}{R} \pm \mathcal{O}(L_2(n,\beta,p))$$
$$= lrd(n,\beta) \frac{|I^n_\alpha| + |I^n_{\alpha+1}|}{|I^n_\beta| + |I^n_{\beta+1}|} \pm \mathcal{O}(M_2(n,\beta,p)) \ .$$

By equality (A.10), for every $0 < i < p$, we have

$$\frac{|I^{n-1}_\alpha| + |I^{n-1}_{\alpha+1}|}{|I^n_\beta| + |I^n_{\beta+1}|} lrd(n,\beta) = \frac{|I^{n-1}_\alpha| + |I^{n-1}_{\alpha+1}|}{|I^n_{\beta+i}| + |I^n_{\beta+i+1}|} lrd(n,\beta+i) \pm \mathcal{O}(M_1(n,\beta,p)) \ .$$

Thus,

$$lrd(n-1,\alpha) = lrd(n,\beta) \frac{|I^n_\alpha| + |I^n_{\alpha+1}|}{|I^n_\beta| + |I^n_{\beta+1}|} \pm \mathcal{O}(M_2(n,\beta,p))$$
$$= lrd(n,\beta+i) \frac{|I^n_\alpha| + |I^n_{\alpha+1}|}{|I^n_{\beta+i}| + |I^n_{\beta+i+1}|} \pm \mathcal{O}(M_2(n,\beta,p)) \ .$$

□

Lemma A.6. *Let $h : I \subset \mathbb{R} \to J \subset \mathbb{R}$ be a homeomorphism and $\mathcal{G}_\Omega$ a grid of the closed interval I. For every level n and every $0 \le i < \Omega(n) - 1$, let $a(n,i)$ and $b(n,i)$ be given by*

$$a(n,i) = \frac{1 + r_h(n,i)}{1 + r(n,i)} \quad \text{and} \quad b(n,i) = \exp(-crd(n,i)) \ .$$

(i) Then, for every $1 \le i < \Omega(n) - 1$, we have

$$a(n,i)a(n,i-1)b(n,i-1) = \frac{r_h(n,i)}{r(n,i)} \ . \qquad \text{(A.22)}$$

(ii) Let $n \geq 1$ and $\beta, p \in \{2, \ldots, \Omega(n)-1\}$ have the following properties:
(a) There is $\varepsilon > 1$ such that $a(n,\beta) \geq \varepsilon$.
(b) There is $\gamma < 1$ such that $\gamma \leq b(n,\beta+i) \leq \gamma^{-1}$, for every $0 \leq i < p$.
Then, for every $1 \leq i \leq p$, we have

$$a(n,\beta+i) \geq 1 + \frac{(\varepsilon-1)\gamma^i}{2} \prod_{k=1}^{i} r(n,\beta+k) + \frac{(\varepsilon-1)\gamma^i B^{-i}}{2} + B(\gamma-1)\frac{1-(B\gamma^{-1})^i}{1-(B\gamma^{-1})} , \quad \text{(A.23)}$$

where $B \geq 1$ is given by the bounded geometry property of the grid.

Proof. Let us prove equality (A.22). By hypotheses, we have

$$\begin{aligned} b(n,i-1) &= \exp(-crd(n,i-1)) \\ &= \frac{1+r(n,i-1)}{1+r_h(n,i-1)} \frac{1+r(n,i)^{-1}}{1+r_h(n,i)^{-1}} \\ &= a(n,i-1)^{-1} \frac{1+r(n,i)}{1+r_h(n,i)} \frac{r_h(n,i)}{r(n,i)} \\ &= a(n,i-1)^{-1} a(n,i)^{-1} \frac{r_h(n,i)}{r(n,i)} . \end{aligned}$$

Thus,

$$b(n,i-1)a(n,i-1)a(n,i) = \frac{r_h(n,i)}{r(n,i)} .$$

Let us prove inequality (A.23). By definition of $a(n,i)$ and by equality (A.22), we have

$$\begin{aligned} a(n,i) &= \frac{1+r_h(n,i)}{1+r(n,i)} \\ b(n,i-1)a(n,i-1)a(n,i) &= \frac{r_h(n,i)}{r(n,i)} \end{aligned}$$

Hence, we get

$$\begin{aligned} a(n,i)(1+r(n,i)) &= 1+r_h(n,i) \\ r_h(n,i) &= b(n,i-1)a(n,i-1)a(n,i)r(n,i) . \end{aligned}$$

Thus,

$$a(n,i)(1+r(n,i)) = 1 + b(n,i-1)a(n,i-1)a(n,i)r(n,i) ,$$

and so

$$a(n,i) = (1 - r(n,i)(b(n,i-1)(a(n,i-1)-1) + b(n,i-1)-1)^{-1} .$$

Therefore, for every $n \geq 1$, $\beta, p \in \{2, \ldots, \Omega(n)-1\}$ and $1 \leq i \leq p$, we get

$$a(n,\beta+i)-1 \geq r(n,\beta+i)(b(n,\beta+i-1)(a(n,\beta+i-1)-1)+b(n,\beta+i-1)-1) .$$

Hence, by induction in $1 \leq i \leq p$, we get

$$\begin{aligned} a(n,\beta+i)-1 \geq{}& (a(n,\beta)-1)\prod_{k=1}^{i} r(n,\beta+k)b(n,\beta+k-1) + \\ &+ r(n,\beta+i)\sum_{k=1}^{i}(b(n,\beta+k-1)-1)\prod_{l=k}^{i-1} r(n,\beta+l)b(n,\beta+l). \end{aligned} \tag{A.24}$$

Using that $B^{-1} < r(n,\beta+k) < B$ by the bounded geometry property of the grid, we get

$$\begin{aligned} (a(n,\beta)-1)\prod_{k=1}^{i} r(n,\beta+k)b(n,\beta+k-1) &\geq (\varepsilon-1)\gamma^i \prod_{k=1}^{i} r(n,\beta+k) \\ &\geq \frac{(\varepsilon-1)\gamma^i}{2}\prod_{k=1}^{i} r(n,\beta+k) + \frac{(\varepsilon-1)\gamma^i B^{-i}}{2} . \end{aligned} \tag{A.25}$$

Furthermore, noting that $\gamma - 1 < 0$, we have

$$\begin{aligned} r(n,\beta+i)&\sum_{k=1}^{i}(b(n,\beta+k-1)-1)\prod_{l=k}^{i-1} r(n,\beta+l)b(n,\beta+l) \\ &\geq B(\gamma-1)\sum_{k=1}^{i}(B\gamma^{-1})^{i-k} \\ &\geq B(\gamma-1)\frac{1-(B\gamma^{-1})^i}{1-(B\gamma^{-1})} . \end{aligned} \tag{A.26}$$

Putting inequalities (A.24), (A.25) and (A.26) together, we obtain that

$$a(n,\beta+i)-1 \geq \frac{(\varepsilon-1)\gamma^i}{2}\prod_{k=1}^{i} r(n,\beta+k) + \frac{(\varepsilon-1)\gamma^i B^{-i}}{2} + B(\gamma-1)\frac{1-(B\gamma^{-1})^i}{1-(B\gamma^{-1})} .$$

□

A.5 Uniformly asymptotically affine (uaa) homeomorphisms

The definition of uniformly asymptotically affine homeomorphism that we introduce in this chapter is more adapted to our problem and, apparently, is stronger than the usual one for symmetric maps, where the constant d of the (uua) condition in Definition 34, below, is taken to be equal to 1. However, in Lemma A.9, we will prove that they are equivalent.

Definition 34 *Let $d \geq 1$ and $\varepsilon : \mathbb{R}_0^+ \to \mathbb{R}_0^+$ be a continuous function with $\varepsilon(0) = 0$. The homeomorphism $h : I \to J$ satisfies the (d, ε)* uniformly asymptotically affine condition, *if*

$$|lrd(I_\beta, I_{\beta'})| \leq \varepsilon(|I_\beta| + |I_{\beta'}|) , \tag{A.27}$$

for all intervals $I_\beta, I_{\beta'} \subset I$ with $d^{-1} \leq |I_{\beta'}|/|I_\beta| \leq d$. The map h is uniformly asymptotically affine (uaa), *if for every $d \geq 1$ there exists ε_d such that h satisfies the (d, ε_d) uniformly asymptotically affine condition.*

We will use the following lemma in the proof of Lemma A.8, below.

Lemma A.7. *Let $\alpha > 1$ and $d \geq 1$. Let $\mathcal{G}_\Omega$ be a (B, M) grid of a compact interval I. Let $x - \delta_1, x, x + \delta_2$ contained in I be such that $\delta_1 > 0$, $\delta_2 > 0$ and $d^{-1} \leq \delta_2/\delta_1 \leq d$. Then, there are intervals L_1, L_2, R_1 and R_2 with the following properties:*

(i)

$$L_1 \subset [x - \delta_1, x] \subset L_2 \quad \text{and} \quad R_1 \subset [x, x + \delta_2] \subset R_2 . \tag{A.28}$$

(ii)

$$\alpha^{-1} < \frac{|L_1|}{\delta_1} < \frac{|L_2|}{\delta_1} < \alpha \quad \text{and} \quad \alpha^{-1} < \frac{|R_1|}{\delta_2} < \frac{|R_2|}{\delta_2} < \alpha . \tag{A.29}$$

(iii) Let $n_0 = n_0(x - \delta_1, x, x + \delta_2, \mathcal{G}_\Omega) \geq 1$ be the biggest integer such that

$$[x - \delta_1, x + \delta_2] \subset I_\beta^{n_0} \cup I_{\beta+1}^{n_0},$$

for some $1 \leq \beta < \Omega(n_0)$. Then, there are integers $n_1 = n_1(\alpha, B, M, d)$ and $n_2 = n_2(\alpha, B, M, d)$ such that

$$\begin{aligned} L_1 &= \cup_{i=l+1}^{m-1} I_i^{n_0+n_1} \quad , \quad & L_2 &= \cup_{i=l}^{m} I_i^{n_0+n_1} , \\ R_1 &= \cup_{i=m+1}^{r-1} I_i^{n_0+n_1} \quad , \quad & R_2 &= \cup_{i=m}^{r} I_i^{n_0+n_1} , \end{aligned}$$

for some l, m, r with the property that $l < m < r$ and $r - l \leq n_2$.

Proof. Let $0 < B_1 = B_1(B, M) < B_2 = B_2(B, M) < 1$ be as in Remark A.1. By construction of n_0, there is $I_\varepsilon^{n_0+1}$ with the property that $I_\varepsilon^{n_0+1} \subset [x - \delta_1, x, x + \delta_2]$. In particular, we have that either $I_\varepsilon^{n_0+1} \subset I_\beta^{n_0}$ or $I_\varepsilon^{n_0+1} \subset I_{\beta+1}^{n_0}$. Thus, using the bounded geometry property of a grid and Remark A.1, we obtain that

$$B^{-1} B_1 |I_\beta^{n_0}| \leq |I_\varepsilon^{n_0+1}| \leq \delta_2 + \delta_1 . \tag{A.30}$$

Since $d^{-1} \leq \delta_2/\delta_1 \leq d$, by inequality (A.30), we get

$$\begin{aligned} \delta_1 &\geq (1+D)^{-1}(\delta_2+\delta_1) \\ &\geq (1+D)^{-1}B^{-1}B_1|I_\beta^{n_0}| \, . \end{aligned} \tag{A.31}$$

Since $[x-\delta_1, x+\delta_2] \subset I_\beta^{n_0} \cup I_{\beta+1}^{n_0}$, by the bounded geometry property of a grid, we obtain that

$$\begin{aligned} \delta_1 &\leq |I_\beta^{n_0}| + |I_{\beta+1}^{n_0}| \\ &\leq (1+B)|I_\beta^{n_0}| \, . \end{aligned} \tag{A.32}$$

By inequalities (A.31) and (A.32), there is $A = A(B_0, B_1, d) > 1$ such that

$$A^{-1}|I_\beta^{n_0}| \leq \delta_1 \leq A|I_\beta^{n_0}| \, . \tag{A.33}$$

Similarly, we have

$$A^{-1}|I_\beta^{n_0}| \leq \delta_2 \leq A|I_\beta^{n_0}| \, . \tag{A.34}$$

Take $0 < \theta(\alpha) < 1$ such that $\alpha^{-1} \leq 1-\theta < 1+\theta \leq \alpha$. Let $n_1 = n_1(B, B_2, A, \theta)$ be the smallest integer such that

$$B_2^{n_1} \leq B^{-1}\theta A^{-1}/2 \, . \tag{A.35}$$

Let $l < m < r$ be such that $x - \delta_1 \in I_l^{n_0+n_1}$, $x \in I_m^{n_0+n_1}$ and $x + \delta_2 \in I_r^{n_0+n_1}$. Then, by the bounded geometry property of a grid, there is $n_2 = 2Mn_1 \geq 1$ such that $r - l \leq n_2$. Hence, the intervals

$$\begin{aligned} L_1 &= \cup_{i=l+1}^{m-1} I_i^{n_0+n_1} \quad , \quad & L_2 &= \cup_{i=l}^{m} I_i^{n_0+n_1} \, , \\ R_1 &= \cup_{i=m+1}^{r-1} I_i^{n_0+n_1} \quad , \quad & R_2 &= \cup_{i=m}^{r} I_i^{n_0+n_1} \end{aligned}$$

satisfy property (i) and property (iii) of Lemma A.7. Let us prove that the intervals L_1, L_2, R_1 and R_2 satisfy property (ii) of Lemma A.7. By the bounded geometry property of a grid and inequality (A.35), we get

$$|I_i^{n_0+n_1}| \leq BB_2^{n_1}|I_\beta^{n_0}| \leq \theta A^{-1}|I_\beta^{n_0}|/2 \, , \tag{A.36}$$

for all $l \leq i \leq r$. Thus, by inequalities (A.33) and (A.36), we get

$$\begin{aligned} |L_1|/\delta_1 &\geq (\delta_1 - |I_l^{n_0+n_1}| - |I_m^{n_0+n_1}|)/\delta_1 \\ &\geq (\delta_1 - \theta A^{-1}|I_\beta^{n_0}|)/\delta_1 \\ &\geq 1-\theta \, . \end{aligned} \tag{A.37}$$

Again, by inequalities (A.33) and (A.36), we get

$$\begin{aligned} |L_2|/\delta_1 &\leq (\delta_1 + |I_l^{n_0+n_1}| + |I_m^{n_0+n_1}|)/\delta_1 \\ &\leq (\delta_1 + \theta A^{-1}|I_\beta^{n_0}|)/\delta_1 \\ &\leq 1+\theta \, . \end{aligned} \tag{A.38}$$

Similarly, using inequalities (A.34) and (A.36), we obtain that

$$|R_1|/\delta_2 \geq 1 - \theta \quad \text{and} \quad |R_2|/\delta_2 \leq 1 + \theta \,. \tag{A.39}$$

Noting that $\alpha^{-1} \leq 1-\theta < 1+\theta \leq \alpha$ and putting together inequalities (A.37), (A.38) and (A.39), we obtain that the intervals L_1, L_2, R_1 and R_2 satisfy property (ii) of Lemma A.7. □

Lemma A.8. *Let $h : I \to J$ be a homeomorphism and I a compact interval. The following statements are equivalent:*

(i) The homeomorphism $h : I \to J$ is (uaa).

(ii) There is a sequence γ_n converging to zero, when n tends to infinity, such that

$$|lrd(n, \beta)| \leq \gamma_n \,, \tag{A.40}$$

for every $n \geq 1$ and every $1 \leq \beta < \Omega(n)$.

Proof. Let us prove that statement (i) implies statement (ii). Let $\mathcal{G}_\Omega$ be a (B, M) grid of I. We have that

$$B^{-1} \leq r(n, \beta) \leq B \,, \tag{A.41}$$

for every level $n \geq 1$ and every $1 \leq \beta < \Omega(n)$. For every level $n \geq 1$ and every $1 \leq \beta < \Omega(n)$, let $x - \delta_1, x, x + \delta_2 \in I$ be such that $I^n_\beta = [x - \delta_1, x]$ and $I^n_{\beta+1} = [x, x + \delta_2]$. Hence,

$$\frac{r_h(n,\beta)}{r(n,\beta)} = \frac{h(x+\delta_2) - h(x)}{h(x) - h(x-\delta_1)} \frac{\delta_1}{\delta_2} \,.$$

Since $h : I \to J$ is (B, ε_B) uniformly asymptotically affine, we get

$$lrd(n, \beta) < \varepsilon_B(|I^n_\beta| + |I^n_{\beta+1}|) \,. \tag{A.42}$$

By Remark A.1, there is $B_2 = B_2(B, M) < 1$ such that $|I^n_\beta| \leq B_2^n |I|$ and $|I^n_{\beta+1}| \leq B_2^n |I|$. Let $\alpha_n = \varepsilon_B(2B_2^n|I|)$. Hence, by inequality (A.42), we have

$$\begin{aligned} lrd(n, \beta) &< \varepsilon_B(|I^n_\beta| + |I^n_{\beta+1}|) \\ &< \varepsilon_B(2B_2^n|I|) \\ &< \alpha_n \,, \end{aligned}$$

for every n and every $1 \leq \beta < \Omega(n)$. Since $\varepsilon_B(0) = 0$ and ε_B is continuous at 0, we get that $\alpha_n = \varepsilon_B(2B_2^n|I|)$ converges to zero, when n tends to infinity.

Let us prove that statement (ii) implies statement (i). Let $\mathcal{G}_\Omega$ be a (B, M) grid of I. Let $d \geq 1$. Let $x - \delta_1, x, x + \delta_2 \in I$, be such that $\delta_1 > 0$, $\delta_2 > 0$ and $d^{-1} \leq \delta_2/\delta_1 \leq d$. For every $\alpha > 1$, let L_1, L_2, R_1 and R_2 be the intervals as constructed in Lemma A.7. By inequality (A.28) and by monotonicity of the homeomorphism h, we obtain that

$$\frac{|h(R_1)|}{|h(L_2)|}\frac{|L_1|}{|R_2|} \le \frac{h(x+\delta_2)-h(x)}{h(x)-h(x-\delta_1)}\frac{\delta_1}{\delta_2} \le \frac{|h(R_2)|}{|h(L_1)|}\frac{|L_2|}{|R_1|} . \tag{A.43}$$

By inequality (A.29),

$$1 \le \frac{|L_2|}{|L_1|}\frac{|R_2|}{|R_1|} \le \alpha^4. \tag{A.44}$$

By inequalities (A.43) and (A.44), we get

$$\alpha^{-4}\frac{|h(R_1)|}{|h(L_2)|}\frac{|L_2|}{|R_1|} \le \frac{h(x+\delta_2)-h(x)}{h(x)-h(x-\delta_1)}\frac{\delta_1}{\delta_2} \le \alpha^4\frac{|h(R_2)|}{|h(L_1)|}\frac{|L_1|}{|R_2|} . \tag{A.45}$$

By Lemma A.7, we obtain that

$$\frac{|h(R_2)|}{|h(L_1)|}\frac{|L_1|}{|R_2|} = \frac{\sum_{i=m-1}^{r-1}\prod_{j=l}^{i} r_h(n_0+n_1,j)}{\sum_{i=m-1}^{r-1}\prod_{j=l}^{i} r(n_0+n_1,j)}\frac{1+\sum_{i=l+1}^{m-2}\prod_{j=l}^{i} r(n_0+n_1,j)}{1+\sum_{i=l+1}^{m-2}\prod_{j=l}^{i} r_h(n_0+n_1,j)} ,$$

$$\frac{|h(R_1)|}{|h(L_2)|}\frac{|L_2|}{|R_1|} = \frac{\sum_{i=m}^{r-2}\prod_{j=l}^{i} r_h(n_0+n_1,j)}{\sum_{i=m}^{r-2}\prod_{j=l}^{i} r(n_0+n_1,j)}\frac{1+\sum_{i=l}^{m-1}\prod_{j=l}^{i} r(n_0+n_1,j)}{1+\sum_{i=l}^{m-1}\prod_{j=l}^{i} r_h(n_0+n_1,j)} . \tag{A.46}$$

By inequality (A.40), there is $C_0 \ge 1$ and there is a sequence γ_n converging to zero, when n tends to infinity, such that

$$\frac{r_h(n_0+n_1,j)}{r(n_0+n_1,j)} = 1 \pm C_0\gamma_{n_0+n_1} , \tag{A.47}$$

for every n_0+n_1 and for every $1 \le j < \Omega(n_0+n_1)$. Without loss of generality, we will consider that γ_n is a decreasing sequence. Hence, by inequalities (A.46) and (A.47), there is $C_1 = C_1(C_0, n_2) > 1$ such that

$$\left|\log\frac{|h(R_1)|}{|h(L_2)|}\frac{|L_2|}{|R_1|}\right| \le C_1\gamma_{n_0+n_1} \quad\text{and}\quad \left|\log\frac{|h(R_2)|}{|h(L_1)|}\frac{|L_1|}{|R_2|}\right| \le C_1\gamma_{n_0+n_1} .$$

Therefore, by inequality (A.45), we obtain that

$$\left|\log\frac{h(x+\delta_2)-h(x)}{h(x)-h(x-\delta_1)}\frac{\delta_1}{\delta_2}\right| \le C_1\gamma_{n_0+n_1} + 4\log(\alpha) .$$

For every $m = 1, 2, \ldots$, let $\alpha_m = \exp(1/8m)$. Hence, we get

$$\left|\log\frac{h(x+\delta_2)-h(x)}{h(x)-h(x-\delta_1)}\frac{\delta_1}{\delta_2}\right| \le C_1\gamma_{n_0+n_1} + 1/(2m) . \tag{A.48}$$

By Lemma A.7, $n_0 = n_0(x-\delta_1, x, x+\delta_2) \ge 1$ is the biggest integer such that $[x-\delta_1, x+\delta_2] \subset I_\beta^{n_0} \cup I_{\beta+1}^{n_0}$. Hence, there is $I_\alpha^{n_0+1} \subset [x-\delta_1, x+\delta_2]$, Thus,

$|I_\alpha^{n_0+1}| \le \delta$, where $\delta = \delta_1 + \delta_2$. By Remark A.1, there is $0 < B_1(B, M) < 1$ such that $|I_\alpha^{n_0+1}| \ge B_1^{n_0+1}|I|$. Hence, we get that $B_1^{n_0+1}|I| \le |I_\alpha^{n_0+1}| \le \delta$, and so

$$n_0 \ge \frac{\log\left(\delta B_1^{-1}|I|^{-1}\right)}{\log(B_1)} .$$

Therefore, there is a monotone sequence $\delta_m > 0$ converging to zero, when m tends to infinity, with the following property: if $\delta_1 + \delta_2 \le \delta_m$, then $n_0 = n_0(x - \delta_1, x, x + \delta_2)$ is sufficiently large such that $C_1\gamma_{n_0+n_1} \le 1/(2m)$. Hence, by inequality (A.48), for every $m \ge 1$ and every $\delta_0 + \delta_1 \le \delta_m$, we have

$$\left|\log \frac{h(x+\delta_2) - h(x)}{h(x) - h(x-\delta_1)} \frac{\delta_1}{\delta_2}\right| \le C_1\gamma_{n_0+n_1} + 1/(2m) \le 1/m . \tag{A.49}$$

Therefore, we define the continuous function $\varepsilon_D : \mathbb{R}^+ \to \mathbb{R}^+$ as follows:

(i) $\varepsilon_d(\delta_m) = 1/(m-1)$ for every $m = 2, 3, \ldots$;
(ii) ε_d is affine in every interval $[\delta_m, \delta_m - 1]$;
(iii) Since I is a compact interval and h is a homeomorphism, there is an extension of ε_d to $[\delta_2, \infty)$ such that inequality (A.27) is satisfied.

By inequality (A.49), we get that ε_d satisfies inequality (A.27). □

Lemma A.9. *If $h : I \to J$ satisfies the (d_0, ε_{d_0}) uniformly asymptotically affine condition, then the homeomorphism h is (uaa).*

Proof. Similarly to the proof that statement (i) implies statement (ii) of Lemma A.8, we obtain that if $h : I \to J$ satisfies the (d_0, ε_{d_0}) uniformly asymptotically affine condition, then satisfies statement (ii) of Lemma A.8 with respect to a symmetric grid (see definition of a symmetric grid in Remark A.1). Since statement (ii) implies statement (i) of Lemma A.2, we get that the homeomorphism h is (uaa). □

Lemma A.10. *Let $h : I \to J$ be a homeomorphism and $\mathcal{G}_\Omega$ a grid of the compact interval I.*

(i) If $h : I \to J$ is (uaa), then there is a sequence α_n converging to zero, when n tends to infinity, such that

$$|crd(n, \beta)| \le \alpha_n ,$$

for every $n \ge 1$ and every $1 \le \beta < \Omega(n) - 1$.
(ii) If there is a sequence α_n converging to zero, when n tends to infinity, such that for every $n \ge 1$ and every $1 \le \beta < \Omega(n) - 1$

$$|crd(n, \beta)| \le \alpha_n ,$$

then, for every closed interval K contained in the interior of I, the homeomorphism h is (uaa) in K.

Proof. Let us prove statement (i). By Lemma A.8, there is a sequence α_n converging to zero, when n tends to infinity, such that

$$|lrd(n,\beta)| \le \gamma_n \ , \tag{A.50}$$

for every $n \ge 1$ and every $1 \le \beta < \Omega(n)$. By inequality (A.9), we have that

$$crd(n,\beta) \in \frac{lrd(n,\beta)}{1+r(n,\beta)^{-1}} - \frac{lrd(n,\beta+1)}{1+r(n,\beta+1)} \pm \mathcal{O}(lrd(n,\beta)^2, lrd(n,\beta+1)^2) \ . \tag{A.51}$$

By the bounded geometry property of a grid, there is $B \ge 1$ such that $B^{-1} \le r(n,\beta) \le B$. Thus, there is $C_0 > 1$ such that

$$C_0^{-1} \le \frac{1}{1+r(n,\beta)^{-1}} \le C_0 \quad \text{and} \quad C_0^{-1} \le \frac{1}{1+r(n,\beta+1)} \le C_0 \ . \tag{A.52}$$

Therefore, putting together inequalities (A.50), (A.51) and (A.52), we obtain that there is $C_1 > 1$ such that $|crd(n,\beta)| \le C_1\gamma_n$, for every level n and every $1 \le \beta < \Omega(n) - 1$.

Let us prove statement (ii). Let us suppose, by contradiction, that there is $\varepsilon_0 > 0$ such that $|lrd(n(j),\beta(j))| > \varepsilon_0$, where $I^{n(j)}_{\beta(j)} \subset K$ and $n(j)$ tends to infinity, when j tends to infinity. Hence, there is a subsequence m_j such that either $lrd(n(m_j),\beta(m_j)) < -\varepsilon_0$ for every $j \ge 1$, or $lrd(n(m_j),\beta(m_j)) > \varepsilon_0$ for every $j \ge 1$. For simplicity of notation, we will denote $n(m_j)$ by n_j, and $\beta(m_j)$ by β_j. It is enough to consider the case where $lrd(n_j,\beta_j) > \varepsilon_0$ (if necessary, after re-ordering all the indices). Thus, there is $\varepsilon = \varepsilon(\varepsilon_0) > 1$ such that, for every $j \ge 1$,

$$\frac{1+r_h(n_j,\beta_j)}{1+r(n_j,\beta_j)} > \varepsilon \ . \tag{A.53}$$

Let $a(n,i)$ and $b(n,i)$ be defined as in Lemma A.6:

$$a(n,i) = \frac{1+r_h(n,i)}{1+r(n,i)} \tag{A.54}$$

$$b(n,i) = \exp(-crd(n,\beta)) = \frac{1+r(n,i)}{1+r_h(n,i)} \frac{1+r(n,i+1)^{-1}}{1+r_h(n,i+1)^{-1}} \ .$$

Hence, we have that $a(n_j,\beta_j) \ge \varepsilon$ for every $j \ge 1$. By hypothesis, the cross-ratio distortion $crd(n,\beta)$ converges uniformly to zero when n tends to infinity. Thus, there is an increasing sequence γ_n converging to one, when n tends to infinity, such that

$$\gamma_n \le b(n,i) \le \gamma_n^{-1} \ , \tag{A.55}$$

for every $1 \le i < \Omega(n) - 1$. Let $\eta = \min\{(\varepsilon-1)/4, 1/2\}$. For every j large enough, let p_j be the maximal integer with the following properties: (i) $\gamma_{n_j}^{p_j} \ge \eta$; (ii) $\gamma_{n_j}^{p_j}(\varepsilon-1)/2 \ge \eta$; and (iii), letting $B \ge 1$ be as given by the bounded geometry property of the grid,

$$\frac{(\varepsilon-1)\gamma^i B^{-p_j}}{2} \geq B(1-\gamma)\frac{1-(B\gamma^{-1})^{p_j}}{1-(B\gamma^{-1})} .$$

Since γ_{n_j} converges to one, when j tends to infinity, we obtain that p_j also tends to infinity, when j tends to infinity. By properties (ii) and (iii) of η and by inequality (A.23), for every j large enough, and for every $1 \leq i \leq p_j$, we have

$$a(n_j, \beta_j + i) \geq 1 + \eta \prod_{k=1}^{i} r(n_j, \beta_j + k) > 1 . \tag{A.56}$$

For every $j \geq 1$, let N_j be the smallest integer such that there are four grid intervals $I^{N_j}_{\alpha_j-1}$, $I^{N_j}_{\alpha_j}$, $I^{N_j}_{\alpha_j+1}$ and $I^{N_j}_{\alpha_j+2}$ such that

$$I^{n_j}_{\beta_j} \subset I^{N_j}_{\alpha_j-1} \quad \text{and} \quad I^{N_j}_{\alpha_j} \cup I^{N_j}_{\alpha_j+1} \cup I^{N_j}_{\alpha_j+2} \subset \cup_{i=1}^{p_j-1} I^{n_j}_{\beta_j+i} .$$

Since the grid intervals $I^{n_j}_{\beta_j}, \ldots, I^{n_j}_{\beta_j+p(j)-1}$ are contained in at most four grid intervals at level $N_j - 1$, we obtain that

$$4M^{n_j-(N_j-1)} \geq p_j ,$$

where $M > 1$ is given by the bounded geometry property of the grid. Thus, $n_j - N_j$ tends to infinity, when j tends to infinity. Let us denote by $RD(j)$ the following ratio:

$$\begin{aligned} RD(j) &= \frac{|I^{N_j}_{\alpha_j}|}{|J^{N_j}_{\alpha_j}|}\frac{|J^{N_j}_{\alpha_j+1}| + |J^{N_j}_{\alpha_j+2}|}{|I^{N_j}_{\alpha_j+1}| + |I^{N_j}_{\alpha_j+2}|} \\ &= \frac{r_h(N_j,\alpha_j)(1+r_h(N_j,\alpha_j+1))}{r(N_j,\alpha_j)(1+r(N_j,\alpha_j+1))} . \end{aligned}$$

By the bounded geometry property of a grid, we have $B^{-1} < r(N_j, \alpha_j + i) < B$ for every $-1 \leq i \leq 3$ and $j \geq 0$. By Lemma A.2 and statement (ii) of Lemma A.4, there is $k_0 > 1$ such that $k_0^{-1} < r_h(N_j, \alpha_j + i) < k_0$ for every $-1 \leq i \leq 3$ and $j \geq 0$. Hence, there is $k = k(B, k_0) > 1$ such that for every $j \geq 0$, we have

$$k^{-1} \leq RD(j) \leq k . \tag{A.57}$$

Now, we are going to prove that $RD(j)$ tends to infinity, when j tends to infinity, and so we will get a contradiction. Let $e_1 < e_2 < e_3 < e_4$ be such that

$$I^{N_j}_{\alpha_j} = \cup_{i=e_1}^{e_2} I^{n_j}_{\beta_j+i} \quad , \quad I^{N_j}_{\alpha_j+1} = \cup_{i=e_2+1}^{e_3} I^{n_j}_{\beta_j+i} \quad , \quad I^{N_j}_{\alpha_j+2} = \cup_{i=e_3+1}^{e_4} I^{n_j}_{\beta_j+i} .$$

Hence, we get

$$RD(j) = \frac{|I^{N_j}_{\alpha_j}|}{|J^{N_j}_{\alpha_j}|}\frac{|J^{N_j}_{\alpha_j+1}| + |J^{N_j}_{\alpha_j+2}|}{|I^{N_j}_{\alpha_j+1}| + |I^{N_j}_{\alpha_j+2}|} = \frac{R_1(j)}{R_{h,1}(j)}\frac{R_{h,2}(j)+R_{h,3}(j)}{R_2(j)+R_3(j)} , \tag{A.58}$$

where

$$R_1(j) = \frac{|I^{N_j}_{\alpha_j}|}{|I^{n_j}_{\beta_j+e_2}|} = 1 + \sum_{q=e_1}^{e_2-2} \prod_{i=q+1}^{e_2-1} r(n_j, \beta_j + i)^{-1}$$

$$R_{h,1}(j) = \frac{|J^{N_j}_{\alpha_j}|}{|J^{n_j}_{\beta_j+e_2}|} = 1 + \sum_{q=e_1}^{e_2-2} \prod_{i=q+1}^{e_2-1} r_h(n_j, \beta_j + i)^{-1}$$

$$R_2(j) = \frac{|I^{N_j}_{\alpha_j+1}|}{|I^{n_j}_{\beta_j+e_2}|} = \sum_{q=e_2}^{e_3-1} \prod_{i=e_2}^{q} r(n_j, \beta_j + i)$$

$$R_{h,2}(j) = \frac{|J^{N_j}_{\alpha_j+1}|}{|J^{n_j}_{\beta_j+e_2}|} = \sum_{q=e_2}^{e_3-1} \prod_{i=e_2}^{q} r_h(n_j, \beta_j + i)$$

$$R_3(j) = \frac{|I^{N_j}_{\alpha_j+2}|}{|I^{n_j}_{\beta_j+e_2}|} = \sum_{q=e_3}^{e_4-1} \prod_{i=e_2}^{q} r(n_j, \beta_j + i)$$

$$R_{h,3}(j) = \frac{|J^{N_j}_{\alpha_j+2}|}{|J^{n_j}_{\beta_j+e_2}|} = \sum_{q=e_3}^{e_4-1} \prod_{i=e_2}^{q} r_h(n_j, \beta_j + i) \; .$$

Hence, by inequalities (A.54) and (A.56), for every $1 \le i \le p_j$, we get

$$\frac{r_h(n_j, \beta_j + i)}{r(n_j, \beta_j + i)} > 1 \; . \tag{A.59}$$

Thus, we deduce that

$$\begin{aligned} R_{h,1}(j) &= 1 + \sum_{q=e_1}^{e_2-2} \prod_{i=q+1}^{e_2-1} r(n_j, \beta_j + i)^{-1} \frac{r(n_j, \beta_j + i)}{r_h(n_j, \beta_j + i)} \\ &\le 1 + \sum_{q=e_1}^{e_2-2} \prod_{i=q+1}^{e_2-1} r(n_j, \beta_j + i)^{-1} \\ &= R_1(j) \; . \end{aligned} \tag{A.60}$$

By inequality (A.59), we obtain

$$\begin{aligned} R_{h,2}(j) &= \sum_{q=e_2}^{e_3-1} \prod_{i=e_2}^{q} r(n_j, \beta_j + i) \frac{r_h(n_j, \beta_j + i)}{r(n_j, \beta_j + i)} \\ &\ge R_2(j). \end{aligned} \tag{A.61}$$

Now, let us bound $R_{h,3}(j)$ in terms of $R_3(j)$. Putting together inequalities (A.22) and (A.56), we obtain

$$\begin{aligned} \frac{r_h(n,i)}{r(n,i)} &= b(n, i-1)a(n,i)a(n,i-1) \\ &\ge b(n, i-1)a(n,i) \; . \end{aligned} \tag{A.62}$$

Noting that $e_3 - e_2 < p_j$, and by inequality (A.55) and property (i) of η, we get

$$\prod_{i=e_2}^{e_3-1} b(n_j, \beta_j + i - 1) \geq \gamma^{p_j} \geq \eta \, .$$

Hence, by inequalities (A.56) and (A.62), we get

$$\begin{aligned}
\prod_{i=e_2}^{e_3-1} \frac{r_h(n_j, \beta_j + i)}{r(n_j, \beta_j + i)} &\geq \prod_{i=e_2}^{e_3-1} b(n_j, \beta_j + i - 1) a(n_j, \beta_j + i) \\
&\geq \eta \prod_{i=e_2}^{e_3-1} \left(1 + \eta \prod_{k=1}^{i} r(n_{m_i}, \beta_j + k)\right) \\
&\geq \eta \left(1 + \eta \sum_{i=e_2}^{e_3-1} \prod_{k=1}^{i} r(n_{m_i}, \beta_j + k)\right) \\
&\geq \eta^2 \frac{|I^{N_j}_{\alpha_j+1}|}{|I^{n_j}_{\beta_j+1}|} \, . \qquad \text{(A.63)}
\end{aligned}$$

Noting that $I^{n_j}_{\beta_j+1} \subset I^{N_j}_{\alpha_j-1} \cup I^{N_j}_{\alpha_j}$ and by the bounded geometry property of the grid, we get

$$\frac{|I^{N_j}_{\alpha_j+1}|}{|I^{n_j}_{\beta_j+1}|} \geq B^{-2} B_2^{N_j - n_j} \, , \qquad \text{(A.64)}$$

where $B_2 < 1$ is given in Remark A.1. Putting together inequalities (A.63) and (A.64), we obtain that

$$\prod_{i=e_2}^{e_3-1} \frac{r_h(n_j, \beta_j + i)}{r(n_j, \beta_j + i)} \geq \eta^2 B^{-2} B_2^{N_j - n_j} \, .$$

Hence,

$$\begin{aligned}
R_{h,3}(j) &= \prod_{i=e_2}^{e_3-1} \frac{r_h(n_j, \beta_j + i)}{r(n_j, \beta_j + i)} r(n_j, \beta_j + i) \sum_{q=e_3}^{e_4-1} \prod_{i=e_3}^{q} \frac{r_h(n_j, \beta_j + i)}{r(n_j, \beta_j + i)} r(n_j, \beta_j + i) \\
&\geq \eta^2 B^{-2} B_2^{N_j - n_j} \prod_{i=e_2}^{e_3-1} r(n_j, \beta_j + i) \sum_{q=e_3}^{e_4-1} \prod_{i=e_3}^{q} r(n_j, \beta_j + i) \\
&= \eta^2 B^{-2} B_2^{N_j - n_j} R_3(j) \, . \qquad \text{(A.65)}
\end{aligned}$$

Noting that $R_2(j) R_3(j)^{-1} = |I^{N_j}_{\alpha_j+1}||I^{N_j}_{\alpha_j+2}|^{-1}$ and by the bounded geometry property of the grid, we obtain

$$B^{-1} \leq R_2(j) R_3(j)^{-1} \leq B \, .$$

Therefore, putting together inequalities (A.60), (A.61) and (A.65), we obtain that

$$
\begin{aligned}
RD(j) &= \frac{R_1(j)}{R_{h,1}(j)} \frac{R_{h,2}(j) + R_{h,3}(j)}{R_2(j) + R_3(j)} \\
&\geq \frac{R_2(j) + \eta^2 B^{-2} B_2^{N_j - n_j} R_3(j)}{R_2(j) + R_3(j)} \\
&\geq \frac{1 + \eta^2 B^{-3} B_2^{N_j - n_j}}{1 + B} .
\end{aligned}
$$

Since $B_2^{N_j-n_j}$ tends to infinity, when j tends to infinity, we get that $RD(j)$ also tends to infinity, when j tends to infinity. However, by inequality (A.57), this is absurd. □

A.6 C^{1+r} diffeomorphisms

Let $0 < r \leq 1$. We say that a homeomorphism $h : I \to J$ is C^{1+r} if its differentiable and its first derivative $dh : I \to \mathbb{R}$ is r-Hölder continuous, i.e. there is $C \geq 0$ such that, for every $x, y \in I$,

$$|dh(y) - dh(x)| \leq C|y - x|^r .$$

In particular, if $r = 1$, then dh is Lipschitz.

Lemma A.11. *Let $h : I \to J$ be a homeomorphism, and let I be a compact interval with a grid $\mathcal{G}_\Omega$.*

(i) For $0 < r \leq 1$, the map h is a C^{1+r} diffeomorphism if, and only if, for every $n \geq 1$ and for every $1 \leq \beta < \Omega(n)$, we have that

$$|lrd(n, \beta)| \leq \mathcal{O}(|I_\beta^n|^r) . \tag{A.66}$$

(ii) The map h is affine if, and only if, for every $n \geq 1$ and every $1 \leq \beta < \Omega(n)$, we have that

$$|lrd(n, \beta)| \leq o(|I_\beta^n|) . \tag{A.67}$$

Proof. By the Mean Value Theorem, if h is a C^{1+r} diffeomorphism, then, for every $n \geq 1$ and for every grid interval I_β^n, we get that $lrd(n, \beta) \in \pm\mathcal{O}(|I_\beta^n|^r)$, and so inequality (A.66) is satisfied. If h is affine, then, for every $n \geq 1$ and for every grid interval I_β^n, we get that $lrd(n, \beta) = 0$, and so inequality (A.67) is satisfied.

Let us prove that inequality (A.66) implies that h is C^{1+r}. For every point $P \in I$, let $I_{\alpha_1}^1, I_{\alpha_2}^2, \ldots$ be a sequence of grid intervals $I_{\alpha_n}^n$ such that $P \in I_{\alpha_n}^n$ and $I_{\alpha_n}^n \subset I_{\alpha_{n-1}}^{n-1}$ for every $n > 1$. Let us suppose that $I_{\alpha_{n-1}}^{n-1} = \cup_{i=0} j I_{\alpha_n+i}^n$ for some $j = j(\alpha_n) \geq 1$. By inequality (A.66) and using the bounded geometry of the grid, we obtain that

$$\begin{aligned}\frac{dh(n-1,\alpha_{n-1})}{dh(n,\alpha_n)} &= \frac{1+\sum_{i=1}^{j}\prod_{k=1}^{i} r_h(n,\alpha_n+k)}{1+\sum_{i=1}^{j}\prod_{k=1}^{i} r(n,\alpha_n+k)} \\ &= \frac{1+\sum_{i=1}^{j}\prod_{k=1}^{i} r(n,\alpha_n+k)(1\pm\mathcal{O}(|I^n_{\alpha_n+k}|)}{1+\sum_{i=1}^{j}\prod_{k=1}^{i} r(n\alpha_n+k)} \\ &= \mathcal{O}(|I^n_{\alpha_n}|^r)\ .\end{aligned}$$

A similar argument to the one above implies that for all $I^n_{\alpha_n} \subset I^{n-1}_{\alpha_{n-1}}$, we have

$$dh(n,\alpha_n) = dh(n-1,\alpha_{n-1}) \pm \mathcal{O}(|I^{n-1}_{\alpha_{n-1}}|^r)\ .$$

Hence, using the bounded geometry property of a grid, for every $m \geq 1$ and for every $n \geq m$, we get

$$dh(n,\alpha_n) = dh(m,\alpha_m) \pm \mathcal{O}(|I^m_{\alpha_m}|^r)\ . \tag{A.68}$$

Thus, the average derivative $dh(n,\alpha_n)$ converges to a value d_P, when n tends to infinity. Let us prove that h is differentiable at P and that $dh(P) = d_P$. Let L be any interval such that the point $P \in L$. Take the largest $m \geq 1$ such that there is a grid interval I^m_γ with the property that $L \subset \cup_{j=-1,0,1} I^m_{\gamma+j}$. By the bounded geometry property of a grid, there is $C \geq 1$, not depending upon P, L and I^m_γ, such that

$$C^{-1} \leq \frac{|I^m_\gamma|}{|L|} \leq C\ . \tag{A.69}$$

Then, using inequality (A.66) and the bounded geometry of the grid, for every $j = \{-1,0,1\}$, we obtain that

$$|ldh(m,\gamma+j) - ldh(m,\gamma)| \leq \mathcal{O}(|L|^r)\ ,$$

and so

$$dh(m,\gamma+j) = dh(m,\gamma) \pm \mathcal{O}(|L|^r)\ . \tag{A.70}$$

For every $n \geq m$, take the smallest sequence of adjacent grid intervals $I^n_{\beta_n},\ldots,I^n_{\beta_n+i_n}$, at level n, such that $L \subset \cup_{i=0}^{i_n} I^n_{\beta_n+i} \subset \cup_{j=-1,0,1} I^m_{\gamma+j}$. By inequalities (A.68) and (A.70), for every $I^n_{\beta_n+i} \subset I^m_{\gamma+j(i)}$ we get that

$$\begin{aligned}dh(m,\beta_n+i) &= dh(m,\gamma+j(i)) \pm \mathcal{O}(|I^m_{\gamma+j(i)}|^r) \\ &= d_P \pm \mathcal{O}(|L|^r)\ .\end{aligned}$$

Hence,

$$\begin{aligned}\frac{h(L)}{L} &= \lim_{n\to\infty}\sum_{i=0}^{i_n}\frac{|I^n_{\beta_n+i}|}{|L|} dh(n,\beta_n+i) \\ &= \lim_{n\to\infty}\sum_{i=0}^{i_n}\frac{|I^n_{\beta_n+i}|}{|L|}\left(d_P \pm \mathcal{O}(|L|^r)\right) \\ &= d_P \pm \mathcal{O}(|L|^r)\ .\end{aligned} \tag{A.71}$$

Therefore, for every $P \in I$, the homeomorphism h is differentiable at P and $dh(P) = d_P$. Let us check that dh is r-Hölder continuous. For every $P, P' \in I$, let L be the closed interval $[P, P']$. Using inequality (A.71), we obtain that

$$\begin{aligned} dh(P') - dh(P) &= \frac{h(L)}{L} - \frac{h(L)}{L} \pm \mathcal{O}(|L|^r) \\ &= \pm\mathcal{O}(|L|^r) , \end{aligned}$$

and so dh is r-Hölder continuous.

Let us prove that inequality (A.67) implies that h is affine. A similar argument to the one above gives us that h is differentiable and that

$$|dh(P') - dh(P)| \leq o(|P' - P|) , \tag{A.72}$$

for every $P, P' \in I$. Hence, we get that

$$\begin{aligned} |dh(P') - dh(P)| &\leq \lim_{n\to\infty} \sum_{i=0}^{n-1} \left| dh\left(P + \frac{(i+1)(P'-P)}{n}\right) \right. \\ &\qquad \left. - dh\left(P + \frac{i(P'-P)}{n}\right) \right| \\ &\leq \lim_{n\to\infty} n\, o\left(\frac{P'-P}{n}\right) = 0 , \end{aligned}$$

and so h is an affine map. □

Lemma A.12. *Let $0 < r \leq 1$. Let $h : I \to J$ be a homeomorphism and $\mathcal{G}_\Omega$ a grid of the compact interval I.*

(i) If $h : I \to J$ is a C^{1+r} diffeomorphism, then, for every $n \geq 1$ and every $1 \leq \beta < \Omega(n) - 1$, we have that

$$|crd(n, \beta)| \leq \mathcal{O}(|I^n_\beta|^r) . \tag{A.73}$$

(ii) If, for every $n \geq 1$ and every $1 \leq \beta < \Omega(n) - 1$, we have that

$$|crd(n, \beta)| \leq \mathcal{O}(|I^n_\beta|^r) , \tag{A.74}$$

then, for every closed interval K contained in the interior of I, the homeomorphism $h|K$ restricted to K is a C^{1+r} diffeomorphism.

Proof. Proof of statement (i). By Lemma A.11, for every $n \geq 1$ and for every $1 \leq \beta < \Omega(n)$, we have that $|lrd(n, \beta)| \leq \mathcal{O}(|I^n_\beta|^r)$. Hence, by the bounded geometry property of a grid and by inequality (A.9), we get $|crd(n, \beta)| \leq \mathcal{O}(|I^n_\beta|^r)$.

Proof of statement (ii). Let K be a closed interval contained in the interior of I. By Lemmas A.8 and A.10, there is a decreasing sequence of positive reals ε_n which converges to 0, when n tends to ∞, such that

$$|lrd(n, \beta)| < |\varepsilon_n| , \tag{A.75}$$

for all $n \geq 1$ and for all grid interval I^n_β intersecting K. For every grid interval I^{n-1}_α intersecting K, let $k_1 = k_1(n, \alpha)$ and $k_2 = k_2(n, \alpha)$ be such that $\cup_{\beta=k_1}^{k_2} I^n_\beta = I^{n-1}_\alpha \cup I^{n-1}_{\alpha+1}$. Let the integers β and i be such that $k_1 \leq \beta \leq k_2$ and $k_1 \leq \beta + i \leq k_2$. By the bounded geometry property of a grid, and by inequalities (A.10) and (A.73), we get

$$lrd(n, \beta + i) = \pm\mathcal{O}\left(|lrd(n, \beta)| + (|I^n_\beta| + |I^n_{\beta+1}|)^r\right) .$$

Therefore,

$$L_2(n, \beta, p) = \pm\mathcal{O}\left(lrd(n, \beta)^2 + (|I^n_\beta| + |I^n_{\beta+1}|)^{2r}\right) . \tag{A.76}$$

By inequalities (A.11) and (A.76), we get

$$lrd(n-1, \alpha) = \frac{|I^{n-1}_\alpha| + |I^{n-1}_{\alpha+1}|}{|I^n_{\beta+i}| + |I^n_{\beta+i+1}|} lrd(n, \beta+i) \pm \mathcal{O}\left(lrd(n, \beta)^2 + (|I^n_\beta| + |I^n_{\beta+1}|)^r\right) . \tag{A.77}$$

Let us suppose, by contradiction, that there is a sequence of grid intervals $I^{n_j}_{\beta_j}$ and a sequence of positive reals $|e_j|$ which tends to infinity, when j tends to infinity, such that

$$lrd(n_j, \beta_j) = e_j(|I^{n_j}_{\beta_j}| + |I^{n_j}_{\beta_j+1}|)^r . \tag{A.78}$$

Using that the number of grid intervals at every level n is finite, we obtain that there exists a subsequence m_j of j such that $I^{n_{m_{j+1}}}_{\beta_{m_{j+1}}} \subset I^{n_{m_j}}_{\beta_{m_j}}$. Therefore, there exists a sequence of grid intervals $I^1_{\alpha_1}, I^2_{\alpha_2}, \ldots$ with the following properties:

(i) for every $i \geq 1$, $I^{i+1}_{\alpha_{i+1}} \subset I^i_{\alpha_i}$;
(ii) for every $i \geq 1$, let a_i be determined such that

$$lrd(i, \alpha_i) = a_i(|I^i_{\alpha_i}| + |I^i_{\alpha_i+1}|)^r . \tag{A.79}$$

Then, there is a subsequence m_j of j such that $|a_i| \leq |a_{m_j}|$ for every $1 \leq i \leq m_j$, and $|a_{m_j}|$ tends to infinity, when j tends to infinity.

Let us denote $|I^i_{\alpha_i}| + |I^i_{\alpha_i+1}|$ by B_i. Using inequality (A.77) inductively, we get

$$lrd(m_j, \beta_{m_j}) = \frac{B_{m_j}}{B_1} lrd(1, \alpha_1) \pm \mathcal{O}\left(\sum_{i=2}^{m_j} \frac{B_{m_j}}{B_i}(lrd(i, \alpha_i)^2 + B_i^r)\right) \tag{A.80}$$

By the bounded geometry property of a grid, there is $0 < \theta < 1$ such that

$$\frac{B_k}{B_i} \leq \theta^{k-i} , \tag{A.81}$$

for every $1 \leq i \leq m_j$ and for every $1 \leq k \leq m_j$. Noting that $|a_1| \leq |a_{m_j}|$, by inequalities (A.79) and (A.81), we get

$$\frac{B_{m_j}}{B_1} lrd(1, \alpha_1) = \frac{a_1 B_1^r B_{m_j}}{B_1}$$
$$= \pm \mathcal{O}\left(|a_{m_j}| B_{m_j}^r \theta^{(1-r)m_j}\right) . \qquad \text{(A.82)}$$

By inequality (A.75), $a_i B_i \leq \varepsilon_i$, and $|a_i| \leq |a_{m_j}|$ for $i \leq m_j$. Hence, by inequalities (A.79) and (A.81), we obtain that

$$\frac{B_{m_j}}{B_i} (lrd(i, \alpha_i)^2 + B_i^r) = \frac{a_i (a_i B_i^r)(B_i^r B_{m_j}) + B_i^r B_{m_j}}{B_i}$$
$$= \pm \mathcal{O}\left((|a_{m_j}| \varepsilon_i + 1) B_{m_j}^r \theta^{(1-r)(m_j - i)}\right) . \qquad \text{(A.83)}$$

Using inequalities (A.82) and (A.83) in inequality (A.80), we get

$$\frac{|lrd(m_j, \beta_{m_j})|}{|a_{m_j}| B_{m_j}^r} \leq \mathcal{O}\left(\theta^{(1-r)m_j} + \sum_{i=2}^{m_j} \left((\varepsilon_i + |a_{m_j}|^{-1}) \theta^{(1-r)(m_j - i)}\right)\right)$$
$$\leq \mathcal{O}\left(\theta^{(1-r)m_j} + \frac{|a_{m_j}|^{-1}}{1 - \theta^{1-r}} + \sum_{i=2}^{m_j} \left(\varepsilon_i \theta^{(1-r)(m_j - i)}\right)\right) . \qquad \text{(A.84)}$$

Since ε_i converges to zero, when i tends to infinity, inequality (A.84) implies that there is $j_0 \geq 0$ such that, for every $j \geq j_0$, we get

$$|lrd(m_j, \beta_{m_j})| < |a_{m_j}| B_{m_j}^r ,$$

which contradicts (A.79). □

A.7 C^{2+r} diffeomorphisms

Let $0 < r \leq 1$. We say that a homeomorphism $h : I \to J$ is C^{2+r} if its twice differentiable and its second derivative $d^2 h : I \to \mathbb{R}$ is r-Hölder continuous. We will state and prove Lemma A.13 which we will use later in the proof of Lemma A.14, below.

Lemma A.13. *Let $\mathcal{G}_\Omega$ be a grid of the closed interval I. Let $h : I \subset \mathbb{R} \to J \subset \mathbb{R}$ be a homeomorphism such that for every $n \geq 1$ and every $1 \leq \beta < \Omega(n) - 1$,*

$$|crd(n, \beta)| \leq \mathcal{O}(|I_\beta^n|^{1+r}) , \qquad \text{(A.85)}$$

where $0 \leq r < 1$. Then, for every closed interval K contained in the interior of I, the logarithmic ratio distortion and the cross-ratio distortion satisfy the following estimates:

(i) There is a constant $C(i) > 0$, not depending upon the level n and not depending upon $1 \leq \beta \leq \Omega(n)$, such that

$$lrd(n, \beta + i) = \frac{|I_{\beta+i}^n| + |I_{\beta+i+1}^n|}{|I_\beta^n| + |I_{\beta+1}^n|} lrd(n, \beta) \pm C(i) |I_\beta^n|^{1+r} . \qquad \text{(A.86)}$$

(ii) Let I_α^{n-1} and $I_{\alpha+1}^{n-1}$ be two adjacent grid intervals. Let I_β^n and $I_{\beta+1}^n$ be grid intervals contained in the union $I_\alpha^{n-1} \cup I_{\alpha+1}^{n-1}$. Then,

$$lrd(n-1,\alpha) = \frac{|I_\alpha^{n-1}| + |I_{\alpha+1}^{n-1}|}{|I_\beta^n| + |I_{\beta+1}^n|} lrd(n,\beta) \pm \mathcal{O}(|I_\beta^n|^{1+r}) . \tag{A.87}$$

Proof. By Lemma A.12, for every $0 < s < 1$, the homeomorphism $h|K$ is C^{1+s}, and so the map $\psi : I \to \mathbb{R}$ is well-defined by $\psi(x) = \log dh(x)$. By bounded geometry property of a grid and by inequality (A.85), for every integer i, there is a positive constant $E_1(i)$ such that

$$|crd(n,\beta+j_1)| \le E_1(i)(|I_\beta^n|^{1+r}) , \tag{A.88}$$

for every grid interval I_β^n and $0 \le j_1 \le i$. Take $s < 1$ such that $2s = 1 + r$ and $0 \le j_2 \le i$. By inequality (A.85) and statement (ii) of Lemma A.12, h is C^{1+s}. Hence, using the bounded geometry property of a grid and statement (i) of Lemma A.11, we obtain that

$$\begin{aligned} |lrd(n,\beta+j_1) lrd(n,\beta+j_2)| &\le \mathcal{O}(|I_{\beta+j_1}^n|^s |I_{\beta+j_2}^n|^s) \\ &\le E_2(i)(|I_\beta^n|^{1+r}), \end{aligned} \tag{A.89}$$

where $E_2(i)$ is a positive constant depending upon i. Using inequalities (A.88) and (A.89) in (A.10), we get inequality (A.86). Furthermore, using inequalities (A.88) and (A.89) in (A.11), we get inequality (A.87). □

Lemma A.14. *Let $0 < r \le 1$. Let $h : I \to J$ be a homeomorphism and $\mathcal{G}_\Omega$ a grid of the compact interval I.*

(i) If $h : I \to J$ is C^{2+r}, then

$$|crd(n,\beta)| \le \mathcal{O}(|I_\beta^n|^{1+r}) ,$$

for every $n \ge 1$ and every $1 \le \beta < \Omega(n) - 1$.

(ii) If, for every $n \ge 1$ and every $1 \le \beta < \Omega(n) - 1$, we have that

$$|crd(n,\beta)| \le \mathcal{O}(|I_\beta^n|^{1+r}) , \tag{A.90}$$

then, for every closed interval K contained in the interior of I, the homeomorphism $h|K$ restricted to K is C^{2+r}.

Proof. Proof of statement (i): Let h be C^{2+r} and let $\psi : I \to \mathbb{R}$ be given by $\psi(x) = \log dh(x)$. For every $n \ge 1$, let $I_\gamma^n = [x,y]$, $I_{\gamma+1}^n = [y,z]$ and $I_{\gamma+1}^n = [z,w]$ be adjacent grid intervals, at level n. By Taylor series, we get

$$\begin{aligned} |h(I_\gamma^n)| &\in |I_\gamma^n| dh(y) + |I_\gamma^n|^2 d^2h(y) \pm \mathcal{O}(|I_\gamma^n|^{2+r}) \\ |h(I_{\gamma+1}^n)| &\in |I_{\gamma+1}^n| dh(y) - |I_{\gamma+1}^n|^2 d^2h(y) \pm \mathcal{O}(|I_{\gamma+1}^n|^{2+r}) \\ |h(I_{\gamma+1}^n)| &\in |I_{\gamma+1}^n| dh(z) + |I_{\gamma+1}^n|^2 d^2h(z) \pm \mathcal{O}(|I_{\gamma+1}^n|^{2+r}) \\ |h(I_{\gamma+2}^n)| &\in |I_{\gamma+2}^n| dh(z) - |I_{\gamma+2}^n|^2 d^2h(z) \pm \mathcal{O}(|I_{\gamma+2}^n|^{2+r}) . \end{aligned}$$

Therefore,

$$\frac{|h(I^n_{\gamma+1})|}{|I^n_{\gamma+1}|}\frac{|I^n_\gamma|}{|h(I^n_\gamma)|} \in \frac{dh(y) - |I^n_{\gamma+1}|d^2h(y) \pm \mathcal{O}(|I^n_{\gamma+1}|^{1+r})}{dh(y) + |I^n_\gamma|d^2h(y) \pm \mathcal{O}(|I^n_\gamma|^{1+r})}$$
$$\in 1 - (|I^n_\gamma| + |I^n_{\gamma+1}|)\frac{d\psi(y)}{2} \pm \mathcal{O}((|I^n_\gamma| + |I^n_{\gamma+1}|)^r) ,$$

and so

$$lrd(n,\gamma) \in -(|I^n_\gamma| + |I^n_{\gamma+1}|)\frac{d\psi(y)}{2} \pm \mathcal{O}((|I^n_\gamma| + |I^n_{\gamma+1}|)^r) .$$

Similarly, we get

$$lrd(n,\gamma+1) \in -(|I^n_{\gamma+1}| + |I^n_{\gamma+2}|)\frac{d\psi(z)}{2} \pm \mathcal{O}((|I^n_{\gamma+1}| + |I^n_{\gamma+2}|)^r).$$

Therefore, by inequality (A.9), the cross-ratio distortion $c(n,\gamma) \in \pm\mathcal{O}(|I^n_\gamma|^r)$.

Proof of statement (ii). We prove statement (ii), first in the case where $0 < r < 1$ and secondly in the case where $r = 1$.
Case $0 < r < 1$. By Lemma A.12, for every $0 < s < 1$, the homeomorphism $h|K$ is C^{1+s}, and so the map $\psi : I \to \mathbb{R}$ is well-defined by $\psi(x) = \log dh(x)$. For every point $P \in I$, let $I^1_{\alpha_1}, I^2_{\alpha_2}, \ldots$ be a sequence of grid intervals $I^n_{\alpha_n}$ such that $P \in I^n_{\alpha_n}$ and $I^n_{\alpha_n} \subset I^{n-1}_{\alpha_{n-1}}$ for every $n > 1$. By the bounded geometry property of a grid and by inequality (A.90), for every grid interval $I^n_\beta \subset \cup_{i=-1,0,1} I^{n-1}_{\alpha_{n-1}+i}$, we have that

$$|crd(n,\beta)| \le \mathcal{O}(|I^n_{\alpha_n}|^{1+r}) . \tag{A.91}$$

By inequality (A.87), we have

$$\frac{lrd(n-1,\alpha_{n-1})}{|I^{n-1}_{\alpha_{n-1}}| + |I^{n-1}_{\alpha_{n-1}+1}|} = \frac{lrd(n,\alpha_n)}{|I^n_{\alpha_n}| + |I^n_{\alpha_n+1}|} \pm \mathcal{O}(|I^n_{\alpha_n}|^r) .$$

Hence, by the bounded geometry property of a grid, for every $m \ge 1$ and for every $n \ge m$, we get that

$$\frac{lrd(n,\alpha_n)}{|I^n_{\alpha_n}| + |I^n_{\alpha_n+1}|} = \frac{lrd(m,\alpha_m)}{|I^m_{\alpha_m}| + |I^m_{\alpha_m+1}|} \pm \mathcal{O}(|I^m_{\alpha_m}|^r) . \tag{A.92}$$

Thus, $lrd(n,\alpha_n)/|I^n_{\alpha_n}| + |I^n_{\alpha_n+1}|$ converges to a value d_P, when n tends to infinity. Let us prove that ψ is differentiable at P and that $d\psi(P) = 2d_P$. Let $L = [x,y]$ be any interval such that the point $P \in L$. Take the largest $m \ge 1$ such that there is a grid interval I^m_γ with the property that $L \subset \cup_{j=-1,0,1} I^m_{\gamma+j}$. By the bounded geometry property of a grid, there is $C \ge 1$, not depending upon P, L and I^m_γ, such that

$$C^{-1} \leq \frac{|I_\gamma^m|}{|L|} \leq C\ . \tag{A.93}$$

For every $n \geq m$, take the smallest sequence of adjacent grid intervals $I^n_{\beta_n}, \ldots, I^n_{\beta_n+i_n}$, at level n, such that $L \subset \cup_{i=0}^{i_n} I^n_{\beta_n+i} \subset \cup_{j=-1,0,1} I^m_{\gamma+j}$. Hence, by definition of the logarithmic ratio distortion, we get

$$\psi(x) = \lim_{n\to\infty} ldh(I^n_{\beta_n})$$

and

$$\psi(y) = \lim_{n\to\infty} ldh(I^n_{\beta_n+i_n})\ .$$

Therefore,

$$\begin{aligned}\frac{\psi(y)-\psi(x)}{y-x} &= \lim_{n\to\infty} \frac{ldh(I^n_{\beta_n+i_n}) - ldh(I^n_{\beta_n})}{y-x} \\ &= \lim_{n\to\infty} \frac{\sum_{i=0}^{i_n-1} lrd(I^n_{\beta_n+i})}{y-x}\ .\end{aligned} \tag{A.94}$$

By inequalities (A.92) and (A.93), for every $I^n_{\beta_n+i} \subset I^m_{\gamma+j(i)}$,we get

$$\begin{aligned} lrd(n,\beta_n+i) &= \left(|I^n_{\beta_n+i}| + |I^n_{\beta_n+i+1}|\right)\left(\frac{lrd(m,\gamma+j(i))}{|I^m_{\gamma+j(i)}| + |I^m_{\gamma+j(i)+1}|} \pm \mathcal{O}(|I^m_{\gamma+j(i)}|^r)\right) \\ &= \left(|I^n_{\beta_n+i}| + |I^n_{\beta_n+i+1}|\right)(d_P \pm \mathcal{O}(|L|^r))\ .\end{aligned} \tag{A.95}$$

Putting together (A.94) and (A.95), we obtain that

$$\begin{aligned}\frac{\psi(y)-\psi(x)}{y-x} &= \lim_{n\to\infty} (d_P \pm \mathcal{O}(|L|^r)) \frac{\sum_{i=0}^{i_n-1} |I^n_{\beta_n+i}| + |I^n_{\beta_n+i+1}|}{y-x} \\ &= \lim_{n\to\infty} (d_P \pm \mathcal{O}(|L|^r)) \frac{|I^n_{\beta_n}| + |I^n_{\beta_n+i_n}| + 2\sum_{i=1}^{i_n-1} |I^n_{\beta_n+i}|}{y-x} \\ &= 2d_P \pm \mathcal{O}(|L|^r)\ .\end{aligned} \tag{A.96}$$

Therefore, for every $P \in I$, the homeomorphism ψ is differentiable at P and $d\psi(P) = 2d_P$. Let us check that $d\psi$ is r-Hölder continuous. For every $P, P' \in I$, let L be the closed interval $[P, P']$. Using (A.96), we obtain that

$$\begin{aligned} d\psi(P') - d\psi(P) &= \frac{\psi(P')-\psi(P)}{P'-P} - \frac{\psi(P')-\psi(P)}{P'-P} \pm \mathcal{O}(|L|^r) \\ &= \pm\mathcal{O}(|L|^r)\ ,\end{aligned}$$

and so $d\psi$ is r-Hölder continuous.

Case $r = 1$. By the above argument, h is C^{2+s} for every $0 < s < 1$ and so, in particular, h is $C^{1+Lipschitz}$. Thus, by Lemma A.11, for every $n \geq 1$ and every $1 \leq \beta \leq \Omega(n) - 1$ we get that

$$|lrd(n,\beta)| \leq \mathcal{O}(|I^n_\beta|) \; ,$$

which implies that inequality (A.87) is also satisfied for $r = 1$. Now, a similar argument to the one above gives that $d\psi$ is Lipschitz. □

A.8 Cross-ratio distortion and smoothness

In this section, we prove the following result.

Theorem A.15. *Let $h : I \to J$ be a homeomorphism between two compact intervals I and J on the real line, and let $\mathcal{G}_\Omega$ be a grid of I.*

(i) If h has the degree of smoothness presented in a line of Table 2, and $dh(x) \neq 0$ for all $x \in I$ (not applicable for quasisymmetric and (uaa) homeomorphisms), then the logarithmic ratio distortion satisfies the bounds presented in the same line with respect to all grid intervals. Conversely, if the logarithmic ratio distortion satisfies the bounds presented in a line of Table 2 with respect to all grid intervals, then $h : I \to J$ has the degree of smoothness presented in the same line, and $dh(x) \neq 0$ for all $x \in I$ (not applicable for quasisymmetric and (uaa) homeomorphisms).

The smoothness of h	*The order of* $lrd\left(I^n_\beta, I^n_{\beta+1}\right)$
Quasisymmetric	$\mathcal{O}(1)$
(uaa)	$o\left(\left\|I^n_\beta\right\|\right)\left\|I^n_\beta\right\|^{-1}$
$C^{1+\alpha}$	$\mathcal{O}\left(\left\|I^n_\beta\right\|^{\alpha}\right)$
$C^{1+Lipschitz}$	$\mathcal{O}\left(\left\|I^n_\beta\right\|\right)$
Affine	$o\left(\left\|I^n_\beta\right\|\right)$

Table 2.

(ii) If h has the degree of smoothness presented in a line of Table 3, and $dh(x) \neq 0$ for all $x \in I$ (not applicable for quasisymmetric and (uaa) homeomorphisms), then the cross-ratio distortion satisfies the bounds presented in the same line with respect to all grid intervals. Conversely, if the cross-ratio distortion satisfies the bounds presented in a line of Table 3 with respect to all grid intervals, then, for every closed interval K contained in the interior of I, the homeomorphism $h|K$ restricted to K has the degree of smoothness presented in the same line, and $dh(x) \neq 0$ for all $x \in I$ (not applicable for quasisymmetric and (uaa) homeomorphisms).

The smoothness of h	*The order of* $crd\left(I^n_\beta, I^n_{\beta+1}, I^n_{\beta+2}\right)$
Quasisymmetric	$\mathcal{O}(1)$
(uaa)	$o\left(\left\|I^n_\beta\right\|\right)\left\|I^n_\beta\right\|^{-1}$
$C^{1+\alpha}$	$\mathcal{O}\left(\left\|I^n_\beta\right\|^{\alpha}\right)$
$C^{2+\alpha}$	$\mathcal{O}\left(\left\|I^n_\beta\right\|^{1+\alpha}\right)$
$C^{2+Lipschitz}$	$\mathcal{O}\left(\left\|I^n_\beta\right\|^{2}\right)$

Table 3.

We point out that some of the difficulties and usefulness of these results come from the fact that (i) we just compute the bounds of the ratio and cross- ratio distortions with respect to a countable set of intervals fixed by a grid, and (ii) we do not restrict the grid intervals, at the same level, to have necessarily the same lengths. In hyperbolic dynamics, these grids are naturally determined by Markov partitions.

Proof of Theorem A.15. The equivalences presented for quasisymmetric homeomorphisms follow from Lemma A.2 with respect to ratio distortion and from Lemma A.4 with respect to cross-ratio distortion, noting that the ratios $r(n,\beta)$ and the cross-ratios $cr(n,\beta)$ are uniformly bounded by the bounded geometry property of the grid. The equivalences presented for uniformly asymptotically affine (uaa) homeomorphisms follow from Lemma A.8 with respect to ratio distortion and from Lemma A.10 with respect to cross- ratio distortion. The equivalences presented for $C^{1+\alpha}$, $C^{1+Lipschitz}$ and affine diffeomorphisms follow from Lemma A.11 with respect to ratio distortion and from Lemma A.12 with respect to cross-ratio distortion. The equivalences presented for $C^{2+\alpha}$ and $C^{2+Lipschitz}$ diffeomorphisms follow from Lemma A.14. □

A.9 Further literature

The quasisymmetric homeomorphisms of the real line extend to quasiconformal homeomorphisms of the upper half-plane, by the Beurling-Ahlfors extension theorem (see Ahlfors and Beurling [3]). The uniformly asymptotically affine (uaa) (or, equivalently, symmetric) homeomorphisms are the boundary values of quasiconformal homeomorphisms of the upper half-plane whose conformal distortion tends to zero at the boundary (see Gardiner and Sullivan [42]). (Uaa) homeomorphisms turn out to be precisely those homeomorphisms which have boundary dilatation equal to one, in the sense of Strebel [216]. The (uaa) homeomorphisms of a circle comprise the closure, in the quasisymmetric topology, of the real-analytic homeomorphisms and this closure contains the set of C^1 diffeomorphisms (see Gardiner and Sullivan [42]). Furthermore, any two C^r expanding circle maps conjugated by a (uaa) homeomorphism are C^r conjugated (see Ferreira and Pinto [38]). In Gardiner and Sullivan [42], Jacobson and Swiatek [51], de Melo and van Strien [99], Pinto and Rand [158] and Pinto and Sullivan [175] other relations are also presented between distinct degrees of smoothness of a homeomorphism of the real line with distinct bounds of ratio and cross-ratio distortions of intervals. This chapter is based on Pinto and Sullivan [175].

B

Appendix B: Classifying C^{1+} structures on Cantor sets

We present a classification of $C^{1+\alpha}$ structures on trees embedded in the real line. This is an extension of the results of Sullivan on embeddings of the binary tree to trees with arbitrary topology and to embeddings without bounded geometry and with contact points.

B.1 Smooth structures on trees

A *tree* consists of a set of vertices of the form $V_T = \cup_{n\geq 0} T_n$, where each T_n is a finite set, together with a directed graph on these vertices such that each $t \in T_n$, $n \geq 1$, has a unique edge leaving it. This edge joins t (the *daughter*) to $m(t) \in T_{n-1}$ (its *mother*). We inductively define $m^p(t) \in T_{n-p}$. We call $m^p(t)$ the *p-ancestor* of t.

Given a tree T, we define the *limit set* or *set of ends* L_T as the set of all sequences $\underline{t} = t_0 t_1 \ldots$ such that $m(t_{i+1}) = t_i$, for all $i \geq 0$. We endow L_T with the metric d where

$$d(s_0 s_1 \ldots, t_0 t_1 \ldots) = 2^{-n},$$

if $s_i = t_i$, for all $0 \leq i \leq n-1$ and $s_n \neq t_n$.

If $\underline{t} = t_0 t_1 \ldots \in L_T$, then by $\underline{t}|n$ we denote the finite words $t_0 \ldots t_{n-1}$. Let $L_{\underline{t}|n}$ denote the set of $\underline{s} \in L_T$ such that $\underline{s}|n = \underline{t}|n$. This is called an *n-cylinder* of the tree. If L is an open subset of L_T containing $L_{\underline{t}|n}$ and $i : L \to \mathbb{R}$ a continuous mapping, then we denote by $C_{\underline{t}|n,i}$ the smallest closed interval in $\mathbb{R}$ that contains $i(L_{\underline{t}|n})$. This is also called an *n-cylinder.* Note that both $L_{\underline{t}|n}$ and $C_{\underline{t}|n,i}$ are determined by t_{n-1}. Therefore, we shall often write these as $L_{t_{n-1}}$ and $C_{t_{n-1},i}$. Say that $\underline{s} \sim \underline{t}$, if $i(\underline{s}) = i(\underline{t})$.

We shall only be interested in mappings i that respect the cylinder structure of L_T in the following way. We demand that if $\underline{s}|n \neq \underline{t}|n$, then

$$\text{int}C_{\underline{s}|n,i} \cap \text{int}C_{\underline{t}|n,i} = \emptyset.$$

The mapping $i : L \to \mathbb{R}$ induces a mapping $L/\sim \to \mathbb{R}$ that we also denote by i.

Definition B.1. *Such a pair (i, L) is a chart of L_T, if L is an open set of L_T with respect to the metric d and the induced map $i : L/\sim \to \mathbb{R}$ is an embedding.*

Two charts (i, L) and (j, K) are *compatible*, if the equivalence relation $\sim$ corresponding to i agrees with that of j on $L \cap K$. They are $C^{1+\alpha}$ *compatible*, if they are compatible and the mapping $j \circ i^{-1}$ from $i(L \cap K)$ to $j(L \cap K)$ has a $C^{1+\alpha}$ extension to a neighbourhood of $i(L \cap K)$ in $\mathbb{R}$.

Definition B.2. *A* structure *on L_T is a set of compatible charts that cover L_T. A $C^{1+\alpha}$* structure *on L_T is a structure such that the charts are $C^{1+\alpha}$ compatible charts that cover L_T. A $C^{1+\alpha^-}$ structure is a structure such that the charts $C^{1+\beta}$ compatible, for all $0 < \beta < \alpha$.*

A finite set of $C^{1+\alpha}$ compatible charts that cover L_T defines a $C^{1+\alpha}$ structure on L_T. Suppose L_T has a smooth structure. We say that $h : L_T \to L_T$ is it structure preserving, if for all charts (i, L) and (i', L') of the structure whenever $t \in L$ and $h(t) \in L'$, then the chart $(i' \circ h, L)$ is compatible with (i, L). Then, we say that a structure-preserving map $h : L_T \to L_T$ is smooth if its representatives in local charts are smooth in the following sense: if $\underline{t} \in L$ and $h(\underline{t}) \in L'$, where (i, L) and (i', L') are charts in the structure, then $i' \circ h \circ i^{-1}$ has a smooth extension to a neighbourhood of $i(\underline{t})$ in $\mathbb{R}$. Similarly, we define smooth maps between different spaces.

Remark B.3. We shall mostly be concerned with situations where either (i) the smooth structure is defined by a single chart or (ii) the structure is defined by a single embedding of $L_T/\sim$ into the circle or into a train-track.

If $\mathcal{S}$ is a $C^{1+\alpha}$ structure on L_T and i is a chart of $\mathcal{S}$, then we say that $\underline{s}|n$ and $\underline{t}|n$ are *adjacent*, if there is no $\underline{u} \in L_T$ such that $C_{\underline{u},i}$ lies between $C_{\underline{s}|n,i}$ and $C_{\underline{t}|n,i}$ and that they are *in contact*, if $C_{\underline{s}|n,i} \cap C_{\underline{t}|n,i} \neq \emptyset$. Note that this conditions are independent of the choice of the chart i of $\mathcal{S}$ that contains $L_{\underline{s}|n}$ and $L_{\underline{t}|n}$ in its domain. It does however depend upon $\mathcal{S}$, so we only use this terminology when we have a specific structure in mind. If $\underline{s}|n = s_0 \dots s_{n-1}$ and $\underline{t}|n = t_0 \dots t_{n-1}$, then we say that s_{n-1} and t_{n-1} are *adjacent* (resp. *in contact*), if $\underline{s}|n$ and $\underline{t}|n$ are.

Definition B.4. *Two $C^{1+\alpha}$ structures $\mathcal{S}$ and $\mathcal{T}$ on L_T are $C^{1+\alpha}$*-equivalent, *if the identity is a $C^{1+\alpha}$ diffeomorphism when it is considered as a map from L_T with one structure to L_T with the other. They are $C^{1+\alpha^-}$*-equivalent, *if the identity is a $C^{1+\beta}$ diffeomorphism, for all $0 < \beta < \alpha$.*

B.1.1 Examples

Standard binary Cantor set

Consider the binary tree T shown in Figure B.1. We can index the vertices of the tree by the finite words $\varepsilon_0 \dots \varepsilon_{n-1}$ of 0s and 1s in such way that the

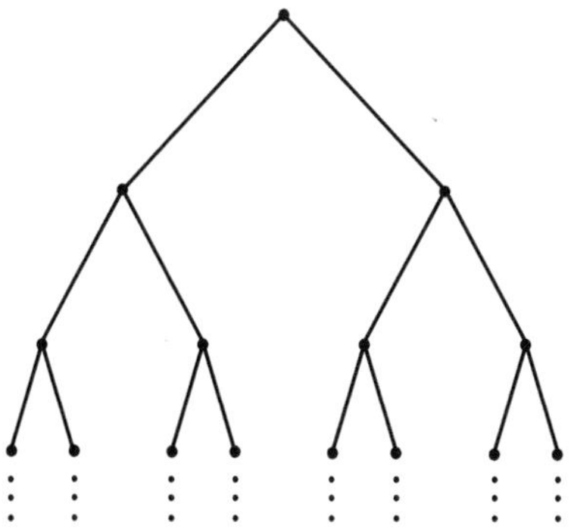

Fig. B.1. A binary tree.

mother of the vertex $t = \varepsilon_0 \dots \varepsilon_n$ is $m(t) = \varepsilon_0 \dots \varepsilon_{n-1}$ and so that $\varepsilon_0 \dots \varepsilon_{n-1}0$ lies to the left of $\varepsilon_0 \dots \varepsilon_{n-1}1$. Now, to each vertex $t = \varepsilon_0 \dots \varepsilon_{n-1}$ associate a closed interval I_t so that $I_t \subset I_{m(t)}$, $I_{\varepsilon_0 \dots \varepsilon_{n-1}0}$ is to the left of $I_{\varepsilon_0 \dots \varepsilon_{n-1}1}$ and

$$I_{\varepsilon_0 \dots \varepsilon_{n-1}} = I_{\varepsilon_0 \dots \varepsilon_{n-1}0} \cup G_{\varepsilon_0 \dots \varepsilon_{n-1}} \cup I_{\varepsilon_0 \dots \varepsilon_{n-1}1},$$

where $G_{\varepsilon_0 \dots \varepsilon_{n-1}}$ is an open interval between $I_{\varepsilon_0 \dots \varepsilon_{n-1}0}$ and $I_{\varepsilon_0 \dots \varepsilon_{n-1}1}$. We assume that the ratios $|G_t|/|I_t|$ are bounded away from 0. Then, the lengths of the intervals $I_{\varepsilon_0 \dots \varepsilon_{n-1}}$ go to 0 exponentially fast as $n \to \infty$, and therefore

$$C = \cap_{n \geq 0} \cup_{\varepsilon_0 \dots \varepsilon_{n-1}} I_{\varepsilon_0 \dots \varepsilon_{n-1}}$$

is a Cantor set.

Let $\Sigma = \{0,1\}^{\mathbb{Z}_{\geq 0}}$ denote the set of infinite right-handed words $\varepsilon_0\varepsilon_1 \dots$ of 0s and 1s. Clearly, L_T can be identified with Σ, since each $\underline{t} = t_0 t_1 \dots \in L_T$ can be identified with a word $\varepsilon_0\varepsilon_1 \dots$ in Σ. The mapping $i : \Sigma \to \mathbb{R}$ defined by

$$i(\varepsilon_0\varepsilon_1 \dots) = \cap_{n \geq 0} I_{\varepsilon_0 \dots \varepsilon_{n-1}}$$

gives an embedding of L_T into $\mathbb{R}$. This is the simplest non-trivial example of an embedded tree. We shall be interested in embedded trees such as this where the analogue of the Cantor set C is generated in one way or another by a dynamical system.

Very often, the set $C = i(L_T)$ will be an invariant set of a hyperbolic dynamical system. For example, there is a map σ defined on L_T above by

$$\sigma(\varepsilon_0\varepsilon_1 \dots) = \varepsilon_1\varepsilon_2 \dots .$$

This induces a map σ' on $C = i(L_T)$ that is candidate for a hyperbolic system. Using our results, we give necessary and sufficient for this map to be smooth in the sense that it has a $C^{1+\alpha}$ extension to $\mathbb{R}$ as a Markov map such as that shown in Figure B.2.

In the above case, the equivalence relation $\sim$ is trivial and there are no contact points. But now consider the case where the tree is embedded in

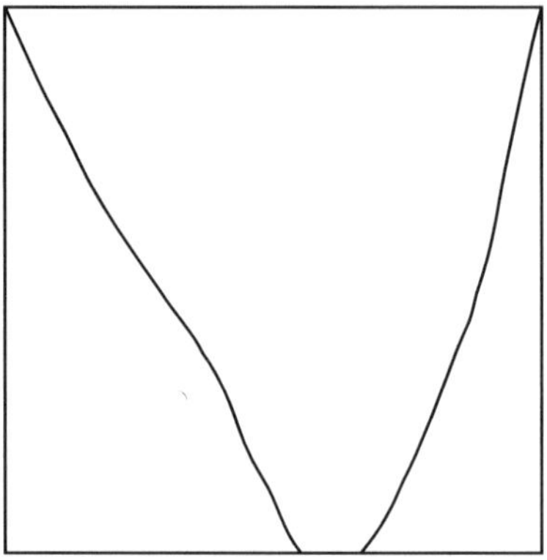

Fig. B.2. A cookie-cutter.

this way but where the gaps G_t are empty. In this case, i maps L_T onto an interval but is not an embedding, because it is not injective. The equivalence relation $\sim$ on L_T is non-trivial: it identifies the points $\varepsilon_0 \ldots \varepsilon_n 1000\ldots$ and $\varepsilon_0 \ldots \varepsilon_n 0111\ldots$. Thus, h is injective on all but a countable set. The space $L_T/\sim$ is homeomorphic to an interval. However, note that L_T has much more structure than an interval, because of the points marked by the cylinder structure. In particular, there are uncountable many smooth structures on L_T, but only one on the interval.

We could regard the vertex set of T as $\cup_{n\geq 0} T_n$, where T_n is the set of intervals $I_{\varepsilon_0\ldots\varepsilon_{n-1}}$ and the edge relation of T is inclusion. In such a case, we say that T is defined by the cylinder structure.

Rotations of the circle

This is another example with contact points. Consider the rotation $R_\alpha(x) = x+\alpha$, where α is an irrational number such that $0<\alpha<1$, represented as the discontinuous mapping

$$R_\alpha = \begin{cases} x+\alpha, & x\in[\alpha-1,0] \\ x+\alpha-1, & x\in[0,\alpha] \end{cases}$$

Let p_n/q_n be the nth rational approximant of α. Consider the orbit $R_\alpha(0), \ldots, R_{(q_n-1)\alpha}(0)$. This partitions the interval $[\alpha-1,\alpha]$ into q_n+1 closed intervals. Let T_n denote the set of such intervals and let T be the tree whose vertex set is $\cup_{n\geq 0} T_n$ and such that the mother of $v \in T_n$ is the interval T_{n-1} that contains v. Thus, T is again defined by the cylinder structure. If $t_o t_1 \ldots \in L_T$, then $i(t_o t_1\ldots) = \cap_{n\geq 0} t_n$ defines an embedding of T with contact points.

Of course, any map that is topologically conjugate to R_α would generate the tree T, but a different embedding. The question of determining whether two such mappings are smoothly conjugate boils down to showing that these embeddings determine the same smooth structure on L_T. The approach used in the theory of renormalization is to show that this tree T can be generated by a Markov family $(F_n)_{n\in\mathbb{Z}_{\geq 0}}$ as defined in Rand [189]. This Markov family and its convergence properties determine the $C^{k+\alpha}$ structure on L_T as is proved in Pinto and Rand [157].

B.2 Basic definitions

We start by introducing some basic definitions.

Gaps

Fix a $C^{1+\alpha^-}$ structure $\mathcal{S}$ on L_T. If s and t are adjacent but not in contact, then there is a *gap* between $i(L_s)$ and $i(L_t)$. We will add a symbol $g_{s,t} = g_{t,s}$ to T_n to stand for this gap, if $m(s) = m(t)$. For the chart (i, L), we let $G_{s,t,i}$ denote the smallest closed interval containing the gap. Let $\tilde{T}_n$ denote the set T_n with all the gap symbols $g_{s,t}$ adjoined. Let $\tilde{V}_T = \cup_{n\geq 1}\tilde{T}_n$. If $m^p(s) = m^p(t)$, then $G_{s,t,i} = G_{m^{p-1}(s),m^{p-1}(t),i}$.

Primary atlas

Suppose that $\mathcal{S}$ is a $C^{1+\alpha^-}$ structure on L_T. Then, clearly there exists $N \geq 0$ such that if $T_N = \{t_1, \ldots, t_q\}$, then there are charts (i_j, U_j) of $\mathcal{S}$, $1 \leq j \leq q$, such that the open subset U_j contains the N-cylinder L_{t_j}. We call such a system of charts a *primary* $N_{\mathcal{I}}$ *atlas* $\mathcal{I}$ with $N_{\mathcal{I}} = N$.

Fix such a primary $N_{\mathcal{I}}$ atlas $\mathcal{I} = \{(i_j, U_j)\}_{j=1,\ldots,q}$. Define $C_{t,\mathcal{I}}$ as the interval C_{t,i_j}, where j is such that $m^r(t) = t_j$, for some $r \geq 1$. Similarly, define $G_{s,t,\mathcal{I}}$ as the gap G_{s,t,i_j}, if s and t are non-contact adjacent points with $m(s) = m(t)$.

If $t, s \in T_n$ are adjacent and in contact, define the scalar

$$d_{s,t,\mathcal{I}} = \frac{1}{2}(|C_{t,\mathcal{I}}| + |C_{s,\mathcal{I}}|).$$

If $t, s \in T_n$ are adjacent but not in contact, let t_2 be the vertex such that $G_{t,s,\mathcal{I}} \subset C_{m(t_2),\mathcal{I}}$ but $G_{t,s,\mathcal{I}}$ is not contained in $C_{t_2,\mathcal{I}}$. If $C_t \neq C_{m(t)}$, then define $t_1 = t$. Otherwise, let t_1 be a descendent of t such that $C_{t_1,\mathcal{I}}$ is adjacent to $G_{t,s,\mathcal{I}}$, $C_{t_1,\mathcal{I}} \neq C_{t,\mathcal{I}}$ but $C_{m(t_1),\mathcal{I}} = C_{t,\mathcal{I}}$. Define the scalar

$$\delta_{t,s,\mathcal{I}} = \frac{1}{2}|C_{t_1,\mathcal{I}}|\frac{|G_{t,s,\mathcal{I}}|}{|C_{t_1,\mathcal{I}}|}.$$

Let t', s' be the vertices such that $m(t') = t$ and $m(s') = s$. Define the scalar

$$e_{t,s,\mathcal{I}} = \delta_{t,s,\mathcal{I}} - \delta_{t',s',\mathcal{I}}.$$

If $t \in T_n$ is in contact, let the connected set $\overline{C}_{t,\mathcal{I}}$ be the union of n cylinders and gaps containing $C_{t,\mathcal{I}}$. The number of n-cylinders and gaps contained in $\overline{C}_{t,\mathcal{I}}$ is bounded independently of t and n.

Scaling tree

(i) The scaling tree $\sigma_{\mathcal{I}}(t)$:

$$\sigma_{\mathcal{I}}(t) = \frac{|C_{t,\mathcal{I}}|}{|C_{m(t),\mathcal{I}}|} \quad \text{and} \quad \sigma_{\mathcal{I}}(g_{t,s}) = \frac{|G_{t,s,\mathcal{I}}|}{|C_{m(t),\mathcal{I}}|}.$$

This defines a function

$$\sigma_{\mathcal{I}} : \bigcup_{n \geq N_{\mathcal{I},\mathcal{J}}} \tilde{T}_n \to [0,1].$$

The fact that it is not necessarily defined for small n is not important.

Ratios distortions

Now, suppose that in addition to the structure $\mathcal{S}$ and its primary $N_{\mathcal{I}}$-atlas $\mathcal{I}$, we have another structure $\mathcal{T}$ and a primary $N_{\mathcal{J}}$-atlas $\mathcal{J}$ for it. Redefine $N_{\mathcal{I},\mathcal{J}} = \max(N_{\mathcal{I}}, N_{\mathcal{J}}) + 1$. To each $t \in \tilde{T}_n$, $n \geq N_{\mathcal{J},\mathcal{J}}$, we associate the following ratios:

(ii) ν_t:

$$\nu_t = \left| 1 - \frac{\sigma_{\mathcal{J}}(t)}{\sigma_{\mathcal{I}}(t)} \right|.$$

(iii) $\nu_{t,s}$: If $t, s \in T_n$ are in contact,

$$\nu_t = \left| 1 - \frac{|C_{t,\mathcal{I}}|}{|C_{s,\mathcal{I}}|} \frac{|C_{s,\mathcal{J}}|}{|C_{t,\mathcal{J}}|} \right|.$$

B.3 $(1+\alpha)$-contact equivalence

Let $\mathcal{S}$ and $\mathcal{T}$ be $C^{1+\alpha^-}$ structures on $L_{\mathcal{T}}$ and let $\mathcal{I}$ (resp. $\mathcal{J}$) be a primary atlas for $\mathcal{S}$ (resp. $\mathcal{T}$). We are going to prove that a sufficient condition for $\mathcal{S}$ and $\mathcal{T}$ to be $C^{1+\alpha^-}$-equivalent is that $\mathcal{I} \overset{\alpha}{\sim} \mathcal{J}$. It is sufficient to prove it locally at each point $\underline{t} \in L_{\mathcal{T}}$. Let $i : U_0 \to \mathbb{R}$ be a chart in $\mathcal{I}$ and $j : V_0 \to \mathbb{R}$ be a chart in $\mathcal{J}$ with $\underline{t} \in U_0 \cap V_0$. Then, it suffices to show that, for some open subsets U and V of $U_0 \cap V_0$ containing $\underline{t}$, the mapping $j \circ i^{-1} : i(U) \to j(V)$ has a $C^{1+\alpha^-}$ extension to $\mathbb{R}$. If this is the case, for all such $\underline{t}$, then the result holds globally. We can restrict our analysis to the case where

(i) the smallest closed interval $\mathbf{I}$ containing $i(U)$ is a cylinder $C_{t,i}$, for some $t \in T_{N_0}$, where $N_0 > N_{\mathcal{I},\mathcal{J}}$ or else is the union of two adjacent cylinders of this form that are in contact; and
(ii) where the smallest closed interval $\mathbf{J}$ containing $j(V)$ consists of the corresponding cylinders for j.

Now, let $\mathbf{I}^n$ (resp. $\mathbf{J}^n$) be the set of endpoints of the cylinders $C_{t,i}$ (resp. $C_{t,j}$), where $t \in T_n$, $n \geq N_0$ and $C_{t,i} \subset \mathbf{I}$ (resp. $C_{t,j} \subset \mathbf{J}$). Then, $j \circ i^{-1}$ maps $\mathbf{I}^n$ onto $\mathbf{J}^n$ and is a homeomorphism of the closure I^∞ of $\cup_{n \geq N_0} I^n$ onto the closure $\mathbf{J}^\infty$ of $\cup_{n \geq N_0} \mathbf{J}^n$. We will construct a sequence of C^∞ mappings L_n such that

(i) L_n agrees with $j \circ i^{-1}$ on $\cup_{N_0 \leq j \leq n} \mathbf{I}^j$;
(ii) L_n is a Cauchy sequence in the space of $C^{1+\varepsilon}$ functions on $\mathbf{I}$, for all $\varepsilon < \alpha$, and, therefore, converges to a $C^{1+\alpha^-}$ function L_∞ on $\mathbf{I}$.
(iii) the mapping L_∞ gives the required smooth extension of $j \circ i^{-1}$ and proves the theorem.

B.3.1 $(1+\alpha)$ scale and contact equivalence

Define the scalar $A_{t,s,\mathcal{I}}$ as follows. Let $t, s \in T_n$ be adjacent vertices, not in contact, such that $m(t) = m(s)$. Define

$$\underline{\Delta}_t = \{z \in \tilde{T}_n : z < t \ \text{ and } \ m(z) = m(t)\}$$

and

$$\overline{\Delta}_t = \{z \in \tilde{T}_n : z > t \ \text{ and } \ m(z) = m(t)\}.$$

We now define the scalars $F_{t,s,\overline{\Delta},\mathcal{I}}$ and $F_{t,s,\underline{\Delta},\mathcal{I}}$. If $s \in \underline{\Delta}_t$, define the scalar $F_{t,s,\overline{\Delta},\mathcal{I}} = \delta_{t,s,\mathcal{I}}$, otherwise define $F_{t,s,\overline{\Delta},\mathcal{I}} = \delta_{t,s,\mathcal{I}} + |C_{t,\mathcal{I}}|$. Similarly, define $F_{t,s,\underline{\Delta},\mathcal{I}}$. Let

$$A_{t,s,\mathcal{I}} = \min_{\Delta \in \{\underline{\Delta}_t, \overline{\Delta}_t\}} \left\{ \sum_{z \in \Delta} \nu_z |C_{z,\mathcal{I}}| + \nu_t F_{t,s,\Delta,\mathcal{I}} \right\}.$$

Roughly, $A_{t,s,\mathcal{I}} - \nu_t F_{t,s,\Delta,\mathcal{I}}$ is given by the weighted average of the cylinder lengths $|C_{z,\mathcal{I}}|$ using weights ν_z.

Definition B.5. *We say that two such primary atlases $\mathcal{I}$ and $\mathcal{J}$ are $(1+\alpha)$-*scale equivalent, *if, for all $0 \leq \varepsilon < \alpha < 1$, there exists a decreasing function $f = f_\varepsilon : \mathbb{Z}_{\geq 0} \to \mathbb{R}$ with the following properties:*

(i) $\sum_{n=0}^{\infty} f(n) < \infty$;
(ii) for all $t \in \tilde{T}_n$, $\nu_t < f(n)$;
(iii) for all $s \in T_n$, adjacent to t but not in contact with it, and all $n > N_{\mathcal{I},\mathcal{J}}$, if $m(s) = m(t)$,

$$A_{t,s,\mathcal{I}} e_{t,s,\mathcal{I}}^{-(1+\varepsilon)} + \nu_t e_{t,s,\mathcal{I}}^{-\varepsilon} < f(n),$$

while if $m(s) \neq m(t)$ *and* $e_{t,s,\mathcal{I}} > 0$*, then*

$$\delta_{t,s,\mathcal{I}} \nu_t e_{t,s,\mathcal{I}}^{-(1+\varepsilon)} < f(n).$$

Definition B.6. *We say that two such primary atlases* $\mathcal{I}$ *and* $\mathcal{J}$ *are* $(1+\alpha)$-contact equivalent, *if, for all* ε *such that* $0 \leq \varepsilon < \alpha < 1$*, there exists a decreasing function* $f = f_\varepsilon : \mathbb{Z}_{\geq 0} \to \mathbb{R}$ *with the following properties:*

(i) $\sum_{n=0}^{\infty} f(n) < \infty$*;*
(ii) for all $t, s \in T_n$*,* $n > N_{\mathcal{I},\mathcal{J}}$ *such that* t *and* s *are in contact,*

$$\nu_{t,s} d_{t,s,\mathcal{I}}^{-\varepsilon} < f(n) \quad \text{and} \quad \nu_t |\overline{C}_{t,\mathcal{I}}^{-\varepsilon} < f(n).$$

By condition (ii) of the Definition B.5, for all $t \in \tilde{T}_n$, $\mathcal{O}(|C_{t,\mathcal{I}}|) = \mathcal{O}(|C_{t,\mathcal{J}}|)$, as easily proven in Lemma B.8. Therefore, Definitions B.5 and B.6 are symmetric in $\mathcal{I}$ and $\mathcal{J}$.

Definition B.7. *We say that two such primary atlases* $\mathcal{I}$ *and* $\mathcal{J}$ *are* $(1+\alpha)$-equivalent *(*$\mathcal{I} \overset{\alpha}{\sim} \mathcal{J}$*), if they are* $(1+\alpha)$*-scale equivalent and* $(1+\alpha)$*-contact equivalent.*

B.3.2 A refinement of the equivalence property

Lemma B.8. $|C_{t,\mathcal{I}}|/|C_{t,\mathcal{J}}|$ *is bounded away from 0 and* ∞*, i.e.* $|C_{t,\mathcal{I}}|/|C_{t,\mathcal{J}}| = \mathcal{O}_t(1)$.

Proof. For all $\underline{t} = t_0 t_1 \ldots \in L_T$, define $Q(t_j) = \ln(|C_{t_j,\mathcal{I}}|/|C_{t_j,\mathcal{J}}|)$, for all $j \geq 0$. By definition of ν_t, $|Q(t_{j-1}) - Q(t_j)| \leq \mathcal{O}(\nu_{t_j})$. By the $(1+\alpha)$-scale equivalence,

$$|Q(t_{N_{\mathcal{I},\mathcal{J}}}) - Q(t_n)| \leq \mathcal{O}\left(\sum_{j=N+1}^{n} \nu_{t_j}\right) < c_1,$$

for some constant c_1. As the set $T_{N_{\mathcal{I},\mathcal{J}}}$ is finite, $|Q(t_n)|$ is bounded above, independently of n and t_n. □

Corollary B.9. *If* $t \in T_n$*,* $n \geq N_{\mathcal{I},\mathcal{J}}$*,*

$$\left| \frac{|C_{t,\mathcal{J}}|}{|C_{t,\mathcal{I}}|} - \frac{|C_{m(t),\mathcal{J}}|}{|C_{m(t),\mathcal{I}}|} \right| \leq \mathcal{O}(\nu_t). \tag{B.1}$$

If $s, t \in T_n$ *are adjacent but not in contact and* $m(s) = m(t)$*, then*

$$\left| \frac{|G_{t,s,\mathcal{J}}|}{|G_{t,s,\mathcal{I}}|} - \frac{|C_{m(t),\mathcal{J}}|}{|C_{m(t),\mathcal{I}}|} \right| \leq \mathcal{O}(\nu_{g_{t,s}}). \tag{B.2}$$

If they are in contact, then

$$\left| \frac{|C_{t,\mathcal{J}}|}{|C_{t,\mathcal{I}}|} - \frac{|C_{s,\mathcal{J}}|}{|C_{s,\mathcal{I}}|} \right| \leq \mathcal{O}(\nu_{t,s}). \tag{B.3}$$

Proof. This follows directly from the definition of ν_t, $\nu_{g_{t,s}}$ and $\nu_{t,s}$ and the boundedness of $|C_{t,\mathcal{I}}|/|C_{t,\mathcal{J}}|$. □

B.3.3 The map L_t

For all $n \geq N_{\mathcal{I},\mathcal{J}}$, and all $t \in T_n$ with adjacent vertices s and r, define the map L_t as the affine map such that $L_t(P_{t,s,\mathcal{I}}) = P_{t,s,\mathcal{J}}$ and $L_t(P_{t,r,\mathcal{I}}) = P_{t,r,\mathcal{J}}$. Therefore, for $z \in \{s,r\}$,

$$L_t(x) = \frac{|C_{t,\mathcal{J}}|}{|C_{t,\mathcal{I}}|}(x - P_{t,z,\mathcal{I}}) + P_{t,z,\mathcal{J}}.$$

To each $s,t \in \tilde{T}_n$, $n \geq N_{\mathcal{I},\mathcal{J}}$, we associate the intervals $C_{t,s,\mathcal{I}}$, $D_{t,s,\mathcal{I}}$ and $E_{t,s,\mathcal{I}}$ that we will use in the construction of the sequence of C^∞ mappings L_n (see Figure B.3).

- $C_{s,t,\mathcal{I}}$, $C_{t,s,\mathcal{I}}$ and $D_{t,s,\mathcal{I}}$: If $t,s \in T_n$ are adjacent and in contact, define $P_{t,s,\mathcal{I}} = P_{s,t,\mathcal{I}}$ as the common point between the closed sets $C_{t,\mathcal{I}}$ and $C_{s,\mathcal{I}}$. Define the closed sets $C_{t,s,\mathcal{I}}$ and $C_{s,t,\mathcal{I}}$, respectively, as the sets obtained from $C_{t,\mathcal{I}}$ and from $C_{s,\mathcal{I}}$, by rescaling them by the factor $1/2$, keeping the point $P_{t,s,\mathcal{I}}$ fixed. Define $D_{t,s,\mathcal{I}} = C_{t,s,\mathcal{I}} \cup C_{s,t,\mathcal{I}}$. Note that $|D_{t,s,\mathcal{I}}| = d_{t,s,\mathcal{I}}$. If $t,s \in T_n$ are adjacent but not in contact, define $P_{t,s,\mathcal{I}}$ and $P_{s,t,\mathcal{I}}$, respectively, as the common points of the closed sets $C_{t,\mathcal{I}}$ and $C_{s,\mathcal{I}}$ with the gap $G_{t,s,\mathcal{I}}$. Define the closed sets $C_{t,s,\mathcal{I}}$ and $C_{s,t,\mathcal{I}}$, respectively, as the intervals contained into the gap $G_{t,s,\mathcal{I}}$, with endpoints $P_{t,s,\mathcal{I}}$ and $P_{s,t,\mathcal{I}}$ and length $\delta_{t,s,\mathcal{I}}$ and $\delta_{s,t,\mathcal{I}}$.
- $E_{t,s,\mathcal{I}}$: Let $t_1, s_1 \in T_{n+1}$ be the adjacent vertices such that $G_{t_1,s_1,\mathcal{I}} = G_{t,s,\mathcal{I}}$. Define $E_{t,s,\mathcal{I}} = C_{t,s,\mathcal{I}} \setminus C_{t_1,s_1,\mathcal{I}}$. Note that $C_{t,\mathcal{I}} = C_{m(t),\mathcal{I}}$ if, and only if, $E_{t,s,\mathcal{I}} = \emptyset$. Moreover, $|E_{t,s,\mathcal{I}}| = e_{t,s,\mathcal{I}}$. Let $t_l, s_l \in T_l$ and $t_j, s_j \in T_j$ be adjacent vertices such that $G_{t_l,s_l,\mathcal{I}} = G_{t_j,s_j,\mathcal{I}}$. Then, $E_{t_l,s_l,\mathcal{I}}$ and $E_{t_j,s_j,\mathcal{I}}$ have the important property that $\mathrm{int}E_{t_l,s_l,\mathcal{I}} \cap \mathrm{int}E_{t_j,s_j,\mathcal{I}} = \emptyset$. This property is used later on in the construction of the map L_n.

Lemma B.10. *(i) For k equal to 0 and 1 and for all $n \geq N_{\mathcal{I},\mathcal{J}}$ and all pairs of adjacent vertices $t,s \in T_n$ that are in contact,*

$$\|L_t - L_s\|_{C^k} \leq \mathcal{O}(\nu_{t,s}|D_{t,s,\mathcal{I}}|^{1-k}) \qquad \text{(B.4)}$$

in the domain $D_{t,s,\mathcal{I}}$.
(ii) For all vertices $t \in T_n$ and all $n \geq N_{\mathcal{I},\mathcal{J}}$,

$$\|L_t - L_{m(t)}\|_{C^0} \leq \mathcal{O}(f_\varepsilon(n)) \qquad \text{(B.5)}$$

in the domain $C_{t,\mathcal{I}}$. For all adjacent vertices s and t not in contact, if $m(s) = m(t)$, one has

$$\|L_t - L_{m(t)}\|_{C^0} \leq \mathcal{O}(A_{t,s,\mathcal{I}}) \qquad \text{(B.6)}$$

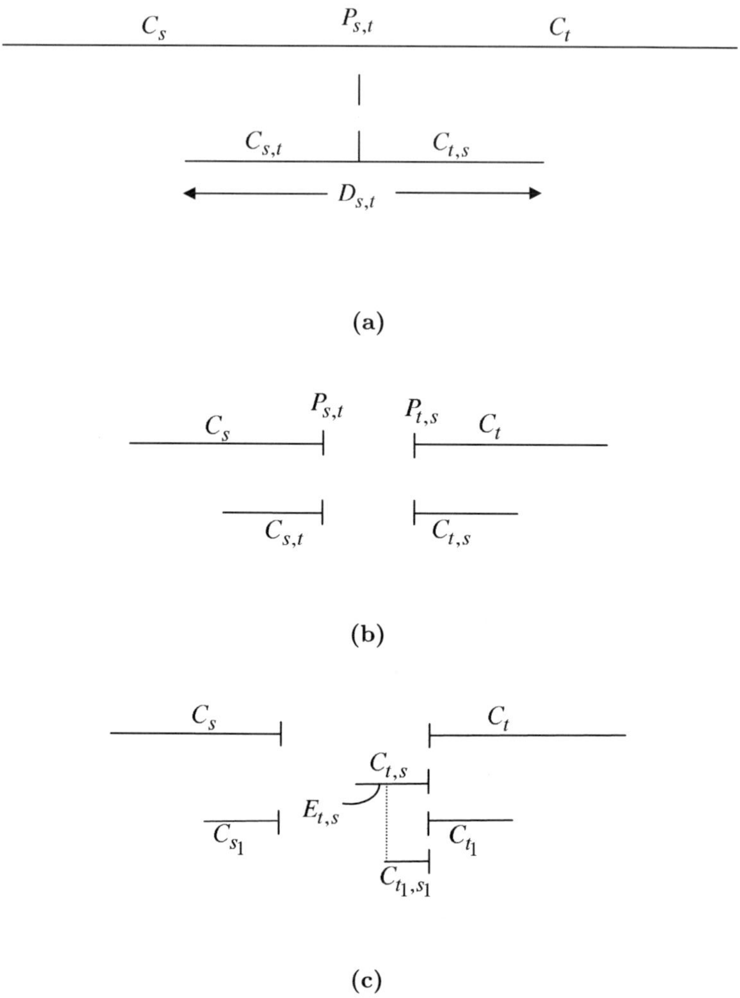

Fig. B.3. The intervals $C_{t,s}$, $C_{s,t}$, $D_{s,t}$ and $E_{t,s}$.

in the domain $C_{t,s,\mathcal{I}}$. *If* $m(s) \neq m(t)$ *and* $E_{t,s,\mathcal{I}} = \emptyset$, *then* $L_t = L_{m(t)}$ *in* $C_{t,s,\mathcal{I}}$. *If* $m(s) \neq m(t)$ *and* $E_{t,s,\mathcal{I}} \neq \emptyset$, *one has*

$$\|L_t - L_{m(t)}\|_{C^0} \leq \mathcal{O}(\nu_t|C_{t,s,\mathcal{I}}|) \tag{B.7}$$

in the domain $C_{t,s,\mathcal{I}}$. *Moreover,*

$$\|dL_t - dL_{m(t)}\|_{C^0} \leq \mathcal{O}(\nu_t) \tag{B.8}$$

in the domains $C_{t,\mathcal{I}}$ *and* $E_{t,s,\mathcal{I}}$.

Proof. Firstly, we prove inequality (B.4). By Corollary B.9 and since $L_t(P_{t,s,\mathcal{I}}) = L_s(P_{t,s,\mathcal{I}}) = P_{t,s,\mathcal{J}} = P_{s,t,\mathcal{J}}$ and $|x - P_{t,s,\mathcal{I}}| \leq \mathcal{O}(|D_{t,s,\mathcal{I}}|)$,

$$|L_t(x) - L_s(x)| = \left| \frac{|C_{t,\mathcal{J}}|}{|C_{t,\mathcal{I}}|} - \frac{|C_{s,\mathcal{J}}|}{|C_{s,\mathcal{I}}|} \right| |x - P_{t,s,\mathcal{I}}| \le \mathcal{O}(\nu_{t,s}|D_{t,s,\mathcal{I}}|) \tag{B.9}$$

and

$$|dL_t - dL_s| = \left| \frac{|C_{t,\mathcal{J}}|}{|C_{t,\mathcal{I}}|} - \frac{|C_{s,\mathcal{J}}|}{|C_{s,\mathcal{I}}|} \right| \le \mathcal{O}(\nu_{t,s}).$$

Let us prove inequality (B.5). Let $v, z, r \in \tilde{T}_n$ be such that $m(v) = m(z) = m(r) = m(t)$ and z is the only vertex between v and r. By definition of $L_{m(t)}$, and as $L_z(P_{z,r,\mathcal{I}}) = L_r(P_{z,r,\mathcal{I}})$, we obtain by Corollary B.9

$$\begin{aligned} |L_{m(t)}(P_{v,z,\mathcal{I}}) - L_z(P_{v,z,\mathcal{I}})| &\le \left| \frac{|C_{m(t),\mathcal{J}}|}{|C_{m(t),\mathcal{I}}|} - \frac{|C_{z,\mathcal{J}}|}{|C_{z,\mathcal{I}}|} \right| |P_{v,z,\mathcal{I}} - P_{z,r,\mathcal{I}}| + \\ &\quad + |L_{m(t)}(P_{z,r,\mathcal{I}}) - L_r(P_{z,r,\mathcal{I}})| \\ &\le \mathcal{O}\left(\nu_z |C_{z,\mathcal{I}}| + |L_{m(t)}(P_{z,r,\mathcal{I}}) - L_r(P_{z,r,\mathcal{I}})|\right). \end{aligned}$$

Let $r_1, v_1 \in \Lambda_1 = \underline{\Lambda}_t$ and $r_2, v_2 \in \Lambda_2 = \overline{\Lambda}_t$ be such that r_1 and r_2 are adjacent to t and v_1 and v_2 have adjacent vertices z_1 and z_2, respectively, such that $m(z_1) \ne m(t) \ne m(z_2)$. Let i be equal to 1 or 2. By definition of $L_{m(t)}$ and L_{v_i},

$$L_{m(t)}(P_{v_i,z_i,\mathcal{I}}) = L_{v_i}(P_{v_i,z_i,\mathcal{I}}). \tag{B.10}$$

By inequalities (B.9) and (B.10),

$$|L_{m(t)}(P_{t,r_i,\mathcal{I}}) - L_{r_i}(P_{t,r_i,\mathcal{I}}) \le \mathcal{O}\left(\sum_{v \in \Lambda_i} \nu_v |C_{v,\mathcal{I}}| \right). \tag{B.11}$$

For all $x \in C_{t,\mathcal{I}}$, by Corollary B.9 and inequality (B.11),

$$\begin{aligned} |L_{m(t)}(x) - L_t(x)| &\le \left| \frac{|C_{m(t),\mathcal{J}}|}{|C_{m(t),\mathcal{I}}|} - \frac{|C_{z,\mathcal{J}}|}{|C_{z,\mathcal{I}}|} \right| |x - P_{t,r_i,\mathcal{I}}| + \\ &\quad + |L_{m(t)}(P_{t,r_i,\mathcal{I}}) - L_{r_i}(P_{t,r_i,\mathcal{I}})| \\ &\le \mathcal{O}\left(\nu_t |C_{t,\mathcal{I}}| + \sum_{v \in \Lambda_i} \nu_v |C_{v,\mathcal{I}}| \right) \\ &\le \mathcal{O}\left(f_\varepsilon(n) |C_{m(t),\mathcal{I}}| \right) \le \mathcal{O}(f_\varepsilon(n)). \end{aligned}$$

Let us prove inequality (B.6). For all $x \in C_{t,s,\mathcal{I}}$, by definition of $A_{t,s,\mathcal{I}}$, by Corollary B.9 and inequality (B.11),

$$\begin{aligned} |L_{m(t)}(x) - L_t(x)| &\le \left| \frac{|C_{m(t),\mathcal{J}}|}{|C_{m(t),\mathcal{I}}|} - \frac{|C_{z,\mathcal{J}}|}{|C_{z,\mathcal{I}}|} \right| |x - P_{t,s,\mathcal{I}}| + \\ &\quad + |L_{m(t)}(P_{t,s,\mathcal{I}}) - L_t(P_{t,s,\mathcal{I}})| \le \mathcal{O}(A_{t,s,\mathcal{I}}). \end{aligned}$$

Let us prove inequality (B.7). By definition, $L_t(P_{t,s,\mathcal{I}}) = L_{m(t)}(P_{t,s,\mathcal{I}})$. For all $x \in C_{t,s,\mathcal{I}}$, by Corollary B.9,

$$|L_t(x) - L_{m(t)}(x)| = \left| \frac{|C_{t,\mathcal{J}}|}{|C_{t,\mathcal{I}}|} - \frac{|C_{m(t),\mathcal{J}}|}{|C_{m(t),\mathcal{I}}|} \right| |x - P_{t,s,\mathcal{I}}|$$
$$\leq \mathcal{O}\left(\nu_t |C_{t,s,\mathcal{I}}|\right).$$

Moreover, inequality (B.8) follows by Corollary B.9, because

$$|dL_t - dL_{m(t)}| = \left| \frac{|C_{t,\mathcal{J}}|}{|C_{t,\mathcal{I}}|} - \frac{|C_{m(t),\mathcal{J}}|}{|C_{m(t),\mathcal{I}}|} \right| \leq \mathcal{O}(\nu_t).$$

□

B.3.4 The definition of the contact and gap maps

Lemma B.11. *For all $\delta \geq 0$, there exists a C^∞ map $\phi : [0, \delta] \to [0, 1]$ such that $\phi(0) = 0$ on $[0, \delta/3]$, $\phi = 1$ on $[2\delta/3, 1]$ and $\|\phi\|_{C^{k+\alpha}} \leq c_k \delta^{-(k+\alpha)}$, where c_k depends only upon $k \in \mathbb{Z}_{\geq 0}$ and not on $\alpha \in (0, 1]$ or δ.*

Proof. Find such a function ϕ_0 for the case $\delta = 1$ and then deduce the general case by letting $\phi(x) = \phi_0(\delta^{-1} x)$. □

If s and t are adjacent vertices in T_n, we use Lemma B.11 to choose functions $\phi_{t,s}$ on $G_{t,s,\mathcal{I}}$ and $\psi_{s,t} = \psi_{t,s}$ on $D_{t,s,\mathcal{I}}$ with the following properties.

(i) $\phi_{t,s} = 0$ (resp. $\psi_{t,s} = 0$) on the left-hand third of $E_{t,s,\mathcal{I}}$ (resp. $D_{t,s,\mathcal{I}}$) and $\phi_{t,s} = 1$ (resp. $\psi_{t,s} = 1$) on the left-hand third of $E_{t,s,\mathcal{I}}$ (resp. $D_{t,s,\mathcal{I}}$).

(ii)
$$\|\phi_{t,s}\|_{C^p} \leq \mathcal{O}\left(|E_{t,s,\mathcal{I}}|^{-p}\right) \tag{B.12}$$
and
$$\|\psi_{t,s}\|_{C^p} \leq \mathcal{O}\left(|D_{t,s,\mathcal{I}}|^{-p}\right), \tag{B.13}$$
for all reals p between 0 and 2 and where the constants are independent of all the data.

Extend $\phi_{t,s}$ to all of the gap $D_{t,s,\mathcal{I}}$ as a smooth map by taking it as constant outside $E_{t,s,\mathcal{I}}$. We call the $\phi_{t,s}$ *gap maps* and the $\psi_{t,s}$ *contact maps*.

Note that, for all $n, m \geq N_{\mathcal{I},\mathcal{J}}$ and all non-contact adjacent vertices $t_1, s_1 \in T_n$ and $t_2, s_2 \in T_m$ such that $\{s_1, t_1\} \neq \{s_2, t_2\}$, the domains of the gap maps where they are different from 0 or 1 do not overlap. For all $n \geq N$ and all contact adjacent vertices $t_3, s_3 \in T_n$ and $t_4, s_4 \in T_m$ such that $\{s_3, t_3\} \neq \{s_4, t_4\}$, the domains of the contact maps do not overlap. Moreover, they do not overlap with any domain of any gap map ϕ_{t_2,s_2}, where $t_2, s_2 \in T_m$ and $m \leq n$.

B.3.5 The map L_n

For all $n \geq N_0$ and all vertices $t \in T_n$, define the map L_n on $C_{t,\mathcal{I}} \subset \mathbf{I}$ as follows. For all vertices s_i in contact with t, $L_n = L_t$ on $C_{t,\mathcal{I}} \setminus \cup_i C_{t,s,\mathcal{I}}$. If s is in contact with t and s is on the left of t, then define L_n on $C_{t,s,\mathcal{I}}$ by

$$L_n = \psi_{t,s} L_t + (1 - \psi_{t,s}) L_s.$$

If s is on the right of t, then define L_n on $C_{t,s,\mathcal{I}}$ by

$$L_n = \psi_{t,s} L_s + (1 - \psi_{t,s}) L_t.$$

Extension of L_n to the gaps

For all $n \geq N_0$ and all non-contact adjacent vertices $t, s \in T_n$, suppose that t os on the left of s. If $E_{t,s,\mathcal{I}} = \emptyset$, define the map L_n on $C_{t,s,\mathcal{I}}$ by $L_n | C_{t,s,\mathcal{I}} = L_t$. If $E_{t,s,\mathcal{I}} \neq \emptyset$, define the map L_n on $C_{t,s,\mathcal{I}}$ by

$$L_n | C_{t,s,\mathcal{I}} = L_{m(t)} \phi_{t,s} + L_t (1 - \phi_{t,s}).$$

If $E_{s,t,\mathcal{I}} = \emptyset$, define the map L_n on $C_{s,t,\mathcal{I}}$ by $L_n | C_{s,t,\mathcal{I}} = L_s$. If $E_{s,t,\mathcal{I}} \neq \emptyset$, define the map L_n on $C_{s,t,\mathcal{I}}$ by

$$L_n | C_{s,t,\mathcal{I}} = L_{m(s)} (1 - \phi_{s,t}) + L_s \phi_{s,t}.$$

Finally, in $G_{t,s,\mathcal{I}} \setminus (C_{t,s,\mathcal{I}} \cup C_{s,t,\mathcal{I}})$, define $L_n = L_{n-1}$.

Let $t_1, s_1 \in T_{n-1}$ be such that $m(t_1) = t$ and $m(s_1) = s$ and $E_{t,s,\mathcal{I}} \neq \emptyset$ and $E_{s,t,\mathcal{I}} \neq \emptyset$. The map L_n is equal to L_t in $C_{t,s,\mathcal{I}} \setminus E_{t,s,\mathcal{I}} = C_{t_1,s_1,\mathcal{I}}$. The map L_n changes smoothly in $E_{t,s,\mathcal{I}}$ to $L_n = L_{m(t)} = L_{n-1}$. The map L_n is equal to L_{n-1} in $G_{t,s,\mathcal{I}} \setminus (C_{t,s,\mathcal{I}} \cup C_{s,t,\mathcal{I}})$. Again the map $L_n = L_{n-1} = L_{m(s)}$ changes smoothly in $E_{s,t,\mathcal{I}}$ such that $L_n = L_s$ in $C_{s,t,\mathcal{I}} \setminus E_{s,t,\mathcal{I}} = C_{s_1,t_1,\mathcal{I}}$. Therefore, the map L_n patches together smoothly in $G_{t,s,\mathcal{I}}$. If $E_{t,s,\mathcal{I}} = \emptyset$, then in $C_{t,s,\mathcal{I}}$, by definition of the map L_{n-1}, $L_{n-1} = L_{m(t)}$ and the map $L_{m(t)} = L_t = L_n$. Therefore, $L_n = L_{n-1}$ in $C_{t,s,\mathcal{I}}$. Similarly, if $E_{s,t,\mathcal{I}} = \emptyset$, then $L_n = L_{n-1}$ in $C_{s,t,\mathcal{I}}$.

This construction builds an infinitely differentiable map L_n that is defined on the closed interval $\mathbf{I}$ and that maps $\mathbf{I}$ diffeomorphically onto $\mathbf{J}$.

B.3.6 The sequence of maps L_n converge

The space of $C^{1+\varepsilon}$ maps on $\mathbf{I}$, for all $0 < \varepsilon < \alpha$, with the $C^{1+\varepsilon}$ norm, is a Banach space. In this section, we present a prove that the sequence $(L_n)_{n > N_0}$ is a Cauchy sequence in this space and therefore converges. First, we prove the following lemma.

Lemma B.12. *Suppose $t \in T_n$ and $n > N_0$. Then, in the three subsets $C_{t,\mathcal{I}} \setminus \cup_s C_{t,s,\mathcal{I}}$, $D_{t,s,\mathcal{I}}$ and $G_{t,s,\mathcal{I}}$,*

$$\|L_n - L_{n-1}\|_{C^{1+\varepsilon}} \leq \mathcal{O}(f_\varepsilon(n-1)).$$

The constants of the inequality only depend upon $\mathcal{I}$ and $\mathcal{J}$.

Proof. We break the proof down into 3 cases corresponding to behavior in the three subsets $C_{t,\mathcal{I}} \setminus \cup_s C_{t,s,\mathcal{I}}$, $D_{t,s,\mathcal{I}}$ and $G_{t,s,\mathcal{I}}$.

(i) For $C_{t,\mathcal{I}} \setminus \cup_s C_{t,s,\mathcal{I}}$, where s is in contact with t. By Lemma B.10,

$$\|L_n - L_{n-1}\|_{C^{1+\varepsilon}} = \|L_t - L_{m(t)}\|_{C^{1+\varepsilon}} \leq \mathcal{O}(f_\varepsilon(n)).$$

(ii) For $D_{t,s,\mathcal{I}} = C_{t,s,\mathcal{I}} \cup C_{s,t,\mathcal{I}}$. Suppose s is on the left of t. We will study $L_n - L_{n-1}$ in the domain $C_{t,s,\mathcal{I}}$. By a similar argument, we have the same result in $C_{s,t,\mathcal{I}}$.

$$L_n - L_t = \psi_{t,s} L_t + (1 - \psi_{t,s}) L_s - L_t = (1 - \psi_{t,s})(L_s - L_t).$$

By inequality (B.4),

$$|L_n - L_t| \leq |1 - \psi_{t,s}||L_s - L_t| \leq \mathcal{O}(\nu_{t,s}|D_{t,s,\mathcal{I}}|).$$

Moreover, by Lemma B.10 and inequality (B.13),

$$|dL_n - dL_t| \leq |d\psi_{t,s}||L_s - L_t| + |\psi_{t,s}||dL_s - dL_t| \leq \mathcal{O}(\nu_{t,s}).$$

and

$$\begin{aligned}\|dL_n - dL_t\|_{C^\varepsilon} \leq\ & \|d\psi_{t,s}\|_{C^\varepsilon}\|L_s - L_t\|_{C^0} + \\ & +\|d\psi_{t,s}\|_{C^0}\|L_s - L_t\|_{C^\varepsilon} + \|d\psi_{t,s}\|_{C^\varepsilon}\|dL_s - dL_t\|_{C^0} \\ \leq\ & \mathcal{O}(\nu_{t,s}|D_{t,s,\mathcal{I}}|^{-\varepsilon}).\end{aligned}$$

Therefore,

$$\|L_n - L_t\|_{C^{1+\varepsilon}} \leq \mathcal{O}(\nu_{t,s}|D_{t,s,\mathcal{I}}|^{-\varepsilon}).$$

If $m(s) \neq m(t)$, then, by Lemma B.10 and the last inequality,

$$\begin{aligned}\|L_n - L_{n-1}\|_{C^{1+\varepsilon}} &\leq \|L_n - L_t\|_{C^{1+\varepsilon}} + \|L_n - L_{m(t)}\|_{C^{1+\varepsilon}} \\ &\quad + \|L_{m(t)} - L_{n-1}\|_{C^{1+\varepsilon}} \\ &\leq \mathcal{O}(\nu_{t,s}|D_{t,s,\mathcal{I}}|^{-\varepsilon}) + \mathcal{O}(f_\varepsilon(n)) + \\ &\quad +\mathcal{O}(\nu_{m(t),m(s)}|D_{m(t),m(s),\mathcal{I}}|^{-\varepsilon}) \\ &\leq \mathcal{O}(f_\varepsilon(n-1)).\end{aligned}$$

If $m(s) = m(t)$, then $L_{m(t)} = L_{n-1}$ or

$$\|L_{m(t)} - L_{n-1}\|_{C^{1+\varepsilon}} \leq \mathcal{O}(\nu_{m(t),z}|D_{m(t),z,\mathcal{I}}|^{-\varepsilon}) \leq \mathcal{O}(f_\varepsilon(n-1)),$$

where z is a contact vertex of $m(t)$. Therefore, in the domain $C_{t,s,\mathcal{I}}$,

$$\|L_n - L_{n-1}\|_{C^{1+\varepsilon}} \le \mathcal{O}(f_\varepsilon(n-1)).$$

By a similar argument, in the domain $C_{s,t,\mathcal{I}}$, we obtain in $D_{t,s,\mathcal{I}}$,

$$\|L_n - L_{n-1}\|_{C^{1+\varepsilon}} \le \mathcal{O}(f_\varepsilon(n-1)).$$

(iii) For $G_{t,s,\mathcal{I}}$. Suppose that t is on the left of s. By definition of the domains of the gap maps, $L_n = L_{n-1}$ in the gap $G_{t,s,\mathcal{I}}$, except in the intervals $C_{t,s,\mathcal{I}}$ and $C_{s,t,\mathcal{I}}$. If $E_{t,s,\mathcal{I}} = \emptyset$, then $L_n = L_{n-1}$ in $C_{t,s,\mathcal{I}}$. If $E_{t,s,\mathcal{I}} \ne \emptyset$, then in $C_{t,s,\mathcal{I}}$

$$L_n - L_{n-1} = L_{m(t)}(\phi_{t,s} - 1) + L_t(1-\phi_{t,s}) = (L_t - L_{m(t)})(1-\phi_{t,s}).$$

If $m(t) = m(s)$, by Lemma B.10 and inequality (B.12),

$$\|L_n - L_{n-1}\|_{C^0} \le |L_t - L_{m(t)}||1-\phi_t| \le \mathcal{O}(\nu_t),$$

$$\begin{aligned}\|dL_n - dL_{n-1}\|_{C^0} &\le |L_t - L_{m(t)}||d\phi_t| + |dL_t - dL_{m(t)}||1-\phi_t| \\ &\le \mathcal{O}\left(A_{t,s,\mathcal{I}}|E_{t,s,\mathcal{I}}|^{-1}\right) + \mathcal{O}(\nu_t)\end{aligned}$$

and

$$\begin{aligned}\|dL_n - dL_{n-1}\|_{C^\varepsilon} &\le \|L_t - L_{m(t)}\|_{C^\varepsilon}\|d\phi_t\|_{C^0} + \|L_t - L_{m(t)}\|_{C^0}\|d\phi_t\|_{C^\varepsilon} + \\ &\quad + \|dL_t - dL_{m(t)}\|_{C^0}\|1-\phi_t\|_{C^\varepsilon} \\ &\le \mathcal{O}\left(\nu_t|E_{t,s,\mathcal{I}}|^{1-\varepsilon-1}\right) + \mathcal{O}\left(A_{t,s,\mathcal{I}}|E_{t,s,\mathcal{I}}|^{-(1+\varepsilon)}\right) + \\ &\quad + \mathcal{O}\left(\nu_t|E_{t,s,\mathcal{I}}|^{-\varepsilon}\right) \\ &\le \mathcal{O}\left(A_{t,s,\mathcal{I}}|E_{t,s,\mathcal{I}}|^{-(1+\varepsilon)}\right) + \mathcal{O}\left(\nu_t|E_{t,s,\mathcal{I}}|^{-\varepsilon}\right).\end{aligned}$$

Similarly, in $C_{s,t,\mathcal{I}}$, if $E_{s,t,\mathcal{I}} = \emptyset$, then $L_n = L_{n-1}$. If $E_{s,t,\mathcal{I}} \ne \emptyset$ and $m(s) = m(t)$, then in $C_{s,t,\mathcal{I}}$,

$$\|L_n - L_{n-1}\|_{C^{1+\varepsilon}} \le \mathcal{O}\left(A_{t,s,\mathcal{I}}|E_{t,s,\mathcal{I}}|^{-(1+\varepsilon)}\right) + \mathcal{O}\left(\nu_s|E_{s,t,\mathcal{I}}|^{-\varepsilon}\right).$$

If $m(t) \ne m(s)$ and $E_{t,s,\mathcal{I}} \ne \emptyset$, we have by Lemma B.10 and inequality (B.12), that in the domain $C_{t,s,\mathcal{I}}$

$$\|L_n - L_{n-1}\|_{C^0} \le |L_t - L_{m(t)}||1-\phi_t| \le \mathcal{O}\left(\nu_t|C_{t,s,\mathcal{I}}|\right),$$

$$\begin{aligned}\|dL_n - dL_{n-1}\|_{C^0} &\le |L_t - L_{m(t)}||d\phi_t| + |dL_t - dL_{m(t)}||1-\phi_t| \\ &\le \mathcal{O}\left(\nu_t|C_{t,s,\mathcal{I}}||E_{t,s,\mathcal{I}}|^{-1}\right)\end{aligned}$$

and

$$\begin{aligned}\|dL_n - dL_{n-1}\|_{C^\varepsilon} &\le \|L_t - L_{m(t)}\|_{C^\varepsilon}\|d\phi_t\|_{C^0} + \|L_t - L_{m(t)}\|_{C^0}\|d\phi_t\|_{C^\varepsilon} + \\ &+\|dL_t - dL_{m(t)}\|_{C^0}\|1-\phi_t\|_{C^\varepsilon} \\ &\le \mathcal{O}\left(\nu_t|C_{t,s,\mathcal{I}}||E_{t,s,\mathcal{I}}|^{-(1+\varepsilon)}\right).\end{aligned}$$

Similarly, in $C_{s,t,\mathcal{I}}$,

$$\|L_n - L_{n-1}\|_{C^{1+\varepsilon}} \le \mathcal{O}\left(\nu_t|C_{s,t,\mathcal{I}}||E_{s,t,\mathcal{I}}|^{-(1+\varepsilon)}\right).$$

□

Lemma B.13. *The sequence of maps $(L_n)_{n>N_0}$ is a Cauchy sequence in the domain **I** with respect to the $C^{1+\varepsilon}$ norm. In fact, $\|L_n - L_{n-1}\|_{C^{1+\varepsilon}} \le \mathcal{O}(f_\varepsilon(n-1))$.*

Proof. For all vertices $t \in T_n$, define P_t as the middle point of $C_{t,\mathcal{I}}$ and for all non-contact vertices $t, s \in T_n$, define $Q_{t,s}$ as the endpoint of $C_{t,s,\mathcal{I}}$ that is not common to $C_{t,\mathcal{I}}$. Denote $dL_n - dL_{n-1}$ by B_n. By inequality (B.8),

$$|B_n(P_t)| \le \mathcal{O}(\nu_t) \text{ and } |dB_n(Q_{t,s})| = 0. \tag{B.14}$$

For all $x, y \in \mathbf{I}$, if the closed interval between x and y is contained in the union of a bounded number of domains of the form $C_{t,\mathcal{I}}$ or $C_{g_{t,s},\mathcal{I}}$, then, by Lemma B.12,

$$\frac{|B_n(y) - B_n(x)|}{|y-x|^\varepsilon} \le \mathcal{O}(f_\varepsilon(n-1)). \tag{B.15}$$

Otherwise, take P_x (resp. P_y) to be the nearest point of the form P_t or $Q_{t,s}$ to x (resp. y) in the closed interval between x and y. Let us consider the case that $P_x = P_t$ and $P_y = P_s$. By inequalities (B.14) and (B.15) and $(1+\alpha)$-contact equivalence,

$$\begin{aligned}\frac{|B_n(y) - B_n(x)|}{|y-x|^\varepsilon} &\le \frac{|B_n(y) - B_n(P_y)|}{|y-P_y|^\varepsilon} + \frac{|B_n(P_y)|}{|\underline{C_{s,\mathcal{I}}}|^\varepsilon} + \\ &+\mathcal{O}(f_\varepsilon(n-1)) + \mathcal{O}\left(\nu_s|\overline{C_{s,\mathcal{I}}}|^{-(\varepsilon)}\right) + \\ &+\mathcal{O}\left(\nu_t|\overline{C_{t,\mathcal{I}}}|^{-(\varepsilon)}\right) + \mathcal{O}(f_\varepsilon(n-1)) \\ &\le \mathcal{O}(f_\varepsilon(n-1)).\end{aligned}$$

Similarly, for the other cases. Therefore, $\|L_n - L_{n-1}\|_{C^{1-\varepsilon}} \le \mathcal{O}(f_\varepsilon(n-1))$ and, by condition (i) of Definition B.5, L_n is a Cauchy sequence. □

B.3.7 The map L_∞

Since the sequence $(L_n)_{n\geq N_0}$ is a Cauchy sequence in $C^{1+\varepsilon}(\mathbf{I})$, it converges to a function $L_\infty \in C^{1+\varepsilon}$.

Lemma B.14. *The map L_∞ is a $C^{1+\alpha^-}$ diffeomorphism of* $\mathbf{I}$ *onto* $\mathbf{J}$ *that extends $i^{-1}\circ j$.*

Proof. By Lemma B.8, for all $t \in T_n$, $|C_{t,\mathcal{J}}|/|C_{t,\mathcal{I}}|$ is bounded away from 0 and ∞ and, by the hypotheses of $(1+\alpha)$-scale equivalence and $(1+\alpha)$-contact equivalence, if $s,t \in T_n$ are adjacent, (i) $A_{t,s,\mathcal{I}}|E_{t,s,\mathcal{I}}|^{-1} \to 0$, (ii) $\nu_t \to 0$ as $n \to \infty$, and (iii) $\nu_{s,t} \to 0$ depending if s is in contact with t or not and if they have the same mother. Thus, there exists $\varepsilon_1 > 0$, $0 < \varepsilon < \varepsilon_1$, and $N_1 > 0$ such that if $n \geq N_1$, then, for all $s,t \in T_n$,

$$\varepsilon_1 < |C_{m(t),\mathcal{J}}|/|C_{m(t),\mathcal{I}}|, \quad \mathcal{O}\left(A_{t,s,\mathcal{I}}|E_{t,s,\mathcal{I}}|^{-1} + \nu_t\right) < \varepsilon \text{ and } \mathcal{O}(\nu_{t,s}) < \varepsilon,$$

when defined.

We break down the proof into four parts corresponding to the sets $C_{t,\mathcal{I}} \setminus (\cup_s C_{t,s,\mathcal{I}})$, $D_{t,s,\mathcal{I}}$, where s is adjacent and in contact with t; $C_{t,s,\mathcal{I}}$, $C_{s,t,\mathcal{I}}$ and $G_{t,s,\mathcal{I}} \setminus (C_{t,s,\mathcal{I}} \cup C_{s,t,\mathcal{I}})$, if s is adjacent and not in contact with t.

(i) In $C_{t,\mathcal{I}} \setminus C_{t,s,\mathcal{I}}$. $dL_t = |C_{t,\mathcal{J}}|/|C_{t,\mathcal{I}}| > \varepsilon_1$.

(ii) In $D_{t,s,\mathcal{I}}$. Suppose that s is on the left of t. Then, in the domain $D_{t,s,\mathcal{I}}$, by the inequalities (B.4) and (B.13),

$$\begin{aligned}
|dL_n| &= |\psi_{t,s}dL_t + d\psi_{t,s}L_t + (1-\psi_{t,s})dL_s - d\psi_{t,s}L_s| \\
&\geq |dL_s| - |d\psi_{t,s}(L_t - L_s) + \psi_{t,s}(dL_t - dL_s)| \\
&\geq |C_{s,\mathcal{J}}|/|C_{s,\mathcal{I}}| - \mathcal{O}(\nu_{t,s}) > \varepsilon_1 - \varepsilon > 0,
\end{aligned}$$

(iii) In $C_{t,s,\mathcal{I}}$. Suppose t is on the left of s. Similarly, if t is on the right of s. Then, in the domain $C_{t,s,\mathcal{I}}$, if $E_{t,s,\mathcal{I}} = \emptyset$, one has $|dL_n| = |dL_t| > \varepsilon_1$. If $E_{t,s,\mathcal{I}} \neq \emptyset$, then, by Lemma B.10 and inequality (B.12),

$$\begin{aligned}
|dL_n| &= |\phi_{t,s}dL_t + d\phi_{t,s}L_t + (1-\phi_{t,s})dL_{m(t)} - d\phi_{t,s}L_{m(t)}| \\
&\geq |dL_{m(t)}| - |d\phi_{t,s}(L_t - L_{m(t)}) + \phi_{t,s}(dL_t - dL_{m(t)})| \\
&\geq |C_{m(t),\mathcal{J}}|/|C_{m(t),\mathcal{I}}| - \mathcal{O}\left(A_{t,s,\mathcal{I}}|E_{t,s,\mathcal{I}}|^{-1} + \nu_t\right) > \varepsilon_1 - \varepsilon > 0,
\end{aligned}$$

(iv) In $G_{t,s,\mathcal{I}} \setminus (C_{t,s,\mathcal{I}} \cup C_{s,t,\mathcal{I}})$. In different subsets of this set, the map $L_n = L_{n-j}$, for some $j \in \mathbb{N}$. We suppose, by induction, that $L_{n-j} > \varepsilon_1 - \varepsilon > 0$. For that take $N_0 = \max\{N_0, N_1\}$.
Therefore, $|dL_n| > \varepsilon_1 - \varepsilon > 0$ in $\mathbf{I}$, for all $n > N_0$, which implies that $|L_\infty| \geq \varepsilon_1 - \varepsilon > 0$.
By construction, $L_n(C_{t,\mathcal{I}}) = C_{t,\mathcal{J}}$, for all $t \in T_m$, $N_0 \leq m \leq n$, and therefore L_∞ equals $i^{-1}\circ j$ on the closure of $\cup_{n\geq N_0}\mathbf{I}^n$.
As $L_\infty(C_{t,\mathcal{I}}) = C_{t,\mathcal{J}}$, for all vertices $t \in T_n$ and all $n > N_0$, L_∞ is a $C^{1+\alpha^-}$ conjugacy between the charts i and j.

□

B.3.8 Sufficient condition for $C^{1+\alpha^-}$-equivalent

Theorem B.15. *Let $\mathcal{S}$ and $\mathcal{T}$ be $C^{1+\alpha^-}$ structures on $L_{\mathcal{T}}$ and let $\mathcal{I}$ (resp. $\mathcal{J}$) be a primary atlas for $\mathcal{S}$ (resp. $\mathcal{T}$). A sufficient condition for $\mathcal{S}$ and $\mathcal{T}$ to be $C^{1+\alpha^-}$-equivalent is that $\mathcal{I} \overset{\alpha}{\sim} \mathcal{J}$.*

Proof. It is sufficient to prove the theorem locally at each point $\underline{t} \in L_{\mathcal{T}}$. Let $i : U_0 \to \mathbb{R}$ be a chart in $\mathcal{I}$ and $j : V_0 \to \mathbb{R}$ be a chart in $\mathcal{J}$ with $\underline{t} \in U_0 \cap V_0$. Then, it suffices to show that, for some open subsets U and V of $U_0 \cap V_0$ containing $\underline{t}$, the mapping $j \circ i^{-1} : i(U) \to j(V)$ has a $C^{1+\alpha^-}$ extension to $\mathbb{R}$. If this is the case, for all such $\underline{t}$, then the result holds globally. We can restrict our analysis to the case where

(i) the smallest closed interval $\mathbf{I}$ containing $i(U)$ is a cylinder $C_{t,i}$, for some $t \in T_{N_0}$, where $N_0 > N_{\mathcal{I},\mathcal{J}}$ or else is the union of two adjacent cylinders of this form that are in contact; and
(ii) where the smallest closed interval $\mathbf{J}$ containing $j(V)$ consists of the corresponding cylinders for j.

Now, let $\mathbf{I}^n$ (resp. $\mathbf{J}^n$) be the set of endpoints of the cylinders $C_{t,i}$ (resp. $C_{t,j}$), where $t \in T_n$, $n \geq N_0$ and $C_{t,i} \subset \mathbf{I}$ (resp. $C_{t,j} \subset \mathbf{J}$). Then, $j \circ i^{-1}$ maps $\mathbf{I}^n$ onto $\mathbf{J}^n$ and is a homeomorphism of the closure I^∞ of $\cup_{n \geq N_0} I^n$ onto the closure $\mathbf{J}^\infty$ of $\cup_{n \geq N_0} \mathbf{J}^n$. By Lemmas B.12 and B.14, there is a sequence of C^∞ mappings L_n such that

(i) L_n agrees with $j \circ i^{-1}$ on $\cup_{N_0 \leq j \leq n} \mathbf{I}^j$;
(ii) L_n is a Cauchy sequence in the space of $C^{1+\varepsilon}$ functions on $\mathbf{I}$, for all $\varepsilon < \alpha$, and, therefore, converges to a $C^{1+\alpha^-}$ function L_∞ on $\mathbf{I}$.
(iii) the mapping L_∞ gives the required smooth extension of $j \circ i^{-1}$ and proves the theorem.

□

The proof of Theorem B.16 is similar to the proof of Theorem B.15, taking ε equal to α.

Theorem B.16. *Let ε be equal to α in Definitions B.5 and B.6. The $C^{1+\alpha^-}$ structures $\mathcal{S}$ and $\mathcal{T}$ are $C^{1+\alpha^-}$-equivalent, if $\mathcal{I} \overset{\alpha}{\sim} \mathcal{J}$.*

B.3.9 Necessary condition for $C^{1+\alpha^-}$-equivalent

Theorem B.15 gave a sufficient condition for $\mathcal{S}$ and $\mathcal{T}$ be $C^{1+\alpha^-}$-equivalent. The following theorem gives a necessary condition that is very closely related.

Theorem B.17. *Let $\mathcal{S}$ and $\mathcal{T}$ be $C^{1+\alpha^-}$ structures on $L_{\mathcal{T}}$ with γ-controlled geometries and $\mathcal{I}$ and $\mathcal{J}$ be, respectively, primary atlases for $\mathcal{S}$ and $\mathcal{T}$. If $\mathcal{S}$ and $\mathcal{T}$ are $C^{1+\alpha^-}$-equivalent, then $\mathcal{I} \overset{\gamma}{\sim} \mathcal{J}$.*

Proof. Suppose that the structures $\mathcal{S}$ and $\mathcal{T}$ are $C^{1+\beta}$-equivalent, for all $0<\beta<\alpha$. Let the respective primary atlas $\mathcal{I}$ and $\mathcal{J}$ have γ-controlled geometry, where $0<\gamma\le\alpha$. This equivalence means that the identity is a $C^{1+\beta}$ diffeomorphism between the two structures. Thus, if (i,U) is a chart of $\mathcal{I}$ and (j,V) is a chart of $\mathcal{J}$ such that $C_{m(z),\mathcal{I}}\subset U$ and $C_{m(z),\mathcal{J}}\subset V$, then there exists a $C^{1+\beta}$ diffeomorphism $h:\mathbb{R}\to\mathbb{R}$ such that $h(C_{m(z),\mathcal{I}})=C_{m(z),\mathcal{J}}$ and $h(C_{t,\mathcal{I}})=C_{t,\mathcal{J}}$, for all descendents t of $m(z)$.

By the Mean Value Theorem, there are points $u,v\in C_{m(t),\mathcal{I}}$ such that

$$|dh(u)|=|C_{m(t),\mathcal{J}}|/|C_{m(t),\mathcal{I}}| \text{ and } |dh(v)|=|C_{t,\mathcal{J}}|/|C_{t,\mathcal{I}}|.$$

Moreover, since h is $C^{1+\beta}$, we have that $|dh(u)-dh(v)|\le\mathcal{O}(|C_{m(t),\mathcal{I}}|^{\beta})$. Therefore,

$$\nu_t=\left|1-\frac{|C_{t,\mathcal{J}}|}{|C_{m(t),\mathcal{J}}|}\frac{|C_{m(t),\mathcal{I}}|}{|C_{t,\mathcal{I}}|}\right|\le\mathcal{O}\left(|C_{m(t),\mathcal{I}}|^{\beta}\right)\le\mathcal{O}(g_{\beta,\varepsilon}(n)). \tag{B.16}$$

By a similar argument,

$$\nu_{g_{t,s}}=\left|1-\frac{|G_{t,s,\mathcal{J}}|}{|C_{m(t),\mathcal{J}}|}\frac{|C_{m(t),\mathcal{I}}|}{|G_{t,s,\mathcal{I}}|}\right|\le\mathcal{O}\left(|C_{m(t),\mathcal{I}}|^{\beta}\right)\le\mathcal{O}(g_{\beta,\varepsilon}(n)). \tag{B.17}$$

Therefore,

$$\begin{aligned}A_{t,s,\mathcal{I}}&\le\mathcal{O}\left(|C_{m(t),\mathcal{I}}|^{\beta}\left(\sum_{z\in\Lambda}|C_{z,\mathcal{I}}|\right)\right)\\&\le\mathcal{O}\left(|C_{m(t),\mathcal{I}}|^{1+\beta}\right)\le\mathcal{O}(g_{\beta,\varepsilon}(n)).\end{aligned}$$

By the hypotheses of Theorem B.17, if $m(t)=m(s)$, then

$$\begin{aligned}A_{t,s,\mathcal{I}}|E_{t,s,\mathcal{I}}|^{-(1+\varepsilon)}+\nu_t|E_{t,s,\mathcal{I}}|^{-\varepsilon}&\le\mathcal{O}\left(|C_{m(t),\mathcal{I}}|^{1+\beta}|E_{t,s,\mathcal{I}}|^{-(1+\varepsilon)}\right)+\\&\quad+\mathcal{O}\left(|C_{m(t),\mathcal{I}}|^{\beta}|E_{t,s,\mathcal{I}}|^{-\varepsilon}\right)\\&\le\mathcal{O}\left(|C_{m(t),\mathcal{I}}|^{1+\beta}|E_{t,s,\mathcal{I}}|^{-(1+\varepsilon)}\right)\\&\le\mathcal{O}(g_{\beta,\varepsilon}(n)).\end{aligned}$$

If $m(t)\ne m(s)$ and $E_{t,s,\mathcal{I}}\ne\emptyset$, then

$$\nu_t|C_{t,s,\mathcal{I}}||E_{t,s,\mathcal{I}}|^{-(1+\varepsilon)}\le\mathcal{O}\left(|C_{m(t),\mathcal{I}}|^{\beta}|C_{t,s,\mathcal{I}}||E_{t,s,\mathcal{I}}|^{-(1+\varepsilon)}\right)\le\mathcal{O}(g_{\beta,\varepsilon}(n)).$$

Thus, the conditions of Definition B.5 are verified, if for $f_\varepsilon(n)$ one takes $cg_{\beta,\varepsilon}(n)$, where $c>0$ is some constant. Therefore, the atlases $\mathcal{I}$ and $\mathcal{J}$ are $(1+\gamma)$-scale equivalent.

If t is in contact, then, by inequality (B.16),

$$\begin{aligned}\nu_t|\overline{C_{t,\mathcal{I}}}|^{-\varepsilon} &\leq \mathcal{O}\left(|C_{m(t),\mathcal{I}}|^{\beta}|\overline{C_{t,\mathcal{I}}}|^{-\varepsilon}\right) \\ &\leq \mathcal{O}(g_{\beta,\varepsilon}(n)).\end{aligned}$$

If s and t are in contact, then, by the Mean Value Theorem, there exist $u \in C_{s,\mathcal{I}}$ and $v \in C_{t,\mathcal{I}}$ such that

$$|dh(u)| = |C_{s,\mathcal{J}}|/|C_{s,\mathcal{I}}| \quad \text{and} \quad |dh(v)| = |C_{t,\mathcal{J}}|/|C_{t,\mathcal{I}}|.$$

Since the map h is $C^{1+\beta}$,

$$|dh(z) - dh(v)| \leq \mathcal{O}\left((|C_{t,\mathcal{I}}| + |C_{s,\mathcal{I}}|)^{\beta}\right) \leq \mathcal{O}\left(|D_{t,s,\mathcal{I}}|^{\beta}\right).$$

Therefore,

$$\nu_{t,s} = \left|1 - \frac{|C_{t,\mathcal{J}}|}{|C_{s,\mathcal{J}}|}\frac{|C_{s,\mathcal{I}}|}{|C_{t,\mathcal{I}}|}\right| \leq \mathcal{O}\left(|D_{t,s,\mathcal{I}}|^{\beta}\right) \tag{B.18}$$

and

$$\frac{\nu_{t,s}}{|D_{t,s,\mathcal{I}}|^{\varepsilon}} \leq \mathcal{O}\left(|D_{t,s,\mathcal{I}}|^{\beta-\varepsilon}\right) \leq \mathcal{O}(g_{\beta,\varepsilon}(n)).$$

The last inequality follows from the hypotheses of the theorem.

Thus, taking $f_\varepsilon(n) = cg_{\beta,\varepsilon}(n)$, the conditions of Definition B.6 are verified. Therefore, the atlases $\mathcal{I}$ and $\mathcal{J}$ are $(1+\gamma)$-contact equivalent. This completes the proof that $\mathcal{I}$ and $\mathcal{J}$ are $(1+\gamma)$-equivalent. □

Lemma B.18. *For $C^{1+\alpha^-}$ structures on L_T with α-controlled geometry, the Definition B.21 is equivalent to Definition B.7.*

Proof. Definition B.7 implies Definition B.21, because, by Theorem B.15, the $C^{1+\alpha^-}$ structures $\mathcal{S}$ and $\mathcal{T}$ are $C^{1+\alpha^-}$-equivalent and by α-controlled geometry Theorem B.17 holds with $\gamma = \alpha$. Therefore, by inequalities (B.16), (B.17) and (B.18), we obtain Definition B.21. Definition B.21 implies Definition B.7 by a straightforward calculation, using the α-controlled geometry property of the structure $\mathcal{S}$. □

B.4 Smooth structures with α-controlled geometry and bounded geometry

The results of the following sections are implied by the general theory on smooth structures that we will present in Section B.3.

Definition B.19. *A $C^{1+\alpha^-}$ structure $\mathcal{S}$ on L_T has γ-*controlled geometry, *if, for some primary atlas $\mathcal{I}$ and for all ε such that $0 < \varepsilon < \gamma \leq \alpha$, there exists β such that $\varepsilon < \beta < \alpha$ and there exists a decreasing function $g = g_{\beta,\varepsilon} : \mathbb{Z}_{n\geq 0} \to \mathbb{R}$ with the following properties:*

(i) $\sum_{n=0}^{\infty} g(n) < \infty$;

(ii) for all $t \in \tilde{T}_n$, $|C_{t,\mathcal{I}}|^\beta < g(n)$;
(iii) for all $t, s \in T_n$, *that are adjacent but not in contact, if* $m(t) = m(s)$, *then*

$$|C_{m(t),\mathcal{I}}|^{1+\beta} e_{t,s,\mathcal{I}}^{-(1+\varepsilon)} < g(n),$$

while if $m(t) \neq m(s)$ *and* $e_{t,s,\mathcal{I}} > 0$, *then*

$$|C_{m(t),\mathcal{I}}|^\beta \delta_{t,s,\mathcal{I}} e_{t,s,\mathcal{I}}^{-(1+\varepsilon)} < g(n);$$

(iv) for all $t, s \in T_n$ *that are in contact, we have that* $d_{t,s,\mathcal{I}}^{\beta-\varepsilon} < g(n)$ *and* $|C_{m(t),\mathcal{I}}|^\beta |\overline{C}_{t,\mathcal{I}}|^{-\varepsilon} < g(n)$.

If the structure $\mathcal{S}$ on L_T does not have gaps, then condition (iii) is trivial satisfied and (ii) follows from (iv). An important example is given by the case of smooth structures generated by smooth circle maps.

Let $\mathcal{I}$ and $\mathcal{J}$ be different primary atlas for $\mathcal{S}$ on L_T. By smoothness of the structure $\mathcal{S}$, there is a constant $c > 0$ such that, for all $t \in \tilde{T}_n$, $\mathcal{O}(|C_{t,\mathcal{I}}|) = \mathcal{O}(|C_{t,\mathcal{J}}|)$. Therefore, Definition B.19 is independent of the atlas considered. Similarly, let $\mathcal{S}$ and $\mathcal{T}$ be C^{1+}-equivalent structures on L_T. Then, $\mathcal{S}$ has γ-controlled geometry if, and only if, $\mathcal{T}$ has γ-controlled geometry.

In Lemma B.25 below, we show that a structure with bounded geometry has γ-controlled geometry, for all $0 < \gamma < 1$.

Lemma B.20. *The structure* $\mathcal{S}$ *has* α*-controlled geometry, if the following condition is verified: The gaps of the structure* $\mathcal{S}$ *have length greater or equal to the cylinders adjacent to it. Let* $l : \mathbb{Z}_{\geq 0} \to \mathbb{R}$ *and* $L : \mathbb{Z}_{\geq 0} \to \mathbb{R}$ *be positive functions such that, for all* $t \in \tilde{T}_n$, $l(n) \leq \sigma_{\mathcal{I}}(t) \leq L(n)$. *Then, for all* $0 < \varepsilon < \gamma$, *there is* $\gamma < \beta < \alpha$ *such that*

$$\sum_{n=1}^{\infty} \left(\prod_{i=1}^{n-1} L(i) \right)^{\beta-\varepsilon} (l(n))^{-(1+\varepsilon)}$$

converges.

Proof. For all $t \in \tilde{T}_n$,

$$l(n) \leq \frac{|C_{t,\mathcal{I}}|}{|C_{m(t),\mathcal{I}}|} \leq L(n). \tag{B.19}$$

For all $t \in \tilde{T}_n$,

$$\prod_{i=1}^{n} l(i) \leq |C_{t,\mathcal{I}}| \leq \prod_{i=1}^{n} L(i). \tag{B.20}$$

Conditions (i), (ii) and (iv) in the definition of γ-controlled geometry are verified by inequality (B.20), for a decreasing function $g = g_{\beta,\varepsilon} : \mathbb{Z}_{\geq 0} \to \mathbb{R}$ such that

$$\mathcal{O}(g(n)) = \mathcal{O}\left((l(n))^{-(1+\varepsilon)} \prod_{i=0}^{n-1} (L(i))^{\beta-\varepsilon} \right).$$

Let us prove that condition (iii) is also verified. For all adjacent vertices $t, s \in T_n$, that are not in contact, we have by definition that

$$\delta_{t,s,\mathcal{I}} = |C_{t,\mathcal{I}}| \frac{|G_{t,s,\mathcal{I}}|}{|C_{t_2,\mathcal{I}}|}.$$

Recall that t_2 is the vertex such that $C_{t_2,\mathcal{I}}$ and $G_{t,s,\mathcal{I}}$ have the same mother and $C_{t_2,\mathcal{I}}$ is an ancestor of $C_{t,\mathcal{I}}$. Therefore, by inequality (B.19), $l(n)|C_{m(t),\mathcal{I}}| \leq |C_{t,\mathcal{I}}| \leq L(n)|C_{m(t),\mathcal{I}}|$. Thus,

$$\delta_{t,s,\mathcal{I}} \leq L(n)|C_{m(t),\mathcal{I}}||G_{t,s,\mathcal{I}}|/|C_{t_2,\mathcal{I}}|$$

and

$$e_{t,s,\mathcal{I}} \geq (1 - L(n))l(n)|C_{m(t),\mathcal{I}}||G_{t,s,\mathcal{I}}|/|C_{t_2,\mathcal{I}}|.$$

By hypotheses $|C_{t_2,\mathcal{I}}|/|G_{t,s,\mathcal{I}}| \leq \mathcal{O}(1)$, thus

$$\begin{aligned} |C_{m(t),\mathcal{I}}|^{1+\beta} e_{t,s,\mathcal{I}}^{-(1+\varepsilon)} &\leq \mathcal{O}\left(l(n)^{-(1+\varepsilon)}|C_{m(t),\mathcal{I}}|^{\beta-\varepsilon}\right) \\ &\leq \mathcal{O}(g(n)). \end{aligned}$$

Hence,

$$\begin{aligned} |C_{m(t),\mathcal{I}}|^{\beta} \delta_{t,s,\mathcal{I}} e_{t,s,\mathcal{I}}^{-(1+\varepsilon)} &\leq \mathcal{O}\left(l(n)^{-(1+\varepsilon)}|C_{m(t),\mathcal{I}}|^{\beta-\varepsilon} L(n)\right) \\ &\leq \mathcal{O}(g(n)). \end{aligned}$$

Therefore, for all $0 < \gamma < 1$, the structure $\mathcal{S}$ has γ-controlled geometry. □

By Lemma B.20, if the structure $\mathcal{S}$ has gaps, the number of vertices with the same mother can increase polynomially or exponentially from level n to level $n+1$ and $\mathcal{S}$ be a structure with γ-controlled geometry. For instance, let $0 < \beta \leq \mu < 1$ and $p_m(n) = a_0 n^m + \ldots$ and $q_m(n) = b_0 n^m + \ldots$ be polynomials of degree m, where $a_0, b_0 > 0$. If $l(n) = \beta^{p_m(n)}$ and $L(n) = \mu^{q_m(n)}$, then the structure $\mathcal{S}$ has α-controlled geometry.

Condition (ii) can easily be modified to allow that a vertex t and its ancestors to at most $m^k(t)$ could define the same cylinders, for some $k \geq 1$ not depending upon the vertex t.

Moreover, γ-controlled geometry include cases, in opposition to Lemma B.20, where the length of the cylinders does not decrease as fast as in the case of bounded geometry. For these cases, γ can be different of α. Therefore, γ-controlled geometry is a concept much more general than bounded geometry.

Definition B.21. *Let $\mathcal{S}$ and $\mathcal{T}$ be $C^{1+\alpha^-}$ structures on L_T with α-controlled geometries and $\mathcal{I}$ and $\mathcal{J}$ be, respectively, primary atlases for $\mathcal{S}$ and $\mathcal{T}$. The structures $\mathcal{S}$ and $\mathcal{T}$ are* $(1+\alpha)$-equivalent *($\mathcal{S} \overset{\alpha}{\sim} \mathcal{T}$), if, for all $0 < \beta < \alpha$ and for all $t \in \tilde{T}_n$, $\nu_t < \mathcal{O}(|C_{m(t),\mathcal{I}}|^{\beta})$ and for all s in contact with t, $\nu_{t,s} < \mathcal{O}\left(d_{t,s,\mathcal{I}}^{\beta}\right)$.*

The following theorem is an immediate consequence of the general theory on smooth structures in Section B.3.

Putting together Lemma B.18 and Theorems B.15 and B.17, we obtain the following result:

Corollary B.22. *Let $\mathcal{S}$ and $\mathcal{T}$ be $C^{1+\alpha^-}$ structures with α-controlled geometries on L_T. The $C^{1+\alpha^-}$ structures on L_T, $\mathcal{S}$ and $\mathcal{T}$ are $C^{1+\alpha^-}$-equivalent if, and only if, $\mathcal{S} \overset{\alpha}{\sim} \mathcal{T}$.*

Putting together Lemma B.18 and Theorem B.16, we get the following result.

Corollary B.23. *Let β be equal to α in the Definitions B.19 and B.21. Then, the $C^{1+\alpha}$ structures $\mathcal{S}$ and $\mathcal{T}$ with α-controlled geometries are $C^{1+\alpha}$-equivalent, if $\mathcal{S} \overset{\alpha}{\sim} \mathcal{T}$.*

An interesting feature of Corollary B.22 is that it gives a balanced equivalence between the scaling of the partition structures and the degree of smoothness between them.

A compatible chart (i, L) with the $C^{1+\alpha^-}$ structure $\mathcal{S}$ can be regarded as a smooth structure $\mathcal{T}$ on L. Let the structure $\mathcal{S}'$ on L be the restriction of the structure $\mathcal{S}$ to L. Then, (i, L) is a compatible $C^{1+\alpha^-}$ chart of $\mathcal{S}$ if, and only if, $\mathcal{T} \overset{\alpha}{\sim} \mathcal{S}'$.

The definitions and results of this section are independent of the primary atlas chosen for the smooth structures on L_T. This is due to the facts that:

(i) the structures have α-controlled geometry and this property is independent of the primary atlas considered;
(ii) by Corollary B.22, the structures $\mathcal{S}$ and $\mathcal{T}$, with primary atlas $\mathcal{I}$ and $\mathcal{J}$, respectively, are $C^{1+\alpha}$-equivalent if, and only if, $\mathcal{I} \overset{\alpha}{\sim} \mathcal{J}$.
(iii) Thus, if $\mathcal{I}$ and $\mathcal{J}$ are different primary atlas for the same structure $\mathcal{S}$, they are $(1+\alpha)$-equivalent, which implies that
(iv) the definition of $(1+\alpha)$-equivalence is independent of the primary atlas considered.
(v) Therefore, Corollary B.22 is independent of the primary atlas considered.

B.4.1 Bounded geometry

Definition B.24. *A structure $\mathcal{S}$ has* bounded geometry, *if, for some primary atlas $\mathcal{I}$, $\sigma_{\mathcal{I}}(t)$ is bounded away from 0, i.e. there exists $0 < \delta < 1$ such that $\sigma_{\mathcal{I}}(t) > \delta$, for all $t \in \tilde{T}_n$, $n \geq N_{\mathcal{I},\mathcal{J}}$. Recall that $\sigma_{\mathcal{I}}(t) = |C_{t,\mathcal{I}}|/|C_{m(t),\mathcal{I}}|$ and $\sigma_{\mathcal{I}}(g_{t,s}) = |G_{t,s,\mathcal{I}}|/|C_{m(t),\mathcal{I}}|$. Moreover, there is $l > 0$ such that, for all $t \in T_n$, if $\sigma_{\mathcal{I}}(t) = 1$, then $\sigma_{\mathcal{I}}(m^l(t)) < 1$.*

The definition of bounded geometry for a smooth structure $\mathcal{S}$ does not depend of the atlas considered, although the constant δ is not necessarily the same for different primary atlas.

Some examples of smooth structures with bounded geometry are the ones generated by smooth circle maps with rotation number of constant type, by the closure of the orbit of the critical point of unimodal maps infinitely renormalizable with bounded geometry and by Markov maps.

Lemma B.25. *A structure $\mathcal{S}$ with bounded geometry has α-controlled geometry, for all $0 < \alpha < 1$.*

Proof. By bounded geometry, for all $t \in \tilde{T}_n$, there is $1 \leq j \leq l$ such that

$$\frac{|C_{t,\mathcal{I}}|}{|C_{m^{j-1}(t),\mathcal{I}}|} = 1 \quad \text{and} \quad \frac{|C_{t,\mathcal{I}}|}{|C_{m^j(t),\mathcal{I}}|} < 1 - \delta. \tag{B.21}$$

Clearly, for all $t \in \tilde{T}_n$,

$$\mathcal{O}(\delta^n) < |C_{t,\mathcal{I}}| < \mathcal{O}((1-\delta)^{n/l}). \tag{B.22}$$

Conditions (i), (ii) and (iv) in the definition of α-controlled geometry are verified, by (B.22), for the decreasing function $g = g_{\beta,\varepsilon} : \mathbb{Z}_{\geq 0} \to \mathbb{R}$ given by

$$g(n) = c((1-\delta)^{n/l})^{\beta-\varepsilon},$$

for some constant $c > 0$. Let us prove that condition (iii) is also verified. For all adjacent vertices $t, s \in \tilde{T}_n$, that are not in contact, we have, by definition, that

$$\delta_{t,s,\mathcal{I}} = \frac{1}{2}|C_{t_1,\mathcal{I}}|\frac{|G_{t,s,\mathcal{I}}|}{|C_{t_2,\mathcal{I}}|}.$$

Recall that t_1 is the vertex such that $C_{t_1,\mathcal{I}} \neq C_{m(t_1),\mathcal{I}} = C_{m(t),\mathcal{I}}$ and t_2 is the vertex such that $C_{t_2,\mathcal{I}}$ and $G_{t,s,\mathcal{I}}$ have the same mother and $C_{t_2,\mathcal{I}}$ is an ancestor of $C_{t,\mathcal{I}}$. Therefore, by (B.21), $|G_{t,s,\mathcal{I}}|/|C_{t_2,\mathcal{I}}| = \mathcal{O}(1)$ and $\mathcal{O}(|C_{t_1,\mathcal{I}}|) = \mathcal{O}(|C_{m(t_1),\mathcal{I}}|) = \mathcal{O}(|C_{m(t),\mathcal{I}}|)$. Thus, $\delta_{t,s,\mathcal{I}} = \mathcal{O}(|C_{m(t),\mathcal{I}}|)$.

Let $t', s' \in \tilde{T}_{n+1}$ be such that $m(t') = t$ and $m(s') = s$ and t'_1 is the vertex such that $C_{t'_1,\mathcal{I}} \neq C_{m(t'_1),\mathcal{I}} = C_{m(t'),\mathcal{I}}$. If $e_{t,s,\mathcal{I}} = \delta_{t,s,\mathcal{I}} - \delta_{t',s',\mathcal{I}} \neq 0$, then, by (B.21),

$$|C_{t_1,\mathcal{I}}| > |C_{t_1,\mathcal{I}}| - |C_{t'_1,\mathcal{I}}| = |C_{m(t'_1),\mathcal{I}}| - |C_{t'_1,\mathcal{I}}| > \delta|C_{m(t'_1),\mathcal{I}}| = \delta|C_{t_1,\mathcal{I}}|.$$

Therefore, if $e_{t,s,\mathcal{I}} \neq 0$, then

$$\mathcal{O}(e_{t,s,\mathcal{I}}) = \mathcal{O}(|C_{t_1,\mathcal{I}}|) = \mathcal{O}(|C_{m(t),\mathcal{I}}|) = \mathcal{O}(\delta_{t,s,\mathcal{I}}),$$

that, together with (B.22), proves condition (iii) of the definition of α-controlled geometry. □

Putting together Lemma B.25 and Corollary B.22, we obtain the following result.

Theorem B.26. *Let $\mathcal{S}$ and $\mathcal{T}$ be $C^{1+\alpha^-}$ structures on $L_{\mathcal{T}}$ with bounded geometry. Then, $\mathcal{S}$ and $\mathcal{T}$ are $C^{1+\alpha^-}$-equivalent if, and only if, $\mathcal{S} \overset{\alpha}{\sim} \mathcal{T}$.*

Definition B.27. *(i) $\mathcal{S}$ is a C^{1+} structure on $L_{\mathcal{T}}$ if, and only if, $\mathcal{S}$ is a $C^{1+\varepsilon}$ structure, for some $\varepsilon > 0$.*
(ii) The structures $\mathcal{S}$ and $\mathcal{T}$ are C^{1+}-equivalent if, and only if, they are $C^{1+\varepsilon}$-equivalent, for some $\varepsilon > 0$.
(iii) The structures $\mathcal{S}$ and $\mathcal{T}$ are $(1+)$-equivalent ($\mathcal{S} \overset{1+}{\sim} \mathcal{T}$) if, and only if, there is $\lambda \in (0,1)$ such that, for all $t \in \tilde{T}_n$, $\nu_t \leq \mathcal{O}(\lambda^n)$ and if s is in contact with t, then $\nu_{t,s} \leq \mathcal{O}(\lambda^n)$.

Theorem B.28. *Let $\mathcal{S}$ and $\mathcal{T}$ be C^{1+} structures on $L_{\mathcal{T}}$ with bounded geometry and $\mathcal{I}$ (resp. $\mathcal{J}$) be primary atlas. For bounded geometry, a necessary and sufficient condition for the C^{1+} structures $\mathcal{S}$ and $\mathcal{T}$ to be C^{1+}-equivalent is that $\mathcal{S} \overset{1+}{\sim} \mathcal{T}$.*

Proof. Let $0 < \varepsilon' < 1$ be such that $\mathcal{S}$ and $\mathcal{T}$ are $C^{1+\varepsilon'}$ structures on $L_{\mathcal{T}}$. Let us prove if, for all $t \in \tilde{T}_n$ and all s in contact with t, $\nu_t \leq \mathcal{O}(\lambda^n)$ and $\nu_{t,s} \leq \mathcal{O}(\lambda^n)$, then there is $0 < \beta < \varepsilon'$ such that $\mathcal{S}$ and $\mathcal{T}$ are $C^{1+\beta}$-equivalent.

Take $0 < \varepsilon < \varepsilon'$ such that $\lambda \leq \delta^{\varepsilon}$. By bounded geometry,

$$\nu_t \leq \mathcal{O}(\lambda^n) \leq \mathcal{O}((\delta^n)^{\varepsilon}) \leq \mathcal{O}(|C_{t_1,\mathcal{I}}|^{\varepsilon})$$

and

$$\nu_{t,s} \leq \mathcal{O}(\lambda^n) \leq \mathcal{O}((\delta^n)^{\varepsilon}) \leq \mathcal{O}(\delta^{\varepsilon}_{t,s,\mathcal{I}}).$$

Therefore, the structures $\mathcal{S}$ and $\mathcal{T}$ are $(1+\varepsilon)$-equivalent, and by Corollary B.26 they are the $C^{1+\beta}$-equivalent for some $0 < \beta < \varepsilon$.

Let us prove that if there is $0 < \beta < \varepsilon'$ such that $\mathcal{S}$ and $\mathcal{T}$ are $C^{1+\beta}$-equivalent, then there is $0 < \lambda < 1$ such that, for all $t \in \tilde{T}_n$, and s in contact with t, $\nu_t \leq \mathcal{O}(\lambda^n)$ and $\nu_{t,s} \leq \mathcal{O}(\lambda^n)$.

Let $0 < \varepsilon < \beta$ and $0 < \lambda < 1$ be such that $\lambda \geq (1-\delta)^{\varepsilon/l}$. By Corollary B.26, the structures $\mathcal{S}$ and $\mathcal{T}$ are $(1+\beta)$-equivalent, and by (B.22) in proof of Lemma B.25,

$$\nu_t \leq \mathcal{O}(|C_{t_1,\mathcal{I}}|^{\varepsilon}) \leq \mathcal{O}\left(((1-\delta)^{n/l})^{\varepsilon}\right) \leq \mathcal{O}(\lambda^n)$$

and

$$\nu_{t,s} \leq \mathcal{O}(\delta^{\varepsilon}_{t,s,\mathcal{I}}) \leq \mathcal{O}\left(((1-\delta)^{n/l})^{\varepsilon}\right) \leq \mathcal{O}(\lambda^n),$$

that proves the theorem. □

B.5 Further literature

This chapter is based on Pinto and Rand [158].

C

Appendix C: Expanding dynamics of the circle

We discuss two questions about degree d smooth expanding circle maps, with $d \geq 2$. (i) We characterize the sequences of asymptotic length ratios which occur for systems with Hölder continuous derivative. The sequence of asymptotic length ratios are precisely those given by a positive Hölder continuous function s (solenoid function) on the Cantor set C of d-adic integers satisfying a functional equation called the matching condition. In the case of the 2-adic integer Cantor set, the functional equation is

$$s(2x+1) = \frac{s(x)}{s(2x)}\left(1 + \frac{1}{s(2x-1)}\right) - 1 \; .$$

We also present a one-to-one correspondence between solenoid functions and affine classes of exponentially fast d-adic tilings of the real line that are fixed points of the d-amalgamation operator. (ii) We calculate the precise maximum possible level of smoothness for a representative of the system, up to diffeomorphic conjugacy, in terms of the functions s and $cr(x) = (1 + s(x))/(1 + (s(x+1))^{-1})$. For example, in the Lipschitz structure on C determined by s, the maximum smoothness is $C^{1+\alpha}$ for $0 < \alpha \leq 1$ if, and only if, s is α-Hölder continuous. The maximum smoothness is $C^{2+\alpha}$ for $0 < \alpha \leq 1$ if, and only if, cr is $(1+\alpha)$-Hölder. A curious connection with Mostow type rigidity is provided by the fact that s must be constant if it is α-Hölder for $\alpha > 1$.

C.1 $C^{1+Hölder}$ structures U for the expanding circle map E

In this section, we present the definition of a $C^{1+Hölder}$ expanding circle map E with respect to a structure U and give its characterization in terms of the ratio distortion of E at small scales with respect to the charts in U.

The *expanding circle map* $E = E(d) : S \to S$, with degree $d \geq 2$, is given by $E(z) = z^d$ in complex notation. Let $p \in S$ be one of the fixed points of the expanding circle map E. The *Markov intervals of the expanding circle map* E are the adjacent closed intervals $I_0, \ldots, I_{d-1}$ with non empty interior such that only their boundaries are contained in the set $\{E^{-1}(p)\}$ of pre-images of the fixed point $p \in S$. Choose the interval I_0 such that $I_0 \cap I_{d-1} = \{p\}$. Let the *branch expanding circle map* $E_i : I_i \to S$ be the restriction of the expanding circle map E to the Markov interval I_i, for all $0 \leq i < d$. Let the interval $I_{\alpha_1 \ldots \alpha_n}$ be $E_{\alpha_n}^{-1} \circ \ldots \circ E_{\alpha_1}^{-1}(S)$. The nth*-level of the interval partition of the expanding circle map* E is the set of all closed intervals $I_{\alpha_1 \ldots \alpha_n} \in S$.

A $C^{1+H\ddot{o}lder}$ diffeomorphism $h : I \to J$ is a $C^{1+\varepsilon}$ diffeomorphism for some $\varepsilon > 0$ (the notion of a quasisymmetric homeomorphism and of a $C^{1+\varepsilon}$ diffeomorphism $h : I \to J$ are the usual ones and are presented in sections A.3 and A.6, respectively.)

Definition 35 The expanding circle map $E : S \to S$ is $C^{1+H\ddot{o}lder}$ with respect to a structure U on the circle S *if for every finite cover U' of U,*

(i) there is an $\varepsilon > 0$ with the property that for all charts $u : I \to \mathbb{R}$ and $v : J \to \mathbb{R}$ contained in U' and for all intervals $K \subset I$ such that $E(K) \subset J$, the maps $v \circ E \circ u^{-1}|u(K)$ are $C^{1+\varepsilon}$ and their $C^{1+\varepsilon}$ norms are bounded away from zero and infinity;
(ii) there are constants $c > 0$ and $\nu > 1$ such that, for every $n > 0$ and every $x \in S$, $|(v \circ E^n \circ u^{-1})'(x)| > c\nu^n$, where $u : I \to \mathbb{R}$ and $v : J \to \mathbb{R}$ are any two charts in U' such that $x \in u(I)$ and $E^n \circ u(x) \in J$.

Remark C.1. The above condition (ii) is equivalent to say that all $C^{1+H\ddot{o}lder}$ expanding maps, that we consider in this chapter, are quasisymmetric conjugated to the affine expanding map $E = E(d) : S \to S$ given by $E(z) = z^d$ in complex notation.

It is well-known that quasisymmetry implies Hölder continuity, but, in general, the opposite is not true. However, in the above remark, condition (ii) is also equivalent to say that the affine expanding map is Hölder conjugated to the $C^{1+H\ddot{o}lder}$ expanding maps that we consider in this chapter.

Lemma C.2. *The expanding circle map $E : S \to S$ is $C^{1+H\ddot{o}lder}$ with respect to a structure U if, and only if, for every finite cover U' of U, there are constants $0 < \mu < 1$ and $b > 1$ with the following property: for all charts $u : J \to \mathbb{R}$ and $v : K \to \mathbb{R}$ contained in U' and for all adjacent intervals $I_{\alpha_1 \ldots \alpha_n}$ and $I_{\beta_1 \ldots \beta_n}$ at level n of the interval partition such that $I_{\alpha_1 \ldots \alpha_n}, I_{\beta_1 \ldots \beta_n} \subset J$ and $E(I_{\alpha_1 \ldots \alpha_n}), E(I_{\beta_1 \ldots \beta_n}) \subset K$, we have that*

$$b^{-1} < \frac{|u(I_{\alpha_1 \ldots \alpha_n})|}{|u(I_{\beta_1 \ldots \beta_n})|} < b \quad \text{and} \quad \left|\log \frac{|u(I_{\alpha_1 \ldots \alpha_n})| \; |v(E(I_{\beta_1 \ldots \beta_n}))|}{|u(I_{\beta_1 \ldots \beta_n})| \; |v(E(I_{\alpha_1 \ldots \alpha_n}))|}\right| \leq \mathcal{O}(\mu^n) \; .$$

Lemma C.2 follows from Theorem A.15 in Section A.2 and Remark C.1.

By using the Mean Value Theorem we obtain the following result for a $C^{1+H\ddot{o}lder}$ expanding circle map $E : S \to S$ with respect to a structure U.

For every finite cover U' of U, there is an $\varepsilon > 0$, with the property that for all charts $u : J \to \mathbb{R}$ and $v : K \to \mathbb{R}$ contained in U' and for all adjacent intervals I and I', such that $I, I' \subset J$ and $E^n(I), E^n(I') \subset K$, for some $n \geq 1$, we have

$$\left| \log \frac{|u(I)||v(E^n(I'))|}{|u(I')||v(E^n(I))|} \right| \leq \mathcal{O}(|v(E^n(I)) \cup v(E^n(I'))|^{\varepsilon}). \tag{C.1}$$

C.2 Solenoids (Ẽ,S̃)

In this section, we introduce the notion of a (thca) solenoid $(\tilde{E}, \tilde{S})$ and we prove that a $C^{1+Hölder}$ expanding circle map E with respect to a structure U determines a unique (thca) solenoid.

The sequence $\mathbf{x} = (\ldots, x_3, x_2, x_1, x_0)$ is *an inverse path of the expanding circle map* E if $E(x_n) = x_{n-1}$, for all $n \geq 1$. The *topological solenoid* $\tilde{S}$ consists of all inverse paths $\mathbf{x} = (\ldots, x_3, x_2, x_1, x_0)$ of the expanding circle map E with the product topology. The *solenoid map* $\tilde{E}$ is the bijective map defined by

$$\tilde{E}(\mathbf{x}) = (\ldots x_0, E(x_0)).$$

The *projection map* $\pi = \pi_S : \tilde{S} \to S$ is defined by $\pi(\mathbf{x}) = x_0$. A *fiber or transversal* over $x_0 \in S$ is the set of all points $\mathbf{x} \in \tilde{S}$ such that $\pi(\mathbf{x}) = x_0$. A fiber is topologically a Cantor set $\{0, \ldots, d-1\}^{\mathbb{N}_0}$. A *leaf* $\mathcal{L} = L_{\mathbf{z}}$ is the set of all points $\mathbf{w} \in \tilde{S}$ path connected to the point $\mathbf{z} \in \tilde{S}$. A *local leaf* $\mathcal{L}'$ is a path connected subset of a leaf. A local leaf $\mathcal{L}'$ is *adjacent* to a local leaf $\mathcal{L}''$, if $\mathcal{L}'$ intersected with $\mathcal{L}''$ is equal to a unique point.

The *monodromy map* $\tilde{M} : \tilde{S} \to \tilde{S}$ is defined such that the local leaf starting on $\mathbf{x}$ and ending on $\tilde{M}(\mathbf{x})$ after being projected by π is an anti-clockwise arc starting on x_0, going around the circle once, and ending on the point x_0. Since the orbit of any point $\mathbf{x} \in \tilde{S}$ under $\tilde{M}$ is dense on its fiber (see Lemma C.5 in Section C.3), we get that all leaves $\mathcal{L}$ of the solenoid $\tilde{S}$ are dense. Hence, the topological solenoid is a compact set and is the twist product of the circle S with the Cantor set $\{0, \ldots, d-1\}^{\mathbb{N}_0}$, where the twist is determined by the monodromy map.

We define a metric m on each transversal as follows: Let $0 < \mu < 1$. For every $\mathbf{x}$ and $\mathbf{y}$ in the same fiber, we define $m(\mathbf{x}, \mathbf{y}) = \mu^n$ if $x_n = y_n$ and $x_{n+1} \neq y_{n+1}$.

Definition 36 The solenoid $(\tilde{E}, \tilde{S})$ is transversely continuous affine (tca) *if (i) every leaf $\mathcal{L}$ has an affine structure; (ii) the solenoid map $\tilde{E}$ preserves the affine structure on the leaves; and (iii) the ratio between the lengths of adjacent leaves, determined by their affine structures, varies continuously along transversals.* The solenoid $(\tilde{E}, \tilde{S})$ is transversely Hölder continuous affine (thca) *if the solenoid is (tca) and the ratio between adjacent leaves determined by their affine structure varies Hölder continuously along transversals.*

We say that $(\mathbf{x}, \mathbf{y}, \mathbf{z})$ *is a triple*, if the points $\mathbf{x}$, $\mathbf{y}$ and $\mathbf{z}$ are distinct and are contained in the same leaf $\mathcal{L}$ of $\tilde{S}$. Let T be the set of all triples $(\mathbf{x}, \mathbf{y}, \mathbf{z})$. A function $r : T \to \mathbb{R}^+$ is *invariant by the action of the solenoid map* $\tilde{E}$ if, and only if, for all triples $(\mathbf{x}, \mathbf{y}, \mathbf{z}) \in T$, we have $r(\mathbf{x}, \mathbf{y}, \mathbf{z}) = r(\tilde{E}(\mathbf{x}), \tilde{E}(\mathbf{y}), \tilde{E}(\mathbf{z}))$. A function $r : T \to \mathbb{R}^+$ *varies Hölder continuously along fibers or, equivalently, transversals* if, and only if, for all triples $(\mathbf{x}, \mathbf{y}, \mathbf{z}), (\mathbf{x}', \mathbf{y}', \mathbf{z}') \in T$ such that $\mathbf{x}$ and $\mathbf{x}'$ are in the same fiber, $\mathbf{y}$ and $\mathbf{y}'$ are in the same fiber, and $\mathbf{z}$ and $\mathbf{z}'$ are in the same fiber, we have

$$|\log(r(\mathbf{x}, \mathbf{y}, \mathbf{z})) - \log(r(\mathbf{x}', \mathbf{y}', \mathbf{z}'))| \leq \max\{m(\mathbf{x}, \mathbf{x}'), m(\mathbf{y}, \mathbf{y}'), m(\mathbf{z}, \mathbf{z}')\} .$$

Definition 37 *A* leaf ratio function $r : T \to \mathbb{R}^+$ *is a continuous function invariant by the action of the solenoid map* $\tilde{E}$ *and satisfying the following* matching condition*: for all triples* $(\mathbf{x}, \mathbf{w}, \mathbf{y}), (\mathbf{w}, \mathbf{y}, \mathbf{z}) \in T$,

$$r(\mathbf{x}, \mathbf{y}, \mathbf{z}) = \frac{r(\mathbf{x}, \mathbf{w}, \mathbf{y}) r(\mathbf{w}, \mathbf{y}, \mathbf{z})}{1 + r(\mathbf{x}, \mathbf{w}, \mathbf{y})} .$$

A Hölder leaf ratio function $r : T \to \mathbb{R}^+$ *is a leaf ratio function varying Hölder continuously along fibers.*

Lemma C.3. *There is a one-to-one correspondence between (thca) solenoids* $(\tilde{E}, \tilde{S})$ *and Hölder leaf ratio functions* $r : T \to \mathbb{R}^+$.

Proof. The affine structures on the leaves of the (thca) solenoid $\tilde{S}$ determine a function $r : T \to \mathbb{R}^+$ that varies continuously along leaves, and satisfies the matching condition. The converse is also true. Moreover, (i) the solenoid map $\tilde{S}$ preserves the affine structure on the leaves if and only if the function $r : T \to \mathbb{R}^+$ is invariant by the action of the solenoid map $\tilde{E}$ and (ii) the ratio between adjacent leaves determined by their affine structure changes Hölder continuously along transversals if and only if the function $r : T \to \mathbb{R}^+$ varies Hölder continuously along fibers. □

Lemma C.4. *A* $C^{1+Hölder}$ *expanding circle map* $E : S \to S$ *with respect to a structure* U *generates a Hölder leaf ratio function* $r_U : T \to \mathbb{R}^+$.

Proof. Let U' be a finite cover of U. For every triple $(\mathbf{x}, \mathbf{y}, \mathbf{z}) \in T$ and every n large enough, let $u_n : J_n \to \mathbb{R}$ be a chart contained in U' such that $x_n, y_n, z_n \in J_n$. Using (C.1), $r_U(\mathbf{x}, \mathbf{y}, \mathbf{z})$ is well-defined by

$$r_U(\mathbf{x}, \mathbf{y}, \mathbf{z}) = \lim_{n \to \infty} \frac{|u_n(y_n) - u_n(z_n)|}{|u_n(x_n) - u_n(y_n)|} .$$

By construction, r_U is invariant by the dynamics of the solenoid map and satisfies the matching condition. Again, using (C.1), we obtain that r_U is a continuous function varying Hölder continuously along transversals. Hence, r_U is a leaf ratio function. □

C.3 Solenoid functions $s : C \to \mathbb{R}^+$

In this section, we will introduce the notion of a solenoid function whose domain is a fiber of the solenoid. We will show that a Hölder leaf ratio function determines a Hölder solenoid function and that a Hölder solenoid function determines an element in the set of sequences $A(d)$ defined below.

Definition 38 *Let the space $A(d)$ be the set of all sequences $\{a_1, a_2, \ldots\}$ of positive real numbers with the following properties:*

(i) there is $0 < \nu < 1$ such that $a_n/a_m \leq \nu^i$ if $n - m$ is divisible by d^i and
(ii) $a_1, a_2, \ldots$ satisfies

$$a_m = \frac{\prod_{i=1}^{d-1} a_{dm-i} \left(\sum_{j=0}^{d-1} \prod_{l=0}^{j} a_{dm+l} \right)}{1 + \sum_{j=1}^{d-1} \prod_{l=j}^{d-1} a_{dm-l}} . \tag{C.2}$$

A geometric interpretation of the sequences contained in the set $A(d)$ is given by the d-adic tilings and grids of the real line defined in Section C.4, below.

Let $\sum_{i=-\infty}^{\infty} a_i d^i$ be a *d-adic number*. The d-adic numbers

$$\sum_{i=-\infty}^{n-1} (d-1)d^i + \sum_{i=n}^{\infty} a_i d^i \quad \text{and} \quad (a_n + 1)d^n + \sum_{i=n+1}^{\infty} a_i d^i$$

such that $a_n + 1 < d$ are *d-adic equivalent. The d-adic set* $\tilde{\Omega}$ is the topological Cantor set $\{0, \ldots, d-1\}^{\mathbb{Z}}$ of all d-adic numbers modulo the above d-adic equivalence. The *product map* $d\times : \tilde{\Omega} \to \tilde{\Omega}$ is the multiplication by d of the d-adic numbers. The *add* 1 *map* $1+ : \tilde{\Omega} \to \tilde{\Omega}$ is the sum of 1 to the d-adic numbers.

Let the map $\tilde{\omega} : \tilde{\Omega} \to \tilde{S}$ be the homeomorphism between the d-adic set $\tilde{\Omega}$ and the solenoid $\tilde{S}$ defined as follows: $\tilde{\omega}(\sum_{i=-\infty}^{\infty} a_i d^i) = \mathbf{x} = (\ldots, x_1, x_0) \in \tilde{S}$, where $x_n = \cap_{i=1}^{\infty} E_{a_{n-1}}^{-1} \circ \ldots \circ E_{a_{n-i}}^{-1} (I_{a_{n-(i+1)}})$ for all $n \geq 0$ (recall that $I_{a_{n-(i+1)}}$ is a Markov interval of the expanding circle map E). Hence $x_n \in I_{a_n}$ for all $n \geq 0$. By construction, the map $\tilde{\omega} : \tilde{\Omega} \to \tilde{S}$ conjugates the product map $d\times : \tilde{\Omega} \to \tilde{\Omega}$ with the solenoid map $\tilde{E} : \tilde{S} \to \tilde{S}$, and conjugates the add 1 map $1+ : \tilde{\Omega} \to \tilde{\Omega}$ with the monodromy map $\tilde{M} : \tilde{S} \to \tilde{S}$.

Lemma C.5. *Every orbit of the monodromy map is dense on its fiber.*

Proof. Since the add 1 map $1+ : \tilde{\Omega} \to \tilde{\Omega}$ is dense on the image $\tilde{\omega}^{-1}(F)$ of every fiber F of the solenoid $\tilde{S}$, the lemma follows. □

Let Ω be the topological Cantor set $\{0, \ldots, d-1\}^{\mathbb{Z}_{\leq 0}}$ corresponding to all d-adic numbers of the form $\sum_{i=-\infty}^{-1} a_i d^i$ modulo the d-adic equivalence. *The projection map* $\pi_\Omega : \tilde{\Omega} \to \Omega$ is defined by $\pi_\Omega \left(\sum_{i=-\infty}^{\infty} a_i d^i \right) = \sum_{i=-\infty}^{-1} a_i d^i$. The map $\omega : \Omega \to S$ is defined by $\omega(\sum_{i=-\infty}^{-1} a_i d^i) = \cap_{i=1}^{\infty} E_{a_{-1}}^{-1} \circ \ldots \circ E_{a_{-i}}^{-1}(I_{a_{-(i+1)}})$. By construction,

$$\omega \circ \pi_\Omega \left(\sum_{i=-\infty}^{\infty} a_i d^i \right) = \pi_S \circ \tilde{\omega} \left(\sum_{i=-\infty}^{\infty} a_i d^i \right),$$

for all $\sum_{i=-\infty}^{\infty} a_i d^i \in \tilde{\Omega}$.

The set C is the topological Cantor set $\{0, \ldots, d-1\}^{\mathbb{Z}_{\geq 0}}$ corresponding to all d-adic integers of the form $\sum_{i=0}^{\infty} a_i d^i$.

Definition 39 *The* solenoid function $s : C \to \mathbb{R}^+$ *is a continuous function satisfying the following* matching condition, *for all* $a \in C$*:*

$$s(a) = \frac{\prod_{i=1}^{d-1} s(da - i) \left(\sum_{j=0}^{d-1} \prod_{l=0}^{j} s(da + l) \right)}{1 + \sum_{j=1}^{d-1} \prod_{l=j}^{d-1} s(da - l)} . \qquad \text{(C.3)}$$

Lemma C.6. *The Hölder leaf ratio function* $r : T \to \mathbb{R}^+$ *determines a Hölder solenoid function* $s_r : C \to \mathbb{R}^+$.

Proof. For all $\sum_{i=0}^{\infty} a_i d^i \in C$, we define

$$s_r \left(\sum_{i=0}^{\infty} a_i d^i \right) = r \left(\tilde{\omega} \left(\sum_{i=0}^{\infty} a_i d^i - 1 \right), \tilde{\omega} \left(\sum_{i=0}^{\infty} a_i d^i \right), \tilde{\omega} \left(\sum_{i=0}^{\infty} a_i d^i + 1 \right) \right) .$$

The matching condition and the Hölder continuity of the leaf ratio function $r : T \to \mathbb{R}^+$ imply the matching condition and the Hölder continuity of the solenoid function $s_r : C \to \mathbb{R}^+$, respectively. □

Lemma C.7. *There is a one-to-one correspondence between Hölder solenoid functions* $s : C \to \mathbb{R}^+$ *and sequences* $\{r_1, r_2, r_3, \ldots\} \in A(d)$.

Proof. Given a Hölder solenoid function $s : C \to \mathbb{R}^+$, for all $i = \sum_{j=0}^{k} a_j d^j \geq 0$, we define r_i by

$$r_i = s \left(\sum_{j=0}^{k} a_j d^j \right) .$$

The matching condition of the solenoid function $s : C \to \mathbb{R}^+$ implies that the ratios $r_1, r_2, \ldots$ satisfy (C.2). The Hölder continuity of the solenoid function $s : C \to \mathbb{R}^+$ implies condition (i). Conversely, for every d-adic integer $a = \sum_{i=0}^{\infty} a_i d^i \in C$, let $\underline{a}_n \in \mathbb{N}_0$ be equal to $\sum_{i=0}^{n} a_i d^i$. Define the value $s(a)$ by

$$s(a) = \lim_{n \to \infty} r_{\underline{a}_n} .$$

Using condition (i) the above limit is well defined and the function $s : C \to \mathbb{R}^+$ is Hölder continuous. Using condition (ii) and the continuity of s we obtain that the function s satisfies the matching condition. □

C.4 d-Adic tilings and grids

In this section, we introduce d-adic tilings of the real line that are fixed points of the d-amalgamation operator and d-adic fixed grids of the real line. We show that their affine classes are in one-to-one correspondence with (thca) solenoids.

A *tiling* $\mathcal{T} = \{I_\beta \subset \mathbb{R} : \beta \in \mathbb{Z}\}$ *of the real line* is a collection of *tiling intervals* I_β with the following properties: (i) The tiling intervals are closed intervals; (ii) The union $\cup_{\beta \in \mathbb{Z}} I_\beta$ of all tiling intervals I_β is equal to the real line; (iii) any two distinct tiling intervals have disjoint interiors; (iv) For every $\beta \in \mathbb{Z}$, the intersection of the tiling intervals I_β and $I_{\beta+1}$ is only an endpoint common to both intervals; (v) There is $B \geq 1$, such that for every $\beta \in \mathbb{Z}$, we have $B^{-1} \leq |I_{\beta+1}|/|I_\beta| \leq B$. The *tiling sequence* $\underline{r} = (r_m)_{m \in \mathbb{Z}}$ is given by $r_m = |I_{m+1}|/|I_m|$. Let $\mathbf{T}$ denote the set of all tiling sequences. The *d-amalgamation operator* $A_d : \mathbf{T} \to \mathbf{T}$ is defined by $A_d(\underline{r}) = \underline{s}$, where

$$s_i = r_{d(i-1)+1,di} \frac{1 + \sum_{m=di+1}^{d(i+1)-1} r_{di+1,m}}{1 + \sum_{m=d(i-1)+1}^{di-1} r_{d(i-1)+1,m}},$$

for all $i \in \mathbb{Z}$.

Definition 40 *A tiling $\mathcal{T}$ is a fixed point of the d-amalgamation operator, if the corresponding tiling sequence is a fixed point of the d-amalgamation operator, i.e. $A_d(\underline{r}) = \underline{r}$. A tiling is* d-adic, *if there is a sequence $\mu_1, \mu_2, \ldots$ converging to zero such that $|r_j - r_k| \leq \mu_i$, when $(j-k)$ is divisible by d^i. A tiling is* exponentially fast d-adic, *if there is $0 < \mu < 1$ such that $|r_j - r_k| \leq \mathcal{O}(\mu^i)$, when $(j-k)$ is divisible by d^i.*

The tilings $\mathcal{T}_1 = \{I_\beta \subset \mathbb{R} : \beta \in \mathbb{Z}\}$ and $\mathcal{T}_2 = \{J_\beta \subset \mathbb{R} : \beta \in \mathbb{Z}\}$ of the real line are in the same affine class, if there is an affine map $h : \mathbb{R} \to \mathbb{R}$ such that $h(I_\beta) = J_\beta$ for every $\beta \in \mathbb{Z}$. We note that a tiling sequence $\underline{r}$ determines an affine class of tilings $\mathcal{T}$ and vice-versa.

Remark C.8. The tiling sequence $\underline{r} = (r_m)_{m \in \mathbb{Z}}$ of an exponentially fast d-adic tiling of the real line that is a fixed point of the d-amalgamation operator determines a sequence $r_1, r_2, \ldots$ in $A(d)$.

A *d-grid* $\mathcal{G}$ of the real line is a collection of intervals I_β^n satisfying properties (i) to (vii) of a (B,d)-grid $\mathcal{G}_\Omega$ (see Appendix A), for some $B \geq 1$, such that every interval I_β^n is the union of d grid intervals at level $n+1$, and $\Omega(n) = \infty$. We note that every level n of a grid forms a tiling of the real line. We say that the grids $\mathcal{G}_1 = \{I_\beta^n\}$ and $\mathcal{G}_2 = \{J_\beta^n\}$ of the real line are in the same affine class, if there is an affine map $h : \mathbb{R} \to \mathbb{R}$ such that $h(I_\beta^n) = J_\beta^n$ for every $\beta \in \mathbb{Z}$ and every $n \in \mathbb{N}$. The *d-grid sequence* $\ldots \underline{r}^2 \underline{r}^1$ is given by $\underline{r}^n = (r_m^n)_{m \in \mathbb{Z}}$, where $r_m^n = |I_{m+1}^n|/|I_m^n|$. The following remark gives a geometric interpretation of the d-amalgamation operator.

Remark C.9. (i) If $\ldots \underline{r}^2\underline{r}^1$ is a d-grid sequence, then $A_d(\underline{r}^{n+1}) = \underline{r}^n$ for every $n \geq 1$.
(ii) If $\ldots \underline{r}^2\underline{r}^1$ is a sequence such that $A_d(\underline{r}^{n+1}) = \underline{r}^n$, then the sequence determines an affine class of d-grids.

Definition 41 *A* fixed *d*-grid $\mathcal{G}$ *of the real line is a d-grid of the real line such that the corresponding grid sequence* $\ldots \underline{r}^2\underline{r}^1$ *is constant, i.e.* $\underline{r}^1 = \underline{r}^n$ *for every* $n \geq 1$*. A* d-adic fixed grid $\mathcal{G}$ *of the real line is a fixed d-grid such that* $\underline{r}^1$ *is a d-adic tiling. An* exponentially fast *d*-adic fixed grid $\mathcal{G}$ *of the real line is a fixed d-grid such that* $\underline{r}^1$ *is an exponentially fast d-adic tiling.*

Hence, all the levels of a *d*-adic fixed grid $\mathcal{G}$ of the real line determine the same *d*-adic tiling of the real line, up to affine equivalence, that is a fixed point of the *d*-amalgamation operator.

Lemma C.10. *There is a one-to-one correspondence between (i) (thca) solenoids; (ii) affine classes of exponentially fast d-adic tilings of the real line that are fixed points of the d-amalgamation operator; (iii) affine classes of exponentially fast d-adic fixed grids of the real line.*

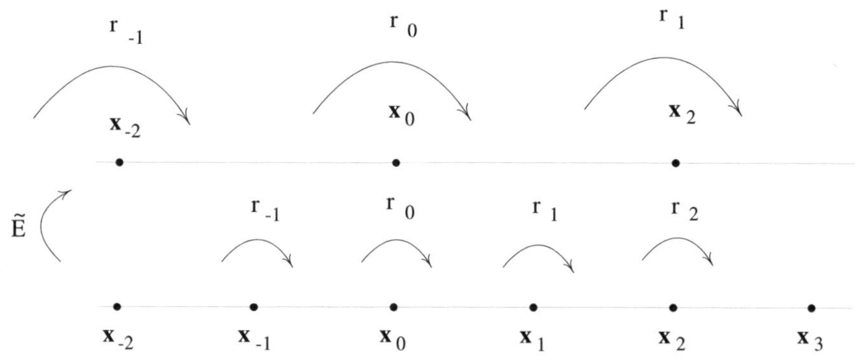

Fig. C.1. The leaf $\mathcal{L}$ fixed by the solenoid map $\tilde{E}$.

Proof. By construction, there is a one-to-one correspondence between (ii) affine classes of exponentially fast *d*-adic quasiperiodic tilings of the real line that are fixed points of the d-amalgamation operator and (iii) affine classes of exponentially fast *d*-adic quasiperiodic fixed grids of the real line. Let us prove that a (thca) solenoid determines canonically an affine class of exponentially fast *d*-adic tilings of the real line that are fixed points of the d-amalgamation operator. Let $\mathcal{L}$ be a leaf of the (thca) solenoid $(\tilde{E}, \tilde{S})$ containing a fixed point $\mathbf{x}_0$ of the solenoid map $\tilde{E}$. The leaf $\mathcal{L}$ is marked by the points $\ldots, \mathbf{x}_{-1}, \mathbf{x}_0, \mathbf{x}_1, \ldots$ that project on the same point of the circle as the fixed point $\mathbf{x}_0$, and such that there is a local leaf $\mathcal{L}_m$ with extreme points $\mathbf{x}_m$ and $\mathbf{x}_{m+1}$ with the property that $\mathcal{L}_m$ does not contain any other point $\mathbf{x}_j$ for $m \neq j \neq m+1$. The affine structure on the leaf $\mathcal{L}$ determines the ratios $r_m = r(\mathbf{x}_{m-1}, \mathbf{x}_m, \mathbf{x}_{m+1})$ of the leaf ratio function $r : T \to \mathbb{R}^+$, for all $m \in \mathbb{Z}$. Since the solenoid map $\tilde{E}$ is affine and $\tilde{E}(\mathcal{L}) = \mathcal{L}$, the sequence of ratios $\underline{r} = (r_m)_{m \in \mathbb{Z}}$ is fixed by the amalgamation operator A_d (see Figure C.1), and so $\underline{r}$ determines an affine class of tilings that are fixed points of the d-amalgamation operator. The Hölder transversality of the solenoid $(\tilde{E}, \tilde{S})$ implies that the sequence $\underline{r}$ determines an affine class of exponentially fast d-adic tilings. Hence, the sequence $\underline{r}$ determines an affine class of exponentially fast d-adic tilings that are fixed points of the d-amalgamation operator, and so the sequence $\underline{r}$ also determines an affine class of exponentially fast d-adic fixed grids of the real line. Conversely, an affine class of exponentially fast d-adic fixed grids of the real line determines uniquely the affine structure of a leaf $\mathcal{L}$ that is fixed by the solenoid map $\tilde{E}$. Since the *grid sequence* $\ldots \underline{r}^2 \underline{r}^1$ is a fixed point of the amalgamation operator, i.e. $A_d(\underline{r}^n) = \underline{r}^{n-1}$, the solenoid map $\tilde{E}$ is affine on the leaf $\mathcal{L}$. By density of the leaf $\mathcal{L}$ on the solenoid $\tilde{S}$ and since the grid g_d is exponentially fast d-adic, the affine structure of the leaf $\mathcal{L}$ extends to an affine structure transversely Hölder continuous on the solenoid $\tilde{S}$ such that the solenoid map $\tilde{E}$ leaves the affine structure invariant. □

C.5 Solenoidal charts for the $C^{1+H\ddot{o}lder}$ expanding circle map E

In this section, we introduce the solenoidal charts which will determine a canonical structure for the expanding circle map.

Definition 42 *Let $\mathcal{L}$ be a local leaf with an affine structure and $\pi_{\mathcal{L}} = \pi_S|\mathcal{L}$ the homeomorphic projection of $\mathcal{L}$ onto an interval $J_{\mathcal{L}}$ of the circle S. Let $\phi_{\mathcal{L}} : \mathcal{L} \to \mathbb{R}$ be a map preserving the affine structure of the leaf $\mathcal{L}$. A* solenoidal chart *$u_{\mathcal{L}} : J_{\mathcal{L}} \to \mathbb{R}$* on the circle *$S$ is defined by $u_{\mathcal{L}} = \phi_{\mathcal{L}} \circ \pi_{\mathcal{L}}^{-1}$ (see Figure C.2).*

Lemma C.11. *The solenoidal charts determined by a (thca) solenoid $(\tilde{E}, \tilde{S})$ produce a canonical structure U such that the expanding circle map E is $C^{1+H\ddot{o}lder}$.*

Proof. Let U' be a finite cover consisting of solenoidal charts. Let $I_{\alpha_1 \ldots \alpha_n}$ and $I_{\beta_1 \ldots \beta_n}$ be adjacent intervals at level n of the interval partition and $u_{\mathcal{L}} : J \to \mathbb{R}$ and $v_{\mathcal{L}'} : K \to \mathbb{R}$ solenoidal charts such that $I_{\alpha_1 \ldots \alpha_n}, I_{\beta_1 \ldots \beta_n} \subset J$ and $I_{\alpha_2 \ldots \alpha_n}, I_{\beta_2 \ldots \beta_n} \subset K$. Let $\mathbf{x}$, $\mathbf{y}$ and $\mathbf{z}$ be the points contained in $\mathcal{L}$ such that $\pi(\mathbf{x})$ and $\pi(\mathbf{y})$ are the endpoints of $I_{\alpha_1 \ldots \alpha_n}$, and $\pi(\mathbf{y})$ and $\pi(\mathbf{z})$ are the endpoints of $I_{\beta_1 \ldots \beta_n}$. Let $\mathbf{x}'$, $\mathbf{y}'$ and $\mathbf{z}'$ be the points contained in $\mathcal{L}'$ such that $\pi(\mathbf{x}')$ and $\pi(\mathbf{y}')$ are the endpoints of $I_{\alpha_2 \ldots \alpha_n}$, and $\pi(\mathbf{y}')$ and $\pi(\mathbf{z}')$ are

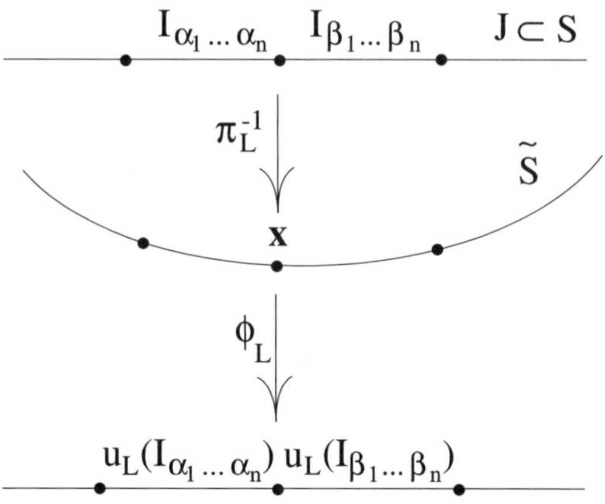

Fig. C.2. The solenoidal chart.

the endpoints of $I_{\beta_2\ldots\beta_n}$ (see Figure C.2). By Lemma C.3, the (thca) solenoid determines a leaf ratio function $r : T \to \mathbb{R}^+$ such that

$$\frac{|u_{\mathcal{L}}(I_{\beta_1\ldots\beta_n})|}{|u_{\mathcal{L}}(I_{\alpha_1\ldots\alpha_n})|}\frac{|v_{\mathcal{L}'}(I_{\alpha_2\ldots\alpha_n})|}{|v_{\mathcal{L}'}(I_{\beta_2\ldots\beta_n})|} = \frac{r(\mathbf{x},\mathbf{y},\mathbf{z})}{r(\mathbf{x}',\mathbf{y}',\mathbf{z}')} \ . \tag{C.4}$$

By Lemma C.6, using that $\tilde{E}$ is affine on leaves, the leaf ratio function $r : T \to \mathbb{R}^+$ determines a solenoid function $s_r : C \to \mathbb{R}^+$ such that

$$\frac{r(\mathbf{x},\mathbf{y},\mathbf{z})}{r(\mathbf{x}',\mathbf{y}',\mathbf{z}')} = \frac{s\left(\tilde{\omega}^{-1}(\tilde{E}^n(\mathbf{y}))\right)}{s\left(\tilde{\omega}^{-1}(\tilde{E}^{n-1}(\mathbf{y}'))\right)} \ . \tag{C.5}$$

By Hölder continuity of the solenoid function,

$$\left|\log \frac{s\left(\tilde{\omega}^{-1}(\tilde{E}^n(\mathbf{x}))\right)}{s\left(\tilde{\omega}^{-1}(\tilde{E}^{n-1}(\mathbf{y}))\right)}\right| \le \mathcal{O}(\mu^n), \tag{C.6}$$

for some $0 < \mu < 1$. Putting (C.4), (C.5) and (C.6) together, and using that C is compact, we obtain that

$$b^{-1} < \frac{|u_{\mathcal{L}}(I_{\alpha_1\ldots\alpha_n})|}{|u_{\mathcal{L}}(I_{\beta_1\ldots\beta_n})|} < b \ \text{ and } \ \left|\log \frac{|u_{\mathcal{L}}(I_{\alpha_1\ldots\alpha_n})|}{|u_{\mathcal{L}}(I_{\beta_1\ldots\beta_n})|}\frac{|v_{\mathcal{L}'}(I_{\beta_2\ldots\beta_n})|}{|v_{\mathcal{L}'}(I_{\alpha_2\ldots\alpha_n})|}\right| \le \mathcal{O}(\mu^n), \tag{C.7}$$

for some $b \ge 1$. Hence, by Lemma C.2, the expanding circle map E is $C^{1+H\ddot{o}lder}$ with respect to the structure U produced by the solenoidal charts. □

Lemma C.12. *The Hölder solenoid function $s : C \to \mathbb{R}^+$ determines a set of solenoidal charts which produce a structure U such that the expanding circle map E is $C^{1+Hölder}$.*

Proof. For every triple $(\mathbf{x}, \mathbf{y}, \mathbf{z})$ such that there are $n \in \mathbb{Z}$ and $a \in C$ with the property that

$$(\tilde{E}^n(\mathbf{x}), \tilde{E}^n(\mathbf{y}), \tilde{E}^n(\mathbf{z})) = (\tilde{\omega}(a-1), \tilde{\omega}(a), \tilde{\omega}(a+1))$$

we define $r(\mathbf{x}, \mathbf{y}, \mathbf{z})$ equal to $s(a)$. Hence, the ratios r are invariant under the solenoid map $\tilde{E}$. Since the solenoid function satisfies the matching condition, the above ratios r determine an affine structure on the leaves of the solenoid. By construction, the solenoidal charts $u_{\mathcal{L}} : J \to \mathbb{R}$ and $v_{\mathcal{L}'} : K \to \mathbb{R}$ determined by this affine structure on the leaves, as in the proof of Lemma C.11 above, satisfy (C.7), and so by Lemma C.2, the expanding circle map E is $C^{1+Hölder}$ with respect to the structure U produced by the solenoidal charts. □

C.6 Smooth properties of solenoidal charts

We will prove that the solenoidal charts maximize the smoothness of the expanding circle map with respect to all charts in the same $C^{1+Hölder}$ structure.

Let U be a $C^{1+Hölder}$ structure for the expanding circle map E. By Lemmas C.3 and C.4, the structure U determines a (thca) solenoid $(\tilde{E}, \tilde{S})_U$.

Lemma C.13. *Let U be a $C^{1+Hölder}$ structure for the expanding circle map E, and let V be the set of all solenoidal charts determined by the (thca) solenoid $(\tilde{E}, \tilde{S})_U$. Then, the set V is contained in U and the degree of smoothness of the expanding circle map E when measured in terms of a cover U' of U attains its maximum when $U' \subset V$.*

Proof. Let the expanding circle map $E : S \to S$ be C^r, for some $r > 1$, with respect to a finite cover U' of the structure U. We shall prove that the solenoidal charts $v_{\mathcal{L}} : I \to \mathbb{R}$ are C^r compatible with the charts contained in U', proving the lemma. Let $\mathcal{L}$ be a local leaf that projects by $\pi_{\mathcal{L}} = \pi_S|\mathcal{L}$ homeomorphically on an interval I contained in the domain J of a chart $u : J \to \mathbb{R}$ of U'. For n large enough, let $u_n : J_n \to \mathbb{R}$ be a chart in U' such that $I_n = \pi_S(\tilde{E}^{-n}(\mathcal{L})) \subset J_n$. Let $\lambda_n : u_n(I_n) \to (0,1)$ be the restriction to the interval $u_n(I_n)$ of an affine map sending the interval $u_n(I_n)$ onto the interval $(0,1)$ (see Figure C.3). Let $e_n : (0,1) \to \mathbb{R}$ be the C^r map defined by $e_n = u \circ E^n \circ u_n^{-1} \circ \lambda_n^{-1}$. The map e_n is the composition of a contraction λ_n^{-1} followed by an expansion $u \circ E^n \circ u_n^{-1}$. Therefore, by the usual blow-down blow-up technique (see the proof of Theorem E.19 and Pinto [150]), the map $e : (0,1) \to \mathbb{R}$ given by $e = \lim_{n\to\infty} e_n$ is a C^r homeomorphism. Hence, the map $v_{\mathcal{L}} : I \to \mathbb{R}$ defined by $e^{-1} \circ u$ is a solenoidal chart and is C^r compatible with the charts contained in U'. □

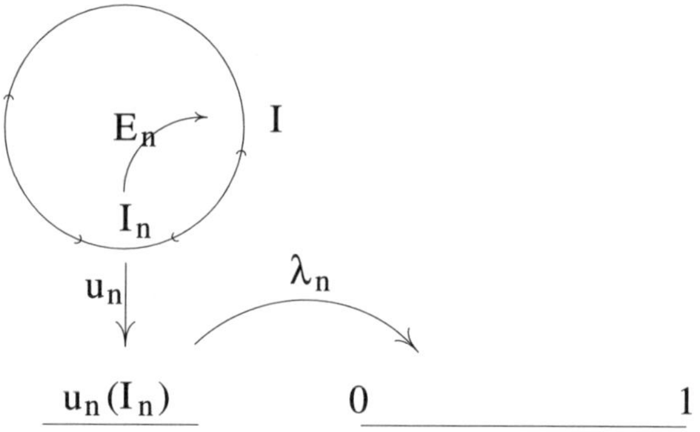

Fig. C.3. The construction of the solenoidal charts from the $C^{1+H\ddot{o}lder}$ structure U.

C.7 A Teichmüller space

Theorem C.14, below, proves the assumptions stated in the first paragraph of the Introduction.

Theorem C.14. *The following sets are canonically isomorphic:*

(i) The set of all $C^{1+H\ddot{o}lder}$ structures U for the expanding circle map $E : S \to S$ of degree $d \geq 2$;
(ii) The set of all (thca) solenoids $(\tilde{E}, \tilde{S})$;
(iii) The set of all Hölder leaf ratio functions $r : T \to \mathbb{R}^+$;
(iv) The set of all Hölder solenoid functions $s : C \to \mathbb{R}^+$;
(v) The set of all sequences $\{r_0, r_1, \ldots\} \in A(d)$;
(vi) The set of all affine classes of exponentially fast d-adic tilings of the real line that are fixed points of the d-amalgamation operator;
(vii) The set of all affine classes of exponentially fast d-adic fixed grids of the real line.

Proof. The proof of this theorem follows from the following diagram, where the implications are determined by the lemmas indicated by their numbers:

$$
\begin{array}{ccccccc}
 & & (i) & \overset{8}{\Leftarrow} & (ii) & \overset{7}{\Leftrightarrow} & (vi),(vii) \\
 & & 9\Uparrow & \overset{3}{\searrow} & \Updownarrow 2 & & \\
(v) & \overset{6}{\Leftrightarrow} & (iv) & \overset{5}{\Leftarrow} & (iii) & &
\end{array}
$$

□

C.8 Sullivan's solenoidal surfaces

We are going to describe Sullivan's one-to-one correspondence between (tca) solenoids and complex structures on a solenoidal surface L. Via this correspondence, the set of all (tca) solenoids is a separable infinite dimensional complex Banach manifold (see Sullivan [232]).

A 2-dimensional *solenoid* is a compact space locally homeomorphic to a (2-ball) product a totally disconnected space. A solenoid is naturally laminated by the path connected components which are called *leaves*. Let $W = \tilde{S} \times \{y : y > 0\}$. Consider the free, properly discontinuous action of the integers generated by the map $(x, y) \to (\tilde{E}(x), 2y)$ on W. The *solenoidal surface* L is the orbit space of this action. Hence, the solenoidal surface L is a 2-dimensional solenoid, since we have a compact fundamental domain $\{(x, y) : a \leq y \leq 2a\}$ for the action considered. Since every leaf of $\tilde{S}$ is dense in $\tilde{S}$, we get that every leaf of L is also dense in L. The periodic leaves under $\tilde{E}$ of $\tilde{S}$ give rise to annuli leaves in L, and the other leaves of L are topological disks. Since the periodic leaves of $\tilde{S}$ are countably many, we obtain that annuli leaves in L are also countably many.

A *complex structure* on solenoidal surface L is a maximal covering of L by lamination charts (disk) $\times$ (transversal) so that overlap homeomorphisms are complex analytic in the disk direction. Two complex structures are *Teichmüller equivalent* if they are related by a homeomorphism which is homotopic to the identity through leaf preserving continuous mappings of L. The set of classes is called the *Teichmüller set* $T(L)$. By Corollary in page 548 of Sullivan [232], the Teichmüller set $T(L)$ can be represented by the smooth conformal structures on L relative to a chosen background smooth structure on L modulo the equivalence relation by diffeomorphisms homotopic to the identity. By Corollary in page 556 of Sullivan [232], the Teichmüller set $T(L)$ has a complex Banach manifold structure.

Let $(\tilde{E}, \tilde{S})$ be a (tca) solenoid. Let $\overline{W} = \tilde{S} \times \{y : y \geq 0\}$. Hence, each leaf l of $\tilde{S}$ has a natural inclusion in $\overline{W}$ as the boundary of a half space H_l. Since the solenoid map $\tilde{E}$ is affine along leaves of $\tilde{S}$, there is a well-defined extension F of $\tilde{E}$ to $\overline{W}$ such that F is a complex affine map when restricted to each half space H_l. Thus, the action of the integers generated by the map F on W determines a orbit space L_F with a natural complex structure. By the Ahlfors-Beurling extension [3], the complex structure of L_F determines a unique element in the Teichmüller set $T(L)$.

Theorem C.15. *(Sullivan [232]) There is a one-to-one correspondence between*

(i) the elements of the Teichmüller set $T(L)$;
(ii) the (tca) solenoids;
(iii) the set of all (uaa) structures U for the expanding circle map E.

See definition of a (uaa) structure U for the expanding circle map E in Section C.9, below. By Theorem C.14, there is a one-to-one correspondence between $C^{1+Hölder}$ expanding circle maps and (thca) solenoids. By Corollary in page 562 of Suulivan [232], the set of elements in the Teichmüller set $T(L)$ corresponding to (thca) solenoids is a dense set of $T(L)$. Similarly, the set of elements in the Teichmüller set $T(L)$ corresponding to (thca) solenoids determined by analytic expanding circle maps is also a dense set of $T(L)$. Furthermore, the set of eigenvalues of the periodic points of $C^{1+Hölder}$ expanding circle maps form a complete set of invariants.

C.9 (Uaa) structures U for the expanding circle map E

In this section, we present the definition of uniformly asymptotically affine (uaa) expanding circle map E, with respect to a structure U, and we define the set $B(d)$. We show a one-to-one correspondence between (uaa) expanding circle map E and the elements in the set $B(d)$.

Definition 43 *The expanding circle map $E : S \to S$ is (uaa) with respect to a structure U if, and only if, for every finite cover U' of U, there is a sequence $\varepsilon_1, \varepsilon_2, \ldots$ converging to zero and a constant $b > 1$ with the following property: for all charts $u : J \to \mathbb{R}$ and $v : K \to \mathbb{R}$ contained in U' and for all adjacent intervals $I_{\alpha_1 \ldots \alpha_n}$ and $I_{\beta_1 \ldots \beta_n}$ at level n of the interval partition, such that $I_{\alpha_1 \ldots \alpha_n}, I_{\beta_1 \ldots \beta_n} \subset J$, and for all $0 \le i \le n$, such that $E^i(I_{\alpha_1 \ldots \alpha_n}), E^i(I_{\beta_1 \ldots \beta_n}) \subset K$, we have that*

$$b^{-1} < \frac{|u(I_{\alpha_1 \ldots \alpha_n})|}{|u(I_{\beta_1 \ldots \beta_n})|} < b \quad \text{and} \quad \left| \log \frac{|u(I_{\alpha_1 \ldots \alpha_n})|\ |v(E^i(I_{\beta_1 \ldots \beta_n}))|}{|u(I_{\beta_1 \ldots \beta_n})|\ |v(E^i(I_{\alpha_1 \ldots \alpha_n}))|} \right| \le \varepsilon_n \ .$$

Using Lemma A.8, in Section A.5, the above definition is equivalent to the one presented in Sullivan [232].

Definition 44 *The space $B(d)$ is the set of all sequences $\{a_1, a_2, \ldots\}$ of positive real numbers with the following properties:*

(i) there is sequence $\nu_1, \nu_2, \ldots$ converging to zero such that $a_n/a_m \le \nu_i$ if $n - m$ is divisible by d^i, and
(ii) $a_1, a_2, \ldots$ satisfies

$$a_m = \frac{\prod_{i=1}^{d-1} a_{dm-i} \left(\sum_{j=0}^{d-1} \prod_{l=0}^{j} a_{dm+l} \right)}{1 + \sum_{j=1}^{d-1} \prod_{l=j}^{d-1} a_{dm-l}} \ .$$

By Sullivan [232], the set of all (uaa) expanding circle maps E is a separable infinite dimensional complex Banach manifold (see Section C.8). Furthermore, this set is the completion of the set of all $C^{1+Hölder}$ expanding circle maps E. Hence, by Theorem C.16 below, the set $B(d)$ inherits a complex Banach structure and it is the closure of $A(d)$ with respect to this structure.

Theorem C.16. *The set $B(d)$ is canonically isomorphic to*

(i) the Teichmüller set $T(L)$;
(ii) the set of all (uaa) structures U for the expanding circle map $E : S \to S$ of degree $d \geq 2$;
(iii) The set of all (tca) solenoids $(\tilde{E}, \tilde{S})$;
(iv) the set of all leaf ratio functions $r : T \to \mathbb{R}^+$;
(v) the set of all solenoid functions $s : C \to \mathbb{R}^+$;
(vi) the set of all affine classes of d-adic tilings of the real line that are fixed points of the d-amalgamation operator;
(vii) the set of all affine classes of d-adic fixed grids of the real line.

The equivalence between (i) and (ii) in Theorem C.16 follows from Theorem C.15. The proof of the other equivalences in Theorem C.16 follows similarly to the proof of Theorem C.14.

C.10 Regularities of the solenoidal charts

In order to state the next theorem, we introduce the following definitions. The *metric* $|\mathbf{u}|_s : C \times C \to \mathbb{R}_0^+$ is defined as follows (see Section C.10 for the geometric interpretation of $|\mathbf{u}|_s$). Let $a = \sum_{m=0}^{\infty} a_m d^m \in C$ and $b = \sum_{m=0}^{\infty} b_m d^m \in C$ be such that $a_n \ldots a_0 = b_n \ldots b_0$ and $a_{n+1} \neq b_{n+1}$. For $0 \leq i \leq n$, let $A_i = \sum_{m=0}^{i} a_m d^m$ and $E_i = \sum_{m=0}^{i} (d-1) d^m$. We define the *metric* by

$$|\mathbf{u}|_s(a,b) = \inf_{0 \leq i \leq n} \left\{ 1 + \sum_{j=A_i}^{E_i} \prod_{l=A_i}^{j} s(l) + \sum_{j=0}^{A_i - 1} \prod_{l=j}^{A_i - 1} s(l) \right\} .$$

In this chapter, the regularities Hölder and Lipschitz have different meaning when written with uppercase or lowercase letters, as we now explain. For $\beta > 0$, we say that a function $f : C \to \mathbb{R}$ is *β-Hölder*, with respect to the metric $|\mathbf{u}| = |\mathbf{u}|_s$, if there is a constant $d \geq 0$ such that $|f(b) - f(a)| \leq d\left(|\mathbf{u}|(a,b)\right)^{\beta}$ for all $a, b \in C$. We say that f is *β-hölder*, with respect to the metric $|\mathbf{u}|$, if there is a continuous function $\varepsilon : \mathbb{R}_0^+ \to \mathbb{R}_0^+$, with $\varepsilon(0) = 0$, such that $|f(b) - f(a)| \leq \varepsilon\left(|\mathbf{u}|(a,b)\right)\left(|\mathbf{u}|(a,b)\right)^{\beta}$ for all $a, b \in C$. By f being *Lipschitz* we mean that f is 1-Hölder. On the real line, with respect to the Euclidean metric, β-Hölder for $\beta > 1$ or lipschitz implies constancy. We define the *solenoid cross-ratio function* $cr(a) : C \to \mathbb{R}^+$ by $cr(a) = (1 + s(a))(1 + (s(a+1))^{-1})$.

Theorem C.17. *For every C^r structure U of the circle S invariant by $E(2)$, the overlap maps and the expanding map $E(2) : S \to S$ attain its maximum of smoothness with respect to the canonical family of solenoid charts $\mathcal{F}_U$ contained in U. Table 1 presents explicit conditions in terms of the solenoid function $s = s_U : C \to \mathbb{R}^+$, determined by the C^r structure U (see Lemmas C.4 and C.6), which give the degree of smoothness of the overlap homeomorphisms and of $E(2)$ in $\mathcal{F}_U$, and vice-versa.*

The regularity of the solenoidal chart overlap maps and $E(2): S \to S$.	*Condition on the functions s and cr, using the metric $\|\mathbf{u}\|_s$ on the Cantor set C.*
have α-Hölder 1st derivative $0 < \alpha \leq 1$	*s is α-Hölder*
have α-Hölder 1st derivative $0 < \alpha \leq 1$	*cr is α-Hölder*
have Lipschitz 1st derivative	*s is Lipschitz*
have α-Hölder 2nd derivative $0 < \alpha \leq 1$	*cr is $(1+\alpha)$-Hölder*
have Lipschitz 2nd derivative	*cr is 2-Hölder*
Affine	*s is lipschitz*

Table 1.

Proof. Let $\mathbf{p}$ be the fixed point of the solenoid map $\tilde{E}$ such that $\pi(\mathbf{p})$ is the fixed point of the expanding circle map chosen in Section C.1 to generate the Markov partition of E. Let $\mathcal{L}_{\mathbf{p}}$ be the local leaf starting on $\mathbf{p}$ and ending on its image $\tilde{M}(\mathbf{p})$ by the monodromy map $\tilde{M}$. Let $z : \pi_S(\mathcal{L}_{\mathbf{p}}) \to (0,1)$ be the corresponding solenoidal chart. Noting that the solenoid function determines a ratio function invariant by the solenoid map and that $|z(J)| = 1$, we obtain the following geometric interpretation of the metric

$$|\mathbf{u}|_s(a,b) = \inf_{0 \leq i \leq n} \{|z(I_{a_i \dots a_0})|\} \ . \tag{C.8}$$

By Lemma C.11, the solenoidal charts determined by a (thca) solenoid $(\tilde{E}, \tilde{S})$ produce a canonical structure U such that the expanding circle map E is $C^{1+H\ddot{o}lder}$. Hence, for every $l \geq 0$ there is a constant $D = D(l) \geq 1$ such that

$$D^{-1}|z(I_{a_{n-l} \dots a_0})| \leq |\mathbf{u}|_s(a,b) \leq D|z(I_{a_{n-l} \dots a_0})| \ , \tag{C.9}$$

for all $a, b \in C$.

Let U be a $C^{1+H\ddot{o}lder}$ structure for the expanding circle map E, and let V be the set of all solenoidal charts determined by the (thca) solenoid $(\tilde{E}, \tilde{S})_U$. By Lemma C.13, the set V is contained in U and the degree of smoothness of the expanding circle map E when measured in terms of a cover U' of U attains its maximum when $U' \subset V$. Let $\mathcal{L}$ and $\mathcal{L}'$ be two local leaves and $u : J = \pi_S(\mathcal{L}) \to \mathbb{R}$ and $v : J' = \pi_S(\mathcal{L}') \to \mathbb{R}$ the corresponding solenoidal charts. If $J \cap J' \neq \emptyset$, let $I_{\beta_1 \dots \beta_n} \subset J \cap J'$ be any interval at any level n of the interval partition. Let the points $\mathbf{x} \in \mathcal{L}$ and $\mathbf{y} \in \mathcal{L}'$ be such that $\pi_S(\mathbf{x}) = \pi_S(\mathbf{y}) \in S$ is the right endpoint of the interval $I_{\beta_1 \dots \beta_n}$. Let a be the point $\tilde{\omega}(\tilde{E}^n(\mathbf{x})) \in C$ and b the point $\tilde{\omega}(\tilde{E}^n(\mathbf{y})) \in C$. Hence, there is a sequence $c_l \dots c_0$, depending only upon $\mathcal{L}$ and $\mathcal{L}'$, such that

$$a_{n+l} \dots a_0 = b_{n+l} \dots b_0 = c_l \dots c_0 \beta_1 \dots \beta_n$$

and $a_{n+l+1} \neq b_{n+l+1}$. Therefore, by (C.9), there is a constant $D = D(l) \geq 1$ such that

$$D^{-1}|z(I_{\beta_1 \dots \beta_n})| \leq |\mathbf{u}|_s(a,b) \leq D|z(I_{\beta_1 \dots \beta_n})| \tag{C.10}$$

with respect to the solenoidal chart $z : \pi_S(\mathcal{L}_\mathbf{p}) \to (0,1)$ defined above. By Lemma C.13, the overlap maps $z \circ u^{-1}$ and $z \circ v^{-1}$ are $C^{1+H\ddot{o}lder}$ smooth, and so there is a constant $D_1 = D_1(\mathcal{L}, \mathcal{L}') \geq 1$ such that

$$D_1^{-1} \leq \frac{|z(I_{\beta_1 \dots \beta_n})|}{|u(I_{\beta_1 \dots \beta_n})|} \leq D_1 \quad \text{and} \quad D_1^{-1} \leq \frac{|z(I_{\beta_1 \dots \beta_n})|}{|v(I_{\beta_1 \dots \beta_n})|} \leq D_1 \, . \tag{C.11}$$

Putting together (C.10) and (C.11), there is a constant $D_2 = D_2(\mathcal{L}, \mathcal{L}') \geq 1$ such that

$$D_2^{-1} \leq \frac{|\mathbf{u}|_s(a,b)}{|u(I_{\beta_1 \dots \beta_n})|} \leq D_2 \quad \text{and} \quad D_2^{-1} \leq \frac{|\mathbf{u}|_s(a,b)}{|v(I_{\beta_1 \dots \beta_n})|} \leq D_2 \, . \tag{C.12}$$

Let $I_{\beta'_1 \dots \beta'_n}$ and $I_{\beta''_1 \dots \beta''_n}$ be adjacent intervals at level n of the interval partition, such that $I_{\beta'_1 \dots \beta'_n}$ is also adjacent to $I_{\beta_1 \dots \beta_n}$. By proof of Lemma C.12,

$$s(a) = \frac{|u(I_{\beta'_1 \dots \beta'_n})|}{|u(I_{\beta_1 \dots \beta_n})|}, \qquad s(a+1) = \frac{|u(I_{\beta''_1 \dots \beta''_n})|}{|u(I_{\beta'_1 \dots \beta'_n})|}, \tag{C.13}$$

and

$$s(b) = \frac{|v(I_{\beta'_1 \dots \beta'_n})|}{|v(I_{\beta_1 \dots \beta_n})|}, \qquad s(b+1) = \frac{|v(I_{\beta''_1 \dots \beta''_n})|}{|v(I_{\beta'_1 \dots \beta'_n})|}. \tag{C.14}$$

The interval partition of the expanding circle map E generates a grid g_u in the set $u(J \cap J')$. Therefore, using (C.12), (C.13), (C.14) and Theorem A.15, the equivalences presented in Tables 2 and 3 imply that the overlap maps $h = v \circ u^{-1} : u(J \cap J') \to v(J \cap J')$ satisfy the equivalences presented in Table 1. □

C.11 Further literature

The scaling and the solenoid functions give a deeper understanding of the smooth structures of one dimensional dynamical systems (cf. Bedford and Fisher [13], Cui et al. [24], Feigenbaum [34], Pinto and Rand [158], Sullivan [230] and Vul et al. [237]). This appendix is based on Pinto and Sullivan [175].

D

Appendix D: Markov maps on train-tracks

One proves that, for any prescribed topological structure, there is a one-to-one correspondence between smooth conjugacy classes of smooth Markov maps and pseudo-Hölder solenoid functions. This gives a characterization of the moduli space for smooth Markov maps.

D.1 Cookie-cutters

Suppose that I_0 and I_1 are two disjoint closed subintervals of the interval I containing the endpoints of $I = [-1, 1]$. A cookie-cutter is a C^{1+} map $F : X \to X$ such that $|dF| > \lambda > l$ and $F(I_0) = F(I_1) = I$. If

$$\Lambda_n = \{x \in I : F^{j-1}(x) \in I_0 \cup I_1, \ 1 \leq j \leq n\},$$

then Λ_n consists of 2^n disjoint closed *n-cylinders*

$$I_{\varepsilon_1 \cdots \varepsilon_{n-1}} = \{x \in I : F^{j-1}(x) \in I_{\varepsilon_j}, \ 1 \leq j \leq n\}.$$

Each cylinder

$$I_{\varepsilon_1 \cdots \varepsilon_{n-2}} = I_{\varepsilon_1 \cdots \varepsilon_{n-1}} \cup G_{\varepsilon_1 \cdots \varepsilon_{n-2}} \cup I_{\varepsilon_1 \cdots \varepsilon'_{n-1}},$$

where $G_{\varepsilon_1 \cdots \varepsilon_{n-2}}$ is an *n-gap*. The invariant Cantor set C of F

$$C = \cap_{n \geq 1} \Lambda_n = \{x \in I : F^j(x) \in I_0 \cup I, \quad \text{for all } j > 0\}$$

is constructed inductively by deleting the n-gaps. The smoothness and the expanding property of F implies that the Cantor set C has bounded geometry.

We can regard this as a Markov map on a train-track X as follows: Let X be the disjoint union of the three closed intervals I_0, $G = I \backslash (I_0 \cup I_1)$ and I_1 quotient by the junctions $J_1 = \{-1\}$, $J_2 = I_0 \cap G$, $J_3 = G \cap I_1$ and $J_4 = \{1\}$. At the junctions J_2 and J_3, we can define the smooth structure by *journeys*.

However, in this case, it is just the smooth structure induced by the inclusion of I_0 and I_1 into I.

The symbolic space Σ is given by $\{0,1\}^{\mathbb{N}}$, the set of infinite right-handed words $\varepsilon_1\varepsilon_2\cdots$ of 0s and 1s. We add a positive or negative sign to 0 and 1 corresponding to the sign of the derivative of the Markov map F in I_0 and I_1, respectively. There are four possible orderings on the symbolic set Σ corresponding to the two different choices of orientation of the cookie-cutter on each of the two intervals I_0 and I_1.

The mapping $h : \Sigma \to \mathbb{R}$ defined by

$$h(\varepsilon_1\varepsilon_2\cdots) = \bigcap_{n\geq 1} I_{\varepsilon_1\cdots\varepsilon_n}$$

gives an embedding of Σ into $\mathbb{R}$. Moreover, the map h is a topological conjugacy between the shift $\phi : \Sigma \to \Sigma$ and the cookie-cutter $F : \Lambda \to \Lambda$ defined on its invariant set.

We use a train-track X to represent the interval I as follows. Let the train-track X be the disjoint union of the closed intervals I_0, $G = I\backslash(I_0 \cup I_1)$ and I_1 quotient by the junctions $J_1 = \{-1\}$, $J_2 = I_0 \cap G$, $J_3 = G \cap I_1$ and $J_4 = \{1\}$. At the junctions J_2 and J_3, we can define the smooth structure by *journeys*. Each journey is just the identity map from any subset of X containing J_2 or J_3 to I.

For this example, we will show that the solenoid function S_F is any Hölder continuous mapping from $\{0,1\}^{\mathbb{Z}_{\geq 0}}$ to the positive reals $\mathbb{R}^+$.

D.2 Pronged singularities in pseudo-Anosov maps

Near a three-pronged singularity the unstable leaves of a pseudo-Anosov map look as in Figure D.1(a). We will carry out the collapsing procedure shown in Figure D.1(b) to obtain a Y-shaped space X. Let λ_1, λ_2 and λ_3 be three transversals as shown in 5(a). The manifold structure of these define charts on X by identification of points on the same unstable manifold. From Figure D.1, these must satisfy the compatibility condition that they agree on the intersection of their domains. To handle the compatibility condition, we will introduce the notion of *turntables*. Each junction in our train-track will contain a stack of turntables. The charts in each turntable satisfy these strong compatibility conditions.

For a one-pronged singularity, we obtain the analogous structures shown in Figure D.2. The unstable manifolds define a map g from λ_1 to itself and the train-track X is naturally identified with the quotient λ_1/g. Some more discussion of these two examples is given in §D.3.1.

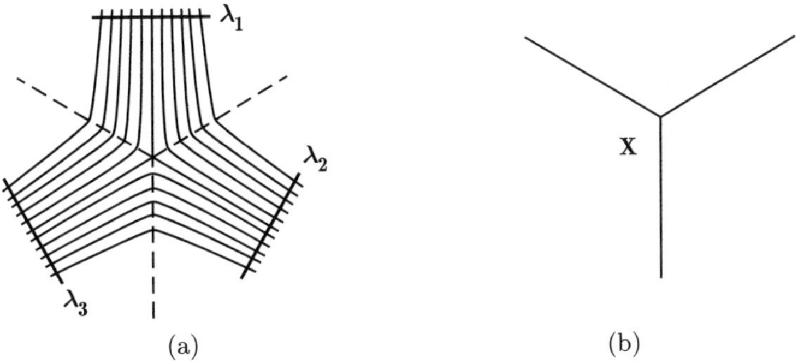

Fig. D.1. (a) The leaves of the unstable foliation of a Pseudo-Anosov diffeomorphism near a three-pronged singularity. The submanifolds λ_1, λ_2 and λ_3 are the transversals used to construct the train-track. (b) The train-track X constructed in this way.

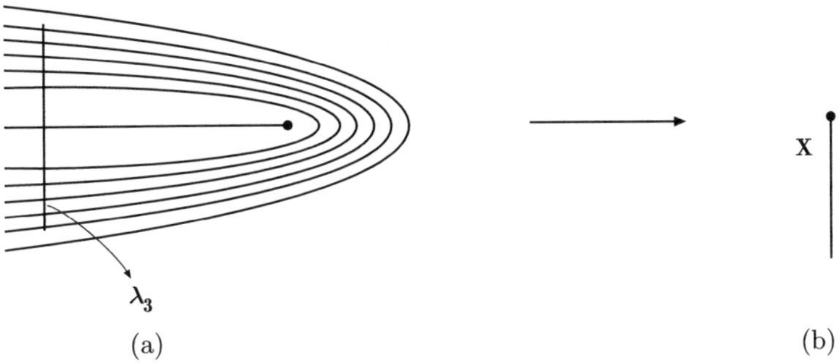

Fig. D.2. (a) The leaves of the unstable foliation of a Pseudo-Anosov diffeomorphism near a one-pronged singularity. The submanifold λ_3 is the transversal used to construct the train-track. (b) The train-track X constructed in this way.

D.3 Train-tracks

The underlying space of a train-track X is a quotient space defined as follows. Consider a finite set of lines $l_1, \ldots, l_m$. Each line is a path connected one-dimensional closed manifold. The endpoints $l_i^{\pm}$ of the line l_i are the *termini* of l_i. A *regular point* is a point in X that is not a terminus. The termini are partitioned into junctions J_α. Then, X is the quotient space obtained from the disjoint union of the lines by identifying termini in the same junction.

A *journey* j is a mapping of an interval $I = (t_0, t_n)$ into X with the following properties:

(i) There is a finite set of times $t_0 < t_1 < \ldots < t_n$ such that $j(t)$ is in a junction if, and only if, $t = t_i$, for some $0 < i < n$;
(ii) j is a local homeomorphism at all regular points;

(iii) For all small $\varepsilon > 0$ and all $0 < i < n$, the map j gives a local homeomorphism of $(t_i - \varepsilon, t_i]$ and $[t_i, t_i + \varepsilon)$ into the lines containing $j(t_i - \varepsilon)$ and $j(t_i + \varepsilon)$.

Suppose that x and y are termini. If there is a journey $j : I \to X$ such that $j(t_i)$ is in the junction and, for some $\varepsilon > 0$, $j(t)$ is in I_x (resp. I_y) when $t \in (t_i - \varepsilon, t_i)$ (resp. $(t_i, t_i + \varepsilon)$), then we write $x \rightleftharpoons y$ and say that there is a *connection* from x to y. If $x \rightleftharpoons x$, then we say that x is *reversible.* An example of a reversible terminus is given by the train-track obtained from a one-pronged singularity in a pseudo-Anosov diffeomorphism (see Figure D.2)

Given journeys j_1 and j_2 such that $j_1(s') = j_2(t')$, let $s(t)$ be the unique function defined on a neighbourhood of t' such that $s(t') = s'$ and $j_1(s(t)) = j_2(t)$. Call $s(t)$ the *timetable conversion* of (j_1, j_2) at x. The journeys j_1 and j_2 are C^r *compatible*, if the timetable conversion is C^r for all common points x.

A C^r structure on X is defined by given a compatible set of journeys that pass through every point of X and through every connection. However, it has to satisfy some extra conditions that we now specify.

As explained in §D.1, we often require extra constrains at junctions. These are described by *turntables.* Associated to every junction is a set (possible empty) of turntables. Each turntable τ is a subset of the junction such that if $x, y \in \tau$, then there is a connection between x and y. We adopt the convection that every subset of a turntable is a turntable. A *maximal turntable* is one that is not contained in any bigger one. The *degree* of a turntable is the number of termini in it, where each reversible terminus is counted twice.

A smooth structure on the train-track X must satisfy the following condition at each turntable τ. If j_1 (resp. j_2) is a is a C^r journey through the connection from x to y (resp. x to z) and $j_1 = j_2$ on the line terminating in x, then $-j_1$ and j_2 define a C^r journey through the connection from y to z. The journey $-j_1$ is the journey j_1 with time reversed.

Definition D.1. *A* C^r structure *on a train-track* X *is defined by a set of journeys* $\{j_\alpha\}$ *such that:*

(i) every point of X *is visited by some journey* j_α*;*
(ii) the journeys $\{j_\alpha\}$ *are* C^r *compatible; and*
(iii) the above turntable condition holds.

When we speak of a smooth metric on X, we just mean on the disjoint union of the lines that is a smooth metric on each line.

D.3.1 Train-track obtained by glueing

A train-track is constructed as follows: We are given a finite number of path connected closed one-manifolds $\lambda_1, \ldots, \lambda_s$ and a set $\mathcal{D}$ of C^r diffeomorphisms whose domain and ranges are each a closed submanifold of the λ_j. The domain and range can be in the same λ_j. From the disjoint union of the λ_j, we form the quotient space X obtained by identifying all pairs of points that are the form $x, g(x)$, where $g(x) \in \mathcal{D}$. The smooth structure is that defined by the set of projections $\pi_j : \lambda_j \to X$. Figure D.3 gives some examples of local train-track structures obtained in this way. Let us use the construction to get a better understanding of the examples given in §D.2 and §D.1 involving pronged singularities of pseudo-Anosov maps. For the three-pronged case, one can use the smooth structure on the Y-shaped space from this figure. We use the glueing construction of §D.1 The unstable leafs define the glueing maps $g_{i,j} : \lambda_i \to \lambda_j$ by holonomy. The smooth structure on the Y is defined by the three charts given by the projection of each λ_i into Y. But from the picture one can see that any two of these determines the third. This is the turntable condition for the Y.

For the single prong or cusp singularity of a pseudo-Anosov map, take λ_1 as shown in Figure D.2. Then, there is a single glueing map $g : \lambda_1 \to \lambda_1$ given by the holonomy on leaves. This is shown in Figure D.3(e).

D.4 Markov maps

We pass now define a smooth Markov map $F : X \to X$ on the train-track X. For such a map, the set of lines is partitioned into the subset of *cylinders* C and the subset of *gaps* $\mathcal{G}$. We let X_0 denote the subspace of X corresponding to the cylinders. The map M does not have to be defined on the gaps. We insist that the lines terminating in a turntable of degree $d > 2$ are all cylinders.

We say that a mapping $F : X \to X$ is *faithful on journeys* at the turntable τ, if every short journey through τ is sent to a journey through the image turntable τ' and the preimage of every short journey through τ' is a journey through τ.

A map $F : X \to X$ is *Markov*, if

(i) F is a local homeomorphism on the interior of each cylinder;
(ii) F maps termini to termini;
(iii) F permutes the turntables of degree $d > 2$ and is faithful on journeys at each of them;
(iv) if τ is a maximal turntable of degree 2, then τ is the image of either a regular point or a maximal turntable which has degree 2; and
(v) for all lines B there exists a cylinder C such that the image of C contains B.

Note that these conditions imply that the image of a cylinder contains all the cylinders that it meets. Suppose that F maps the turntable τ into the turntable τ'. We say that F is a C^r diffeomorphism at τ, if F maps every C^r journey through τ' C^r diffeomorphically onto a C^r journey through τ'.

A C^r Markov map $F : X \to X$, with $r > 1$, is a Markov map such that:

(i) at every regular point F is a local C^r diffeomorphism;

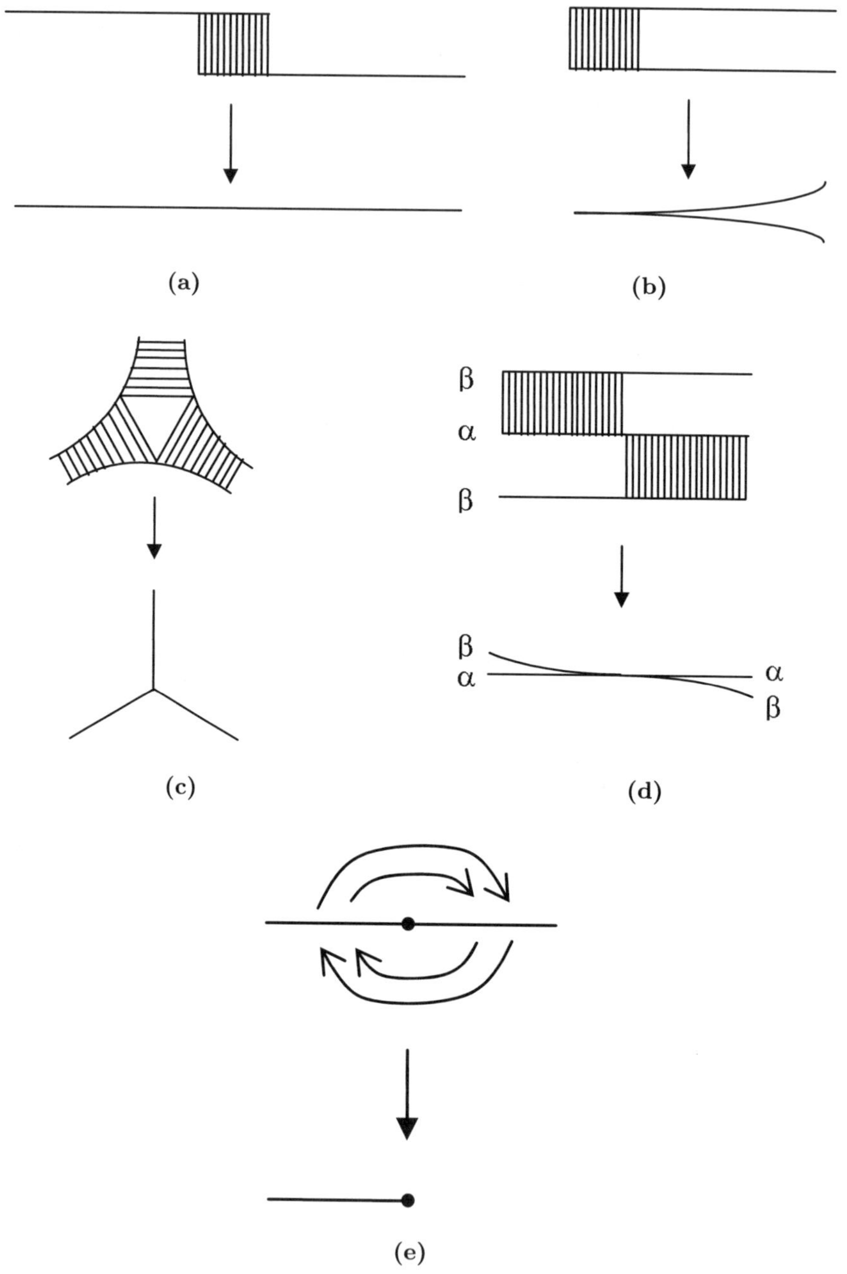

Fig. D.3. Some local train- tracks obtained by glueing. Note that those shown in (c) and (d) have no embedding into Euclidean space.

(ii) F is a C^r diffeomorphism in each turntable of degree $d > 2$;
(iii) if τ is a maximal turntable of degree 2, then τ is the C^r diffeomorphic image of either a regular point or a maximal turntable τ' of degree 2; and
(iv) there exists $\lambda > 1$ and a smooth metric on X such that at every regular point x, $\|dF(x)\| \geq \lambda$.

Let us suppose that the cylinders are indexed by a set S. Thus, we denote them by C_a, $a \in S$. A point $x \in \cup_{a\in S} C_a$ is *captured*, if $F^m(x) \in \cup_{a\in S} C_a$, for all $m > 0$. The set of all captured points in $\cup_{a\in S} C_a$ is denoted by $\Lambda = \Lambda_F$. The set of intervals $\{C_a\}$ is called the *Markov partition* of F.

The Markov maps F and G are *topologically conjugate*, if there exists a homeomorphism $h : \Lambda_F \to \Lambda_G$ such that $G \circ h = h \circ F$ on Λ_F. If the map h has a C^r extension to X, then we say that the map h is a C^r *conjugacy*.

Suppose that h is a mapping of a closed subset X of $\mathbb{R}^n$ into $\mathbb{R}^m$. We say that h is C^r, if h has a C^r extension to some open neighbourhood of X. Moreover, we say that a map h is C^{l+}, if h is $C^{1+\varepsilon}$, for some $0 < \varepsilon < 1$.

Theorem D.2. *Two C^r Markov maps F and G on a C^r train-track are C^r conjugate if, and only if, they are in the same C^{1+} conjugacy class.*

Proof. This theorem is proved by using a blow-down blow-up technique as in the case where X is a one-manifold (e.g. see Theorem E.19). □

Symbolic Dynamics

Given a Markov map F, let $\Sigma_s = \Sigma_s^F$ denote the symbolic set of infinite right-handed words $\underline{\varepsilon} = \varepsilon_1\varepsilon_2\cdots$ such that: (i) for all $m \geq 1$, $\varepsilon_m \in S$ and (ii) there exists $x_{\underline{\varepsilon}} \in C$ with the property that $F^m(x_{\underline{\varepsilon}}) \in C_{\varepsilon_m}$, for all $m \geq 1$. We call these words *admissible*.

Endow Σ_s with the usual topology. Let $\Sigma = \Sigma_F$ be the space obtained from Σ_s by identifying $\underline{\varepsilon}$ with $\underline{\varepsilon}'$, if $x_{\underline{\varepsilon}}$ is equal to $x'_{\underline{\varepsilon}}$. If $x_{\underline{\varepsilon}}$ and $x'_{\underline{\varepsilon}}$ are in the same junction of X, then $\underline{\varepsilon}$ and $\underline{\varepsilon}'$ are defined to be in the same junction of Σ. If $\{x_{\varepsilon_1}, \ldots, x_{\varepsilon_n}\}$ is a turntable for X, then $\{\varepsilon_1, \ldots, \varepsilon_n\}$ is defined to be a turntable for Σ.

Define the shift $\Phi = \Phi_F : \Sigma \to \Sigma$ by $\Phi(\varepsilon_1\varepsilon_2\cdots) = \varepsilon_2\varepsilon_3\cdots$. The Markov map F on Λ_F is topologically conjugate to Φ_F on Σ_F.

Two Markov maps can give rise to the same shift map even though they are not topologically conjugate, because Σ does not take in account the order of the points in each set $C_a \subset X$. Therefore, we order the points on Σ using the ordering on the corresponding points in the sets $C_a \subset X$. The *ordered symbolic dynamical system* is the ordered set Σ with the shift $\Phi : \Sigma \to \Sigma$.

Remark D.3. The correspondence $F \to \Phi_F$ induces a one-to-one correspondence between topological conjugacy classes of Markov maps and ordered symbolic dynamical systems.

Cylinder structures

For all $\underline{\varepsilon} \in \Sigma_F$, let $t = \varepsilon_1 \cdots \varepsilon_l$ and define the *l-cylinder* $C_t = C_t^F$ as the closed interval consisting of all $x \in C$ such that, for all $1 \leq j \leq l$, $F^j(x) \in C_{\varepsilon_j}$. Suppose that C_t and C_s are two l-cylinders such that:

(i) C_t and C_s are contained in the same 1-cylinder C_a;
(ii) there is no other l-cylinder between C_t and C_s in C_a;
(iii) in the interior of the cylinder C_a, $C_t \cap C_s = \emptyset$.

We define the *l-gap* $C_g = C_{g_{t,s}}$ to be the closed interval between C_t and C_s. A *l-line* is defined to be a l-cylinder or a l-gap. This defines the *cylinder structure* of F.

We say that a cylinder structure has *bounded geometry*, if there are constants $c > 0$ and $m > 0$ such that:

(i) for all $l > 1$ if D is a l-line and E is the $(l-1)$-cylinder that contains D, then $|D|/|E| > c$; and
(ii) if F is the $(l-m)$-cylinder that contains D, then $D \neq F$.

D.5 The scaling function

For the special case of C^{1+} Markov maps that do not have connections, one proves the one-to-one correspondence between Hölder scaling functions and C^{1+} conjugacy classes of C^{1+} Markov maps without connections.

Let us consider the cylinder structure generated by a C^{1+} Markov map F. Define the set $\Omega_n = \Omega_n^F$ as the set of all symbols t corresponding to the n-cylinders and n-gaps. Let $\Omega = \Omega^F$ be the union $\cup_{n \geq 1} Omega_n$.

We also keep a record of other basic topological information as follows: (1) the topological order of all n-cylinders within each 1-cylinder; (ii) which endpoints of each n-cylinders are junctions and which junction they are.

For all $t \in \Omega_{n+1}$, the *mother* of t is the symbol $m(t) \in \Omega_n$ that has the property that $C_t \subset C_{m(t)}$.

A *pre-scaling function (or scaling tree)* is a function $\sigma : \Omega \to \mathbb{R}^+$ such that for all $t \in \Omega$

$$\sum_{m(s)=t} \sigma(s) = 1 \,. \tag{D.1}$$

The *pre-scaling function (or scaling tree)* $\sigma_F : \Omega \to \mathbb{R}^+$ *determined by a Markov map* F is the pre-scaling function $\sigma : \Omega \to \mathbb{R}^+$ given by

$$\sigma_F(t) = \lim_{n \to \infty} |C_t| / \left| C_{m(t)} \right| .$$

Define the set $\overline{\Omega} = \overline{\Omega}^F$ as the set of all infinite left-handed words $\bar{t} = \cdots t_n \cdots t_1$ such that, for all $n \geq 1$, $t_n \in \Omega_n$ and $F(C_{t_{n+1}}) = C_{t_n}$.

Define the metric $d : \overline{\Omega} \times \overline{\Omega} \to \mathbb{R}^+$ as follows: Choose $0 < \mu < 1$. For all $\bar{t}, \bar{s} \in \overline{\Omega}$, the distance $d(\bar{t}, \bar{s})$ is equal to μ^n if, and only if, $t_n = s_n$ and $t_{n+1} \neq s_{n+1}$.

The pair $(t_n, s_n) \in \Omega_n \times \Omega_n$ is an *adjacent symbol* if, and only if, the lines C_{t_n} and C_{s_n} have a common point or one of the endpoints x of C_{t_n} and one of the endpoints y of C_{s_n} are a connection $\{x, y\}$. The set $\underline{\Omega}_n \subset \Omega_n \times \Omega_n$ is the set of all adjacent symbols.

For all $\bar{t} \in \overline{\Omega}$, define the *set of children* $C_{\bar{t}}$ of $\bar{t}$ as the set of all infinite left symbols $\bar{s}$ such that t_n is the mother of s_{n+1}, for all $n \geq 1$:

$$C_{\bar{t}} = \{\bar{s} : m(s_{n+1}) = t_n, \text{ for all } n \geq 1\}.$$

Definition D.4. *A* scaling function *is a function* $\overline{\sigma} : \overline{\Omega} \to \mathbb{R}^+$ *such that for all* $\bar{t} \in \overline{\Omega}$

$$\sum_{\bar{s} \in C_{\bar{t}}} \overline{\sigma}(\bar{s}) = 1 \; . \tag{D.2}$$

We say that the scaling function $\overline{\sigma}$ *is* Hölder, *if* $\overline{\sigma}$ *is Hölder continuous in the above metric* d.

Lemma D.5. *Let* F *be a* C^{1+} *Markov map. The Hölder scaling function* $\overline{\sigma}_F : \overline{\Omega} \to \mathbb{R}^+$ *determined by* F *is well-defined by*

$$\overline{\sigma}_F(\bar{t}) = \lim_{n \to \infty} |C_{t_n}| / \left|C_{m(t_n)}\right| .$$

For simplicity of notation, we will denote $\overline{\sigma}$ by σ and $\overline{\sigma}_F$ by σ_F.

The Markov partition of F has the (1+)-*scaling property*, if there is $0 < \lambda < 1$ such that, for all $\bar{t} = \cdots t_1 \in \overline{\Omega}$,

$$\left| 1 - \frac{\sigma_F(t_n)}{\sigma_F(t_{n-1})} \right| \leq \mathcal{O}(\lambda^n).$$

Proof of Lemma D.5. We are going to prove that if the Markov map F is C^{1+}, then $\sigma_F : \overline{\Omega} \to \mathbb{R}^+$ is a Hölder scaling function. By Theorem B.28, the Markov partition of F has the (1+)-*scaling property*. Therefore, the limit $\sigma_F(\bar{t})$ is well defined and

$$\frac{\sigma_F(\bar{t})}{\sigma_F(t_n)} \in 1 \pm \mathcal{O}(\lambda^n). \tag{D.3}$$

By smoothness of the Markov map F and the expanding nature of F, there is $\delta > 0$ such that, for all $t \in \Omega = \cup_{n \geq 1} \Omega_n$, $\sigma_F(t) > \delta$. Therefore, for all $\bar{t} \in \overline{\Omega}$,

$$\sigma(\bar{t}) = \lim_{n \to \infty} \sigma_F(t_n) > \delta.$$

Let $0 < \varepsilon \leq 1$ be such that $\lambda \leq \mu^{\varepsilon}$. For all $\bar{t}, \bar{s} \in \overline{\Omega}$ such that $t_n = s_n$ and $t_{n+1} \neq s_{n+1}$ we have, by (D.3),

$$|\sigma_F(\bar{t}) - \sigma_F(\bar{s})| \leq \mathcal{O}(\lambda^n) \leq \mathcal{O}\left((\mu^\varepsilon)^n\right) \leq \mathcal{O}\left((d(\bar{t},\bar{s}))^\varepsilon\right).$$

Therefore, the function σ_F is Hölder continuous in $\overline{\Omega}$. For all $\bar{t} \in \overline{\Omega}$, the set $C_{\bar{t}}$ is bounded and

$$\sum_{\bar{s}\in C_{\bar{t}}} \sigma(\bar{s}) = \lim_{n\to\infty} \sum_{\bar{s}\in C_{\bar{t}}} \sigma(s_{n+1}) = 1.$$

Therefore, $\sigma_F : \overline{\Omega} \to \mathbb{R}^+$ is a Hölder scaling function.□

Lemma D.6. *Let F and G be two C^r Markov maps in the same topological conjugacy class with $r > 1$. The C^r Markov maps F and G are C^r conjugate if, and only if, the scaling function σ_F is equal to σ_G.*

By Lemma D.6, the Hölder scaling function $\sigma : \overline{\Omega} \to \mathbb{R}^+$ is a complete invariant of the C^{1+} conjugacy classes of C^{1+} Markov maps.

Since F and G are topologically conjugate, F and G define the same set $\overline{\Omega} = \overline{\Omega}_F = \overline{\Omega}_G$.

The cylinder structures of F and G are (1+)-*scale equivalent*, if there is $0 < \lambda < 1$ such that

$$\left|1 - \frac{\sigma_F(t_n)}{\sigma_G(t_n)}\right| \leq \mathcal{O}(\lambda^n) \tag{D.4}$$

for all $\bar{t} = \cdots t_1 \in \overline{\Omega}$. The cylinder structures F and G are $(l+)$-*connection equivalent*, if

$$\frac{|C_{t_n}|\,|D_{s_n}|}{|C_{s_n}|\,|D_{t_n}|} \in 1 \pm \mathcal{O}(\lambda^n)$$

for all $(t_n, s_n) \in \underline{\Omega}_n$.

Proof of Lemma D.6. If the Markov maps F and G are C^{1+} conjugate, then by Theorem D.2, they are C^r-conjugate . By Theorem B.28, two C^{1+} Markov maps F and G are C^{1+} conjugate if, and only if, the cylinder structures of F and G are (1+)-scale equivalent and (1+)-connection equivalent. Therefore, we will prove that the cylinder structures of F and G are (1+)-scale equivalent and (1+)-connection equivalent if, and only if, the scaling function $\sigma_F : \overline{\Omega}_F \to \mathbb{R}^+$ is equal to the scaling function $\sigma_G : \overline{\Omega}_G \to \mathbb{R}^+$. By Theorem B.28 and smoothness of the Markov maps F and G, their cylinder structures have the (1+)-scale property and the (1+)-connection property. Therefore, they satisfy (D.3). Let us prove that, if the cylinder structures of F and G are (1+)-scale equivalent and (1+)-connection equivalent, then they define the same scaling function. By (D.3) and (D.4), for all $\bar{t} = \cdots t_1 \in \overline{\Omega}$ and for all $n > 0$,

$$\begin{aligned}\frac{\sigma_F(\bar{t})}{\sigma_G(\bar{t})} &= \frac{\sigma_F(\bar{t})}{\sigma_F(t_n)}\frac{\sigma_F(t_n)}{\sigma_G(t_n)}\frac{\sigma_G(t_n)}{\sigma_G(\bar{t})} \\ &\in (1 \pm \mathcal{O}(\lambda^n))(1 \pm \mathcal{O}(\lambda^n))(1 \pm \mathcal{O}(\lambda^n)) \\ &\subset 1 \pm \mathcal{O}(\lambda^n).\end{aligned}$$

On letting n converge to infinity, we obtain that the scaling functions $\sigma_F : \overline{\Omega}_F \to \mathbb{R}^+$ and $\sigma_G : \overline{\Omega}_G \to \mathbb{R}^+$ are equal. Let us prove that if the scaling functions $\sigma_F : \overline{\Omega}_F \to \mathbb{R}^+$ and $\sigma_G : \overline{\Omega}_G \to \mathbb{R}^+$ are equal, then the cylinder structures of F and G are (1+)-scale equivalent and (1+)-connection equivalent. For all $t_n \in \Omega_n$, choose $\bar{t} = \cdots t_n \cdots t_1 \in \overline{\Omega}$. Since $\sigma_F(\bar{t}) = \sigma_G(\bar{t})$ and by (D.3),

$$\begin{aligned}\frac{\sigma_F(t_n)}{\sigma_G(t_n)} &= \frac{\sigma_F(t_n)}{\sigma_F(\bar{t})}\frac{\sigma_F(\bar{t})}{\sigma_G(\bar{t})}\frac{\sigma_G(\bar{t})}{\sigma_G(t_n)} \\ &\in (1 \pm \mathcal{O}(\lambda^n))(1 \pm \mathcal{O}(\lambda^n)) \\ &\subset 1 \pm \mathcal{O}(\lambda^n). \end{aligned} \tag{D.5}$$

The cylinder structures of F and G are (1+)-scale equivalent. For all $t \in \Omega_n$, denote the cylinders C_t^F by C_t and the cylinders C_t^G by D_t. Let us prove that the cylinder structures F and G are (1+)-connection equivalent. For all adjacent symbols $(t_n, s_n) \in \underline{\Omega}_n$, choose

$$\bar{t} = \cdots t_n \cdots t_1, \bar{s} = \cdots s_n \cdots s_1 \in \overline{\Omega}$$

such that (i) $(t_l, s_l) \in \underline{\Omega}_l$, (ii) there is $0 < k \le n$ such that $m^k(t_l) = m^k(s_l)$, for all l large enough. Denote $m^i(t_l)$ by t^i and $m^i(s_l)$ by s^i for all $i = 0, \ldots, k$. Let H be the Markov map F or G. By the definition of *(l+)-connection property* of the cylinder structure of F and G, there is $0 < \lambda < 1$ such that, for all $l > n$,

$$\left| 1 - \frac{|C_{t_n}^H|}{|C_{s_n}^H|}\frac{|C_{s_l}^H|}{|C_{t_l}^H|} \right| \le \mathcal{O}(\lambda^n). \tag{D.6}$$

By (D.5) and (D.6),

$$\begin{aligned}\frac{|C_{t_n}|}{|C_{s_n}|}\frac{|D_{s_n}|}{|D_{t_n}|} &= \frac{|C_{t_n}|}{|C_{s_n}|}\frac{|C_{s_l}|}{|C_{t_l}|}\frac{|C_{t_l}|}{|C_{t^k}|}\frac{|C_{s^k}|}{|C_{s_l}|}\frac{|C_{t^k}|}{|C_{s^k}|} \\ &\quad \frac{|D_{s^k}|}{|D_{t^k}|}\frac{|D_{s_l}|}{|C_{s^k}|}\frac{|D_{t^k}|}{|D_{t_l}|}\frac{|D_{t_l}|}{|D_{s_l}|}\frac{|D_{s_n}|}{|D_{t_n}|} \\ &\in (1 \pm \mathcal{O}(\lambda^n)) \prod_{i=0}^{k} \left(\frac{\sigma_F(t^i)}{\sigma_G(t^i)}\frac{\sigma_G(s^i)}{\sigma_F(s^i)} \right) \\ &\subset (1 \pm \mathcal{O}(\lambda^n))(1 \pm \mathcal{O}(\lambda^{l-k})). \end{aligned}$$

On letting l tend to infinity, we obtain that the cylinder structures of F and G are (1+)-connection equivalent. □

Theorem D.7. *Given a Hölder scaling function $\sigma : \overline{\Omega} \to \mathbb{R}^+$ with domain $\overline{\Omega}$ corresponding to a topological Markov map without connections, there is a C^{1+} Markov map F with scaling function $\sigma_F = \sigma$.*

Putting together theorems D.2 and D.7 and lemmas D.5 and D.6, we obtain the following result.

Corollary D.8. *There is a one-to-one correspondence between C^{1+} conjugacy classes of C^{1+} Markov maps without connections and Hölder scaling functions $\sigma : \overline{\Omega} \to \mathbb{R}^+$ with domain $\overline{\Omega}$ corresponding to topological Markov maps without connections. Furthermore, if F and G are C^r Markov maps in the same C^{1+} conjugacy class of C^{1+} Markov maps, then they are C^r conjugate.*

The cylinder structure has the (1+)-*scaling property*, if there is $0 < \mu < 1$ such that

$$\left|1 - \frac{\sigma(s_n)}{\sigma(s_{n-1})}\right| \leq \mathcal{O}(\mu^n)$$

for all $n > 1$ and for all $s \in \Omega_n$.

Proof of Theorem D.7. We are going to prove that given a scaling function $\sigma : \overline{\Omega} \to \mathbb{R}^+$ corresponding to a topological Markov map without connections, then there is also a C^{1+} Markov map without connections with scaling function σ. By Theorem B.28, given a cylinder structure without connections and with (1+)-scale property and bounded geometry, there is a C^{1+} Markov map F without connections that generates this cylinder structure. Define the pre-scaling function $\sigma : \Omega \to \mathbb{R}^+$ determined by a scaling function $\sigma : \overline{\Omega} \to \mathbb{R}^+$ as follows. For all $n > 1$ and for all $t_{n-1} \in \Omega_{n-1}$, choose $\bar{t} = \cdots t_{n-1} \cdots t_1 \in \overline{\Omega}$. For all $\bar{s} \in \mathcal{C}_{\bar{t}}$ define $\sigma(s_n) = \sigma(\bar{s})$. Since the scaling function is bounded from zero, trivially the pre-scaling function is bounded from zero. Thus, the cylinder structure corresponding to the pre-scaling function $\sigma : \Omega \to \mathbb{R}^+$ has bounded geometry. We are going to prove that this cylinder structure has the (1+)-scaling property. For all $n > 1$ and for all $s_n \in \Omega_n$, choose $\bar{s} = \cdots s_n s_{n-1} \cdots s_1 \in \overline{\Omega}$. Since the scaling function is Hölder continuous and it is bounded away from zero, we have that

$$\begin{aligned} \frac{\sigma(s_n)}{\sigma(s_{n-1})} &\in \frac{\sigma(\bar{s}) \pm \mu^n}{\sigma(\bar{s}) \pm \mu^{n-1}} \\ &\subset 1 \pm \mathcal{O}(\mu^n). \end{aligned}$$

By Theorem B.28, there is a C^{1+} Markov map F that generates a cylinder structure with a pre-scaling function equal to $\sigma : \Omega \to \mathbb{R}^+$. By construction of the C^{1+} Markov map F, the scaling function $\sigma_F : \overline{\Omega}_F \to \mathbb{R}^+$ of F is equal to the scaling function $\sigma : \Omega \to \mathbb{R}^+$. □

D.5.1 A Hölder scaling function without a corresponding smooth Markov map

If we wish to find a moduli space for expanding maps of the circle, we are naturally led to the question of which scaling functions occur, for a given class of Markov maps. Sometimes, is difficult to characterize which scaling functions are realizable by these Markov maps

To illustrate this, we consider the following simple class of Markov maps of the interval $I = [0, 1]$. We consider expanding maps $f : I \to I$ such that, for some $0 < r < 1$, $f_0 = f|_{[0,r]}$ (resp. $f_1 = f|_{[r,1]}$) is a C^{1+} diffeomorphism of $[0, r]$ (resp. $[r, 1]$) onto I. The map f determines a Hölder scaling function $s_f : \overline{\Omega} = \{0,1\}^{\mathbb{Z}_{>0}} \to \mathbb{R}$. Moreover, Theorem D.7 asserts that every such function occurs in this way.

We now consider the subclass of such mappings that correspond to degree two expanding mappings of the circle. By the above comments, the scaling functions for these are a complete invariant of C^{1+} conjugacy. Moreover, we have the following fact.

Lemma D.9. *There is a Hölder scaling function $\sigma : \overline{\Omega} \to \mathbb{R}^+$ of the above form such that no C^{1+} expanding map of the circle has σ as its scaling function.*

Proof. If F defines a C^{1+} expanding map of the circle, then $\sigma_F(\cdots 00) = \sigma_F(\cdots 11)$. However,there are Hölder scaling functions in the above class which do not satisfy this property. □

D.6 Smoothness of Markov maps and geometry of the cylinder structures

In this section, we give an equivalence between the geometry of the cylinder structures corresponding to the Markov maps F and G and the smoothness of the conjugacy h between the Markov maps F and G.

D.6.1 Solenoid set

Let F be a topological Markov map. A *connection preorbit* $\bar{c}$ of a connection $c_1 = \{c_1^-, c_1^+\} \in C$ is a sequence $\bar{c} = \cdots c_2 c_1$ such that (i) for all $m > 1$, $c_m = \{c_m^-, c_m^+\}$ is a connection or $c_m^- = c_m^+$; (ii) $F(c_m^-) = c_{m-1}^-$ and $F(c_m^+) = c_{m-1}^+$. Given $n \in \mathbb{N}$, the pair $(\bar{c}, n)$ determines sequences $E^-(\bar{c}, n) = \cdots E_{n+1}^- E_n^-$ and $E^+(\bar{c}, n) = \cdots E_{n+1}^+ E_n^+$ of lines as follows: E_{n+m}^- (resp. E_{n+m}^+) is the $(n+m-1)$-line with c_m^- (resp. c_m^+) as an endpoint. We call the pair $(E^-(\bar{c}, n), E^+(\bar{c}, n))$ a *two-line preorbit*. If c_1 is a preimage of a connection, then the scaling structure of its two-line preorbits $(E^-(\bar{c}, n), E^+(\bar{c}, n))$ with $n > 1$ is determined by those of its image. In such a case, we just need to keep track of the scaling for the two-line preorbits $(E^-(\bar{c}, n), E^+(\bar{c}, n))$ with $n = 1$. On the other hand, if c_1 has no preimage, then we must study the scaling of its two-line preorbits $(E^-(\bar{c}, n), E^+(\bar{c}, n))$, for all $n \geq 1$. Let the set of all *preimage connections* PC be equal to the set of all connections that are C^{1+} preimages of either a connection or a regular point. Therefore, by definition of a Markov map, if a connection c is contained in a turntable of degree $d > 2$, then the connection c is a preimage connection. Let the set GC of all *gap connections* be the set of all connections $\{x, y\}$ such that x or y is an endpoint of a gap. Let the set $\mathcal{A} = \mathcal{A}_F$ of F be equal to

$$\mathcal{A} = \{(\bar{c}, n) : (c_1, n) \in C \times \{1\} \text{ or } (c_1, n) \in C \backslash (PC \cup GC) \times \mathbb{N}\},$$

where $\bar{c}$ is a connection preorbit of $c_1 \in C$.

Let $(E^-(\bar{c}, n), E^+(\bar{c}, n)) = (\cdots E^-_{n+1} E^-_n, \cdots E^+_{n+1} E^+_n)$ be a two-line preorbit. Let a_m (resp. b_m) be the label of the m-line that contains E^-_{n+m} (resp. E^+_{n+m}). Therefore, $a = a(\bar{c}, n) = \cdots a_2 a_1$ and $b = b(\bar{c}, n) = \cdots b_2 d_1$ are contained in the set $\overline{\Omega}$. The *solenoid set* $\mathcal{S} = \mathcal{S}_F$ is the set

$$\mathcal{S} = \{(a(\bar{c}, n), b(\bar{c}, n), n) : (\bar{c}, n) \in \mathcal{A}\}.$$

For all $(\bar{t}, n) = (\bar{a}, \bar{b}, n) \in \mathcal{S}_F$, adjoin to the symbols a_n and b_n all the order information and all the topological information on the endpoints of the $(m+n)$-lines $D_{a_m,n} = E^-_{n+m}$ and $D_{b_m,n} = E^+_{n+m}$. This information codes the order of the n-lines in the 1-lines and which lines and points are in which junction.

Let $\underline{\Omega}^s_n$ be the set of all adjacent symbols (t, s) such that the cylinders C_t and C_s have a common regular point or one of the endpoints x of C_t and one of the endpoints y of C_s are a preimage connection $\{x, y\}$. Let $\underline{\Omega}^g_n$ be the set of all adjacent symbols (t, s) such that C_t or C_s is a gap and C_t and C_s have a common point. Let $\underline{\Omega}_n = \underline{\Omega}^s_n \cup \underline{\Omega}^g_n$ and $\underline{\Omega} = \cup_{n \geq 1} \underline{\Omega}_n$.

The solenoid set $\mathcal{S} = \mathcal{S}_F$ also corresponds to the set of all pairs

$$(\bar{t}, \bar{s}) = (\ldots t_1, \ldots s_1) \in \overline{\Omega} \times \overline{\Omega}$$

of two-lines preorbits such that there exists $N_{\bar{t},\bar{s}} > 0$ with the property that, for all $n \geq N_{\bar{t},\bar{s}}$, $(t_n, s_n) \in \underline{\Omega}^s_n \cup \underline{\Omega}^g_n$ if, and only if, $n \geq N_{\bar{t},\bar{s}}$. The set $\mathcal{S}_{GC} \subset \mathcal{S}$ is the set of all $(\bar{t}, \bar{s}) \in \mathcal{S}$ such that $N_{\bar{t},\bar{s}} = 1$.

The sets $\mathcal{A}$ and $\mathcal{S}$ are isomorphic. By Remark D.3, we obtain the following correspondence.

Remark D.10. The correspondence $F \to \mathcal{S}_F$ induces a one-to-one correspondence between topological conjugacy classes of Markov maps and solenoid sets.

D.6.2 Pre-solenoid functions

Let F be a Markov map. Define the *pre-solenoid function* $\mathrm{s} = \mathrm{s}_F : \mathcal{S} \times \mathbb{N} \to \mathbb{R}^+$ *determined by* F by

$$\mathrm{s}_F(\bar{a}, \bar{b}, n, p) = \frac{|D_{a_p,n}|}{|D_{b_p,n}|}.$$

Equivalently,

$$\mathrm{s} = \mathrm{s}_F : \underline{\Omega} \to reals^+$$

is given by

$$\mathrm{s}(t, t') = \frac{|C_{t_i}|}{|C_{t_{i+1}}|}.$$

For all $t \in \Omega$, define the set $\mathcal{B}_t$ of *brothers* of t as the set of all symbols $t \in \Omega$ such that t and s have the same mother. The pre-scaling function $\sigma : \Omega \to \mathbb{R}^+$ and the pre-solenoid function $\mathrm{s} = \mathrm{s}_F : \mathcal{S} \times \mathbb{N} \to \mathbb{R}^+$ are related by

$$(\sigma(t))^{-1} = \sum_{s \in \mathcal{B}_t} \mathrm{s}(s,t).$$

D.6.3 The solenoid property of a cylinder structure

The cylinder structure generated by the Markov map F has the α*-solenoid property* (resp. α*-strong solenoid property*) if, and only if, for all $0 < \beta < \alpha$ (resp. $0 < \beta \leq \alpha$), there are constants $c, c_\beta > 0$ such that for all $(\bar{t}, n, p) = (\bar{a}, \bar{b}, n, p) \in \mathcal{S} \times \mathbb{N}$, (i) $\mathrm{s}_F(\bar{t}, 1, p) > c$; (ii)

$$\left| 1 - \frac{\mathrm{s}_F(\bar{t}, n, p)}{\mathrm{s}_F(\bar{t}, n, p+1)} \right| < c_\beta (|D_{a_p,n}| + |D_{b_p,n}|)^\beta.$$

Lemma D.11. *A $C^{1+\alpha^-}$ Markov map F generates a cylinder structure with the α-solenoid property. A cylinder structure with the α-solenoid property generates a Markov map G such that $\Lambda_G = \Lambda_F$ and $G|_{\Lambda_G} = F|_{\Lambda_F}$.*

The cylinder structure of F has the $(1+\alpha)$*-scaling property* if

$$\left| \frac{\sigma(t)}{\sigma(\phi(t))} - 1 \right| \leq \mathcal{O}\left(|C_t|^\beta \right),$$

for all $0 < \beta < \alpha$ and for all $t \in \Omega$. The cylinder structure of F has the $(1+\alpha)$*-connection property* if

$$\left| \frac{|C_t|}{|C_s|} \frac{|C_{\phi(s)}|}{|C_{\phi(t)}|} - 1 \right| \leq \mathcal{O}\left((|C_t| + |C_s|)^\beta \right),$$

for all $0 < \beta < \alpha$ and for all $(t, s) \in \underline{\Omega}_n^s$.

Proof of Lemma D.11. Let us prove that the Markov map F is $C^{1+\alpha^-}$ if, and only if, the cylinder structure of F has the α-solenoid property. By Theorem B.26, the Markov map F is $C^{1+\alpha^-}$ if, and only if, the cylinder structure of F has the $(1+\alpha)$-scaling property, the $(1+\alpha)$-connection property and has bounded geometry. Property (i) of the α-solenoid property implies that the cylinder structure of F has bounded geometry and vice-versa. Therefore, we will prove that the cylinder structure of F has α-solenoid property if, and only if, the cylinder structure of F has the $(1+\alpha)$-scaling property and the $(1+\alpha)$-connection property.

We now prove that if the cylinder structure of F has the α-solenoid property, then it has the $(1+\alpha)$-scaling property and the $(1+\alpha)$-connection property. Let $\phi(t) \in \Omega$ be such that $F(C_t) = C_{\phi(t)}$. Since the cylinder structure of F has the α-solenoid, we have that

$$
\begin{aligned}
\frac{\sigma(t)}{\sigma(\phi(t))} &= \frac{\sum_{s\in\mathcal{B}_t} \mathrm{s}(\phi(s),\phi(t))}{\sum_{s\in\mathcal{B}_t} \mathrm{s}(s,t)} \\
&\in \frac{\sum_{s\in\mathcal{B}_t} s(s,t)\left(1\pm\mathcal{O}\left(|C_{m(t)}|^\beta\right)\right)}{\sum_{s\in\mathcal{B}_t} s(s,t)} \\
&\subset 1\pm\mathcal{O}\left(|C_{m(t)}|^\beta\right) \\
&\subset 1\pm\mathcal{O}\left(|C_t|^\beta\right).
\end{aligned}
$$

By the α-solenoid property, the cylinder structure of F has the $(1+\alpha)$-connection property. Let us prove that if the cylinder structure of F has the $(1+\alpha)$-scaling property and the $(1+\alpha)$-connection property, then the cylinder structure of F has the α-solenoid property. By the $(1+\alpha)$-connection property of the cylinder structure of F, for all $0<\beta<\alpha$ and for all $(t,s)\in\underline{\Omega}^s_n$, we have that

$$
\frac{\mathrm{s}(t,s)}{\mathrm{s}(\phi(t),\phi(s))} \in 1\pm\mathcal{O}\left(\left(|C_t|+|C_s|\right)^\beta\right).
$$

By the $(1+\alpha)$-scaling property and by bounded geometry, for all $0<\beta<\alpha$ and for all $(t,s)\in\underline{\Omega}^g_n$, we have that

$$
\begin{aligned}
\frac{\mathrm{s}(t,s)}{\mathrm{s}(\phi(t),\phi(s))} &= \frac{\sigma(t)}{\sigma(\phi(t))}\frac{\sigma(\phi(s))}{\sigma(s)} \\
&\in 1\pm\mathcal{O}\left(\left(|C_{m(t)}|\right)^\alpha\right) \\
&\subset 1\pm\mathcal{O}\left(\left(|C_t|+|C_s|\right)^\alpha\right).
\end{aligned}
$$

Therefore, the cylinder structure of F has the α-solenoid property. □

Lemma D.12. *A cylinder structure with the α-strong solenoid property generates a $C^{1+\alpha}$ Markov map G such that $\Lambda_G=\Lambda_F$ and $G|_{\Lambda_G}=F|_{\Lambda_F}$.*

Proof of Lemma D.12. The proof follows in a similar way to the proof of Lemma D.11, using $\beta=\alpha$ in the definitions of $(1+\alpha)$-scaling property and $(1+\alpha)$-connection property and using Theorem B.26. □

The cylinder structure generated by the Markov map F has the *solenoid property* if, and only if, there are constants $c_1,c_2>0$ and $0<\lambda<1$ such that for all $(\bar{t},n,p)\in\mathcal{S}\times\mathbb{N}$, (i)$\mathrm{s}_F(\bar{t},1,p)>c_1$; (ii)

$$
\left|1-\frac{\mathrm{s}_F(\bar{t},n,p)}{\mathrm{s}_F(\bar{t},n,p+1)}\right| < c_2\lambda^{n+p}.
$$

Theorem D.13. *A C^{1+} Markov map F generates a cylinder structure with the solenoid property and vice-versa.*

The cylinder structure of F has the $(1+)$-*scale property* if, and only if, there is $0<\lambda<1$ such that, for all $t\in\Omega$,

$$
\left|\frac{\sigma(t)}{\sigma(\phi(t))}-1\right| \le \mathcal{O}(\lambda^n).
$$

The cylinder structure of F has the $(1+)$-*connection property* if, and only if, there is $0<\lambda<1$ such that, for all $(t,s)\in \underline{\Omega}^s_n$,

$$\left|\frac{|C_t|}{|C_s|}\frac{|C_{\phi(s)}|}{|C_{\phi(t)}|}-1\right|\leq \mathcal{O}\left(\lambda^n\right).$$

Proof of Theorem D.13. Let us prove that the Markov map F is C^{1+} if, and only if, the cylinder structure of F has the solenoid property. By Corollary B.23, the Markov map F is C^{1+} if, and only if, the cylinder structure of F has the (1+)-scale property, the (1+)-connection property and has bounded geometry. Property (i) of the solenoid property is equivalent to the cylinder structure of F to have bounded geometry. Therefore, we will prove that the cylinder structure of F has the (1+)-scale property and the (1+)-connection property. We now prove that if the cylinder structure of F has the solenoid property, then it has the (1+)-scale property and the (1+)-connection property. Since the cylinder structure of F has the solenoid property, we have that

$$\begin{aligned}\frac{\sigma(t)}{\sigma(\phi(t))}&=\frac{\sum_{s\in\mathcal{B}_t}\mathrm{s}(\phi(s),\phi(t))}{\sum_{s\in\mathcal{B}_t}\mathrm{s}(s,t)}\\&\in\frac{\sum_{s\in\mathcal{B}_t}\mathrm{s}(s,t)\left(1\pm\mathcal{O}c\lambda^n\right)}{\sum_{s\in\mathcal{B}_t}\mathrm{s}(s,t)}\\&\subset 1\pm\mathcal{O}\left(\lambda^n\right).\end{aligned}$$

Since the cylinder structure of F has the solenoid property, it has the (1+)-connection property. Let us prove that if the cylinder structure of F has the (1+)-scale property and the (1+)-connection property, then it has the solenoid property. By the (1+)-connection property of the cylinder structure of F, there is $0<\lambda<1$ such that, for all $(t,s)\in\underline{\Omega}^s_n$,

$$\frac{\mathrm{s}(t,s)}{\mathrm{s}(\phi(t),\phi(s))}\in 1\pm\mathcal{O}\left(\lambda^n\right).$$

By the (1+)-scale property and by bounded geometry, for all $(t,s)\in\underline{\Omega}^g_n$,

$$\begin{aligned}\frac{\mathrm{s}(t,s)}{\mathrm{s}(\phi(t),\phi(s))}&=\frac{\sigma(t)}{\sigma(\phi(t))}\frac{\sigma(\phi(s))}{\sigma(s)}\\&\in 1\pm\mathcal{O}\left(\lambda^n\right).\end{aligned}$$

Therefore, the cylinder structure of F has the solenoid property. □

D.6.4 The solenoid equivalence between cylinder structures

Let F and G be two topologically conjugate Markov maps. The sets $\mathcal{S}_F\times\mathbb{N}$ and $\mathcal{S}_G\times\mathbb{N}$ can isomorphic. Hence, we can identify these sets and denote both of them by $\mathcal{S}\times\mathbb{N}$.

Definition D.14. *The cylinder structures generated by the C^{1+} Markov maps F and G are* solenoid equivalent *if, and only if, there are constants $c_1, c_2 > 0$ and $0 < \lambda < 1$ such that for all $(\bar{t}, n, p) \in \mathcal{S} \times \mathbb{N}$, (i) $\mathrm{s}_F(\bar{t}, 1, p) > c_1$; (ii)*

$$\left| 1 - \frac{\mathrm{s}_F(\bar{t}, n, p)}{\mathrm{s}_G(\bar{t}, n, p)} \right| < c_2 \lambda^{n+p}.$$

Theorem D.15. *Let F and G be C^r Markov maps in the same topological conjugacy class. The conjugacy between F and G is C^r if, and only if, the cylinder structures generated by F and G are solenoid equivalent.*

Proof of Theorem D.15. By Theorem B.28, the Markov maps F and G are C^{1+} conjugate if, and only if, the cylinder structures of F and G are (1+)-scale equivalent and (1+)-connection equivalent. Therefore, we will prove that the cylinder structures of F and G are solenoid equivalent if, and only if, they are (1+)-scale equivalent and (1+)-connection equivalent. We will prove that if the cylinder structures of F and G are solenoid equivalent, then they are (1+)-scale equivalent and (1+)-connection equivalent. Since the cylinder structures of F and G are solenoid equivalent, we obtain that

$$\begin{aligned} \frac{\sigma_F(t)}{\sigma_G(t)} &= \frac{\sum_{s \in \mathcal{B}_t} \mathrm{s}_G(s,t)}{\sum_{s \in \mathcal{B}_t} \mathrm{s}_F(s,t)} \\ &\in \frac{\sum_{s \in \mathcal{B}_t} \mathrm{s}_F(s,t)\left(1 \pm \mathcal{O} c \lambda^n\right)}{\sum_{s \in \mathcal{B}_t} \mathrm{s}_F(s,t)} \\ &\subset 1 \pm \mathcal{O}\left(\lambda^n\right). \end{aligned}$$

Now, we prove that the cylinder structures of F and G are (1+)-connection equivalent. For all $t \in \Omega$, denote the cylinders C_t^F by C_t and the cylinders C_t^F by D_t. Since the cylinder structures of F and G are solenoid equivalent, they are (1+)-connection equivalent. Let us prove that if the cylinder structures of F and G (1+)-scale equivalent and (1+)-connection equivalent, then they are solenoid equivalent. Since they are (1+)-connection equivalent, there is $0 < \lambda < 1$ such that, for all $n > 1$ and for all $(t, s) \in \underline{\Omega}_n^s$,

$$\frac{\mathrm{s}_F(t,s)}{\mathrm{s}_G(t,s)} \in 1 \pm \mathcal{O}\left(\lambda^n\right).$$

By the (1+)-scale equivalence, for all $(t, s) \in \underline{\Omega}_n^g$,

$$\begin{aligned} \frac{\mathrm{s}_F(t,s)}{\mathrm{s}_G(t,s)} &= \frac{\sigma_F(t)}{\sigma_G(t)} \frac{\sigma_G(s)}{\sigma_F(s)} \\ &\in 1 \pm \mathcal{O}\left(\lambda^n\right). \end{aligned}$$

Therefore, the cylinder structures of F and G are solenoid equivalent. □

D.7 Solenoid functions

We define a metric d on the solenoid set $\mathcal{S}$ as follows. The distance between $(\bar{s}, n)$ and $(\bar{z}, n)$ in $\mathcal{S}$ is equal to μ^{m+n} if, and only if, $s_m = z_m$ and $s_{m+1} = z_{m+1}$, where $0 < \mu < 1$. Otherwise, the distance between two elements of $\mathcal{S}$ is equal to infinity.

A function $\mathrm{s} : \mathcal{S} \to \mathbb{R}^+$ is *pseudo-Hölder continuous* if there is a constant $c > 0$ and $0 < \alpha < 1$ such that

$$\left|1 - \frac{\mathrm{s}(\bar{s}, n)}{\mathrm{s}(\bar{z}, n)}\right| < c(d((\bar{s}, n), (\bar{z}, n)))^\alpha,$$

for all $(\bar{s}, n), (\bar{z}, n) \in \mathcal{S}$.

Lemma D.16. *Let F be a C^{1+} Markov map. The function $\mathrm{s} = \mathrm{s}_F : \mathcal{S} \to \mathbb{R}^+$ is well-defined by*

$$\mathrm{s}(\bar{a}, \bar{b}, n) = \lim_{m \to \infty} \frac{|D_{a_m,n}|}{|D_{b_m,n}|}.$$

Furthermore, s *is pseudo-Hölder continuous.*

Proof of Lemma D.18. By Theorem D.13 and by the smoothness of the Markov map F, the cylinder structure of F has the solenoidal property. Thus, there is $0 < \lambda < 1$ such that, for all $(\bar{t}, \bar{s}) \in \mathcal{S}$, and for all $n \geq N_{\bar{t},\bar{s}}$,

$$\left|1 - \frac{\mathrm{s}(t_n, s_n)}{\mathrm{s}(t_{n+1}, s_{n+1})}\right| \leq \mathcal{O}(\lambda^n).$$

Thus, for all $p, q > n \geq N_{\bar{t},\bar{s}} \geq 1$, we have that

$$\frac{\mathrm{s}(t_p, s_p)}{\mathrm{s}(t_q, s_q)} \in 1 \pm \mathcal{O}(\lambda^n).$$

Therefore, the function $s : \mathcal{S} \to \mathbb{R}^+$ is well defined and

$$\frac{\mathrm{s}(\bar{t}, \bar{s})}{\mathrm{s}(t_n, s_n)} \in 1 \pm \mathcal{O}(\lambda^n). \tag{D.7}$$

Let $0 < \alpha \leq 1$ be such that $\lambda \leq \nu^\alpha$. Let $(\bar{t}, \bar{s}), (\bar{t}', \bar{s}') \in \mathcal{S}$ be such that $N_{\bar{t},\bar{s}} = N_{\bar{t}',\bar{s}'}$ and posses the property that $t_n = t'_n$, $s_n = s'_n$ and $t_{n+1} \neq t'_{n+1}$ or $s_{n+1} \neq s'_{n+1}$, for some $n \geq N_{\bar{t},\bar{s}}$. By (D.7), the function $\mathrm{s} : \mathcal{S} \to \mathbb{R}^+$ is pseudo-Hölder continuous

$$\begin{aligned}\left|1 - \frac{\mathrm{s}(\bar{t}, \bar{s})}{\mathrm{s}(\bar{t}', \bar{s}')}\right| &\leq \left|1 - \frac{\mathrm{s}(\bar{t}, \bar{s})}{\mathrm{s}(t_n, s_n)} \frac{\mathrm{s}(t'_n, s'_n)}{\mathrm{s}(\bar{t}', \bar{s}')}\right| \\ &\leq \mathcal{O}\left((d((\bar{t}, \bar{s}), (\bar{t}', \bar{s}')))^\alpha\right).\end{aligned}$$

□

D.7.1 Turntable condition

Let $(\bar{c}, n)$ be a two-line preorbit such that the points c_m^- and c_m^+ belong to a turntable s_m. Let $e_1^m, \ldots, e_d^m$ be any d connections through the turntables s_m such that the exit $e_j^{+.m}$ of e_j^m is the entrance $e_{j+1}^{-.m}$ of e_{j+1}^m, for all $1 \le j < d$, and the exit $e_d^{+.m}$ of e_d^m is the entrance $e_1^{-.m}$ of e_1^m. For all $1 \le j \le d$, let D_j be an l-line with endpoints $e_j^{-.m}$ and $e_j^{+.m}$. Then, the product of the ratios is equal to

$$\prod_{j=1}^{d-1} \frac{|D_{j+1}|}{|D_j|} = \frac{|D_d|}{|D_1|}.$$

Hence, the solenoid function $\mathrm{s} : \mathcal{S} \to \mathbb{R}^+$ satisfies the following *turntable condition*:

$$\prod_{j=i}^{d} \mathrm{s}(\bar{\varepsilon}_j, n) = 1,$$

where $\bar{\varepsilon}_j = \cdots \varepsilon_j^2 \varepsilon_j^1$.

D.7.2 Matching condition

The ratio between two cylinders D_1 and D_2 at level n is determined by the ratios of all cylinders contained in the union $D_1 \cup D_2$ which will impose the matching condition that we now describe.

For all $(\bar{t}, n) = (\bar{a}, \bar{b}, n) \in \mathcal{S}_F$ define

$$\mathcal{C}_{(\bar{t},n)'} = \{\bar{z} \in \overline{\Omega} : D_{z_{m+n+1}} \subset D_{a_m,n} \text{ for all } m \ge 1\}$$

and

$$\mathcal{C}_{(\bar{t},n)''} = \{\bar{z} \in \overline{\Omega} : D_{z_{m+n+1}} \subset D_{b_m,n} \text{ for all } m \ge 1\}.$$

For all $\bar{u}, \bar{v} \in \mathcal{C}_{(\bar{t},n)'} \cup \mathcal{C}_{(\bar{t},n)''}$ define

$$\mathrm{s}(\bar{u}, \bar{v}) = \lim_{m \to \infty} \frac{|D_{v_{m+n+1}}|}{|D_{u_{m+n+1}}|}.$$

Hence, the solenoid function $\mathrm{s} : \mathcal{S} \to \mathbb{R}^+$ satisfies the following *matching condition*:

$$\frac{\sum_{\bar{v} \in \mathcal{C}_{(\bar{t},n)'}} \mathrm{s}(\bar{u}, \bar{v})}{\sum_{\bar{v} \in \mathcal{C}_{(\bar{t},n)''}} \mathrm{s}(\bar{u}, \bar{v})} = \mathrm{s}(\bar{t}, n).$$

for all $(\bar{t}, n) \in \mathcal{S}$ and for all $\bar{u} \in \mathcal{C}_{(\bar{t},n)'} \cup \mathcal{C}_{(\bar{t},n)''}$,

Now, we give the following abstract definition of a solenoid function.

Definition D.17. *A function* $\mathrm{s} : \mathcal{S} \to \mathbb{R}^+$ *is a* solenoid function *if,* s *satisfies the matching and the turntable conditions.*

Lemma D.18. *Let F be a C^{1+} Markov map. The function* $\mathrm{s}_F : \mathcal{S} \to \mathbb{R}^+$ *defined by*

$$\mathrm{s}(\bar{a}, \bar{b}, n) = \lim_{m \to \infty} \frac{|D_{a_m,n}|}{|D_{b_m,n}|}$$

is a pseudo-Hölder solenoid.

Proof of Lemma D.18. By Lemma D.16, the function s_F is well-defined and is pseudo-Hölder continuous. In the previous two sections, we proved that the function $\mathrm{s} : \mathcal{S} \to \mathbb{R}^+$ satisfies the matching and the turntable condition. □

D.8 Examples of solenoid functions for Markov maps

We will give some examples of solenoid sets and solenoid functions for Markov maps. The examples we give of solenoid functions are very simple, usually they have a much richer structure. For example, the scaling functions related to renormalizable structures usually have a lot of self-similarities (see [150]).

Cookie-cutters. Suppose that C_0 and C_1 are two disjoint closed subintervals of the interval C containing the endpoints of $C = [-1, 1]$. Let the train-track X be the disjoint union of the closed intervals I_0, $G = I \backslash (I_0 U I_1)$ and I_1 with junctions $J_1 = \{-1\}$, $J_2 = I_0 \cap G$, $J_3 = G \cap I_1$ and $J_4 = \{1\}$. The set of all connections is equal to $\{J_2, J_3\}$. Let $F : X \to X$ be a cookie-cutter. Let $S_F = \{0, 1\}$ and G be the 1-gap between C_0 and C_1. Add the symbol 0 to the gap point $g_0 = C_0 \cap G$ and associate the symbol 1 to the gap point $g_1 = C_1 \cap G$. Associate the information that the Markov branches F_0 and F_1 are orientation preserving or orientation reversing to the symbols 0 and 1. Let the symbol sequence $\cdots \varepsilon_2 \varepsilon_1 \in \{0, 1\}^{\mathbb{N}}$ represent the image of the gap point g_{ε_1} by the inverse Markov branches $F_{\varepsilon_m}^{-1} \circ \cdots \circ F_{\varepsilon_2}^{-1}$ for all $m > 1$. Then, the solenoid set $\mathcal{S}$ is represented by the set

$$\mathcal{S} = \{0, 1\}^{\mathbb{N}}.$$

Let F be the following cookie-cutter:

$$F(x) = \begin{cases} 2x & x \in [0, \frac{1}{2}] \\ 3x - 2 & x \in [\frac{2}{3}, 1] \end{cases}$$

The solenoid function $\mathrm{s}_F : \mathcal{S} \to \mathbb{R}^+$ is defined as follows. For all $\cdots \varepsilon_2 \varepsilon_2 \in \mathcal{S}$,

$$\mathrm{s}(\cdots \varepsilon_2 0) = \frac{|F_{\varepsilon_m}^{-1} \circ \cdots \circ F_{\varepsilon_2}^{-1}(C_0)|}{|F_{\varepsilon_m}^{-1} \circ \cdots \circ F_{\varepsilon_2}^{-1}(G)|} = 3$$

and

$$\mathrm{s}(\cdots \varepsilon_2 1) = \frac{|F_{\varepsilon_m}^{-1} \circ \cdots \circ F_{\varepsilon_2}^{-1}(C_1)|}{|F_{\varepsilon_m}^{-1} \circ \cdots \circ F_{\varepsilon_2}^{-1}(G)|} = 2$$

Tent maps defined on an interval. Let X be the train-track containing the cylinders $C_0 = [a_0, b_0]$ and $C_1 = [a_1, b_1]$ with the junctions $p_2 = \{a_0, a_0\}$, $p_3 = \{b_0, a_1\}$ and $p_4 = \{b_1, b_1\}$. Let the set of all connections be equal to $\{p_3\}$. Let $T : X \to X$ be the C^{1+} *tent map* such that: (i) $dT > \lambda > 1$ in C_0 and $dT < \lambda < -1$ in C_1; (ii) $T(C_0) = T(C_1) = I$. Let the set S_T be equal to $\{0, 1\}$. Therefore, the set of all preimage connections PC is equal to the set of all connections. Associate the information that the Markov branches F_0 is orientation preserving to the symbol 0 and that F_1 is orientation reversing to the symbol 1.

Let the symbol sequences

$$(\cdots \varepsilon_1, n) \in \{0, 1\}^{\mathbb{N}} \times \mathbb{N}$$

represent the image of the two n-cylinders with connection p_3 by the inverse Markov branches $T_{\varepsilon_m}^{-1} \circ \cdots \circ T_{\varepsilon_1}^{-1}$, for all $m > 1$. The solenoid set $\mathcal{S}$ is represented by the set

$$\mathcal{S} = \{0, 1\}^{\mathbb{N}} \times \mathbb{N}.$$

We are going to give two examples T_1 and T_2 of tent maps. In the first example, the solenoid function is constant and so Hölder continuous. In the second example, the solenoid function is pseudo-Hölder continuous but it is not Hölder continuous.

Let T_1 be the following tent map:

$$T_1(x) = \begin{cases} 3x & x \in [0, \frac{1}{3}] \\ -\frac{3}{2}x + \frac{3}{2} & x \in [\frac{1}{3}, 1] \end{cases}.$$

The solenoid function $s_{T_1} : \mathcal{S} \times \mathbb{N} \to \mathbb{R}^+$ is the constant function $s_{T_1} = 2$. Let T_2 be the following tent map:

$$T_2(x) = \begin{cases} 3x & x \in [0, \frac{1}{3}] \\ \frac{3}{2}x - \frac{1}{2} & x \in [\frac{1}{3}, 1] \end{cases}.$$

For all $(\cdots \varepsilon_1, n) \in \mathcal{S}$ the solenoid function $s_{T_2}(\cdots \varepsilon_1, n) = 2^n$. Therefore, the solenoid function $s_{T_2} : \mathcal{S} \to \mathbb{R}^+$ is pseudo-Hölder continuous, but it is not Hölder continuous.

D.8.1 The horocycle maps and the diffeomorphisms of the circle.

Let $H : X \to X$ be the horocycle (Markov) map (as defined in Chapter 13) such that (i) $H_a(C_a) = (C_b)$ and the endpoints a_1 and a_2 of C_a are sent into $H(a_1) = b_2$ and $H(a_2) = b_2$ (ii) $H_b(C_b) = X$ and the endpoints b_1 and b_2 of C_a are sent into $H(b_1) = a_2$ and $H(b_2) = b_1$. The set of all connections $p_1 = \{a_1, b_2\}$, $p_2 = \{a_2, b_1\}$ and $p_3 = \{b_1, b_2\}$ is equal to the set of all preimage connections.

Let the sequence $\cdots \varepsilon_2 \varepsilon_1 b p_1$ correspond to the preimage

$$H_{\varepsilon_m}^{-1} \cdots H_{\varepsilon_2}^{-1} H_{\varepsilon_1}^{-1} H_b^{-1}(p_1)$$

of p_1, for all $m \geq 1$. Let the sequence $\cdots \varepsilon_2 \varepsilon_1 b p_1 p_3$ correspond to the preimage

$$H_{\varepsilon_m}^{-1} \cdots H_{\varepsilon_2}^{-1} H_{\varepsilon_1}^{-1} H_b^{-1} H^{-1}(p_3)$$

of p_3, for all $m \geq 1$. Let the sequence $\cdots \varepsilon_2 \varepsilon_1 b p_1 p_3 p_2 \cdots p_2$ correspond to the preimage

$$H_{\varepsilon_m}^{-1} \cdots H_{\varepsilon_2}^{-1} H_{\varepsilon_1}^{-1} H_b^{-1} H^{-1} \cdots H^{-1}(p_2)$$

of p_2, for all $m \geq 1$. The solenoid set $\mathcal{S}$ can be represented as the subset of all sequences

$$\cdots \varepsilon_2 \varepsilon_1 \in \mathcal{S} \subset \{a, b, p_1, p_2, p_3\}^{\mathbb{N}} \times \mathbb{N}$$

such that a is followed by b; b is followed by a or b or p_1; p_1 is followed by p_3; p_3 is followed by p_2; p_2 is followed by p_2.

Let H be the horocycle map corresponding to the rigid golden rotation R_g, where g is the golden number:

$$H(x) = \begin{cases} -gx + 1 & x \in C_b = [0, \frac{1}{g}] \\ -gx + g & x \in C_a = [\frac{1}{g}, 1] \end{cases} .$$

For this example, the solenoid function $s_H : \mathcal{S} \to \mathbb{R}^+$ is the following quasi-constant function. For all $\cdots \varepsilon_2 \varepsilon_1 b p_1 \in \mathcal{S}$,

$$s_H(\cdots \varepsilon_2 \varepsilon_1 b p_1) = \frac{|H_{\varepsilon_m}^{-1} \cdots H_{\varepsilon_2}^{-1} H_{\varepsilon_1}^{-1} H_b^{-1}(C_a)|}{|H_{\varepsilon_m}^{-1} \cdots H_{\varepsilon_2}^{-1} H_{\varepsilon_1}^{-1} H_b^{-1}(C_b)|} = \frac{1}{g}.$$

For all $\cdots \varepsilon_2 \varepsilon_1 b p_1 p_3 \in \mathcal{S}$,

$$s_H(\cdots \varepsilon_2 \varepsilon_1 b p_1 p_3) = \frac{|H_{\varepsilon_m}^{-1} \cdots H_{\varepsilon_2}^{-1} H_{\varepsilon_1}^{-1} H_b^{-1} H^{-1}(C_a)|}{|H_{\varepsilon_m}^{-1} \cdots H_{\varepsilon_2}^{-1} H_{\varepsilon_1}^{-1} H_b^{-1} H^{-1}(C_b)|} = 1.$$

For all $\cdots \varepsilon_2 \varepsilon_1 b p_1 p_3 p_2 \cdots p_2 \in \mathcal{S}$,

$$s_H(\cdots \varepsilon_2 \varepsilon_1 b p_1 p_2 \cdots p_2) = \frac{|H_{\varepsilon_m}^{-1} \cdots H_{\varepsilon_2}^{-1} H_b^{-1} H^{-1} \cdots H^{-1}(C_a)|}{|H_{\varepsilon_m}^{-1} \cdots H_{\varepsilon_2}^{-1} H_b^{-1} H^{-1} \cdots H^{-1}(C_b)|} = \frac{1}{g} .$$

D.8.2 Connections of a smooth Markov map.

The connection $c = C_a \cap C_b$ between two cylinders C_a and C_b expresses the existence of a smooth structure on a neighbourhood of $c = C_a \cap C_b$ in $C_a \cup C_b$. We give an example which illustrates the importance of the connection property.

Let $F : I \to I$ be the C^{1+} Markov map defined by

$$F(x) = \begin{cases} -\frac{3}{2}x + 3 & x \in [0, 2] \\ 2x - 1 & x \in [2, 3] \\ \frac{3}{2}x - \frac{9}{2} & x \in [3, 5] \end{cases}$$

with Markov partition $C_0 = [0,2]$, $C_1 = [2,3]$ and $C_2 = [3,5]$. The set of connections is $C_F = \{2\}$ and the set of preimage connections PC_F is equal to the empty set.

Define the homeomorphism $h : [0,5] \to J$ such that: (i) h is equal to a smooth map h_1 in the set $[0,3]$ and to a smooth map h_2 in the set $[3,5]$; (ii) h is not smooth at the point 3.

Let $G = h \circ F \circ h^{-1} : J \to J$ be the smooth Markov map with Markov partition $B_0 = [h(0), h(2)]$, $B_1 = [h(2), h(3)]$ and $B_2 = [h(3), h(5)]$. The set of connections is $C_G = \{h(2)\}$ and the set of preimage connections PC_G is equal to the empty set.

Since the point 3 is not a connection of F and the point $h(3)$ is not a connection of G, the map h is a smooth conjugacy between the smooth Markov map F and the smooth Markov map G, even if h is not smooth at the point 3 in the usual sense.

The scaling function $\sigma_F : \overline{\Omega}_F \to \mathbb{R}^+$ is equal to $\sigma_G : \overline{\Omega}_G \to \mathbb{R}^+$ and the solenoid function $\mathrm{s}_F : \mathcal{S} \to \mathbb{R}^+$ is equal to $\mathrm{s}_G : \mathcal{S} \to \mathbb{R}^+$.

D.9 α-solenoid functions.

A map F is $C^{1+\alpha^-}$ *smooth* if, and only if, F is $C^{1+\beta}$, for all $0 < \beta < \alpha \leq 1$.

The Lipschitz metric $d_L = d_L(F)$. Let F be a topological Markov map. For all $(\bar{t}, n) = (\bar{a}, \bar{b}, n)$ and $(\bar{s}, n) \in \mathcal{S}$, the distance $d_L((\bar{t}, n), (\bar{s}, n))$ is equal to

$$d_L((\bar{t}, n), (\bar{s}, n)) = |D_{a_m,n}| + |D_{b_m,n}|$$

if $t_m = s_m$ and $t_{m+1} \neq s_{m+1}$. Otherwise, the distance between two elements of $\mathcal{S}$ is infinity.

The solenoid function $\mathrm{s} : \mathcal{S} \to \mathbb{R}^+$ is β pseudo-Hölder continuous, with respect to the metric d_L, if, and only if, there is a constant $c_\beta > 0$ such that for

$$\left| 1 - \frac{\mathrm{s}(\bar{t}, n)}{\mathrm{s}(\bar{s}, n)} \right| < c_\beta (d_L((\bar{t}, n), (\bar{s}, n)))^\beta ,$$

all $(\bar{t}, n), (\bar{s}, n) \in \mathcal{S}$.

Definition D.19. *An* α-solenoid function $\mathrm{s} : \mathcal{S} \to \mathbb{R}^+$, *with respect to the metric* d_L, *is a solenoid function* $\mathrm{s} : \mathcal{S} \to \mathbb{R}^+$ *that is* β *pseudo-Hölder continuous, with respect to the metric* d_L, *for all* $0 < \beta \leq \alpha \leq 1$.

Lemma D.20. *Given a* $C^{1+\alpha^-}$ *Markov map* F *the solenoid function* $\mathrm{s} : \mathcal{S} \to \mathbb{R}^+$ *is* β *pseudo-Hölder continuous, with respect to the metric* $d_L(F)$, *for all* $0 < \beta < \alpha$.

Proof of Lemma D.20. By Lemma D.11, the cylinder structure of F has the α-solenoid property. Thus, for all $(\bar{t}, \bar{s}) \in \mathcal{S}$ and for all $n \geq N_{\bar{t},\bar{s}}$, we have thet

$$\left|1 - \frac{s(t_n, s_n)}{s(t_{n+1}, s_{n+1})}\right| \leq \mathcal{O}\left((|C_{t_n}| + |C_{s_n}|)^\beta\right).$$

By the expanding property of the Markov map F, for all $p, q > n \geq N_{\bar{t},\bar{s}} \geq 1$, we have that

$$\frac{s(t_p, s_p)}{s(t_q, s_q)} \in 1 \pm \mathcal{O}\left((|C_{t_n}| + |C_{s_n}|)^\beta\right).$$

Therefore,

$$\frac{s(\bar{t}, \bar{s})}{s(t_n, s_n)} \in 1 \pm \mathcal{O}\left((|C_{t_n}| + |C_{s_n}|)^\beta\right). \tag{D.8}$$

Let $(\bar{t}, \bar{s}), (\bar{u}, \bar{v}) \in \mathcal{S}$ be such that $N_{\bar{t},\bar{s}} = N_{\bar{u},\bar{v}}$ and have the property that $t_n = u_n$, $s_n = v_n$ and $t_{n+1} \neq u_{n+1}$ or $s_{n+1} \neq v_{n+1}$, for some $n \geq N_{\bar{t},\bar{s}} = N_{\bar{u},\bar{v}}$. By (D.8), we have that

$$\begin{aligned}\left|1 - \frac{s(\bar{t}, \bar{s})}{s(\bar{u}, \bar{v})}\right| &\leq \left|1 - \frac{s(\bar{t}, \bar{s})}{s(t_n, s_n)} \frac{s(u_n, v_n)}{s(\bar{u}, \bar{v})}\right| \\ &\leq \mathcal{O}\left((d((\bar{t}, \bar{s}), (\bar{u}, \bar{v})))^\beta\right).\end{aligned}$$

Therefore, the solenoid function $s : \mathcal{S} \to \mathbb{R}^+$ is an α-solenoid function. □

D.10 Canonical set C of charts

By Remark D.10, the solenoid function $s : \mathcal{S} \to \mathbb{R}^+$ defines a symbolic set Σ_F corresponding to a topological Markov map F. We construct a set **C** of canonical charts with domains contained in the symbolic set Σ_F of F F such that:

(i) for all $x \in \Sigma_F$ and for the shift $\sigma(x) \in \Sigma_F$, there are charts $c : \Sigma_c \to \mathbb{R}^+$ and $e : \Sigma_e \to \mathbb{R}^+$ in a neighbourhood of x and in a neighbourhood of $F(x)$, respectively, such that the Markov map F is affine with respect to the charts c and e;
(ii) the solenoid function s_F is equal to the solenoid function s;
(iii) the composition map $d \circ c^{-1}$ between any two charts c and d is a smooth map, whenever defined.

We define the canonical charts $c : \Sigma_c \to \mathbb{R}^+$ by the respective pre-solenoid functions $s_c : \underline{\Omega}_c \to \mathbb{R}^+$ up to affine transformations as follows.

Let $c = (\bar{t}, 1) = (\bar{a}, \bar{b}, 1) \in \mathcal{S}$. The *pre-solenoid set* $\underline{\Omega}_c = \cup_{n \geq 1} \underline{\Omega}_{c,n}$ is the set of all points $(\bar{s}, m, p) = (\bar{v}, \bar{z}, m, p) \in \mathcal{S} \times \mathbb{N}$ such that the cylinders $D_{v_{p+j-1},m}, D_{z_{p+j-1},m} \subset D_{a_{j,1}} \cup D_{b_{j,1}}$, for all $j \geq 1$. Let $\underline{\Omega}_{c,n} \subset \underline{\Omega}_c$ be the set of all symbols $(\bar{s}, m, p) \in \underline{\Omega}_c$ such that $m + j + p = n$. The *pre-solenoid function* $s_c : \underline{\Omega}_c \to \mathbb{R}^+$ determined by the solenoid function s is defined by $s_c(\bar{s}, m, p) = s(\bar{s}, m)$.

The domain Σ_c of the canonical chart c is the set of all symbols $\varepsilon_1 \varepsilon_2 \cdots \in \Sigma_F$ such that ε_1 is equal to a_1 or b_1.

The matching condition implies that the pre-solenoid functions $s_c : \underline{\Omega}_c \to \mathbb{R}^+$ define a cylinder structure, i.e. the set of all extreme points of the cylinders at level n is contained in the set of all extreme points of the cylinders at level $n+1$, for all $n \geq 1$. The turntable condition implies the existence of turntable journeys in the neighbourhood of each turntable.

Theorem D.21. *Given a solenoid function* $s : \mathcal{S} \to \mathbb{R}^+$, *the composition map* $c_d \circ c_c^{-1}$ *between two canonical charts* c *and* d *is a* C^{1+} *smooth map, whenever defined. If* $s : \mathcal{S} \to \mathbb{R}^+$ *is an* α*-solenoid function, then the composition map* $c_d \circ c_c^{-1}$ *is a* $C^{1+\alpha^-}$ *smooth map whenever defined.*

Proof of Theorem D.21. By Theorem B.28, the composition map $c \circ d^{-1}$ is a smooth map if, and only if, the cylinder structures of the canonical charts c and d are (1+)-scale equivalent and (1+)-connection equivalent. Similarly to the proof of Theorem D.15, the (1+)-scale equivalence and (1+)-connection equivalence are equivalent to the following solenoid equivalence defined by (D.9). For all $n \geq 1$ and for all $(w_m, v_n) \in \underline{\Omega}_{c,n} \cap \underline{\Omega}_{d,n}$, let $(\overline{w} = \ldots w_n \ldots w_1, \overline{v} = \ldots v_n \ldots v_1) \in \mathcal{S}$. By the pseudo-Hölder continuity of the solenoid function, there is $o < \lambda < 1$ and there are constants $c_1, c_2 > 0$ such that

$$\begin{aligned} \frac{s_c(w_n, v_n)}{s_d(w_n, v_n)} &\in \frac{s(\overline{w}, \overline{v})(1 \pm c_1 \lambda^n)}{s(\overline{w}, \overline{v})(1 \pm c_1 \lambda^n)} \\ &\subset 1 \pm c_1 \lambda^n \end{aligned} \tag{D.9}$$

and $s_c(w_n, v_n) > c_2$. Therefore, the cylinder structures of the canonical charts c and d are solenoid equivalent. By Theorem B.28, the composition map $c \circ d^{-1}$ is a $C^{1+\alpha^-}$ map if, and only if, the cylinder structures of the canonical charts c and d are $(1+\alpha)$-scale equivalent and $(1+\alpha)$-connection equivalent. Similarly to the proof of Lemma D.11 the $(1+\alpha)$-scale equivalence and $(1+\alpha)$-connection equivalence are equivalent to the following α-solenoid equivalence defined by (D.10). For all $n \geq 1$ and for all $(w_m, v_n) \in \underline{\Omega}_{c,n} \cap \underline{\Omega}_{d,n}$, let $(\overline{w} = \ldots w_n \ldots w_1, \overline{v} = \ldots v_n \ldots v_1) \in \mathcal{S}$. By the β-pseudo-Hölder continuity of the solenoid function, for all $0 < \beta < \alpha$, there are constants $c, c_\beta > 0$ such that

$$\begin{aligned} \frac{s_c(w_n, v_n)}{s_d(w_n, v_n)} &\in \frac{s(\overline{w}, \overline{v})(1 \pm c_\beta(|C_{w_n} + C_{v_n}|))}{s(\overline{w}, \overline{v})(1 \pm c_\beta(|C_{w_n} + C_{v_n}|))} \\ &\subset 1 \pm c_\beta(|C_{w_n} + C_{v_n}|) \end{aligned} \tag{D.10}$$

and $s_c(w_n, v_n) > c$. Therefore, the cylinder structures of the canonical charts c and d are α-solenoid equivalent. □

Theorem D.22. *Given a solenoid function* $s : \mathcal{S} \to \mathbb{R}^+$, *there is a* C^{1+} *Markov map* F *such that* $s_F = s : \mathcal{S} \to \mathbb{R}^+$.

Proof of Theorem D.22. Let $F : \Sigma_F \to \Sigma_F$ be the topological Markov map corresponding to the symbolic solenoid set $\mathcal{S}$. By the turntable condition of the solenoid function and by Theorem D.21, the canonical charts give a smooth structure of the set Σ_F. For all $c = (\bar{t}, \bar{s}) \in \mathcal{S}_{GC}$, define $e = (\bar{v}, \bar{z}) \in \mathcal{S}_{GC}$ by $F(C_{t_n}) \subset F(C_{v_{n-1}})$ and $F(C_{s_n}) \subset F(C_{z_{n-1}})$, for all $n > 1$. By the construction of the canonical charts $c : \Sigma_c \to \mathbb{R}^+$ and $e : \Sigma_e \to \mathbb{R}^+$, the Markov map F is an affine map in Σ_c with respect to the charts c and e. Therefore, the Markov map F is a C^{1+} Markov map. By construction of the canonical charts, the solenoid function $\mathrm{s}_F : \mathcal{S} \to \mathbb{R}^+$ coincides with the solenoid function $\mathrm{s} : \mathcal{S} \to \mathbb{R}^+$. □

Putting together theorems D.21 and D.22, we obtain the following result.

Corollary D.23. *Given an α-solenoid function $\mathrm{s} : \mathcal{S} \to \mathbb{R}^+$ there is a $C^{1+\alpha^-}$ Markov map F such that $\mathrm{s}_F = \mathrm{s} : \mathcal{S}^F \to \mathbb{R}^+$.*

D.11 One-to-one correspondences

Theorem D.24. *The correspondence $F \to \mathrm{s}_F$ induces a one-to-one correspondence between C^{1+} conjugacy classes of C^{1+} Markov maps and pseudo-Hölder solenoid functions. Moreover, if F and G are C^r Markov maps in the same C^{1+} conjugacy class of Markov maps, then they are C^r conjugate.*

By compactness of the subset S_{GC} of all $(\bar{z}, 1) \in \mathcal{S}$, the solenoid function restricted to S_{GC} is bounded from zero and infinity which corresponds to the bounded geometry of the cylinder structure of the Markov map. The matching condition of the solenoid function corresponds to the existence of a topological Markov map. The turntable condition corresponds to the existence of journeys through the turntables. The pseudo Hölder property of the solenoid function corresponds to the existence of a C^{1+} Markov map.

If the solenoid function is bounded from zero, then a pseudo-Hölder solenoid function is Hölder continuous. Otherwise, the solenoid function is just pseudo-Holder continuous, see §D.8.

If the set of all preimage connections PC is equal to the set of all connections for a C^{1+} Markov map F, then the corresponding solenoid function is Hölder continuous. That is the case of cookie-cutters, tent maps on train-tracks, expanding circle maps, horocycle maps and Markov maps generated by pseudo-Anosov maps, for example.

Proof of Theorem D.24. By Lemma D.18 and Theorem D.22, a C^{1+} Markov map F defines a solenoid function s_F and vice-versa. By Theorem D.2, if the C^r Markov maps F and G are C^{1+} conjugate, then they are C^r conjugate. Therefore, we have to prove that the smooth Markov maps F and G are C^{1+} conjugate if, and only if, the solenoid function $\mathrm{s}_F : \mathcal{S}_F \to \mathbb{R}^+$ is equal to the solenoid function $\mathrm{s}_G : \mathcal{S}_G \to \mathbb{R}^+$. By Theorem D.15, the smooth Markov maps F and G are C^{1+} conjugate if, and only if, the cylinder structures of F and G are solenoid equivalent. Therefore, we will prove that the cylinder structures of F and G are solenoid equivalent if, and only if, $\mathrm{s}_F = \mathrm{s}_G$. By smoothness of the Markov maps F and G and by Theorem D.13, the cylinder structures of F and G have the solenoid property. Let us prove that if the cylinder structures of F and G are solenoid equivalent, then the solenoid function $\mathrm{s}_F : \mathcal{S}_F \to \mathbb{R}^+$ is equal to the solenoid function $\mathrm{s}_G : \mathcal{S}_G \to \mathbb{R}^+$. Since the Markov maps F and G are topologically conjugate, the solenoid set $\mathcal{S}_F$ and $\mathcal{S}_F$ are equal. Since the cylinder structures of F and G are solenoid equivalent and have the solenoid property, there is $0 < \lambda < 1$ such that, for all $(\bar{t} = \cdots t_1, \bar{s} = \cdots s_1) \in \mathcal{S}$ and for all $n \geq N_{\bar{t},\bar{s}} \geq 1$,

$$\begin{aligned}\frac{\mathrm{s}_F(\bar{t},\bar{s})}{\mathrm{s}_G(\bar{t},\bar{s})} &= \frac{\mathrm{s}_F(\bar{t},\bar{s})}{\mathrm{s}_F(t_n,s_n)} \frac{\mathrm{s}_F(t_n,s_n)}{\mathrm{s}_G(t_n,s_n)} \frac{\mathrm{s}_G(t_n,s_n)}{\mathrm{s}_G(\bar{t},\bar{s})} \\ &\in (1 \pm \mathcal{O}(\lambda^n))(1 \pm \mathcal{O}(\lambda^n))(1 \pm \mathcal{O}(\lambda^n)) \\ &\subset 1 \pm \mathcal{O}(\lambda^n).\end{aligned}$$

On letting n converge to infinity, we obtain that the solenoid functions $\mathrm{s}_F : \mathcal{S}_F \to \mathbb{R}^+$ and $\mathrm{s}_G : \mathcal{S}_G \to \mathbb{R}^+$ are equal. Now, we prove that if the solenoid functions $\mathrm{s}_F : \mathcal{S}_F \to \mathbb{R}^+$ and $\mathrm{s}_G : \mathcal{S}_G \to \mathbb{R}^+$ are equal, then the cylinder structures of F and G are solenoid equivalent. For all $(t_n, s_n) \in \underline{\Omega}^s_n \times \underline{\Omega}^g_n$, choose $(\bar{t} = \cdots t_n \cdots t_1, \bar{s} = \cdots s_n \cdots s_1) \in \mathcal{S}$ such that $N_{\bar{t},\bar{s}} \leq n$. Since the cylinder structures of F and G have the solenoid property and $\mathrm{s}_F(\bar{t},\bar{s}) = \mathrm{s}_G(\bar{t},\bar{s})$, we have that

$$\begin{aligned}\frac{\mathrm{s}_F(t_n,s_n)}{\mathrm{s}_G(t_n,s_n)} &= \frac{\mathrm{s}_F(t_n,s_n)}{\mathrm{s}_F(\bar{t},\bar{s})} \frac{\mathrm{s}_F(\bar{t},\bar{s})}{\mathrm{s}_G(\bar{t},\bar{s})} \frac{\mathrm{s}_G(\bar{t},\bar{s})}{\mathrm{s}_G(t_n,s_n)} \\ &\in (1 \pm \mathcal{O}(\lambda^n))(1 \pm \mathcal{O}(\lambda^n)) \\ &\subset 1 \pm \mathcal{O}(\lambda^n).\end{aligned}$$

Therefore, the cylinder structures of F and G are solenoid equivalent. □

Lemma D.25. *The definition of the α-solenoid function $\mathrm{s} : \mathcal{S} \to \mathbb{R}^+$ is independent of the C^{1+} Markov map F used to define the metric $d_L(F)$, if $\mathrm{s}_F = \mathrm{s}$.*

Proof of Lemma D.25. Let F and G be two C^{1+} Markov maps such that $\mathrm{s}_F = \mathrm{s}_G = \mathrm{s}$. By Theorem D.24, the Markov maps F and G are C^{1+} conjugate. Thus, for all $t \in \Omega_F = \Omega_G$, $|C^F_t| = \mathcal{O}(|C^G_t|)$. Therefore, for all $0 < \beta < \alpha$, if the solenoid function $\mathrm{s} : \mathcal{S} \to \mathbb{R}^+$ is β pseudo-Hölder continuous with respect to the metric d defined by using the Markov map F, then the solenoid function $\mathrm{s} : \mathcal{S} \to \mathbb{R}^+$ is pseudo-Hölder continuous with respect to the metric d defined by using the Markov map G. □

Theorem D.26. *The correspondence $F \to \mathrm{s}_F$ induces a one-to-one correspondence between $C^{1+\alpha^-}$ conjugacy classes of $C^{1+\alpha^-}$ Markov maps and α-solenoid functions, with respect to the metric $d_L(F)$.*

Proof of Theorem D.26. By Lemma D.20 and Corollary D.23, a $C^{1+\alpha^-}$ Markov map F defines an α-solenoid function s_F and vice-versa. By Theorem D.24, the smooth Markov maps F and G are C^{1+} conjugate by h if, and only if, the solenoid functions $\mathrm{s}_F : \mathcal{S}_F \to \mathbb{R}^+$ and $\mathrm{s}_G : \mathcal{S}_G \to \mathbb{R}^+$ are equal. By Theorem D.2, the conjugacy h is a $C^{1+\alpha^-}$ map. □

Lemma D.27. *If the Markov map F defines an α-solenoid function $\mathrm{s} : \mathcal{S} \to \mathbb{R}^+$, with respect to the metric $d_L(F)$, that is not $\eta > \alpha$ pseudo-Hölder continuous, then the Markov map F is not $C^{1+\eta}$.*

Proof of Lemma D.27. Let $\eta > \alpha$ and suppose that the Markov map F is $C^{1+\eta}$. By Theorem D.26, the solenoid function $\mathrm{s} : \mathcal{S} \to \mathbb{R}^+$ is η pseudo-Hölder continuous which is absurd. □

D.12 Existence of eigenvalues for (uaa) Markov maps

Let $M : T \to T$ be a Markov map and Λ_M its invariant set. Let $I, J \subset T$ be two intervals such that $M_{IJ} = M_{IJ}^{-1}$ is a homeomorphism, for some $p \geq 1$. Let $M_{JI} = M_{IJ}^{-1} : J \to I$ be the inverse map of the map M_{IJ}.

The Markov map M is *(uaa)* with respect to an atlas $\mathcal{A}$ if, and only if, there is a constant $c > 1$ and a continuous function $\varepsilon_c : \mathbb{R}_0^+ \to \mathbb{R}_0^+$ with $\varepsilon_c(0) = 0$ and with the property that for all maps M_{IJ} as above, for all charts $(i, I' \supset I)$, $(j, J' \supset J) \in A$ and for all $x, y, z \in J$ so that $0 < j(y) - j(x)$, $j(z) - j(y) < 5$ and $c^{-1} < (j(z) - j(y))/(j(y) - j(x)) < c$, we have that

$$\left| \log \frac{(i \circ M_{JI})(y) - (i \circ M_{JI})(x)}{(i \circ M_{JI})(z) - (i \circ M_{JI})(y)} \frac{j(z) - j(y)}{j(y) - j(x)} \right| < \varepsilon_c(\delta).$$

We note that the (uaa) definition of a Markov map is stronger than just to say that a map is (uaa). Roughly, a Markov map is (uaa), if every inverse composition is (uaa) with the same constant c and the function ε_c.

The Markov maps $M : T \to T$ and $N : P \to P$ are *topologically conjugate*, if there exists a homeomorphism $h : T \to P$ such that

(i) $h \circ M|_{\Lambda_M} = N \circ h|_{\Lambda_M}$ and
(ii) the singularities x and $h(x)$ have the same order.

If the homeomorphismn h is (uaa), then we say that the Markov maps M and N are (uaa) conjugate.

Let $M : T \to T$ be a (uaa) Markov map with respect to an atlas $\mathcal{A}$ on T. Let $p \in T$ be a periodic point with period q. Consider a local chart $(i, I) \in A$ such that $p \in I$. The eigenvalue $e(p)$ of p is well defined, if the following limit exists and it is independent of the chart considered:

$$e(p) = \lim_{z \to p} \frac{(i \circ M^q)(z) - (i \circ M^q)(p)}{i(z) - i(p)}.$$

Theorem D.28. *Let $M : T \to T$ be a (uaa) Markov map with respect to an atlas $\mathcal{A}$ on T and p a periodic point of the Markov map M. The eigenvalue $e(p)$ is well defined and the set of all the eigenvalues is an invariant of the (uaa) conjugacy class.*

Proof. We are going to prove, in two lemmas, the existence of eigenvalues for (uaa) Markov maps and that they are invariants of the (uaa) conjugacy classes. We prove in Lemma D.29 and Lemma D.30, for a fixed point p, that the eigenvalue $e(p)$ is well defined and it is an invariant of the (uaa) conjugacy class. The proof for a periodic point p' of period q follows in the same way as for the fixed point p using the composition M^q of the Markov map M. □

Lemma D.29. *Let p be a fixed point of the (uaa) Markov map $M : T \to T$ with respect to the atlas $\mathcal{A}$. Let $i : I \to I'$ be a chart in $\mathcal{A}$, such that $p \in I$. Then, the eigenvalue $e(p)$ is well defined by*

$$e(p) = \lim_{z \to p} \frac{(i \circ M)(z) - (i \circ M)(p)}{i(z) - i(p)}.$$

Proof. We consider two different cases: the first case when the Markov map M is orientation reversing and the second case when the Markov map M is orientation preserving. We prove both cases in two steps. First, we prove that given a sequence of points $q_n = M(q_{n+1})$ converging to p, they define a candidate $\underline{e}(p)$ for the eigenvalue $e(p)$. Secondly, we show that for any z converging to p, we obtain that

$$\lim_{z \to p} \frac{(i \circ M)(z) - (i \circ M)(p)}{i(z) - i(p)} = \underline{e}(p).$$

Let $q_n \in T$ be a sequence of points q_n such that

(i) the point $M(q_n)$ is equal to the point q_{n-1} and
(ii) the point q_n is close of the point p, for all $n \geq 0$.

Let r_n be equal to the ratio between the distances of $|i(q_{n-1}) - i(p)|$ and $|i(q_n) - i(p)|$ (see Figure D.4).

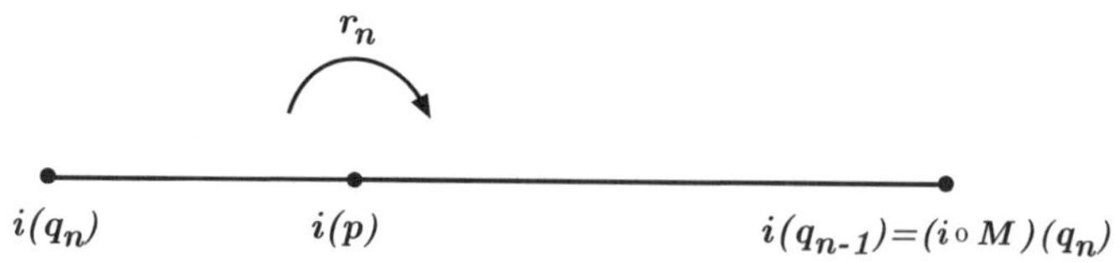

Fig. D.4. The ratio r_n.

Since the Markov map M is (uaa), the limit $r = \lim r_n$ exists and

$$\left|\frac{r}{r_n} - 1\right| \leq \varepsilon_c\left(|i(q_{n-1}) - i(q_n)|\right) .$$

Therefore,

$$|\underline{e}(p)| = \lim\left|\frac{(i \circ M)(q_n) - (i \circ M)(p)}{i(q_n) - i(p)}\right| = r . \tag{D.11}$$

First case. The Markov map M is orientation reversing, $\underline{e}(p) = -r$.

For all point z converging to p, let the point $q_n \in [p, z]$ be such that the ratio between the distance $|i(z) - i(q_n)|$ and the distance $|i(q_n) - i(p)|$ is bounded away from zero and infinity. Let s_n be equal to the ratio between the distances of $|i(z) - i(q_n)|$ and $|i(q_n) - i(p)|$. Let s_{n-1} be equal to the ratio between the distances of $|(i \circ M)(z) - i(q_{n-1})|$ and $|i(q_{n-1}) - i(p)|$ (see Figure D.5).

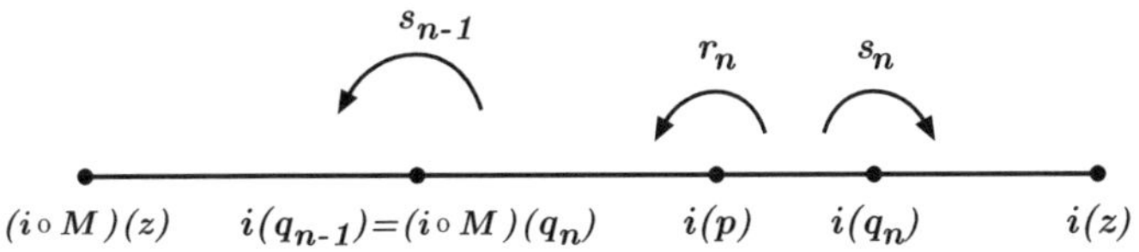

Fig. D.5. The ratios s_n and s_{n-1}.

By equality (D.11),

$$\begin{aligned}\frac{(i \circ M)(z) - (i \circ M)(p)}{i(z) - i(p)} &\in \underline{e}(p)\left(1 \pm \varepsilon_c\left(|i(q_{n-1}) - i(q_n)|\right)\right)\frac{1 + s_{n-1}}{1 + s_n} \\ &\subset \underline{e}(p)\left(1 \pm \varepsilon_c\left(|i(z) - (i \circ M)(z)|\right)\right) . \end{aligned} \tag{D.12}$$

Therefore,

$$\lim_{z \to p}\frac{(i \circ M)(z) - (i \circ M)(p)}{i(z) - i(p)} = \underline{e}(p) .$$

Second case. The Markov map M is orientation preserving, $\underline{e}(p) = r$.

For all point z converging to p, either there is a point q_n between p and z or there is not. In the case where there is a point q_n between p and z, we get a similar condition to (D.12). Otherwise, we consider a point q_n such that the ratio between $|i(q_n) - i(p)|$ and $|i(z) - i(p)|$ is bounded away from zero and infinity.

Let s_n be equal to the ratio between the distances of $|i(q_n) - i(p)|$ and $|i(z) - i(p)|$. Let s_{n-1} be equal to the ratio between the distance $|i(q_{n-1}) - i(p)|$ and the distance $|(i \circ M)(z) - (i \circ M)(p)|$ (see Figure D.6).

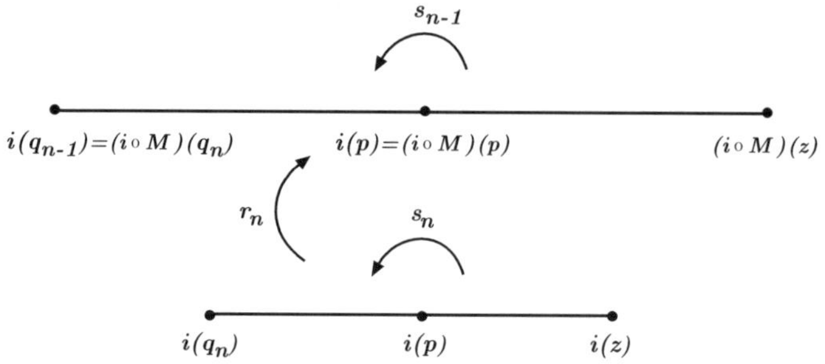

Fig. D.6. The ratios s_n and s_{n-1}.

By equality (D.11),

$$\frac{(i \circ M)(z) - (i \circ M)(p)}{i(z) - i(p)} \in \underline{e}(p)\,(1 \pm \varepsilon_c\,(|i(p) - i(q_{n-1})|))\,\frac{s_n}{s_{n-1}}$$
$$\subset \underline{e}(p)\,(1 \pm \varepsilon_c\,(|(i \circ M)(z) - i(q_{n-1})|))\,.$$

□

Lemma D.30. *The eigenvalue e(p) is an invariant of the (uaa) conjugacy class of the Markov map $M : T \to T$ with respect to the atlas $\mathcal{A}$.*

Proof. Let M and N be (uaa) Markov maps with respect to the atlases $\mathcal{A}$ and $\mathcal{B}$, respectively. Let h be the conjugate map between M and N,

$$h \circ M = N \circ h\ .$$

Let p and $p' = h(p)$ be fixed points of M and N, respectively. Let q_n and $q'_n = h(q_n)$ be two sequences of points converging to p and p', respectively, such that $M(q_n) = q_{n_1}$, for all $n \geq 0$. Let t_n be equal to the ratio between the distances of $|i(q_{n_1}) - i(q_n)|$ and $|i(q_n) - i(p)I|$ and t'_n be equal to the ratio between the distances of $|j(q'_{n-1}) - j(q'_n)|$ and $j(q'_n) - j(p')|$ (see Figure D.7), where $i \in \mathcal{A}$ and $j \in \mathcal{B}$.

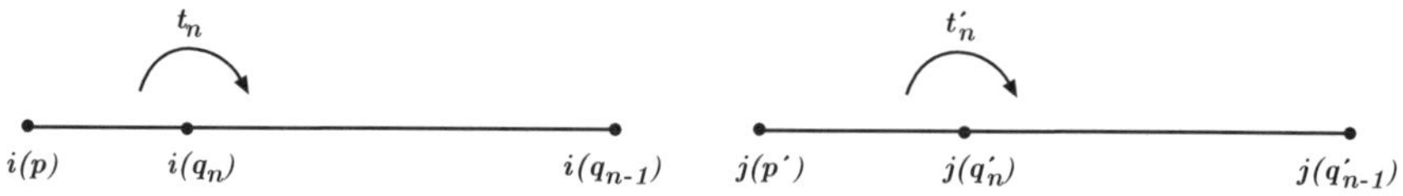

Fig. D.7. The ratios t_n and t'_n.

By Lemma D.29,

$$e_M(p) = \pm \lim(1 + t_n) \quad \text{and} \quad e_N(p') = \pm \lim(1 + t'_n) \ .$$

Since the map h is (uaa),

$$\frac{t_n}{t'_n} \in 1 \pm \varepsilon_c \left(|i(q_{n-1}) - i(q_n)| \right) \ .$$

Since the map h preserves the order, we have

$$e_M(p) = e_N(p').$$

□

D.13 Further literature

This appendix is based on Ferreira and Pinto [36] and Pinto and Rand [158].

E

Appendix E: Explosion of smoothness for Markov families

For uniformly asymptotically affine (uaa) Markov maps on train-tracks, we establish the following type of rigidity result: if a topological conjugacy between them is (uaa) at a point in the train-track then the conjugacy is (uaa) everywhere. In particular, our methods apply to the case in which the domains of the Markov maps are Cantor sets. We also present similar statements for (uaa) and C^r Markov families.

E.1 Markov families on train-tracks

E.1.1 Train-tracks

Let $\tilde{T} = \sqcup \tilde{C}_i/_{\sim}$ be the disjoint union of closed intervals $\tilde{C}_i$ of $\mathbb{R}$ with an equivalence relation $\sim$ on the endpoints of the intervals $\tilde{C}_i$. A set $\tilde{I} \subset \tilde{T}$ is an *open segment of* $\tilde{T}$ if, for every $x \in \tilde{I}$, $\mathrm{cl}(\tilde{I})\setminus\{x\}$ has two connected components $\tilde{I}_1$ and $\tilde{I}_2$. A *closed segment* $\tilde{J} \subset \tilde{T}$ is the closure $\mathrm{cl}(\tilde{I})$ of an open segment $\tilde{I}$. The boundary of an (open or closed) segment $\tilde{I}$ is $\partial \tilde{I} = \mathrm{cl}(\tilde{I}) \setminus \mathrm{int}\, \tilde{I}$. We say that $\tilde{S}$ is an *admissible set of open segments of* $\tilde{T}$ if it satisfies the following properties: (i) if $I \in \tilde{S}$ then I is an open segment of $\tilde{T}$; (ii) for all $x \in \tilde{T}$ there exists $I \in \tilde{S}$ which contains x; (iii) if I is an open segment of $\tilde{T}$ and I is contained in an union of segments in $\tilde{S}$ then I is also in $\tilde{S}$. Let T be a (compact and proper) subset of $\tilde{T}$, and $\tilde{S}$ an admissible set of open segments of $\tilde{T}$. We say that Δ_O is an admissible set of open segments of T if there is an admissible set $\tilde{S}$ of open segments of $\tilde{T}$ such that $\Delta_O = \{\tilde{I} \cap T : \tilde{I} \in \tilde{S}\}$. We say that J is a *closed segment of* T if there is an open segment $I \subset \Delta_O$ such that $J = \mathrm{cl}(I)$. Let Δ be the set of all open and closed segments of T determined by Δ_O. The *boundary* ∂I *of a segment* I of T is the boundary of the smallest segment $\tilde{I} \subset \tilde{T}$ such that $I = \tilde{I} \cap T$. The *interior* $\mathrm{int}\, I$ *of a segment* I of T is $\mathrm{int}\, I = I \setminus \partial I$. The triple $(T, \tilde{T}, \Delta)$ forms a *train-track* T_Δ. Let $T_\Delta = (T, \tilde{T}, \Delta)$ be a train-track. A *chart* (i, I) is a map $i : I \to \mathbb{R}$ which is the restriction of an injective and continuous map $\tilde{i} : \tilde{I} \to \mathbb{R}$, where $\tilde{I}$ is an open segment of $\tilde{T}$ and $\tilde{I} \cap T = I \in \Delta_O$. An *atlas* A *on* T_Δ is a set of charts with the property that for every $x \in T$ and $J \in \Delta_O$ with $x \in J$, there exists a chart (i, I) such that $I \cap J$ contains an open segment K with $x \in K$. We note that (for simplicity of exposition) if (i, I) is in A we will consider that $(i|I', I')$ is also in A for every interval $I' \subset I$. Two charts (i, I) and (j, J) with $I, J \subset T$ are *(uaa) compatible* if the overlap map $i \circ j^{-1} : j(I \cap J) \to i(I \cap J)$ is (uaa) when $I \cap J \neq \emptyset$. An *(uaa) atlas* A *on* T_Δ is an atlas formed by charts which are (uaa) compatible. Let $T_\Delta = (T, \tilde{T}, \Delta)$ and $P_\Gamma = (P, \tilde{P}, \Gamma)$ be train-tracks. The map $h : I \subset T \to J \subset P$ is a homeomorphism if there are connected sets $\tilde{I} \subset \tilde{T}$ and $\tilde{J} \subset \tilde{P}$ with $I = \tilde{I} \cap T$ and $h(I) = \tilde{J} \cap P$ such that h extends to a homeomorphism $\tilde{h} : \tilde{I} \to \tilde{J}$ and the image of every segment in $\tilde{I}$ is a segment in I, and vice-versa. Let A and B be atlases on T_Δ and on P_Γ, respectively. The homeomorphism $h : I \subset T \to J \subset P$ is (aa) at $x \in T$ if for every chart $(i, I') \in A$ with $x \in I' \subset I$ and every chart $(j, J') \in B$ with $h(x) \in J' \subset J$ we have that $j \circ h \circ i^{-1}|i(I' \cap h^{-1}(J'))$ is (aa) at $i(x)$ with modulus of continuity not depending upon the charts considered. The homeomorphism $h : I \subset T \to J \subset P$ is (uaa) at $x \in T$ if for every chart $(i, I') \in A$ with $x \in I' \subset I$ and every chart $(j, J') \in B$ with $h(x) \in J' \subset J$ we have that $j \circ h \circ i^{-1}|i(I' \cap h^{-1}(J'))$ is (aa) at $i(x)$ with modulus of continuity not depending upon the charts considered. The homeomorphism h is (uaa) if h is (uaa) at every point $x \in I$ with modulus of continuity χ_c not depending upon the point x.

E.1.2 Markov families

For every $n \in \mathbb{Z}$, let $T^n_\Delta = (T^n, \tilde{T}^n, \Delta^n)$ be a train-track and $M_n : T^n \to T^{n+1}$ a map. A *Markov partition of* $(M_n, T^n_\Delta)_{n \in \mathbb{Z}}$ is a collection $\left(C^n_1, \cdots, C^n_{m(n)}\right)_{n \in \mathbb{Z}}$ of closed and proper segments in Δ^n with the following properties for every $n \in \mathbb{Z}$:

(i) $T^n = \bigcup_{i=1}^{m(n)} C^n_i$, and the constant $m(n)$ is bounded away from infinity independently of n;
(ii) $\operatorname{int} C^n_i \bigcap \operatorname{int} C^n_j = \emptyset$ if $i \neq j$;
(iii) $M_n|\operatorname{int} C^n_i$ is a homeomorphism onto its image;
(iv) If $x \in \operatorname{int} C^n_i$ and $M_n(x) \in C^{n+1}_j$ then $M_n(C^n_i)$ contains C^{n+1}_j;
(v) For every $C^{n+1}_j \subset T^{n+1}$, there exists a C^n_i such that $M_n(C^n_i)$ contains C^{n+1}_j;
(vi) Let

$$C^n_{\varepsilon_1 \varepsilon_2 \ldots \varepsilon_m} = \{x \in C^n_{\varepsilon_1} : (M_{n+j-1} \circ \ldots \circ M_n)(x) \in C^{n+j}_{\varepsilon_j}, j = 1, 2, \ldots, m-1\}$$

be an *m-cylinder* if $C^n_{\varepsilon_1 \varepsilon_2 \ldots \varepsilon_m} \neq \emptyset$. For every sequence $C^n_{\varepsilon_1}, C^n_{\varepsilon_1 \varepsilon_2}, \ldots$ of cylinders, $\lim_{i \to \infty} \bigcap_{m=1}^{i} C^n_{\varepsilon_1 \varepsilon_2 \ldots \varepsilon_m}$ is a single point;
(vii) For every C^n_i, there exists $l = l(i, n)$ such that $T^{n+l} = M^l_n(C^n_i)$, where $l(i, n)$ is bounded away from infinity independently of i and n;

(viii) For every open segment K and $x \in K$, there is an open segment I such that $M_n(I) \subset K$ and $x \in M_n(I)$.

An *m-gap* G^n is a closed segment contained in an $(m-1)$-cylinder with the property that G^n is equal to two points which are endpoints of two m-cylinders (in particular, G^n is equal to its boundary).

Definition 45 *A* Markov family $(M_n, T^n_\Delta)_{n\in\mathbb{Z}}$ *is a sequence of train-tracks* $T^n_\Delta = (T^n, \tilde{T}^n, \Delta^n)$ *and maps* $M_n : T^n \to T^{n+1}$ *with a Markov partition. A* Markov map (M, T_Δ) *is a Markov family* $(M_n, T^n_\Delta)_{n\in\mathbb{Z}}$*, where* $M_n = M$ *and* $T^n_\Delta = T_\Delta$ *for every* $n \in \mathbb{Z}$*.*

E.1.3 (Uaa) Markov families

A local homeomorphism $\phi : I \subset \mathbb{R} \to \mathbb{R}$ is *uniformly asymptotically affine (uaa) at a point* $x \in I$ if for all $c \geq 1$ there is a continuous function $\chi_c : \mathbb{R}^+_0 \to \mathbb{R}^+_0$ satisfying $\chi_c(0) = 0$ such that for all points $y_1, y_2, y_3 \in I$ with $c^{-1} \leq (y_3 - y_2)/(y_2 - y_1) \leq c$, we have

$$\left| \log \frac{\phi(y_2) - \phi(y_1)}{\phi(y_3) - \phi(y_2)} \frac{y_3 - y_2}{y_2 - y_1} \right| < \chi_c(\max\{|y_3 - x|, |y_1 - x|\}). \tag{E.1}$$

We call χ_c the *modulus of continuity of* ϕ. The left hand-side of (E.1) is called the *ratio distortion* of ϕ at the points y_1, y_2 and y_3. The local homeomorphism $\phi : I \to \mathbb{R}$ is *(uaa)* if ϕ is (uaa) at every point $x \in I$ with modulus of continuity χ_c not depending upon the point x. We say that $\phi : I \to \mathbb{R}$ is *asymptotically affine (aa) at a point* $x \in I$ if ϕ satisfies inequality (E.1) in the case where $y_2 = x$. The classical definition of an (uaa) or symmetric function ϕ is given by taking $c = 1$ (see also Appendix A). Here, we consider in the definition all $c \geq 1$ because I does not have to be an interval. For instance I can be a Cantor set. However, by the following remark these two conditions are equivalent if I is an interval.

Remark E.1. If I is an interval and if, for $c = 1$, ϕ satisfies inequality (E.1) for all $x \in I$ then ϕ satisfies that inequality for all $c > 1$.

Proof. Follows similarly to the proof of Remark E.2. □

Definition 46 *Let* $(M_n, T^n_\Delta)_{n\in\mathbb{Z}}$ *be a Markov family,* $(A_n)_{n\in\mathbb{Z}}$ *a family of atlas* A_n *on* T^n_Δ*, and* (i, I) *a chart in the atlas* A_n*. For all distinct points* $x, y, z \in I$ *with* $i(y)$ *lying between* $i(x)$ *and* $i(z)$*, we define the* ratio $r_i(x, y, z)$ *by*

$$r_i(x, y, z) = \frac{i(z) - i(y)}{i(y) - i(x)}.$$

For every segment $K \subset I$ *we denote by* $|K|_i$ *the length of the smallest interval which contains* $i(K)$*. For simplicity of notation, we will use* $r(x, y, z)$ *and* $|K|$ *instead of* $r_i(x, y, z)$ *and* $|K|_i$*, respectively, when it is clear which is the chart that we are considering.*

We note that the set of all ratios $r_i(x,y,z)$ determines the chart (i,I) up to affine composition. Let $(M_n, T^n_\Delta)_{n\in\mathbb{Z}}$ be a Markov family and $(A_n)_{n\in\mathbb{Z}}$ a family of atlas A_n on T^n_Δ. Given two open segments $I \subset T^m$ and $J \subset T^n$ we denote by $M_{IJ} : I \to J$ the map $M_{IJ} = M_{n-1} \circ \cdots \circ M_m|I$ if M_{IJ} is a homeomorphism. We say that $(M_n, T^n_\Delta, A_n)_{n\in\mathbb{Z}}$ is an *(uaa) Markov family* if it satisfies the following properties:

(i) For every $c \geq 1$, there exists a continuous function $\chi_c : \mathbb{R}^+_0 \to \mathbb{R}^+_0$ with $\chi_c(0) = 0$ such that for all homeomorphisms $M_{IJ} : I \to J$, for all charts $(i,I) \in A_m$ and $(j,J) \in A_n$, and for all points $x,y,z \in J$ with $c^{-1} \leq r(x,y,z) \leq c$, we have

$$\left|\log \frac{r\left(M_{IJ}^{-1}(x), M_{IJ}^{-1}(y), M_{IJ}^{-1}(z)\right)}{r(x,y,z)}\right| < \chi_c(|z-x|); \tag{E.2}$$

(ii) For every closed segment I which is a 1-cylinder or an union of two 1-cylinders with a common endpoint, there is a chart $(i,I') \in A_n$ such that $I \subset I'$. There exists a constant $b > 1$ such that for every 2-cylinder or 2-gap I, $b^{-1} < |I|_i < b$ for every chart (i,I') with $I \subset I'$.

We call χ_c the *modulus of continuity of* $(M_n, T^n_\Delta, A_n)_{n\in\mathbb{Z}}$. An *(uaa) Markov map* (M, T_Δ, A) is an (uaa) Markov family $(M_n, T^n_\Delta, A_n)_{n\in\mathbb{Z}}$, where $M_n = M$, $T^n_\Delta = T_\Delta$ and $A_n = A$ for every $n \in \mathbb{Z}$. We note that condition (ii) is a technical assumption easily fulfilled in the case of a Markov map (by refining the Markov partition if necessary).

Remark E.2. Let $(M_n, T^n_\Delta, A_n)_{n\in\mathbb{Z}}$ be a Markov family such that $T^n = \tilde{T}^n$ for every $n \in \mathbb{Z}$. If $(M_n, T^n_\Delta, A_n)_{n\in\mathbb{Z}}$ satisfies property (ii) in the case where $c = 1$ then also satisfies property (ii) for every $c > 1$.

Proof. Let us prove that, for every $c \geq 1$ and for all small $\varepsilon > 0$, there exists $\delta = \delta(c,\varepsilon)$ such that, for all maps $M_{IJ} : I \to J$, for all charts $(i,I') \in A_m$ and $(j,J') \in A_n$ with $I \subset I'$ and $J \subset J'$, there exists $\delta_0 = \delta_0(c,\varepsilon)$ such that, for all $\delta < \delta_0$ and for all points $x,y,z \in J$ with $c^{-1} \leq r(x,y,z) \leq c$ and $0 < j(y) - j(x), j(z) - j(y) < \delta$, we have

$$\left|\log \frac{r\left(M_{IJ}^{-1}(x), M_{IJ}^{-1}(y), M_{IJ}^{-1}(z)\right)}{r(x,y,z)}\right| < \varepsilon, \tag{E.3}$$

and so M is (uaa). Let us denote by $[t]$ the integer part of $t \geq 0$. There exists $k = k(c,\varepsilon)$ such that, for every pair of adjacent intervals $L, R \subset \mathbb{R}$ with $c^{-1} < |L|/|R| < c$, there are adjacent intervals $P_1, \ldots, P_k$ and a constant $l = l(L,R)$ with the following properties (see Figure E.1):

(i) $\bigcup_{i=1}^{l-1} P_i \subset L$, $\bigcup_{i=l+1}^{k} P_i \subset R$ and $\bigcup_{i=1}^{k} P_i = L \cup R$;
(ii) $\left|\log |L| / \left|\bigcup_{i=1}^{l} P_i\right|\right| < \frac{\varepsilon}{3}$ and $\left|\log |R| / \left|\bigcup_{i=l+1}^{k} P_i\right|\right| < \frac{\varepsilon}{3}$.

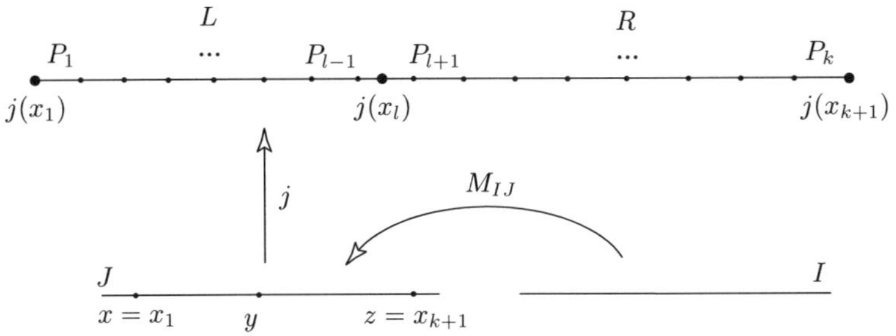

Fig. E.1. The intervals P_i.

Thus, there exist constants $k = k(c, \varepsilon)$ and $l = l(j(x), j(y), j(z))$ and points $x_1, \ldots, x_{k+1} \in J$ with the following properties:

(i) $x_1 = x$ and $x_{k+1} = z$;
(ii) the intervals $[j(x_1), j(x_2)], \ldots, [j(x_k), j(x_{k+1})]$ have the same length and pairwise disjoint interiors;
(iii)

$$\left|\log \frac{j(x_l) - j(x_1)}{j(y) - j(x)}\right| < \frac{\varepsilon}{3} \text{ and } \left|\log \frac{j(x_{k+1}) - j(x_{l+1})}{j(z) - j(y)}\right| < \frac{\varepsilon}{3}. \qquad \text{(E.4)}$$

For simplicity of notation, let us denote the map $i \circ M_{IJ}^{-1}$ by f. Since, by hypotheses $(M_n, T_\Delta^n, A_n)_{n \in \mathbb{Z}}$ satisfies property (i) with $c = 1$ in the definition of an (uaa) Markov family, there is a continuous function $\chi_1 : \mathbb{R}_0^+ \to \mathbb{R}_0^+$ satisfying $\chi_1(0) = 0$ such that, for all $1 < p < k + 1$,

$$\left|\log \frac{f(x_p) - f(x_{p-1})}{f(x_{p+1}) - f(x_p)}\right| < \chi_1(\delta).$$

Thus, for all $1 \leq n < k + 1$ and $1 < m \leq k + 1$,

$$\left|\log \frac{f(x_m) - f(x_{m-1})}{f(x_{n+1}) - f(x_n)}\right| < |m - n|\chi_1(\delta) < k(c, \varepsilon)\chi_1(\delta),$$

and so there exists a constant $c_1 > 0$ such that

$$\begin{aligned}(1 - c_1 k(c, \varepsilon)\chi_1(\delta)) \left|f(x_n) - f(x_{n-1})\right| &< \left|f(x_m) - f(x_{m-1})\right| \\ &< (1 + c_1 k(c, \varepsilon)\chi_1(\delta)) \left|f(x_n) - f(x_{n-1})\right|.\end{aligned}$$

Therefore, there exists a constant $c_2 > 0$ such that

$$\left|\log \frac{\sum_{p=1}^{l} (f(x_{p+1}) - f(x_p))}{\sum_{p=l+1}^{k} (f(x_{p+1}) - f(x_p))} \frac{k - l}{l}\right| \leq c_2 k(c, \varepsilon)\chi_1(\delta). \qquad \text{(E.5)}$$

Let us choose $\delta_0 > 0$ such that, for all $\delta < \delta_0$ we get $c_2 k(c, \varepsilon)\chi_1(\delta) < \varepsilon/3$. By inequalities (E.4) and (E.5), we obtain that

$$\left|\log \frac{r\left(M_{IJ}^{-1}(x), M_{IJ}^{-1}(y), M_{IJ}^{-1}(z)\right)}{r(x,y,z)}\right| < \frac{\varepsilon}{3} + \frac{\varepsilon}{3} + \frac{\varepsilon}{3} = \varepsilon,$$

which ends the proof. □

E.1.4 Bounded Geometry

We note that without loss of generality, we can take the modulus of continuity $\chi_c : \mathbb{R}_0^+ \to \mathbb{R}_0^+$ as being an increasing continuous function. Hence, for simplicity of the arguments in this section, we always consider that this is the case. Let $(M_n, T_\Delta^n, A_n)_{n\in\mathbb{Z}}$ be a Markov family. Let $C, D \subset T^n$ be m-cylinders or m-gaps. We say that the sets C and D are *adjacent* if they have a common endpoint.

Lemma E.3. *Let $(M_n, T_\Delta^n, A_n)_{n\in\mathbb{Z}}$ be an (uaa) Markov family. There exists a constant $d > 1$ such that, for all m-cylinders or m-gaps $C, D \subset T^n$ which are adjacent and contained in the domain I of a chart $(i, I) \in A_n$,*

$$d^{-1} < \frac{|C|_i}{|D|_i} < d.$$

Proof. Let $C' = M_{n+m-2} \circ \ldots \circ M_n(C)$, $D' = M_{n+m-2} \circ \ldots \circ M_n(D)$, and (j, J) be a chart in the atlas A_{n+m-1} such that $C', D' \subset J$. Let $b > 1$ be as considered in the definition of an (uaa) Markov family. Then

$$b^{-2} < |C'|_j/|D'|_j < b^2. \tag{E.6}$$

Take $c > b^2$. Using inequaliy (E.2) and that χ_c is an increasing function, we obtain

$$\left|\log \frac{|C|_i}{|D|_i} \frac{|D'|_j}{|C'|_j}\right| < \chi_c(b). \tag{E.7}$$

Now, Lemma E.3 follows from inequalities (E.6) and (E.7). □

Lemma E.4. *Let $(M_n, T_\Delta^n, A_n)_{n\in\mathbb{Z}}$ be an (uaa) Markov family. There exist constants $d > 1$ and $0 < \alpha, \beta < 1$ with the property that, for every m-cylinder or m-gap $C \subset T^n$, and for all charts $(i, I) \in A_n$ such that $C \cap I \neq \emptyset$, we have $|C| < d\beta^m$. If $C \subset I$ then $|C| > d^{-1}\alpha^m$.*

Proof. Since the number of Markov intervals contained in T^n is bounded independently of $n \in \mathbb{Z}$, Lemma E.4 follows from Lemma E.3. □

Lemma E.5. *If $(M_n, T^n_\Delta, A_n)_{n\in\mathbb{Z}}$ and $(N_n, P^n_\Gamma, B_n)_{n\in\mathbb{Z}}$ are two (uaa) Markov families topologically conjugate by $(h_n)_{n\in\mathbb{Z}}$ then they are C^α conjugate, for some $\alpha > 0$, i.e. there exist constants $d > 1$ and $\alpha > 0$ with the property that for every chart $(i, I) \in A_n$, for all $x, y \in I$, and for every chart $(j, J) \in B_n$ with $h_n(x), h_n(y) \in J$, we have*

$$|h_n(y) - h_n(x)|_j < d|y - x|_i^\alpha \quad \text{and} \quad |y - x|_i < d|h_n(y) - h_n(x)|_j^\alpha. \tag{E.8}$$

Proof. Let (i, I) be a chart in the atlas A_n, and for all $x, y \in I$ let (j, J) be a chart in B_n such that $h_n(x), h_n(y) \in J$. Then, choose the smallest m with the property that there are adjacent m-cylinders or m-gaps C and D, and an $(m+1)$-cylinder or $(m+1)$-gap E such that (i) $x, y \in C \cup D$, and (ii) the interval $K \subset I$ with endpoints x and y contains E. By Lemma E.4, there exist constants $d_1 > 1$ and $0 < \alpha_1, \beta_1 < 1$ such that

$$2d_1^{-1}\alpha_1^{m+1} < |E|_i \le |y - x|_i \le |(C \cup D) \cap I|_i < 2d_1\beta_1^m.$$

Similarly, there exist constants $d_2 > 1$ and $0 < \alpha_2, \beta_2 < 1$ such that

$$2d_2^{-1}\alpha_2^{m+1} < |h_n(E)|_j \le |h_n(y) - h_n(x)|_j \le |h_n((C \cup D) \cap I)|_j < 2d_2\beta_2^m.$$

Therefore, there exist constants $d > 1$ and $\alpha > 0$ such that (E.8) follows. □

E.2 (Uaa) conjugacies

The Markov families $(M_n, T^n_\Delta, A_n)_{n\in\mathbb{Z}}$ and $(N_n, P^n_\Gamma, B_n)_{n\in\mathbb{Z}}$ are *topologically conjugate* if there exists a *conjugacy family* $(h_n)_{n\in\mathbb{Z}}$ of homeomorphisms $h_n : T^n \to P^n$ such that $h_{n+1} \circ M_n = N_n \circ h_n$ for all $n \in \mathbb{Z}$. The conjugacy family $(h_n)_{n\in\mathbb{Z}}$ is (uaa) if for every n, the homeomorphisms h_n and h_n^{-1} are (uaa) and the modulus of continuity χ_c of h_n and h_n^{-1} do not depend upon n.

Definition 47 *Two (uaa) Markov families $(M_n, T^n_\Delta, A_n)_{n\in\mathbb{Z}}$ and $(N_n, P^n_\Gamma, B_n)_{n\in\mathbb{Z}}$ are* (uaa) conjugate *if there exists an (uaa) conjugacy family between $(M_n, T^n_\Delta, A_n)_{n\in\mathbb{Z}}$ and $(N_n, P^n_\Gamma, B_n)_{n\in\mathbb{Z}}$.*

An *orbit* $(w_n)_{n\in\mathbb{Z}}$ of the Markov family $(M_n, T^n_\Delta)_{n\in\mathbb{Z}}$ is a sequence of points $w_n \in T^n$ such that $M_n(w_n) = w_{n+1}$ for every $n \in \mathbb{Z}$. A *sub-orbit* $(w_{n_i})_{i\in\mathbb{Z}}$ is a subsequence of $(w_n)_{n\in\mathbb{Z}}$ (where $(n_i)_{i\in\mathbb{Z}}$ is an increasing sequence of integers).

Theorem E.6. *Let $(M_n, T^n_\Delta, A_n)_{n\in\mathbb{Z}}$ and $(N_n, P^n_\Gamma, B_n)_{n\in\mathbb{Z}}$ be (uaa) Markov families, and let $(h_n)_{n\in\mathbb{Z}}$ be a topological conjugacy family between $(M_n, T^n_\Delta, A_n)_{n\in\mathbb{Z}}$ and $(N_n, P^n_\Gamma, B_n)_{n\in\mathbb{Z}}$. If, for every point w_{n_i} of a sub-orbit $(w_{n_i})_{i\in\mathbb{Z}}$, h_{n_i} is (aa) at w_{n_i} and the modulus of continuity does not depend upon i, then $(h_n)_{n\in\mathbb{Z}}$ is an (uaa) conjugacy.*

Proof. We are going to prove that the homeomorphism $h_0 : T^0 \to P^0$ is (uaa). Then it follows, in a similar way, that h_n is (uaa) for all $n \in \mathbb{Z}$. For simplicity of exposition, we are also going to consider the case in which the conjugacy is (aa) in an orbit $(w_m)_{m \in \mathbb{Z}}$. The proof for the case where the conjugacy is (aa) just in a sub-orbit follows similarly to this one. Let (i, I) be a chart in A_0, and $x, y, z \in I$ any three points such that $c^{-1} \leq r(x, y, z) \leq c$. Take a sequence of charts $(i_m, I'_m) \in A_m$ such that for some $M < 0$ and all $m < M$ it has the following properties: (i) there are intervals I_m and J_m such that $I_m \subset J_m \subset I'_m$, and the maps

$$M_{I_m I} = M_{-1} \circ \cdots \circ M_m : I_m \to I \quad \text{and} \quad M_{J_m J_M} = M_{M-1} \circ \cdots \circ M_m : J_m \to J_M$$

are homeomorphisms; (ii) $w_m \in J_m \setminus I_m$ (see Figure E.2). Let x_M, y_M and z_M be the preimages by $M_{I_M I}$ of x, y and z, respectively. Take a point $p_M \in J_M$ and a constant $\underline{c} = \underline{c}(x_M, y_M, z_M, w_M, p_M) > 1$ such that

$$\underline{c}^{-1} < r(x_M, w_M, p_M) \ , \quad r(y_M, w_M, p_M) \ , \quad r(z_M, w_M, p_M) < \underline{c}$$

(see Figure E.2). Let $x_m, y_m, z_m, p_m \in J_m$ be the preimages by $M_{J_m J_M}$ of x_M, y_M, z_M and p_M, respectively.

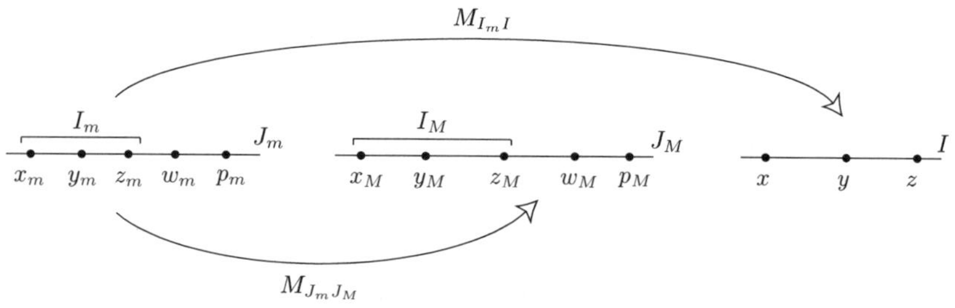

Fig. E.2. The maps $M_{I_m I}$ and $M_{J_m J_M}$.

Since the Markov family $(M_n, T^n_\Delta, A_n)_{n \in \mathbb{Z}}$ is (uaa),

$$\left| \log \frac{r(x, y, z)}{r(x_m, y_m, z_m)} \right| < \chi_c(|z - x|). \tag{E.9}$$

Let $(u, U) \in B_0$ and $(u_m, U_m) \in B_m$ be charts such that $h_0(I) \subset U$ and $h_m(J_m) \subset U_m$. Since the Markov family $(N_n, B_n)_{n \in \mathbb{Z}}$ is (uaa) and by Lemma E.5, there exist constants $d > 1$ and $0 < \alpha \leq 1$ such that

$$\left| \log \frac{r(h_0(x), h_0(y), h_0(z))}{r(h_m(x_m), h_m(y_m), h_m(z_m))} \right| < \chi_c\left(|h_0(z) - h_0(x)|_u\right) < \chi_c\left(d(|z - x|_i)^\alpha\right). \tag{E.10}$$

By hypothesis, the conjugacy h_m is (aa) at w_m, which implies that

$$\left|\log \frac{r(h_m(x_m), h_m(w_m), h_m(p_m))}{r(x_m, w_m, p_m)}\right| < \chi_{\underline{c}}\left(|p_m - x_m|\right),$$

$$\left|\log \frac{r(h_m(y_m), h_m(w_m), h_m(p_m))}{r(y_m, w_m, p_m)}\right| < \chi_{\underline{c}}\left(|p_m - y_m|\right),$$

$$\left|\log \frac{r(h_m(z_m), h_m(w_m), h_m(p_m))}{r(z_m, w_m, p_m)}\right| < \chi_{\underline{c}}\left(|p_m - z_m|\right).$$

The last three inequalities imply that

$$\log \frac{r(x_m, y_m, z_m)}{r(h_m(x_m), h_m(y_m), h_m(z_m))} \to 0, \text{ when } m \to -\infty. \tag{E.11}$$

By (E.9), (E.10) and (E.11), there is a continuous function $\chi_c' : \mathbb{R}_0^+ \to \mathbb{R}_0^+$ satisfying $\chi_c'(0) = 0$, and such that

$$\left|\log \frac{r(h_0(x), h_0(y), h_0(z))}{r(x, y, z)}\right| < \chi_c'(|z - x|).$$

Therefore, the conjugacy h_0 is (uaa). □

A *generating set* $\mathcal{G}$ of $(T^n)_{n\in\mathbb{Z}}$ is a set of points $a \in T^{l(a)}$ with $l(a) \in \mathbb{Z}$, and with the property that, for every $n \in \mathbb{Z}$, we have

$$T^n = \text{cl}\left(\{w = M_{n-1} \circ \cdots \circ M_{l(a)}(a) : a \in \mathcal{G} \text{ and } l(a) \le n\}\right).$$

Theorem E.7. *Let $(M_n, T_\Delta^n, A_n)_{n\in\mathbb{Z}}$ and $(N_n, P_\Gamma^n, B_n)_{n\in\mathbb{Z}}$ be (uaa) Markov families, and let $(h_n)_{n\in\mathbb{Z}}$ be a topological conjugacy family between $(M_n, T_\Delta^n, A_n)_{n\in\mathbb{Z}}$ and $(N_n, P_\Gamma^n, B_n)_{n\in\mathbb{Z}}$. If, for every point a of a generating set $\mathcal{G}$, $h_{l(a)}$ is (aa) at a and the modulus of continuity does not depend upon a, then $(h_n)_{n\in\mathbb{Z}}$ is an (uaa) conjugacy.*

Proof. We are going to prove that the homeomorphism $h_0 : T^0 \to P^0$ is (uaa). It follows, in a similar way, that h_n is (uaa), for all $n \in \mathbb{Z}$. Let (i, I) be a chart in A_0, and $x, y, z \in I$ any three points such that $c^{-1} \le r(x, y, z) \le c$. By construction of the set $\mathcal{G}$, there is a sequence $(w_k)_{k\in\mathbb{Z}}$ of points $w_k = \left(M_{-1} \circ \cdots \circ M_{l(a_k)}\right)(a_k) \in I$ such that (i) $a_k \in \mathcal{G}$, (ii) $i(x) < i(w_k) < i(z)$, and (iii) $\lim w_k = y$. Take a sequence of charts (i_k, I_k') in $A_{l(a_k)}$ such that for some $K > 0$ and all $k > K$ it has the following properties: (i) there are points $x_k, y_k, z_k, a_k \in I_k'$ whose images by $M_{-1} \circ \cdots \circ M_{l(a_k)}$ are the points $x, y, z, w_k \in I$, respectively; (ii) the interval $I_k \subset I_k'$ with endpoints x_k and z_k contains the points y_k and a_k; (iii) I_k is sent injectively by $M_{-1} \circ \cdots \circ M_{l(a_k)}$ in the interval with endpoints x and z. Since the Markov family $(M_n, T_\Delta^n, A_n)_{n\in\mathbb{Z}}$ is (uaa), for k large enough, we get

$$\left|\log \frac{r(x, w_k, z)}{r(x_k, a_k, z_k)}\right| < \chi_c(|z - x|). \tag{E.12}$$

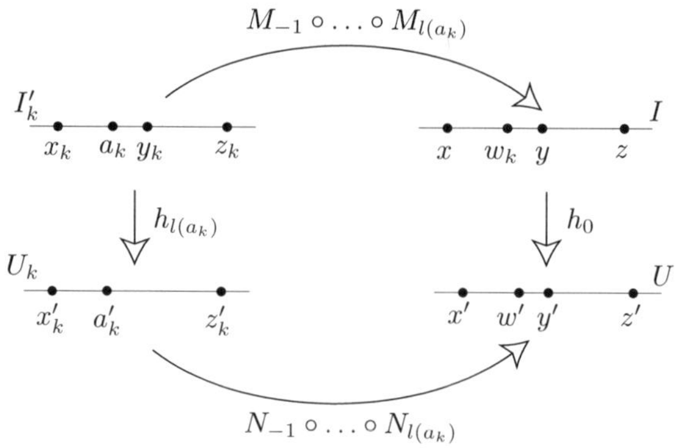

Fig. E.3. The points in the ratios of the proof of Theorem E.7.

Set $x' = h_0(x)$, $y' = h_0(y)$, $w'_k = h_0(w_k)$, $z' = h_0(z)$ and $x'_k = h_{l(a_k)}(x_k)$, $a'_k = h_{l(a_k)}(a_k)$, $z'_k = h_{l(a_k)}(z_k)$ (see Figure E.3).

Let $(u, U) \in B_0$ and $(u_k, U_k) \in B_{l(a_k)}$ be charts such that $h_0(I) \subset U$ and $h_{l(a_k)}(I_k) \subset U_k$. Since the Markov family $(N_n, P^n_{\Gamma}, B_n)_{n\in\mathbb{Z}}$ is (uaa) and by Lemma E.5, there exist constants $d > 1$ and $0 < \alpha \le 1$ such that

$$\left|\log \frac{r(x', w'_k, z')}{r(x'_k, a'_k, z'_k)}\right| < \chi_c\left(|z' - x'|_u\right) < \chi_c\left(d(|z - x|_i)^{\alpha}\right). \tag{E.13}$$

Since the conjugacy $h_{l(a_k)}$ is (aa) at the point a_k,

$$\left|\log \frac{r(x'_k, a'_k, z'_k)}{r(x_k, a_k, z_k)}\right| < \chi_c\left(|z_k - x_k|\right). \tag{E.14}$$

Note that $\chi_c\left(|z_k - x_k|\right)$ converges to zero, when k tends to infinity. Therefore, by (E.12), (E.13) and (E.14), there is a continuous function $\chi'_c : \mathbb{R}^+_0 \to \mathbb{R}^+_0$ satisfying $\chi'_c(0) = 0$, and such that

$$\left|\log \frac{r(x', w'_k, z')}{r(x, w_k, z)}\right| < \chi'_c(|z - x|). \tag{E.15}$$

By continuity of the ratios, we obtain

$$\lim_{k\to\infty} r(x', w'_k, z') = r(x', y', z') \text{ and } \lim_{k\to\infty} r(x, w_k, z) = r(x, y, z). \tag{E.16}$$

Therefore, by (E.15) and (E.16), we conclude

$$\left|\log \frac{r(x', y', z')}{r(x, y, z)}\right| \le \chi'_c(|z - x|),$$

and so h_0 is (uaa). □

A *sub-sequence* $(w_{n_i})_{i\in\mathbb{Z}}$ is any sequence of points $w_{n_i} \in T^{n_i}$ (where $(n_i)_{i\in\mathbb{Z}}$ is an increasing sequence of integers).

Theorem E.8. *Let $(M_n, T^n_\Delta, A_n)_{n\in\mathbb{Z}}$ and $(N_n, P^n_\Gamma, B_n)_{n\in\mathbb{Z}}$ be (uaa) Markov families, and let $(h_n)_{n\in\mathbb{Z}}$ be a topological conjugacy family between $(M_n, T^n_\Delta, A_n)_{n\in\mathbb{Z}}$ and $(N_n, P^n_\Gamma, B_n)_{n\in\mathbb{Z}}$. If, for every point w_{n_i} of a subsequence $(w_{n_i})_{i\in\mathbb{Z}}$, h_{n_i} is (uaa) at w_{n_i} and the modulus of continuity does not depend upon i, then $(h_n)_{n\in\mathbb{Z}}$ is an (uaa) conjugacy.*

Proof. We are going to prove that the homeomorphism $h_0 : T^0 \to P^0$ is (uaa). It follows, in a similar way, that h_n is (uaa) for all $n \in \mathbb{Z}$. Let (i, I) be a chart in A_0, and $x, y, z \in I$ any three points such that $c^{-1} \le r(x, y, z) \le c$. By conditions (v) and (vii) of the definition of Markov partition, there exists $L > 0$ such that for all $n > L$ and all $(n-L)$-cylinders C, we have $(M_{-1} \circ \ldots \circ M_n)(C) = T^0$. Hence, by Lemma E.4 there is n_k sufficiently large and there is a chart $(i_k, I_k) \in A_{n_k}$ such that (i) $w_{n_k} \in I_k$; (ii) $|I_k|_{i_k} < |z-x|_i$; (iii) $(M_{-1} \circ \ldots \circ M_{n_k})(I_k) = I$; and (iv) $M_{I_k I} = M_{-1} \circ \ldots \circ M_{n_k} : I_k \to I$ is a homeomorphism. Set $x_k = M^{-1}_{I_k I}(x)$, $y_k = M^{-1}_{I_k I}(y)$ and $z_k = M^{-1}_{I_k I}(z)$ (see Figure E.4).

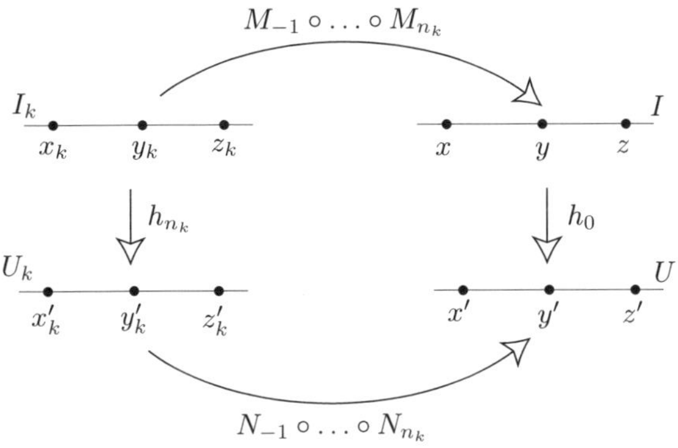

Fig. E.4. The points in the ratios of the proof of Theorem E.8.

Since the Markov family $(M_n, T^n_\Delta, A_n)_{n\in\mathbb{Z}}$ is (uaa),

$$\left|\log \frac{r(x,y,z)}{r(x_k,y_k,z_k)}\right| < \chi_c(|z-x|). \tag{E.17}$$

Set $x' = h_0(x)$, $y' = h_0(y)$ and $z' = h_0(z)$ and $x'_k = h_{n_k}(x_k)$, $y'_k = h_{n_k}(y_k)$ and $z'_k = h_{n_k}(z_k)$. Let $(u, U) \in B_0$ and $(u_k, U_k) \in B_k$ be charts such that $h_0(I) \subset U$ and $h_{n_k}(I_k) \subset U_k$. Since the Markov family $(N_n, P^n_\Gamma, B_n)_{n\in\mathbb{Z}}$ is (uaa) and by Lemma E.5, there exist constants $d > 1$ and $0 < \alpha \le 1$ such that

$$\left|\log \frac{r(x',y',z')}{r(x'_k,y'_k,z'_k)}\right| < \chi_c(|z'-x'|_u) < \chi_c\left(d(|z-x|_i)^\alpha\right). \tag{E.18}$$

Since the conjugacy h_{n_k} is (uaa) at the point w_{n_k} and $|z_k - x_k|_{j_k} < |z - x|_i$, we have

$$\left|\log \frac{r(x'_k, y'_k, z'_k)}{r(x_k, y_k, z_k)}\right| < \chi_c(|z - x|). \tag{E.19}$$

Therefore, by (E.17), (E.18) and (E.19), there is a continuous function $\chi'_c : \mathbb{R}_0^+ \to \mathbb{R}_0^+$ satisfying $\chi'_c(0) = 0$, and such that

$$\left|\log \frac{r(x', y', z')}{r(x, y, z)}\right| < \chi'_c(|z - x|). \tag{E.20}$$

Therefore, h_0 is (uaa). □

E.3 Canonical charts

Given an (uaa) Markov family $(M_n, T^n_\Delta, A_n)_{n\in\mathbb{Z}}$, we define a *canonical chart* (c_0, J_0) with $J_0 \subset T^0$ containing a 1-cylinder as follows (see also [158] and [175]).

Let $I_0, I_{-1}, \ldots$ be segments such that $I_0 \subset J_0$, $I_m \subset T^m$, $M_m|I_m$ is a homeomorphism onto its image and $M_m(I_m) = I_{m+1}$. Let $K_m : I_m \to J_0$ be the homeomorphism given by $K_m = M_{-1} \circ \ldots \circ M_m|I_m$ for every $m < 0$. Let us denote by j_l and j_r the endpoints of J_0. Take a chart $(i_m, I'_m) \in A_m$ such that $I_m \subset I'_m$. Let $L_m : i_m(I_m) \to (0, 1)$ be the map determined uniquely by $L_m(K_m^{-1}(j_l)) = 0$, $L_m(K_m^{-1}(j_r)) = 1$, and L_m has an affine extension to $\mathbb{R}$. Let $d_m : J_0 \to (0, 1)$ be the chart defined by $d_m = L_m \circ i_m \circ K_m^{-1}$ (see Figure E.5). By Lemma E.9 below, the sequence $(d_m)_{m\in\mathbb{Z}}$ converges when m tends to minus infinity. We define the canonical chart $c_0 : J_0 \to \mathbb{R}$ as being this limit $c_0 = \lim_{m\to-\infty} d_m$. The canonical charts (c_0, J_0) with $J_0 \subset T^0$ form the canonical atlas $C_{A,0}$ on T^0. Similarly, for every $n \in \mathbb{Z}$, we define the canonical charts (c_n, J_n) with $J_n \subset T^n$ containing a 1-cylinder which form the canonical atlas $C_{A,n}$ on T^n.

Lemma E.9. *The canonical charts $c_0 : J_0 \to \mathbb{R}$ are well-defined by*

$$c_0 = \lim_{m\to-\infty} d_m .$$

The canonical charts (c_0, J_0) with $J_0 \subset T^0$ form the canonical atlas $C_{A,0}$ on T^0. Similarly, for every $n \in \mathbb{Z}$, we define the canonical charts (c_n, J_n) with $J_n \subset T^n$ containing a 1-cylinder which form the canonical atlas $C_{A,n}$ on T^n.

Lemma E.10. *$(M_n, T^n_\Delta, C_{A,n})_{n\in\mathbb{Z}}$ is an (uaa) Markov family, and it is (uaa) conjugate to $(M_n, T^n_\Delta, A_n)_{n\in\mathbb{Z}}$.*

Proof of Lemmas E.9 and E.10. Let us begin proving that the canonical chart $(c_{A,0}, J_0)$ with $J_0 \subset T^0$ is well-defined by $c_{A,0} = \lim_{m\to-\infty} d_{A,m}$, where the charts $(d_{A,m}, J_0)$ are as introduced in §E.3. Let x, y, z be any three points in J_0

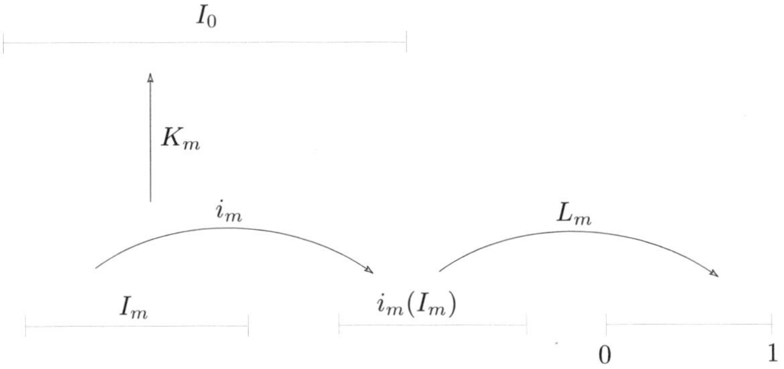

Fig. E.5. The chart $d_m = L_m \circ i_m \circ K_m^{-1}$.

such that $c^{-1} < r_{d_{A,0}}(x, y, z) < c$. Since the Markov family $(M_n, T_\Delta^n, A_n)_{n \in \mathbb{Z}}$ is (uaa), the ratios $r_{d_{A,m}}(x, y, z)$ converge to a unique limit $r(x, y, z)$ when m tends to minus infinity. Furthermore,

$$\left| \log \frac{r_{d_{A,0}}(x, y, z)}{r(x, y, z)} \right| < \chi_c \left(|z - x|_{d_{A,0}} \right) . \tag{E.21}$$

Thus, the ratio $r(x, y, z)$ varies continuously with x, y and z, and there exists a constant $c_1 > 1$ such that $c_1^{-1} < r(x, y, z) < c_1$. Let j_l and j_r be the endpoints of the interval J_0. For every point $y \in J_0$, there is a sequence of pairwise distinct points $x_0, \ldots, x_p, \ldots, x_q \in J_0$ such that $x_0 = j_l$, $x_p = y$, $x_q = j_r$ and $c^{-1} < r_{d_{A,0}}(x_i, x_{i+1}, x_{i+2}) < c$. Hence, writing $r(j_l, y, j_r)$ in terms of the ratios $r(x_i, x_{i+1}, x_{i+2})$ we get that the ratio $r(j_l, y, j_r)$ varies monotonically and continuously with $y \in J_0$. Thus,

$$c_{A,0}(y) = \lim_{m \to -\infty} d_{A,m}(y) = \lim_{m \to -\infty} \frac{1}{1 + r_{d_{A,m}}(j_l, y, j_r)} = \frac{1}{1 + r(j_l, y, j_r)} ,$$

which implies that $c_{A,0}$ is a bijection and topologically compatible with $d_{A,0}$. Hence, the canonical chart $(c_{A,0}, J_0)$ is well-defined. Therefore, the set of canonical charts $(c_{A,0}, J_0)$ with $J_0 \subset T^0$ form a topological atlas $C_{A,0}$. Moreover, by inequality (E.21), the canonical charts $(c_{A,0}, J_0)$ are (uaa) compatible with the charts in A_0. By a similar construction, for every $n \in \mathbb{Z}$, we obtain that the canonical charts in $C_{A,n}$ are (uaa) compatible with the charts in A_n, and the modulus of continuity does not depend upon the charts considered and upon $n \in \mathbb{Z}$. Therefore, using that the Markov family $(M_n, T_\Delta^n, A_n)_{n \in \mathbb{Z}}$ is (uaa), we get that the Markov family $(M_n, T_\Delta^n, C_{A,n})_{n \in \mathbb{Z}}$ is also (uaa). □

E.4 Smooth bounds for C^r Markov families

For $r = k + \alpha$, where $k \in \mathbb{N}$ and $0 < \alpha \leq 1$, a function $h : I \to \mathbb{R}$ defined on an interval I is C^r if the kth derivative of h is α-*Hölder* continuous. We say that a function $h : J \to \mathbb{R}$ defined on a set $J \subset \mathbb{R}$ is C^r if h has a C^r extension to an interval $I \supset J$ of $\mathbb{R}$. An atlas A on a train-track T_Δ is C^r if the overlap map between any two charts in A is C^r and its C^r norm is bounded away from infinity independently of the charts considered. A C^r *Markov family* $(M_n, T^n_\Delta, A_n)_{n\in\mathbb{Z}}$ is a Markov family with the following properties:

(i) The atlases A_n are C^r, and locally the maps M_n with respect to any pair of charts are C^r diffeomorphisms with C^r norm bounded away from infinity independently of the charts considered and of $n \in \mathbb{Z}$;
(ii) There exist constants $c > 0$ and $\lambda > 1$ such that, for all $x \in T^n$ and $p \geq 0$, we have

$$\left|\left(j \circ M_{n+p} \circ \cdots \circ M_n \circ i^{-1}\right)' (i(x))\right| > c\lambda^p$$

where $(i, I) \in A_n$, $(j, J) \in A_{n+p+1}$ and there is an open segment $I' \subset I$ such that $x \in I'$ and $M_{n+p} \circ \cdots \circ M_n(I') \subset J$;
(iii) The property (i) of the definition of (uaa) Markov family is also satisfied.

A *Markov map* (M, T_Δ, A) *is* C^r if there is a C^r Markov family $(M_n, T^n_\Delta, A_n)_{n\in\mathbb{Z}}$ with $M_n = M$, $T^n_\Delta = T_\Delta$ and $A_n = A$ for all $n \in \mathbb{Z}$. Let $I_0, I_{-1}, \ldots$ be segments such that $I_0 \subset J_0$, $I_{-n} \subset T^{-n}$, $M_{-n}|I_{-n}$ is a homeomorphism onto its image and $M_{-n}(I_{-n}) = I_{-n+1}$. Take a chart $(i_{-n}, I'_{-n}) \in A_{-n}$ such that $I_{-n} \subset I'_{-n}$. Let F_{-n} be the inverse map of $i_{-n+1} \circ M_{-n} \circ i_{-n}$. Let $f_{-n} = F_{-n} \circ F_{-1}$.

Lemma E.11. *Let F be a $C^{k+\alpha}$ Markov family. Then, for all $r \in \{1, \ldots, k-1\}$,*

$$d^r \ln df_n = \sum_{l=0}^{r-1} \sum_{i=0}^{n-1} \left(\left(d^{r-l} \ln dF_{-(i+1)} \circ f_i\right)\right.$$
$$\left.(df_i)^{r-l} E^r_l \left(d \ln df_i, \ldots, d^l \ln df_i\right)\right),$$

where E^r_l is a polynomial of order l and the coefficients are independent of $n, i \geq 0$. For $i = 0$, we define the map f_i equal to the identity map.

Proof. We will prove it by induction in the degree of smoothness r.
Case $r = 1$. By differentiation,

$$\ln df_n = \sum_{i=0}^{n-1} \ln dF_{-(i+1)} \circ f_i.$$

Therefore,

$$d \ln df_n = \sum_{i=0}^{n-1} \left(d \ln dF_{-(i+1)} \circ f_i\right) df_i.$$

Thus, the formula is valid for $r = 1$, with $E^1_0 = 1$.

Induction step. Let us suppose by induction hypothesis that the formula is valid for r and let us prove that it is valid for $r+1$. First, we differentiate separately the three terms of the formula in Lemma E.11.

The derivative of the first term is

$$d\left(d^{r-l}\ln dF_{-(i+1)}\circ f_i\right) = \left(d^{r+1-l}\ln dF_{-(i+1)}\circ f_i\right) df_i.$$

The derivative of the second term is

$$d\left(\left(df_i\right)^{r-l}\right) = (r-l)\left(df_i\right)^{r-l}\left(d\ln df_i\right).$$

The derivative of the third term is

$$dE_l^r\left(d\ln df_i,\ldots,d^l\ln df_i\right) = F_l^r\left(d\ln df_i,\ldots,d^{l+1}\ln df_i\right),$$

where F_l^r has degree l and coefficients independent of i and n. We define the polynomial

$$G_{l+1}^r(x_1,\ldots,x_{l+1}) = F_l^r(x_1,\ldots,x_{l+1}) + (r-l)x_1E_l^r(x_1,\ldots,x_l).$$

The polynomial G_{l+1}^r has degree $l+1$ and the coefficients are independent of i and n. Therefore,

$$\begin{aligned}d^{r+1}\ln df_n = &\sum_{l=0}^{r-1}\sum_{i=0}^{n-1}\Big(\left(d^{r+1-l}\ln dF_{-(i+1)}\circ f_i\right)\\ &\left(df_i\right)^{r+1-l}E_l^r\left(d\ln df_i,\ldots,d^l\ln df_i\right)\Big)\\ &+\sum_{l=0}^{r-1}\sum_{i=0}^{n-1}\Big(\left(d^{r-l}\ln dF_{-(i+1)}\circ f_i\right)\\ &\left(df_i\right)^{r-l}G_{l+1}^r\left(d\ln df_i,\ldots,d^{l+1}\ln df_i\right)\Big).\end{aligned}$$

Replacing $l+1$ by l in the second term, we have

$$E_0^{r+1}(x_1,\ldots,x_{l+1}) = E_0^r(x_1,\ldots,x_l) = 1.$$

Define $E_r^r = 0$. For $l = 1,\ldots,r$, $E_l^{r+1}(x_1,\ldots,x_l)$ is equal to

$$F_{l-1}^r(x_1,\ldots,x_l) + (r-l+1)x_1E_{l-1}^r(x_1,\ldots,x_{l-1}) + E_l^r(x_1,\ldots,x_l).$$

□

Lemma E.12. *Let F be a $C^{k+\alpha}$ Markov family. Then, for all $x,y\in C^{F_0}$,*

$$\left|\ln\frac{df_n(y)}{df_n(x)}\right| \le c|x-y|^{\beta},$$

where $\beta=\alpha$ if $k=1$, or $\beta=1$ if $k>1$. Moreover,

$$df_n(y)\in\exp(\pm c_3)df_n(x).$$

Proof. By property (i) and (ii) of a $C^{k+\alpha}$ Markov family, for all $x, y \in C^{F_0}$, there is $z_{x,y} \in C^{F_0}$ such that

$$\begin{aligned}\left|\ln \frac{df_n(y)}{df_n(x)}\right| &\leq \sum_{i=0}^{n-1} \left(\left|\ln \left|dF_{-(i+1)} \circ f_i(y)\right| - \ln \left|dF_{-(i+1)} \circ f_i(x)\right|\right|\right) \\ &\leq c_1 \sum_{i=0}^{n-1} |f_i(y) - f_i(x)|^\beta \leq c_1 \sum_{i=0}^{n-1} (df_i(z_{x,y}))^\beta \, |y - x|^\beta \\ &\leq c|y - x|^\beta \leq c_3,\end{aligned}$$

for some constant $c_3 > 0$. Therefore,

$$df_n(y) \in \exp(\pm c_3) df_n(x).$$

□

Lemma E.13. *Let F be a $C^{k+\alpha}$ Markov family. Then,*

$$\| \ln df_n \|_{C^{k-1+\alpha}} \leq b_k.$$

Proof. The case $k = 1$ is proved by Lemma E.12. For $k \geq 2$, we will prove by induction in r that $d^r \ln df_n$ is bounded in the C^0 norm, independent of n, for all $r = 1, \ldots, k-1$. After, we prove that $d^{k-1} \ln df_n$ is α-Hölder continuous with constant independent of n.
Case $r = 1$. By Lemma E.12 and as $k \geq 2$,

$$\left|\ln \frac{df_n(y)}{df_n(x)}\right| \leq c|x - y|.$$

Therefore, $d \ln df_n$ is bounded in the C^0 norm, independent of n.
Induction step. By induction hypotheses, we suppose that the maps $d \ln df_n$, $\ldots$, $d^{r-1} \ln df_n$ are bounded in the C^0 norm, independent of n. We will prove that $d \ln df_n$ is bounded in the C^0 norm, independent of n.

By Lemma E.11,

$$\begin{aligned} d^r \ln df_n = \sum_{l=0}^{r-1} \sum_{i=0}^{n-1} &\left(\left(d^{r-l} \ln dF_{-(i+1)} \circ f_i\right) \right. \\ &\left. (df_i)^{r-l} \, E_l^r \left(d \ln df_n, \ldots, d^l \ln df_n\right)\right), \end{aligned}$$

where the coefficients of the polynomial E_l^r are independent of n and i, for all $r = 1, \ldots, k-1$.

By property (i) of a $C^{k+\alpha}$ Markov family, there is $b > 0$ such that $\left|dF_{-(i+1)}\right| > b$. Since the first $r+1$ derivatives of the map $F_{-(i+1)}$ are bounded independent of i,

$$\left|d^{r-l} \ln dF_{-(i+1)} \circ f_i\right| \leq b_{r,l}, \tag{E.22}$$

for all $l = 0, \ldots, r-1$, $i = 0, \ldots, n-1$ and $n \in \mathbb{N}$.

By property (ii) of a $C^{k+\alpha}$ Markov family F,

$$\left| \sum_{i=o}^{n-1} (df_i)^{r-l} \right| \leq b_r, \tag{E.23}$$

for all $l = 0, \ldots, r-1$, $i = 0, \ldots, n-1$ and $n \in \mathbb{N}$.

The induction hypotheses implies

$$\left| E_l^r \left(d \ln df_i, \ldots, d^l \ln df_i \right) \right| \leq b_{r,l}, \tag{E.24}$$

for all $l = 0, \ldots, r-1$, $i = 0, \ldots, n-1$ and $n \in \mathbb{N}$.

By Lemma E.11 and inequalities (E.22), (E.23) and (E.24), we have that

$$|d^r \ln df_n| \leq b_r.$$

Let us prove that the map $d^{k-1} \ln df_n$ is α-Hölder continuous with constant independent of n. The map $d^{k-1-l} \ln dF_{-(i+1)}$ is α-Hölder continuous for $l = 0$ and it is Lipschitz for $l = 1, \ldots, k-2$. By property (i) of a $C^{k+\alpha}$ Markov family, α-Hölder (resp. Lipschitz) constant is independent of $i \geq 0$, i.e

$$\left\| d^{k-1-l} \ln dF_{-(i+1)} \right\|_{C^\alpha \text{ or } C^{Lipschitz}} \leq c,$$

for some constant $c > 0$. Thus, the map $d^{k-1-l} \ln F_{-(i+1)} \circ f_i$ is Lipschitz if $l > 0$, or α-Hölder continuous if $l = 0$. Therefore, the Lipschitz (resp. α-Hölder) constant of the map $d^{k-1-l} \ln F_{-(i+1)} \circ f_i$ converges exponentially fast to zero, when i tends to infinity.

The map $(df_i)^{(k-1-l)}$ is Lipschitz, where the Lipschitz constant converges exponentially fast to zero, when i tends to infinity because it has bounded nonlinearity and it is exponentially contracting.

The map

$$E_l^{k-1} \left(d \ln df_i, \ldots, d^l \ln df_i \right)$$

is Lipschitz with constant independent of i because it is an l-product of maps bounded in the C^1 norm, independently of i, as proved by induction.

Therefore, the map

$$\left(d^{k-1-l} \ln dF_{-(i+1)} \circ f_i \right) (df_i)^{k-1-l} E_l^{k-1} \left(d \ln df_i, \ldots, d^l \ln df_i \right)$$

is a product of α-Hölder and Lipschitz maps with constants bounded independent of $i = 0, \ldots, n$ and $n \in \mathbb{N}$. The map $(df_i)^{(k-1-l)}$ converges exponentially fast to zero in the $C^{Lipschitz}$ norm, when i tends to infinity. Therefore, the product of the three maps above is α-Hölder continuous, where the α-Hölder constant converges exponentially fast to zero, when i tends to infinity. Therefore, the map $d^{k-1} \ln df_n$ is α-Hölder continuous. $\square$

E.4.1 Arzelà-Ascoli Theorem

A subset of a topological space is called *conditionally compact*, if its closure is compact in its relative topology.

Theorem E.14. Arzelà-Ascoli. *If S is a compact set, then a set in the space of continuous functions with domain S is conditionally compact if, and only if, it is bounded and equicontinuous.*

We say that the map f is bounded in the $C^{k+\alpha^-}$ norm, if, for all $0 < \varepsilon < \alpha$, f is bounded in the $C^{k+\varepsilon}$ norm. A sequence $(f_n)_{n\geq 0}$ converge in the $C^{k+\alpha^-}$ norm, if, for all $0 < \varepsilon < \alpha$, the sequence $(f_n)_{n\geq 0}$ converge in the $C^{k+\varepsilon}$ norm.

Lemma E.15. *Let $(f_n)_{n\geq 0}$ be a sequence of $C^{k+\alpha}$ smooth functions f_n defined in an interval $I = [a, c]$, where $k > 0$ and $\alpha \in (0, 1]$. If $\|f_n\|_{C^{k+\alpha}} \leq b$, for all $n \geq 0$, then there is a subsequence $(f_{n_i})_{i\geq 0}$ converging to a $C^{k+\alpha}$ smooth function f in the $C^{k+\alpha^-}$ norm.*

Corollary E.16. *The set of all functions $f \in C^{k+\alpha}$ defined in an interval I such that $\|f\|_{C^{k+\alpha}} \leq b$ is a compact set with respect to the $C^{k+\alpha-\varepsilon}$ norm, for all small $\varepsilon > 0$.*

Proof of Lemma E.15 . As the sequence of maps f_n is bounded in the $C^{k+\alpha}$ norm, we have that

$$\left|d^k f_n(x) - d^k f_n(y)\right| \leq b|x - y|^\alpha,$$

for all $n \geq 0$. Therefore, $\left(d^k f_n\right)_{n\geq 0}$ is an equicontinuous family of functions. By the Arzela-Ascoli theorem, there is a subsequence $\left(d^k f_{n_i}\right)_{i\geq 0}$ converging to a function h in the C^0 norm. In other words, there is a sequence $(l_i)_{i\geq 0}$ converging to zero such that

$$\left|d^k f_{n_i} - h\right| \leq l_i.$$

As the function h is continuous, it is integrable. Let us show that the sequence $\left(d^{k-m} f_{n_i}\right)_{i\geq 0}$ converges to m-times the integral of h in the C^0 norm, for all $m = 1, \ldots, k$.

$$\left|d^{k-m} f_{n_i} - \int_a^x \cdots \int_a^x h\right| \leq \left|\int_a^x \cdots \int_a^x \left(d^k f_{n_i} - h\right)\right| \leq l_i|c - a|^m.$$

Therefore, the sequence $(f_{n_i})_{i\geq 0}$ converges to k-times the integral of h in the C^k norm.

Let us prove that the subsequence $(f_{n_i})_{i\geq 0}$ converges in the $C^{k+\varepsilon}$ norm to k-times the integral of h, for all $\varepsilon < \alpha$. Define the map $H = H_{m,j} = d^k f_{n_m} - d^k f_{n_j}$. As the subsequence $(f_{n_i})_{i\geq 0}$ is contained in a Banach space with respect to the $C^{k+\varepsilon}$ norm, we have to prove that

$$\frac{|H(y) - H(x)|}{|y - x|^\varepsilon}$$

tends to zero, when j tends to infinity, for all $m \geq j$.

If $|x - y| > l_j$, then

$$\frac{|H(y) - H(x)|}{|y - x|^\varepsilon} \leq \frac{|H(y)|}{|y - x|^\varepsilon} + \frac{|H(x)|}{|y - x|^\varepsilon} \leq \frac{4l_j}{(l_j)^\varepsilon} \leq 4(l_j)^{1-\varepsilon}.$$

If $|x - y| \leq l_j$, then

$$\begin{aligned}\frac{|H(y) - H(x)|}{|y - x|^\varepsilon} &\leq \frac{\left|d^k f_{n_m}(x) - d^k f_{n_m}(y)\right| + \left|d^k f_{n_j}(x) - d^k f_{n_j}(y)\right|}{|y - x|^\varepsilon} \\ &\leq \frac{2b||y - x|^\alpha}{|y - x|^\varepsilon} \leq 2b(l_j)^{\alpha - \varepsilon}.\end{aligned}$$

Therefore, the sequence of functions $(f_{n_i})_{i \geq 0}$ converges to a function f in the $C^{k+\varepsilon}$ norm.

The function f is $C^{k+\varepsilon}$, because

$$\begin{aligned}\left|d^k f(x) - d^k f(y)\right| &\leq \left|d^k f(x) - d^k f_{n_i}(x)\right| + \left|d^k f_{n_i}(x) - d^k f_{n_i}(y)\right| \\ &\quad + \left|d^k f_{n_i}(y) - d^k f(y)\right| \\ &\leq 2l_i + c|x - y|^\alpha,\end{aligned}$$

and as the sequence $(l_i)_{i \in \mathbb{N}}$ tends to zero, when i tends to infinity, we obtain that $\left|d^k f(x) - d^k f(y)\right| \leq c|x - y|^\alpha$. □

E.5 Smooth conjugacies

The C^r Markov families $(M_n, T^n_\Delta, A_n)_{n \in \mathbb{Z}}$ and $(N_n, P^n_\Gamma, B_n)_{n \in \mathbb{Z}}$ are C^r *conjugate* if there is a family $(h_n)_{n \in \mathbb{Z}}$ of C^r diffeomorphisms $h_n : T^n \to P^n$ such that $h_{n+1} \circ M_n = N_n \circ h_n$, and the C^r norms of the maps h_n and h_n^{-1} are bounded away from infinity independently of $n \in \mathbb{Z}$.

Lemma E.17. *Let $(M_n, T^n_\Delta, A_n)_{n \in \mathbb{Z}}$ be a $C^{k+\delta}$ Markov family, where $k \in \mathbb{N}$ and $\delta > 0$. Let $(C_{A,n})_{n \in \mathbb{Z}}$ be the family of canonical atlas determined by the family $(A_n)_{n \in \mathbb{Z}}$. Then $(M_n, T^n_\Delta, C_{A,n})_{n \in \mathbb{Z}}$ is a $C^{k+\delta}$ Markov family, and $(M_n, T^n_\Delta, C_{A,n})_{n \in \mathbb{Z}}$ is $C^{k+\delta}$ conjugate to $(M_n, T^n_\Delta, A_n)_{n \in \mathbb{Z}}$.*

Proof. We are going to prove that the canonical charts (c_n, J_n) with $J_n \subset T^n$ are $C^{k+\delta}$ compatible with the charts contained in A_n. Furthermore, the overlap maps have $C^{k+\delta}$ norm bounded away from infinity, independently of the charts considered, and of $n \in \mathbb{Z}$. Let (c_0, J_0) be the canonical chart in $C_{A,0}$ defined by $c_0 = \lim_{m \to -\infty} d_m$, where the charts (d_m, J_0) are given by

$d_m = L_m \circ i_m \circ K_m^{-1}$, and the maps L_m, i_m and K_m are as introduced in §E.3. The map $d_m \circ i_0^{-1}$ is $C^{k+\delta}$ and it is the composition of a contraction $i_m \circ K_m^{-1} \circ i_0^{-1}$ followed by an expansion L_m. By Lemma E.13, the $C^{k+\delta}$ norm of the maps $d_m \circ i_0^{-1}$ is uniformly bounded. Hence, by Lemma E.15, there is a subsequence of maps $d_{m_l} \circ i_0^{-1}$ converging in the $C^{k+\delta-\varepsilon}$ norm to a $C^{k+\delta}$ map ψ. Moreover, the $C^{k+\delta}$ norm of ψ is bounded away from infinity independently of the charts (c_0, J_0) and (d_m, J_0) considered. By Lemma E.9, the map ψ is equal to $c_0 \circ i_0^{-1}$, where c_0 is the canonical chart. By the same argument, the map $(d_m \circ i_0^{-1})^{-1}$ has a subsequence converging in the $C^{k+\delta-\varepsilon}$ norm to a $C^{k+\delta}$ map ϕ, and the $C^{k+\delta}$ norm of ϕ is bounded away from infinity, independently of the charts (c_0, J_0) and (d_m, J_0) considered. By Lemma E.9, the map ϕ is equal to $\psi^{-1} = (c_0 \circ i_0^{-1})^{-1}$. Thus, the chart c_0 is $C^{k+\delta}$ compatible with i_0, and the norm of the overlap map ϕ is bounded away from infinity, independently of the charts c_0 and i_0 considered. Similarly, we obtain that the charts (c_n, J_n) with $J_n \subset T^n$ are $C^{k+\delta}$ compatible with the charts contained in A_n and the norm of the overlap maps is bounded away from infinity, independently of the charts considered and of $n \in \mathbb{Z}$. Therefore, using that $(M_n, T_\Delta^n, A_n)_{n\in\mathbb{Z}}$ is a $C^{k+\delta}$ Markov family, we obtain that $(M_n, T_\Delta^n, C_{A,n})_{n\in\mathbb{Z}}$ is also a $C^{k+\delta}$ Markov family, and that $(M_n, T_\Delta^n, C_{A,n})_{n\in\mathbb{Z}}$ is $C^{k+\delta}$ conjugate to $(M_n, T_\Delta^n, A_n)_{n\in\mathbb{Z}}$. □

Proposition E.18. *The Markov family $(M_n, T_\Delta^n, C_{A,n})_{n\in\mathbb{Z}}$ attains the maximum possible smoothness in the (uaa) conjugacy class of the Markov family $(M_n, T_\Delta^n, A_n)_{n\in\mathbb{Z}}$. Moreover, the family $(C_{A,n})_{n\in\mathbb{Z}}$ is canonical in the following sence: if $(M_n, T_\Delta^n, A_n)_{n\in\mathbb{Z}}$ and $(N_n, P_\Gamma^n, B_n)_{n\in\mathbb{Z}}$ are (uaa) conjugate by a conjugacy family $(h_n)_{n\in\mathbb{Z}}$ then, for every chart $(c_{A,n}, J_{A,n}) \in C_{A,n}$, there is a chart $(c_{B,n}, J_{B,n}) \in C_{B,n}$ with $J_{B,n} = h_n(J_{A,n})$ such that*

$$r_{c_{A,n}}(x, y, z) = r_{c_{B,n}}(h_n(x), h_n(y), h_n(z))$$

for all distinct points $x, y, z \in J_{A,n}$, or equivalently $c_{B,n} \circ h_n \circ c_{A,n}^{-1}$ has an affine extension to the reals.

Proof. Let $(M_n, T_\Delta^n, A_n)_{n\in\mathbb{Z}}$ and $(N_n, P_\Gamma^n, B_n)_{n\in\mathbb{Z}}$ be (uaa) Markov families which are (uaa) conjugated by $(h_n)_{n\in\mathbb{Z}}$. Let $(c_{A,0}, J_{A,0})$ with $J_{A,0} \subset T^0$ be a canonical chart contained in $C_{A,0}$ defined by

$$c_{A,0} = \lim_{m\to-\infty} L_{A,m} \circ i_{A,m} \circ K_{A,m}^{-1}$$

where $L_{A,m}$ is an affine map, $(i_{A,m}, I'_{A,m})$ is a chart contained in A_m, and $K_{A,m} = M_{-1} \circ \ldots \circ M_m$ is as defined in §E.3. Similarly, let $(c_{B,0}, J_{B,0})$ with $J_{B,0} = h_0(J_{A,0})$ be a canonical chart contained in $C_{B,0}$ defined by

$$c_{B,0} = \lim_{m\to-\infty} L_{B,m} \circ i_{B,m} \circ K_{B,m}^{-1}$$

where $L_{B,m}$ is an affine map, $(i_{B,m}, I'_{B,m})$ is a chart contained in B_m, and $K_{B,m} = N_{-1} \circ \ldots \circ N_m$ is as defined in §E.3. For all distinct points

$x, y, z \in J_{A,0}$, let us denote $K_{A,m}^{-1}(x)$, $K_{A,m}^{-1}(y)$ and $K_{A,m}^{-1}(z)$ by x_m, y_m and z_m, respectively. By construction of the charts $c_{A,0}$ and $c_{B,0}$, we have that

$$r_{c_{A,0}}(x, y, z) = \lim_{m \to -\infty} r_{i_{A,m}}(x_m, y_m, z_m) \tag{E.25}$$

and

$$r_{c_{B,0}}(h_0(x), h_0(y), h_0(z)) = \lim_{m \to -\infty} r_{i_{B,m}}(h_m(x_m), h_m(y_m), h_m(z_m)). \tag{E.26}$$

Since the family $(h_n)_{n \in \mathbb{Z}}$ is (uaa), there is $\chi_c : \mathbb{R}_0^+ \to \mathbb{R}_0^+$ satisfying $\chi_c(0) = 0$, and such that

$$\left| \log \frac{r_{i_{B,m}}(h_m(x_m), h_m(y_m), h_m(z_m))}{r_{i_{A,m}}(x_m, y_m, z_m)} \right| < \chi_c\left(|z_m - x_m|_{i_{A,m}}\right). \tag{E.27}$$

Putting together (E.25), (E.26) and (E.27), we get

$$r_{c_{A,0}}(x, y, z) = r_{c_{B,0}}(h_0(x), h_0(y), h_0(z)),$$

and so $c_{B,0} \circ h_0 \circ c_{A,0}^{-1}$ has an affine extension to the reals. Similarly, for every $n \in \mathbb{Z}$ and for all canonical charts $(c_{A,n}, J_{A,n})$ with $J_{A,n} \subset T^n$ and $(c_{B,n}, J_{B,n})$ with $J_{B,n} = h_n(J_{A,n})$, we obtain that $c_{B,n} \circ h_n \circ c_{A,n}^{-1}$ has an affine extension to the reals. Hence, for all distinct points $x, y, z \in J_{A,n}$, $r_{c_{A,n}}(x, y, z) = r_{c_{B,n}}(h_n(x), h_n(y), h_n(z))$. Let us suppose that the Markov family $(N_n, P_\Gamma^n, B_n)_{n \in \mathbb{Z}}$ is C^r, for $r > 1$. By Lemma E.17, the Markov family $(N_n, P_\Gamma^n, C_{B,n})_{n \in \mathbb{Z}}$ is C^s, for $s \geq r$. Since the maps $c_{B,n} \circ h_n \circ c_{A,n}^{-1}$ are affine, we obtain that the Markov family $(M_n, T_\Delta^n, C_{A,n})_{n \in \mathbb{Z}}$ is also C^s. Therefore, $(M_n, T_\Delta^n, C_{A,n})_{n \in \mathbb{Z}}$ attains the maximum possible smoothness in the (uaa) conjugacy class of $(M_n, T_\Delta^n, A_n)_{n \in \mathbb{Z}}$. □

Theorem E.19. *Let $(M_n, T_\Delta^n, A_n)_{n \in \mathbb{Z}}$ and $(N_n, P_\Gamma^n, B_n)_{n \in \mathbb{Z}}$ be C^r Markov families and let $(h_n)_{n \in \mathbb{Z}}$ be a topological conjugacy between $(M_n, T_\Delta^n, A_n)_{n \in \mathbb{Z}}$ and $(N_n, P_\Gamma^n, B_n)_{n \in \mathbb{Z}}$. If $(h_n)_{n \in \mathbb{Z}}$ is (uaa) then $(M_n, T_\Delta^n, A_n)_{n \in \mathbb{Z}}$ and $(N_n, P_\Gamma^n, B_n)_{n \in \mathbb{Z}}$ are C^r conjugate.*

Proof. By Proposition E.18, the Markov families $(M_n, T_\Delta^n, C_{A,n})_{n \in \mathbb{Z}}$ and $(N_n, P_\Gamma^n, C_{B,n})_{n \in \mathbb{Z}}$ are at least C^r and the conjugacy family between them is as smooth as the Markov families. By Lemma E.17, the Markov families $(M_n, T_\Delta^n, A_n)_{n \in \mathbb{Z}}$ and $(M_n, T_\Delta^n, C_{A,n})_{n \in \mathbb{Z}}$ are C^r conjugate, and the Markov families $(N_n, P_\Gamma^n, B_n)_{n \in \mathbb{Z}}$ and $(N_n, P_\Gamma^n, C_{B,n})_{n \in \mathbb{Z}}$ are C^r conjugate. Therefore, $(M_n, T_\Delta^n, A_n)_{n \in \mathbb{Z}}$ and $(N_n, P_\Gamma^n, B_n)_{n \in \mathbb{Z}}$ are C^r conjugate. □

E.6 Further literature

The results for Markov maps presented in Appendix D have a natural extension to Markov families. The results presented in this Appendix have a natural extension to non uniformly expanding multimodal maps as presented in Alves, Pinheiro and Pinto [6]. This chapter is based in Bedford and Fisher [13], Ferreira and Pinto [38], Pinto [150] and Pinto and Rand [169].

References

1. Abraham, R., Robbin, J.: Transversal Mappings and Flows. W.A. Benjamin, Inc. New York (1967).
2. Abraham, R., Smale, S.: 'Nongenericity of Ω-stability. Global Analysis, Proc. of Symposia in Pure Mathematics', AMS, **14**, 5–8 (1970).
3. Ahlfors, L.V., Beurling, A.: 'The boundary correspondence under quasiconformal mappings'. Acta Math., **96**, 125–142 (1956).
4. Almeida, J.P.: Pinto's golden tilings. Proceedings of the conference DYNA 2008, International Conference in honour of M. M. Peixoto and D. A. Rand. (2008) M. M. Peixoto, A. A. Pinto and D. A. Rand (Eds.) Springer Verlag.
5. Almeida L, Cruz J, Ferreira, H., Pinto, A. Plato's allegory applied to the theory of planed behaviour through game theory. Proceedings of the conference DYNA 2008, International Conference in honour of M. M. Peixoto and D. A. Rand. (2008) M. M. Peixoto, A. A. Pinto and D. A. Rand (Eds.) Springer Verlag.
6. Alves, J.F., Pinheiro, V., Pinto, A.A.: Explosion of smoothness for multimodal maps. In preparation.
7. Anosov, D.V.: Tangent fields of transversal foliations in U-systems. Math. Notes Acad. Sci., USSR, **2** (5) (1967).
8. Antezana, J., Pujals, E.R., Stojanoff, D.: Covergences of iterated Aluthge transform sequence for diagonalizable matrices. Advances in Mathematics (New York), v. **216**, 255–278, (2007).
9. Araujo, V., Pacifico, M.J., Pujals, E.R., Viana, M: Singular-hyperbolic attractors are chaotic. Transactions AMS, p. 1-11, 2007 (to appear).
10. Arnol'd, V.I.: Small denominators I: on the mapping of a circle into itself. Investijia Akad. Nauk Math, **25**, 1, 21–96 (1961). Transl. A.M.S., 2nd series, 46, 213–284.
11. Avez, A.: Anosov diffeomorphims. In: Gottschalk, W. and Auslander, J. (eds) Topological Dynamics. An International Symposium. Benjamin N.Y. 17–51 (1968).
12. Barreira, L.: Dimension and Recurrence in Hyperbolic Dynamics. Progress in Mathematics **272**, Birkhäuser (2008).
13. Bedford T., Fisher, A.M.: Ratio geometry, rigidity and the scenery process for hyperbolic Cantor sets, Ergod. Th. & Dynam. Sys., **17**, 531–564 (1997).
14. Bonatti, C., Diaz, L., Pujals, E.R.: C^1-generic dichotomy for diffeomorphisms: Weak forms of hyperbolicity or infinitely many sinks or sources.. Annals of Mathematics, v. **158**, 355–418 (2003).

15. Birkhoff, G., Martens, M., Tresser, C.P.: On the scaling structure for period doubling. Asterisque volume dedicated to Jacob Palis 60th birthday.
16. Bochi, J.: Generecity of zero Lyapunov exponents. Ergod. Th. & Dynam. Sys., **22**, 1667–1696 (2002).
17. Bowen, R.: Equilibrium States and the Ergodic Theory of Axiom A Diffeomorphisms. Lecture Notes in Mathematics, No. 470, Springer-Verlag, Ney York (1975).
18. Burroughs, N.J., Oliveira, B.M.P.M., Pinto, A.A.: Regulatory T cell adjustment of quorum growth thresholds and the control of local immune responses. Journal of Theoretical Biology **241**, 134–141 (2006).
19. Burroughs, N.J., Oliveira, B.M.P.M., Pinto, A.A., Sequeira, H.J.T.: Sensibility of the quorum growth thresholds controlling local immune responses. Special issue Mathematical Methods and Modeling of Biophysical Phenomena of the Journal Mathematical and Computer Modelling **47**, 714–725 (2008).
20. Carvalho, A., Peixoto, M.M., Pinheiro, D., Pinto A.A.: Universal pendulum focal decomposition. Submitted for publication.
21. Cawley, E.: The Teichmüller space of an Anosov diffeomorphism of T^2. Inventiones Mathematicae, **112**, 351–376 (1993).
22. Coelho, Z., Pinto, A.A., Rand, D.A.: Rigidity result for comuting pairs of the circle. In preparation.
23. Coullet, P., Tresser, C.: Itération d'endomorphismes et groupe de renormalisation. J. de Physique, **C5**, 25 (1978).
24. Cui, G., Gardiner, F.P., Jiang, Y.: Scaling functions of degree 2 circle endomorphisms, Ahlfors-Bers Colloquium at Storrs, November (2001).
25. Denjoy, A.: Sur les courbes définies par les équations différentielles á la surface du tore. J. Math. Pure et Appl., **11**, série 9, 333-375 (1932).
26. Diaz, L., Pujals, E.R., Ures, R.: Partial hyperbolicity and robust transitivity. Acta Mathematica, v. **183**, n. **1**, p. 1-43 (1999).
27. Falconer, K.J.: The geometry of fractal sets. CUP (1985).
28. de Faria, E.: Quasi-symmetric distortion and rigidity of expanding endomorphisms of S^1. Proceedings of the American Mathematical Society, **124**, 1949–1957 (1996).
29. de Faria, E., de Melo, W., Pinto, A.A.: Global hyperbolicity of renormalization for C^r unimodal mappings. Annals of Mathematics **164**, 731–824 (2006).
30. Federer, H.: Geometric Measure Theory. New York: Springer (1969).
31. Feigenbaum, M.J.: Presentation functions, fixed points, and a theory of scaling function dynamics. J. Stat. Phys., **52**, 527–569 (1988).
32. Feigenbaum, M.J.: Presentation functions, and scaling function theory for circle maps. Nonlinearity, **1**, 577–602 (1988).
33. Feigenbaum, M.J.: The universal metric properties of nonlinear transformations. J. Stat. Phys., **21**, 669–706 (1979).
34. Feigenbaum, M.J.: Quantitative universality for a class of nonlinear transformations. J. Stat. Phys., **19**, 25–52 (1978).
35. Ferreira, F.: Diferenciabilidade das Aplicações de Markov não diferenciáveis. M. Sc. Thesis, FCUP, Porto, Portugal (1994).
36. Ferreira, F., Pinto, A.A.: Existence of eigenvalues for (uaa) Markov maps. Portugaliae Mathematica, **54** (4), 421–429 (1997).
37. Ferreira, F., Pinto, A.A.: Explosion of smoothness from a point to everywhere for conjugacies between diffeomorphisms on surfaces. Ergod. Th. & Dynam. Sys., **23**, 509–517 (2003).

38. Ferreira, F., Pinto, A.A.: Explosion of smoothness from a point to everywhere for conjugacies between Markov families. Dyn. Sys., **16** (2), 193–212 (2001).
39. Flaminio, L., Katok, A.: Rigidity of sympletic Anosov diffeomorphisms on low dimensional tori. Ergod. Th. & Dynam. Sys., **11**, 427–440 (1991).
40. Franks, J.: Anosov diffeomorphisms. In: Smale, S. (ed) Global Analysis. AMS Providence, 61–93 (1970).
41. Franks, J.: Anosov diffeomorphisms on tori. Trans. Amer. Math. Soc., **145**, 117–124 (1969).
42. Gardiner, F.P., Sullivan, D.P.: Symmetric strucutres on a closed curve. American Journal of Mathematics, **114**, 683–736 (1992).
43. Gaspar Ruas, A.A.: Atratores hiperbólicos de codimensão um e classes de isotopia em superficies. Tese de Doutoramento. IMPA, Rio de Janeiro (1982).
44. Ghys, E.: Rigidité différentiable des groupes Fuchsiens. Publ. IHES **78**, 163–185 (1993).
45. Harrison, J.: An introduction to fractals. Proceedings of Symposia in Applied Mathematics, **39**, American Mathematical Society, 107–126 (1989).
46. Hasselblatt, B.: Regularity of the Anosov splitting and of horospheric foliations. Ergod. Th. & Dynam. Sys., **14**, 645–666 (1994).
47. Herman, M.R.: Sur la conjugaison différentiable des difféomorphismes du cercle á des rotations. Publ. IHES, **49**, 5–233 (1979).
48. Hirsch, M., Pugh, C.: Stable manifolds and hyperbolic sets. Proc. Symp. Pure Math., Amer. Math. Soc., **14**, 133–164 (1970).
49. Hirsch, M., Pugh, C., Shub, M.: Invariant manifolds. Lecture notes in mathematics, **583**, Springer (1977).
50. Hurder, S., Katok, A.: Differentiability, rigidity and Godbillon-Vey classes for Anosov flows. Publ. Math. IHES, **72**, 5–61 (1990).
51. Jacobson, M.V., Swiatek, G.: Quasisymmetric conjugacies between unimodal maps. I. Induced expansion and invariant measures. Stony Brook preprint (1991).
52. Jiang, Y.: Generalized Ulam-von Neumann transformations. Ph.D. Thesis, Graduate School of CUNY and UMI publication, (1990).
53. Jiang, Y.: Asymptotic differentiable structure on Cantor set. Comm. in Math. Phys., **155** (3), 503–509 (1993).
54. Jiang, Y.: Renormalization and Geometry in One-Dimensional and Complex Dynamics. Advanced Series in Nonlinear Dynamics, Vol. **10**, World Scientific Publishing Co. Pte. Ltd., River Edge, NJ (1996).
55. Jiang, Y.: Smooth classification of geometrically finite one-dimensional maps. Trans. Amer. Math. Soc., **348** (6), 2391–2412 (1996).
56. Jiang, Y.: On rigidity of one-dimensional maps. Contemporary Mathematics, AMS Series, **211**, 319–431 (1997).
57. Jiang, Y.: Metric invariants in dynamical systems. Journal of Dynamics and Differentiable Equations, Vol. **17** (1), 51–71 (2005).
58. Jiang, Y.: Teichmüller Structures and Dual Geometric Gibbs Type Measure Theory for Continuous Potentials, preprint (2008).
59. Jiang, Y., Cui, G., Gardiner, F.: Scaling functions for degree two circle endomorphisms. Contemporary Mathematics, AMS Series, **355**, 147–163 (2004).
60. Jiang, Y., Cui, G., Quas, A.: Scaling functions, Gibbs measures, and Teichmuller space of circle endomorphisms. Discrete and Continuous Dynamical Systems, **5**, no. 3, 535–552 (1999).

61. Jonker, L., Rand, D.A.: The periodic orbits and entropy of certain maps of the unit interval. J. Lond. Math. Soc. **22**, 175–181 (1980).
62. Jonker, L., Rand, D.A.: Bifurcations in one-dimension. I. The nonwandering set. Inventiones Matematicae **62**, 347–365 (1981).
63. Jonker, L., Rand, D.A.: Bifurcations in one-dimension. II. A versal model for bifurcations. Inventiones Matematicae **63**, 1–15 (1981).
64. Journé, J.L.: A regularity lemma for functions of several variables. Revista Math. Iberoamericana, **4**, 187–193 (1988).
65. Katok, A., Hasselblatt, B.: Introduction to The Modern Theory of Dynamical Systems. CUP (1994).
66. Kollmer, H.: On the structure of axiom-A attractors of codimension one. XI Colóquio Brasileiro de Matemática, vol. II, 609–619 (1997).
67. Kra, B., Schmeling, J.: Diophantine classes, dimension and Denjoy maps. Acta Arith., **105**, 323–340 (2002).
68. Lanford, O.: Renormalization group methods for critical circle mappings with general rotation number, VIIIth International Congress on Mathematical Physics (Maresille, 1986), *World Sci. Publishing*, Singapore, 532–536 (1987).
69. Livšic, A.N., Sinai, Ja.G.: Invariant measures that are compatible with smoothness for transitive C-systems. (Russian) Dokl. Akad. Nauk SSSR, **207**, 1039–1041 (1972).
70. Llave, R.: Invariants for Smooth conjugacy of hyperbolic dynamical systems II. Commun. Math. Phys., **109**, 369–378 (1987).
71. Llave, R., Marco J.M., Moriyon, R.: Canonical perturbation theory of Anosov systems and regularity results for the Livsic cohomology equations. Annals of Math., **123**, 537–612 (1986).
72. Mackay, R.S., Pinto, A.A., van Zeijts, J.B.J.: Coordinate change eigenvalues for bimodal period doubling renormalization. Phys. Lett. **A 190**, 412–416 (1994).
73. Mañé, R.: Ergodic Theory and Differentiable Dynamics. Springer-Verlag Berlin (1987).
74. Manning, A.: There are no new Anosov diffeomorphisms on tori. Amer. J. Math., **96**, 422 (1974).
75. Marco, J.M., Moriyon, R.: Invariants for Smooth conjugacy of hyperbolic dynamical systems I. Commun. Math. Phys., **109**, 681–689 (1987).
76. Marco, J.M., Moriyon, R.: Invariants for Smooth conjugacy of hyperbolic dynamical systems III. Commun. Math. Phys., **112**, 317–333 (1987).
77. Martens, M.: Distortion results and invariant Cantor sets for unimodal maps. Erg. Th and Dyn. Sys, **14**, 331–349 (1994).
78. Martens, M.: The periodic points of renormalization. Ann. of Math., **147**, 543–584 (1998).
79. Martens, M., Nowicki, T.: Invariant Measures for Typical Quadratic Maps. Asterisque **261**, 243–256 (1999).
80. Masur, H.: Interval exchange transformations and measured foliations. The Annals of Mathematics. 2nd Ser., **115** (1), 169–200 (1982).
81. de Melo, W.: Structural Stability of Diffeomorphisms on two-manifolds. Inventiones Mathematicae **21**, 233–246 (1973).
82. de Melo, W.: Stability and optimization of several functions. Topology **15**, 1–12 (1976).
83. de Melo, W.: Moduli of stability of two-dimensional diffeomorphisms. Topology **19** vol. 1, 9–21 (1980).

84. de Melo, W.: A Finiteness problem for one dimensional maps. Proc. AMS, **101** (4), 721–727 (1987).
85. de Melo, W.: On the Cyclicity of Recurrent Flows on Surfaces. Nonlinearity, **10** n. **2**, 311–319 (1997).
86. de Melo, W., Aoki, N., Shiraiwa, K., Takahashi, Y.: Full Families of Circle Endomorphisms. Conf. Dynamical Systems and Chaos, Tokyo v. **1**, 25-27, (1995).
87. de Melo, W., Avila, A., Lyubich, M.: Regular or stochastic dynamics in real analytic families of unimodal maps. Inventiones Mathematicae, **3** v. **154**, 451–550 (2003).
88. de Melo, W., Dumortier, F.: A type of moduli for saddle connections of planar diffeomorphisms. J. Diff. Eq., v. **75**, n. **11**, 88–102 (1988).
89. de Melo, W., de Faria, E.: Rigidity of critical circle mappings I. J. Eur. Math. Soc., **1** n. **4**, 339–392 (1999)
90. de Melo, W., de Faria, E.: Rigidity of critical circle mappings II. J. Am. Math. Soc., **13** n. **2**, 343–370 (2000).
91. de Melo, W., Martens, M.: The Multipliers of Periodic Point in One-Dimensional Dynamics. Nonlinearity, **12** n. **2**, 217–227 (1999).
92. de Melo, W., Martens, M.: Universal Models for Lorenz Maps. Erg.Th. and Dyn. Sys., **21** n. **2**, 343–370 (2000).
93. de Melo, W., Palis, J.: Geometric Theory of Dynamical Systems. Springer-Verlag (1980).
94. de Melo, W., Palis, J., van Strien, S.: Characterizing diffeomorphisms with modulus of stability one. Proc. Symp. Dynamical Systems and Turbulence. Warwick and Springer Lectures Notes, **818**, 266–285 (1981).
95. de Melo, W., Pinto, A.A.: Rigidity of C^2 infinitely renormalizable unimodal maps. Commun. Math. Phys. **208**, 91–105 (1999).
96. de Melo, W., Reis, G.L. Mendes, P.: Equivariant diffeomorphisms with simple recurrences on two-manifolds. Trans. AMS, v. **289** n. **12**, 793–807 (1985).
97. de Melo, W., van Strien, S.: One dimensional dynamics: The Schwarzian derivatives and beyond. Bull. AMS, v. **18**, n. **2**, 159–162 (1988).
98. de Melo, W., van Strien, S.: A structure theorem in one dimensional dynamics. Annals of Mathematics, v. **129**, 519–546 (1989).
99. de Melo, W., van Strien, S.: One-dimensional Dynamics. A series of Modern Surveys in Mathematics. Springer-Verlag, New York (1993).
100. de Melo, W., van Strien, S., Martens, M.: Julia-Fatou-Sullivan theory for real one-dimensional dynamics. Acta Math., **168**, 273–318 (1992).
101. Morales, C., Pacifico, M.J., Pujals, E.R.: Singular Hyperbolic Sets. Proceedings of the AMS, v. **127**, 3393–3400 (1999).
102. Morales, C., Pujals, E.R.: Singular strange atractors on the boundary of Morse-Smale systems. Annales Scientifiques de L' Ecole Normale Superieure, v. **30**, 693–717 (1997).
103. Newhouse, S.: On codimension one Anosov diffeomorphisms. Amer. J. Math., **92**, 671–762 (1970).
104. Newhouse, S., Palis, J.: Hyperbolic nonwandering stes on two-dimension manifolds. In: Peixoto, M. (ed) Dynamical Systems (1973).
105. Norton, A.: Denjoy's Theorem with exponents. Proceedings of the American Mathematical Society, **127** (10), 3111–3118 (1999).
106. Ostlund, S., Pandit, R., Rand, D.A., Schellnhuber, H., Siggia, E.: The 1-dimensional Schrodinger equation with an almost-periodic potential. Phys. Rev. Letts **50**, 1873–1876 (1983).

107. Ostlund, S., Rand, D.A., Sethna, J., Siggia, E.: Universal properties of the transition from quasi-periodicity to chaos. Physica D, **8**, 1–54 (1983).
108. Pacifico, M.J., Pujals, E.R., Sambarino, M., Vieitez, J.L.: Robustly expansive codimension-one homoclinic classes are hyperbolic. Ergodic Theory & Dynamical Systems, v. **30**, 233–247 (2008).
109. Pacifico, M.J., Pujals, E.R., Vieitez, J.L.: Robustly expansive homoclinic classes. Ergodic Theory and Dynamical Systems, v. **25**, n. **1**, 271–300 (2004).
110. Palis, J.: On Morse-Smale diffeomorphisms. Bulletin American Math. Society, **741**, 985–988 (1968).
111. Palis, J.: On Morse-Smale dynamical systems. Topology, **19**, 385–405 (1969).
112. Palis, J.: Ω-explosions for flows. Proceedings American Math. Society, vol. **27**, 85–90 (1971).
113. Palis, J.: Vector fields generate few diffeomorphisms. Bulletin American Mathematical Society, **80**, 503–505, (1974).
114. Palis, J.: On the continuity of Hausdorff dimension and limit capacity for horseshoes. Dynamical Systems, Lecture Notes in Mathematics, Springer-Verlag, vol. **1331**, 150–160 (1988).
115. Palis, J.: On the C^1 omega-stability conjecture. Publications Math. Institut Hautes Etudes Scientifiques, vol. **66**, 210-215, (1988).
116. Palis, J.: On the contribution of Smale to dynamical systems. From Topology to Computation, volume in honour of Stephen Smale, Springer-Verlag, 165–178 (1993).
117. Palis, J.: From dynamical stability and hyperbolicity to finitude of ergodic attractors. Proceedings of the Third World Academy of Sciences, 11th General Conference, Italy, 1996.
118. Palis, J., Camacho, C., Kuiper, N.: The topology of holomorphic flows near a singularity. Publications Math. Institut Hautes Études Scientifiques, vol. **48**, 5–38 (1978).
119. Palis, J., Carneiro, M.J.: Bifurcations and global stability families of gradient. Publications Mathematiques Institut Hautes Études Scientifiques, vol. **70**, 103–168 (1989).
120. Palis, J., Carneiro, M.J.: On the codimension of gradients with umbilic singularity. Publications Math. Institut Hautes Études Scientifiques, vol. **70**, 103–168 (1989).
121. Palis, J., Hirsch, M., Pugh, C., Shub, M.: Neighborhoods of hyperbolic sets. Inventiones Mathematicae. **9**, 212–234 (1970).
122. Palis, J., de Melo, W.: Moduli of stability for diffeomorphisms. Proc. Symp. Dyn. Systems, Lecture Notes in Mathematics, Springer-Verlag, vol. **819**, 318–339 (1980).
123. Palis, J., Newhouse, S.: Bifurcations of Morse-Smale dynamical systems. Dynamical Systems, Academic Press, 303–366, (1973).
124. Palis, J., Newhouse, S.: Hyperbolic nonwandering sets on two-dimensional manifolds. Dynamical Systems, Academic Press, 293–302, (1973).
125. Palis, J., Newhouse, S., Takens, F.: Stable arcs of diffeomorphisms. Bulletin American Mathematical Society, vol. **82**, 499–502 (1976).
126. Palis, J., Newhouse, S., Takens, F.: Bifurcations and stability of families of diffeomorphisms. Publications Math. Institut Hautes Études Scientifiques, vol. **57**, 5–72, (1983).

127. Palis, J., Pugh, C., Shub, M., Sullivan, D.: Genericity theorems in topological dynamics. Lecture Notes in Mathematics, Springer-Verlag, vol. **468**, 241–251 (1975).
128. Palis, J., Smale, S.: Structural stability theorems. Proc. of the Inst. on Global Analysis, American Math. Society, vol. **XIV**, 223–232 (1970).
129. Palis, J., Takens, F.: Topological equivalence of normally hyperbolic vector fields. Topology, vol. **16**, 335–345 (1977).
130. Palis, J., Takens, F.: Stability of parameterized families of gradient vector fields. Annals of Mathematics, vol. **118**, 383–421 (1983).
131. Palis, J., Takens, F.: Cycles and Measure of Bifurcation Sets for Two-Dimensional Diffeomorphisms. Inventiones Mathematicae, vol. **82**, 397–422 (1985).
132. Palis, J., Takens, F.: Hyperbolicity and creation of homoclinic orbits. Annals of Mathematics, vol. **125**, 337–374 (1987).
133. Palis, J., Takens, F.: Hyperbolicity and sensitive chaotic dynamics at homoclinic bifurcations. Cambridge University Press (1993).
134. Palis, J., Viana, M.: High dimension diffeomorphisms displaying infinitely many sinks. Annals of Mathematics, vol. **140**, 207–250 (1994).
135. Palis, J., Yoccoz, J.C.: Rigidity of Centralizeres of Diffeomorphisms. Ann. Scient. École Normale Superieure, vol. **22**, 81–98 (1989).
136. Palis, J., Yoccoz, J.C.: Homoclinic tangencies for hyperbolic sets of large Hausdorff dimension. Acta Mathematica, vol. **172**, 91–136 (1994).
137. Paterson, S.J.: The limit set of a Fuchsian group. Acta. Math., **136**, 241–273 (1976).
138. Peixoto, M.: Generalized convex functions and second order differential inequalities. Bulletin of the American Mathematical Society, **55** (6), 563–572 (1949).
139. Peixoto, M.: On structural stability. Annals of Mathematics, **69**, 199–222 (1959).
140. Peixoto, M.: Some examples on n-dimensional structural stability. Proc. Nat. Acad. Sci, **45**, 633–636 (1959).
141. Peixoto, M.: Structural stability on two-dimensional manifolds. Topology, **1**, 101–120 (1962).
142. Peixoto, M.: Structural stability on two-dimensional manifolds - a further remark. Topology, **2**, 179–180 (1963).
143. Peixoto, M.: On an approximation theorem of Kupka and Smale. Journal of Differential Equations, **3**, 214–227 (1967).
144. Peixoto, M. and Peixoto, M.C.: Structural stability in the plane with enlarged boundary conditions. Anais da Academia Brasileira de Ciências, **31**, 135–160 (1959).
145. Peixoto, M., Pugh, C.C.: Structurally stable systems on open manifolds are never dense. Annals of Mathematics, **87**, 423–430 (1968).
146. Peixoto, M., Rocha, A.C., Sutherland, S., Veerman, P.: On Brillouin zones. Communications in Mathematical Physics, **212** , 725–744 (2000).
147. Peixoto, M., Thom, R.: Le point de vue ênumératif dans les problèmes aux limites pour les équations différentielles ordinaires. I Quelques exemples. C.R. Acad. Sc. Paris, **303**, 629–633 (1986). Erratum, **307**, 197–198 (1988)
148. Peixoto, M., Thom, R.: Le point de vue ênumératif dans les problèmes aux limites pour les équations différentielles ordinaires. II. Le théorème. C.R. Acad. Sc. Paris, **303**, 693–698 (1986).

149. Penner, R.C., Harer, J.L.: Combinatorics of Train Tracks. Princeton University Press, Princeton, New Jersey (1992).
150. Pinto, A.A.: Convergence of Renormalisation and Rigidity of Dynamical Systems. Ph.D. Thesis, University of Warwick (1991).
151. Pinto, A.A.: Hyperbolic and minimal sets. Prooceedings of the 12th International Conference on Difference Equations and Applications, Lisbon 2008. World Scientific. Accepted for publication (2008).
152. Pinto, A.A.: A stable manifold theorem for pseudo-Anosov diffeomorphisms with C^r pseudo-smooth structures. Submitted for publication (2008).
153. Pinto, A.A., Almeida, J.P., Fisher, A.: Anosov diffeomorphisms and real tilings. In preparation.
154. Pinto, A.A., Almeida, J.P., Portela, A.: Golden tilings. Submitted for publication (2008).
155. Pinto, A.A., Pujals, E.R.: Pseudo-Anosov diffeomorphisms versus Pujals's non-uniformly hyperbolic diffeomorphisms. In preparation.
156. Pinto, A.A., Pujals, E.R.: C^r pseudo-Anosov diffeomorphisms and non-uniformly hyperbolic diffeomorphisms. In preparation.
157. Pinto, A.A., Rand, D.A.: Global phase space universality, smooth conjugacies and renormalisation: 2. The $C^{k+\alpha}$ case using rapid convergence of Markov families. Nonlinearity, **5**, 49–79 (1992).
158. Pinto, A.A., Rand, D.A.: Classifying C^{1+} structures on dynamical fractals: 1 The moduli space of solenoid functions for Markov maps on train-tracks. Ergod. Th. & Dynam. Sys., **15** 697–734 (1995).
159. Pinto, A.A., Rand, D.A.: Classifying C^{1+} structures on dynamical fractals: 2 Embedded trees. Ergod. Th. & Dynam. Sys., **15** 969–992 (1995).
160. Pinto, A.A., Rand, D.A.: Characterising rigidity and flexibility of hyperbolic surface dynamics, Warwick preprint, 1–53 (1995).
161. Pinto, A.A., Rand, D.A.: Anosov diffeomorphisms with an invariant measure absolutely continuous with respect to Lebesgue, Warwick preprint, 1–23 (1996).
162. Pinto, A.A., Rand, D.A.: Existence, uniqueness and ratio decomposition for Gibbs states via duality. Ergod. Th. & Dynam. Sys., **21**, 533–544 (2001).
163. Pinto, A.A., Rand, D.A.: Teichmüller spaces and HR structures for hyperbolic surface dynamics. Ergod. Th. & Dynam. Sys., **22**, 1905–1931 (2002).
164. Pinto, A.A., Rand, D.A.: Smoothness of holonomies for codimension 1 hyperbolic dynamics. Bull. London Math. Soc., **34**, 341–352 (2002).
165. Pinto, A.A., Rand, D.A.: Rigidity of hyperbolic sets on surfaces. J. London Math Soc., **2**, 1–22 (2004).
166. Pinto, A.A., Rand, D.A.: Geometric measures for hyperbolic sets on surfaces. SUNY, Stony Brook preprint (2006).
167. Pinto, A.A., Rand, D.A.: Solenoid functions for hyperbolic sets on surfaces. Recent Progress in Dynamics. MSRI Publications, **54**, 145–178 (2007).
168. Pinto, A.A., Rand, D.A.: Train tracks with C^{1+} self-renormalisable structures. Submitted for publication (2008).
169. Pinto, A.A., Rand, D.A.: Explosion of smoothness for Markov families. Submitted for publication (2008).
170. Pinto, A.A., Rand. D.A., Ferreira, F.: Hausdorff dimension bounds for smoothness of holonomies for codimension 1 hyperbolic dynamics. J. Differential Equations **243**, 168–178 (2007).

171. Pinto, A.A., Rand. D.A., Ferreira, F.: Cantor exchange systems and renormalization. J. Differential Equations, **243**, 593–616 (2007).
172. Pinto, A.A., Rand, D.A., Ferreira F.: Arc exchange systems and renormalization. Journal of Difference Equations and Applications (2008). pp. 1-30 Accepted for publication.
173. Pinto, A.A., Rand, D.A., Ferreira F.: C^{1+} Self-renormalisable structures. Part 12 of the book Communications of the Laufen Colloquium on Science 2007. A. Ruffing, A. Suhrer, J. Suhrer (Eds.) Shaker Verlag.
174. Pinto, A.A., Rand, D.A., Ferreira, F., Almeida, J.P.: Fine Structures of Hyperbolic Diffeomorphisms. Proceedings of the conference DYNA 2008, International Conference in honour of M.M. Peixoto and D.A. Rand. (2008) M.M. Peixoto, A.A. Pinto and D.A. Rand (Eds.) Springer Verlag.
175. Pinto, A.A., Sullivan, D.: The circle and the solenoid. Dedicated to Anatole Katok On the Occasion of his 60th Birthday. DCDS-A, **16** (2), 463–504 (2006).
176. Pinto, A.A., Viana, M.: Mañé duality for pseudo-Anosov diffeomorphisms. In preparation.
177. Pugh, C., Shub, M., Wilkinson, A.: Hölder foliations. Duke Math. J., **86**, 517–546 (1997).
178. Pujals, E.R., Morales, C.: Strange attractors containing a singularity with two positive multipliers. Comm. Math. Phys, v. **196**, 671–679, (1998).
179. Pujals, E.R., Pacifico, M.J., Morales, C.: Robust transitive singular sets for 3-flows are partially hyperbolic attractors or repellers. Annals of Mathematics, v. **160**, n. **2**, 375–432 (2003).
180. Pujals, E.R., Robert, L., Shub, M.: Expanding maps of the circle rerevisited: Positive Lyapunov exponents in a rich family. Ergodic Theory & Dynamical Systems, v. **26**, 1931–1937, (2006).
181. Pujals, E.R., Sambarino, M.: Homoclinic tangencies and hyperbolicity for surface diffeomorphisms. Annals of Mathematics, **151**, 961–1023 (2000).
182. Pujals, E.R., Sambarino, M.: On homoclinic tangencies, hyperbolicity, creation of homoclinic orbits and varation of entropy. Nonlinearity, v. **13**, n. **3**, 921–926 (2000).
183. Pujals, E.R., Sambarino, M.: A sufficient condition for robustly minimal foliations. Ergodic Theory and Dynamical Systems, **26**, 281–289 (2006).
184. Pujals, E.R., Sambarino, M.: On the dynamics of dominated splitting. Annals of Mathematics, v. **166**, (2006).
185. Pujals, E.R., Shub, M.: Dynamics of two-dimensional Blaschke products. Ergodic Theory & Dynamical Systems, v. **28**, 575–585 (2008).
186. Rand, D.A.: Exotic phenomena in games and duopoly models. J. of Math. Economics **5**, 173–184 (1978).
187. Rand, D.A.: Fractal bifurcation sets, renormalization strange sets and their universal invariants. Proc. R. Soc. Lond. A **413**, 45–61 (1987). Proceedings of the Royal Society Discussion Meeting on Dynamical Chaos. December, (1986).
188. Rand, D.A.: Universality and renormalisation in dynamical systems. In: Bedford, T., Swift, J. (eds) New Directions in Dynamical Systems, CUP (1988).
189. Rand, D.A.: Global phase space universality, smooth conjugacies and renormalisation: 1. The $C^{1+\alpha}$ case. Nonlinearity, **1**, 181–202 (1988).
190. Rand, D.A.: The singularity spectrum $f(\alpha)$ for cookie-cutters. Ergod. Th. & Dynam. Sys. **9**, 527–541 (1989).

191. Rand, D.A.: Existence, non-existence and universal breakdown of dissipative golden invariant tori. I: Golden critical circle maps. Nonlinearity **5** (3), 639–662 (1992).
192. Rand, D.A.: Existence, non-existence and universal breakdown of dissipative golden invariant tori. II. Convergence of renormalisation for mappings of the annulus. Nonlinearity **5** (3), 663–680 (1992).
193. Rand, D.A.: Existence, non-existence and universal breakdown of dissipative golden invariant tori. III. Invariant circles for mappings of the annulus. Nonlinearity **5** (3), 681–706 (1992).
194. Rand, D.A.: Correlation equations and pair approximations for spatial ecologies. In Advanced Ecological Theory. Principles and Applications (ed. J. McGlade), Blackwell Science, 100–142 (1999).
195. Rand, D.A., Bohr, T.: The entropy function for characteristic exponents. Physica D **25**, 387–398 (1987).
196. Rand, D.A., Ostlund, S., Sethna, J., Siggia, E.: Universal transition from quasiperiodicity to chaos in dissipative systems. Phys. Rev. Letters **49** (2), 132–135 (1982).
197. Schmeling, J., Siegmund-Schultze Ra: Hölder continuity of the holonomy maps for hyperbolic basic sets. I. Ergodic theory and related topics, III, Güstrow, (1990) Lecture Notes in Mathematics, 1514), Springer, Berlin, 174–191 (1992).
198. Shub, M.: Endomorphisms of compact differentiable manifolds. Amer. J. of Math., (1969).
199. Shub, M.: Global Stability of Dynamical Systems, Springer-Verlag (1987).
200. Sinai, Ya: Markov Partitions and C-diffeomorphisms. Anal. and Appl., **2**, 70–80 (1968).
201. Smale, S.: Structurally stable systems are not dense. American Journal of Mathematics, **88**, 491–496 (1966).
202. Smale, S.: Dynamical systems on n-dimensional manifolds. Differential Equations and Dynamical Systems, 483–486 (1967).
203. Smale, S.: Differentiable dynamical systems. Bulletin of the American Mathematical Society, **73** , 747–817 (1967).
204. Smale, S.: What is global analysis? American Math. Monthly, 4–9 (1969).
205. Smale, S.: Stability and genericity in dynamical systems. Seminaire Bourbaki, Springer, Berlin, 177–186 (1969–70).
206. Smale, S.: Notes on differential dynamical systems. Global Analysis, Proc. of Symposia in Pure Mathematics, AMS, **14**, 277–287 (1970).
207. Smale, S.: The Ω-stability theorem, Global Analysis. Proc. of Symposia in Pure Mathematics, AMS, **14**, pp. 289–297 (1970).
208. Smale, S.: Topology and mechanics, I., Inventiones Mathematicae, **10**, 305–331 (1970).
209. Smale, S.: Topology and mechanics, II., Inventiones Mathematicae, **11**, 45–64 (1970).
210. Smale, S.: Problems on the nature of relative equilibria in celestial mechanics, Proc. of Conference on Manifolds, Amsterdam, Springer, Berlin, 194–198 (1970).
211. Smale, S.: On the mathematical foundations of electrical circuit theory, Journal of Differential Geometry, **7**, 193–210 (1972).
212. Smale, S., Shub, M.: Beyond hyperbolicity. Annals of Mathematics, **96**, 591–597 (1972).

213. Smale, S.: Personal perspectives on mathematics and mechanics, Statistical Mechanics: New Concepts, New Problems, New Applications, Stuart A. Rice, Karl F. Freed, and John C. Light (Eds.), Univ. of Chicago Press, pp. 3–12 (1972).
214. Smale, S.: Stability and isotopy in discrete dynamical systems, Dynamical Systems, M.M. Peixoto (Ed.), Academic Press, New York (1973).
215. Stollenwerk, N., Martins, J., Pinto, A.A.: The phase transition lines in pair approximation for the basic reinfection model SIRI. Phys. Lett. **A 371**, 379–388 (2007).
216. Strebel, K.: On the existence of extremal Teichmüller mappings. J. Anal. Math., **30**, 441–447 (1976).
217. van Strien, S.: Normal hyperbolicity and linearisability. Inventiones Mathematicae, **87**, 377–384 (1987).
218. van Strien, S.: Hyperbolicity and invariant measures for general C^2 interval maps satisfying the Misiurewicz conditions. Comm. in Math. Physics. **128**, 437–496 (1990).
219. van Strien, S., Bruin H., Luzzatto, S.: Decay of correlations in one-dimensional dynamics. Annales de l'ENS. 4e série, **36**, 626–646 (2003).
220. van Strien, S., Bruin H., Shen, W.: Invariant measures exist without a growth condition. Commun. Math. Phys. **241**, 287–306 (2003).
221. van Strien, S., Galeeva, R.: Which families of ℓ-modal maps are full? Trans. of the A.M.S. **348**, 3215–3221 (1996).
222. van Strien, S., Levin, G.: Locally connected Julia sets of real polynomials. Annals of Math., **147**, 471–541 (1998).
223. van Strien, S., Levin, G.: Bounds, ergodicity and negative Schwarzian for multimodal maps, Bounds for interval maps with one inflection point II. Inventiones Mathematicae, **141**, 399–465 (2000).
224. van Strien, S., Nowicki, T.: Absolutely continuous measures for Collet-Eckmann maps without Schwarzian derivative conditions. Inventiones Mathematicae **93**, 425–437 (1988).
225. van Strien, S., Nowicki, T.: Hyperbolicity properties of multimodal Collet-Eckmann maps without the Schwarzian derivative conditions. Trans. A.M.S., **321**, 793–810 (1990).
226. van Strien, S., Nowicki, T.: Invariant measures for unimodal maps under a summability condition. Inventiones Mathematicae, **105**, 123–136 (1991).
227. van Strien, S., Vargas, E.: Real Bounds, ergodicity and negative Schwarzian for multimodal maps. J. Amer. Math. Soc. **17**, 749–782 (2004).
228. van Strien, S., Bruin, H., Keller, G., Nowicki, T.: Wild Cantor attractors exist. Annals of Math., **143**, 97–130 (1996).
229. Sullivan, D.: Conformal dynamical systems. Lecture Notes in Mathematics, **1007**, Springer-Verlag, New-York, 725–752 (1983).
230. Sullivan, D.: Differentiable structures on fractal-like sets determined by intrinsic scaling functions on dual Cantor sets. Proceedings of Symposia in Pure Mathematics, **48**, American Mathematical Society (1988).
231. Sullivan, D.: Bounds, quadratic differentials, and renormalization conjectures. AMS Centennial Publications. Volume 2: Mathematics into the Twenty-first Century (1988 Centennial Symposium, August 8-12), American Mathematical Society, Providence, RI (1991).

232. Sullivan, D.: Linking the universalities of Milnor-Thurston, Feigenbaum and Ahlfors-Bers. In: Golberg, L., Phillips, A. (eds) Topological Methods in Modern Mathematics, Publish or Perish, 543–563 (1993).
233. Thurston, W.: On the geometry and dynamics of diffeomorphisms of surfaces. Bull. Amer. Math. Soc., **19**, 417–431 (1988).
234. Thurston, W.: The Geometry and topology of three-manifolds. Princeton University Press, Princeton (1978).
235. Veech, W.: Gauss measures for transformations on the space of interval exchange maps. The Annals of Mathematics, 2nd Ser., **115** (2), 201–242 (1982).
236. Viana, M., Bonatti, C., Diaz, L.J.: Dynamics Beyond Uniform Hyperbolicity: A global geometric and probabilistic perspective. Heidelberg: Springer Verlag (2004).
237. Vul, E.B., Sinai, Ya.G., Khanin, K.M.: Feigenbaum universality and the thermodynamical formalism, Uspekhi Mat. Nauk, **39**, 3–37 (1984).
238. Williams, R.F.: Expanding attractors. Publ. I.H.E.S., **43**, 169–203 (1974).
239. Yoccoz, J.C.: Conjugaison différentiable des difféomorphismes du cercle dont le nombre de rotation vérifie une condition diophantienne. Ann. scient. Éc. Norm. Sup., 4 série, t, **17**, 333–359 (1984).

Index

Printing: Krips bv, Meppel, The Netherlands
Binding: Stürtz, Würzburg, Germany